BEHAVIORAL GENETICS

BEHAVIORAL · GENETICS

Sixth Edition

Robert Plomin
Institute of Psychiatry
London

John C. DeFries
University of Colorado
Boulder

Valerie S. Knopik
Rhode Island Hospital and Brown University

Jenae M. Neiderhiser
Penn State University

Worth Publishers
New York

VICE PRESIDENT, EDITORIAL AND PRODUCTION: Catherine Woods
PUBLISHER: Kevin Feyen
SENIOR ACQUISITIONS EDITOR: Christine M. Cardone
SENIOR MARKETING MANAGER: Katherine Nurre
MARKETING MANAGER: Jennifer Bilello
MARKETING ASSISTANT: Stephanie Ellis
EXECUTIVE MEDIA EDITOR: Andrea Musick
SUPPLEMENTS EDITOR: Anthony Casciano
EDITORIAL ASSISTANT: Eric Dorger
PHOTO EDITOR: Christine Buese
SENIOR PROJECT EDITOR: Liz Geller
TEXT DESIGN: Dreamit Inc.
COVER DESIGN: Kevin Kall
PRODUCTION MANAGER: Sarah Segal
COMPOSITION AND ILLUSTRATIONS: Northeastern Graphic, Inc.
PRINTING AND BINDING: RR Donnelley

COVER PHOTOS: Identical twin students in class, Julian Winslow/Corbis; Line DNA, Josh Westrich/Corbis; DNA double helix, Corbis/Veer.

Library of Congress Control Number: 2012945229

ISBN-13: 978-1-4292-4215-8
ISBN-10: 1-4292-4215-9

Printed in the United States of America
First printing

Worth Publishers
41 Madison Avenue
New York, NY 10010
www.worthpublishers.com

Contents

● CHAPTER 4

DNA: The Basis of Heredity — 39

● CHAPTER 5

Animal Models in Behavioral Genetics — 50

● CHAPTER 6

Nature, Nurture, and Human Behavior — 73

● CHAPTER 7

Estimating Genetic and Environmental Influences   86

● CHAPTER 8

The Interplay between Genes and Environment   105

● CHAPTER 11

Cognitive Disabilities 163

● CHAPTER 12

General Cognitive Ability 186

● CHAPTER 13

Specific Cognitive Abilities   210

● CHAPTER 14

Schizophrenia   230

● CHAPTER 15

Other Adult Psychopathology   242

ABOUT THE AUTHORS

ROBERT PLOMIN is MRC Research Professor of Behavioral Genetics at the Social, Genetic and Developmental Psychiatry Centre at the Institute of Psychiatry in London. He received his doctorate in psychology from the University of Texas, Austin, in 1974, one of the few graduate programs in psychology that offered a specialty in behavioral genetics at that time. He then became an assistant professor at the Institute for Behavioral Genetics at the University of Colorado, Boulder, where he began working with John DeFries. Together, they created the longitudinal Colorado Adoption Project of behavioral development, which has continued for more than 30 years. Plomin worked at Penn State University from 1986 until 1994, when he moved to the Institute of Psychiatry in London to help launch the Social, Genetic and Developmental Psychiatry Centre. The goal of his research is to bring together genetic and environmental research strategies to investigate behavioral development. Plomin is now conducting a study of all twins born in England during the period 1994 to 1996, focusing on developmental delays in childhood. He is a past president of the Behavior Genetics Association (1989–1990) and has received lifetime achievement awards from the Behavior Genetics Association (2002), American Psychological Society (2005), the Society for Research in Child Development (2005), and the International Society for Intelligence Research (2011).

JOHN C. DEFRIES is professor of psychology and faculty fellow of the Institute for Behavioral Genetics, University of Colorado, Boulder. After receiving his doctorate in agriculture (with specialty training in quantitative genetics) from the University of Illinois in 1961, he remained on the faculty of the University of Illinois for six years. In 1962, he began research on mouse behavioral genetics, and the following year he was a research fellow in genetics at the University of California, Berkeley. After returning to Illinois in 1964, DeFries initiated an extensive genetic analysis of open-field behavior in laboratory mice. Three years later, he joined the Institute for Behavioral Genetics, and he served as its director from 1981 to 2001. DeFries and Steve G. Vandenberg founded the journal *Behavior*

Genetics in 1970, and DeFries and Robert Plomin founded the Colorado Adoption Project in 1975. For over three decades, DeFries's major research interest has concerned the genetics of reading disabilities, and he founded the Colorado Learning Disabilities Research Center with Richard K. Olson in 1990. He served as president of the Behavior Genetics Association in 1982 and 1983, receiving the association's Th. Dobzhansky Award for Outstanding Research in 1992; and he became a Fellow of the American Association for the Advancement of Science (Section J, Psychology) in 1994 and the Association for Psychological Science in 2009.

VALERIE S. KNOPIK is Director of the Division of Behavioral Genetics at Rhode Island Hospital and Associate Professor in the Departments of Psychiatry & Human Behavior and Behavioral & Social Sciences at the Warren Alpert School of Medicine at Brown University. She received her doctorate in Psychology from the University of Colorado, Boulder in 2000, where she worked with John DeFries and conducted research in the Colorado Learning Disabilities Research Center. She subsequently completed a fellowship in psychiatric genetics and genetic epidemiology at Washington University School of Medicine in St. Louis from 2000 to 2002 and continued as junior faculty for two years. She joined the faculty at Brown University in 2004 and holds an Adjunct Associate Professor appointment at Washington University School of Medicine in St. Louis. Knopik's primary area of interest is the joint effect of genetic and environmental (specifically prenatal and early postnatal) risk factors on child and adolescent externalizing behavior, associated learning and cognitive deficits, and later substance use. She serves as Associate Editor of *Behavior Genetics* and Field Chief Editor of *Frontiers in Behavioral and Psychiatric Genetics*. Her work has been recognized by the Research Society for Alcoholism, which chose her as a finalist for the Enoch Gordis Research Recognition Award; the NIDA Genetics Workgroup; and the Behavior Genetics Association, from whom she received the Fuller and Scott Early Career Award in 2007.

JENAE M. NEIDERHISER is Liberal Arts Research Professor of Psychology at Penn State University. After receiving her Ph.D. in Human Development and Family Studies from Penn State University in 1994, she joined the faculty of the Center for Family Research, Department of Psychiatry and Behavioral Science, at The George Washington University in Washington, D.C., advancing from Assistant Research Professor to Professor from 1994 to 2007. In 2007 she joined the Department of Psychology at Penn State University, and she also holds the appointment of Professor of Human Development and affiliate scientist at the Oregon Social Learning Center. Neiderhiser's work has focused on how genes and environments work together throughout the lifespan. She has had a particular focus on genotype-environment correlation and how individuals shape their own environments, especially within the family. In her pursuit of this question she has collaborated on developing a number of novel or underutilized research designs, including the Extended Children of Twins and an ongoing prospective adoption study, the Early Growth and Development Study. Neiderhiser is an associate editor for the *Journal of Research on Adolescence* and *Frontiers in Behavioral and Psychiatric Genetics* and is on the editorial board of several developmental psychology journals.

PREFACE

Some of the most important scientific accomplishments of the twentieth century occurred in the field of genetics, beginning with the rediscovery of Mendel's laws of heredity and ending with the first draft of the complete DNA sequence of the human genome. The pace of discoveries has continued to accelerate in the first part of the twenty-first century. One of the most dramatic developments in the behavioral sciences during the past few decades is the increasing recognition and appreciation of the important contribution of genetic factors to behavior. Genetics is not a neighbor chatting over the fence with some helpful hints—it is central to the behavioral sciences. In fact, genetics is central to all the life sciences and gives the behavioral sciences a place in the biological sciences. Genetic research includes diverse strategies, such as twin and adoption studies (called quantitative genetics), which investigate the influence of genetic and environmental factors, as well as strategies to identify specific genes (called molecular genetics). Behavioral geneticists apply these research strategies to the study of behavior in biopsychology, clinical psychology, cognitive psychology, developmental psychology, educational psychology, neuroscience, psychopharmacology, and social psychology, and increasingly in other areas of the social sciences such as behavioral economics and political science.

The goal of this book is to share with you our excitement about behavioral genetics, a field in which we believe some of the most important discoveries in the behavioral sciences have been made in recent years. This sixth edition continues to emphasize what we know about genetics in the behavioral sciences rather than how we know it. Its goal is not to train students to become behavioral geneticists but rather to introduce students in the behavioral, social, and life sciences to the field of behavioral genetics.

This sixth edition presages a passing of the baton to the next generation. Two new and younger authors (Knopik and Neiderhiser) have joined forces with two authors from the previous editions (Plomin and DeFries), which has brought new energy and ideas that help to capture developments in this fast-moving and highly interdisciplinary field. In addition to updating research with more than 550 new references, this edition represents a substantial reorganization. One feature of this edition is that it highlights the value of behavioral genetics for understanding the environment (Chapter 7) and its interplay with genetics (Chapter 8). At first, chapters on the environment might seem odd in a textbook on genetics, but in fact the environment is crucial at every step in the pathways between genes, brain, and behavior. One of the oldest controversies in the behavioral sciences, the so-called nature (genetics) versus nurture (environment) controversy, has given way to a view that both nature and nurture are important for complex behavioral traits. Moreover, genetic research has made important discoveries about how the environment affects behavioral development.

We have also expanded our coverage of gene expression and especially epigenetics as pathways between genes and behavior (Chapter 10). Our review of cognitive abilities includes the new area of brain imaging genetics (Chapter 13). Coverage of psychopathology has been expanded (Chapters 14, 15, and 16), a new section on behavioral economics has been included (Chapter 17), and a new chapter on substance use disorders has been added (Chapter 18), reflecting the enormous growth of genetic research in these areas.

We begin with an introductory chapter that will, we hope, whet your appetite for learning about genetics in the behavioral sciences. The next few chapters present the basic rules of heredity, its DNA basis, and the methods used to find genetic influence and to identify specific genes. The rest of the book highlights what is known about genetics in the behavioral sciences. The areas about which the most is known are cognitive disabilities and abilities, psychopathology, personality, and substance abuse. We also consider areas of behavioral sciences that were introduced to genetics more recently, such as health psychology, aging, and evolutionary psychology. Throughout these chapters, quantitative genetics and molecular genetics are interwoven. One of the most exciting developments in behavioral genetics is the ability to begin to identify specific genes that influence behavior. The last chapter, "The Future of Behavioral Genetics," has been reviewed by 30 of the world's top behavioral geneticists and represents a consensus statement about the future of the field.

Because behavioral genetics is an interdisciplinary field that combines genetics and the behavioral sciences, it is complex. We have tried to write about it as simply as possible without sacrificing honesty of presentation. Although our coverage is representative, it is by no means exhaustive or encyclopedic. History and methodology are relegated to boxes and an appendix to keep the focus on what we now know about genetics and behavior. The appendix, by Shaun Purcell, presents an overview of statistics, quantitative genetic theory, and a type of quantitative genetic analysis called model fitting. In this edition we have retained an interactive Web site that brings the appendix to life with demonstrations: http://pngu.mgh.harvard.edu/purcell/bgim/. The Web site was designed and written by Shaun Purcell. A list of other useful Web sites, including those of relevant associations, databases, and other resources, is included after the appendix. Following the Web sites list is a glossary; the first time each glossary entry appears in the text it is shown in boldface type.

We thank the following individuals, who gave us their very helpful advice for this new edition: Arpana Agrawal, *Washington University*; Ros Arden, *King's College London*; Dorret Boomsma, *VU University Amsterdam*; S. Alexandra Burt, *Michigan State University*; John Crabbe, *Oregon Health & Science University*; Oliver Davis, *King's College London*; Lisbeth DiLalla, *Southern Illinois University*; Bruce Dudek, *State University of New York, Albany*; Richard Duhrkopf, *Baylor University*; Gary Dunbar, *Central Michigan University*; Thalia Eley, *King's College London*; Cathy Fernandes, *King's College London*; Jonathan Flint, *University of Oxford*; Sarah Francazio, *Providence College*; Corina Greven, *Radboud University Medical Centre, Nijmegen*; Claire Haworth, *King's College London*; Christina Hewitt, *University of Colorado, Boulder*; Crystal Hill-Chapman, *Francis Marion University*; Ken

Kendler, *Virginia Commonwealth University, Richmond;* Christopher Kliethermes, *University of California, San Francisco;* Bob Krueger, *University of Minnesota;* Paul Lichtenstein, *Karolinska Institute, Stockholm;* Clare Llewellyn, *University College London;* Nick Martin, *Queensland Institute of Medical Research,* John McGeary, *Providence VA Medical Center;* Alison Pike, *University of Sussex;* Elise Robinson, *Harvard University;* Kathryn Roecklein, *University of Pittsburgh;* Angelica Ronald, *Birkbeck, University of London;* Frank Spinath, *Saarland University;* Jeanette Taylor, *Florida State University;* Anita Thapar, *Cardiff University School of Medicine;* Essi Viding, *University College London;* and Irwin Waldman, *Emory University.*

We also gratefully acknowledge the important contributions of the co-authors of the previous editions of this book: Gerald E. McClearn, Michael Rutter, and Peter McGuffin. We especially wish to thank Neil Harvey, who helped us organize the revision and references and prepare the final manuscript. Finally, we thank our editor at Worth Publishers, Christine M. Cardone, and the Senior Project Editor, Liz Geller, who helped support our efforts in this new edition, as well as Marketing Assistant Stephanie Ellis.

CHAPTER ONE

Overview

Some of the most important recent discoveries about behavior involve genetics. For example, autism (Chapter 16) is a rare but severe disorder beginning early in childhood in which children withdraw socially, not engaging in eye contact or physical contact, with marked communication deficits and stereotyped behavior. Until the 1980s, autism was thought to be environmentally caused by cold, rejecting parents or by brain damage. But genetic studies comparing the risk for identical twins, who are identical genetically (like clones), and fraternal twins, who are only 50 percent similar genetically, indicate substantial genetic influence. If one member of an identical twin pair is autistic, the risk that the other twin is also autistic is very high, about 60 percent. In contrast, for fraternal twins, the risk is low. Molecular genetic studies are attempting to identify individual **genes*** that contribute to the genetic susceptibility to autism.

Later in childhood, a very common concern, especially in boys, is a cluster of attention-deficit and disruptive behavior problems called attention-deficit hyperactivity disorder (ADHD) (Chapter 16). Several twin studies have shown that ADHD is highly heritable (genetically influenced). ADHD is one of the first behavioral areas in which specific genes have been identified. Although many other areas of childhood psychopathology show genetic influence, none are as heritable as autism and ADHD. Some behavior problems, such as childhood anxiety and depression, are only moderately heritable, and others, such as antisocial behavior in adolescence, show little genetic influence.

More relevant to college students are personality traits such as risk-taking (often called sensation-seeking) (Chapter 17), drug use and abuse (Chapter 18), and learning abilities (Chapters 12 and 13). All these domains have consistently shown substantial genetic influence in twin studies and have recently begun to yield clues concerning individual genes that contribute to their **heritability**. These domains are also examples of an important general principle: Not only do genes contribute to disorders such as autism and ADHD, they also play an important role in normal variation. For

---

**Boldface indicates the first appearance in the text of a word or phrase that is in the Glossary.*

example, you might be surprised to learn that differences in weight are almost as heritable as differences in height (Chapter 19). Even though we can control how much we eat and are free to go on crash diets, differences among us in weight are much more a matter of nature (genetics) than nurture (environment). Moreover, normal variation in weight is as highly heritable as overweight or obesity. The same story can be told for behavior. Genetic differences do not just make some of us abnormal; they contribute to differences among all of us in normal variation for mental health, personality, and cognitive abilities.

One of the greatest genetic success stories involves the most common behavioral disorder in later life, the terrible memory loss and confusion of Alzheimer disease, which strikes as many as one in five individuals in their eighties (Chapter 11). Although Alzheimer disease rarely occurs before the age of 65, some early-onset cases of dementia run in families in a simple manner that suggests the influence of single genes. Three genes have been found to be responsible for many of these rare early-onset cases.

These genes for early-onset Alzheimer disease are not responsible for the much more common form of Alzheimer disease that occurs after 65 years of age. Like most behavioral disorders, late-onset Alzheimer disease is not caused by just a few genes. Still, twin studies indicate genetic influence. If you have a twin who has late-onset Alzheimer disease, your risk of developing it is twice as great if you are an identical twin rather than a fraternal twin. These findings suggest genetic influence.

Even for complex disorders like late-onset Alzheimer disease, it is now possible to identify genes that contribute to the risk for the disorder. For example, a gene has been identified that predicts risk for late-onset Alzheimer disease far better than any other known risk factor. If you inherit one copy of a particular form (**allele**) of the gene, your risk for Alzheimer disease is about four times greater than if you have another allele. If you inherit two copies of this allele (one from each of your parents), your risk is much greater. Finding these genes for early-onset and late-onset Alzheimer disease has greatly increased our understanding of the brain processes that lead to dementia.

Another example of recent genetic discoveries involves mental retardation (Chapter 11). The single most important cause of mental retardation is the inheritance of an entire extra **chromosome** 21. (Our **DNA**, the basic hereditary molecule, is packaged as 23 pairs of chromosomes, as explained in Chapter 4.) Instead of inheriting only one pair of chromosomes 21, one from the mother and one from the father, an entire extra chromosome is inherited, usually from the mother. Often called Down syndrome, trisomy-21 is one of the major reasons why women worry about pregnancy later in life. Down syndrome occurs much more frequently when mothers are over 40 years old. The extra chromosome can be detected after 15 weeks of pregnancy by a procedure called **amniocentesis** or earlier by chorionic villus sampling.

Another gene has been identified that is the second most common cause of mental retardation, called *fragile X retardation*. The gene that causes the disorder is on the X chromosome. Fragile X mental retardation occurs nearly twice as often in males as in females because males have only one X chromosome. If a boy has the fragile X allele on his X chromosome, he will develop the disorder. Females have two X chromosomes, and it is necessary to inherit the fragile X allele on both X chromosomes in order to develop the disorder. However, females with one fragile X allele can also be affected to some extent. The fragile X gene is especially interesting because it involves a type of genetic defect in which a short sequence of DNA mistakenly repeats hundreds of times. This type of genetic defect is now also known to be responsible for several other previously puzzling diseases (Chapter 3).

Genetic research on behavior goes beyond just demonstrating the importance of genetics to the behavioral sciences and allows us to ask questions about how genes influence behavior. For example, does genetic influence change during development? Consider cognitive ability, for example; you might think that as time goes by we increasingly accumulate the effects of Shakespeare's "slings and arrows of outrageous fortune." That is, environmental differences might become increasingly important during one's life span, whereas genetic differences might become less important. However, genetic research shows just the opposite: Genetic influence on cognitive ability increases throughout the individual's life span, reaching levels later in life that are nearly as great as the genetic influence on height (Chapter 12). This finding is an example of developmental behavioral genetic research.

School achievement and the results of tests you took to apply to college are influenced almost as much by genetics as are the results of tests of cognitive abilities such as intelligence (IQ) tests (Chapters 12 and 13). Even more interesting, the substantial overlap between such achievement and the ability to perform well on tests is nearly all genetic in origin. This finding is an example of what is called **multivariate genetic analysis.**

Genetic research is also changing the way we think about the environment (Chapters 7 and 8). For example, we used to think that growing up in the same family makes brothers and sisters similar psychologically. However, for most behavioral dimensions and disorders, it is genetics that accounts for similarity among siblings. Although the environment is important, environmental influences can make siblings growing up in the same family different, not similar. This genetic research has sparked an explosion of environmental research looking for the environmental reasons why siblings in the same family are so different.

Recent genetic research has also shown a surprising result that emphasizes the need to take genetics into account when studying the environment: Many environmental measures used in the behavioral sciences show genetic influence! For example, research in developmental psychology often involves measures of parenting that are, reasonably enough, assumed to be measures of the family environment. However, genetic research has convincingly shown genetic influence on parenting measures. How can this be? One way is that genetic differences among parents influence their behavior toward their children. Genetic differences among children can also make a contribution. For example, parents who have more books in their home have children

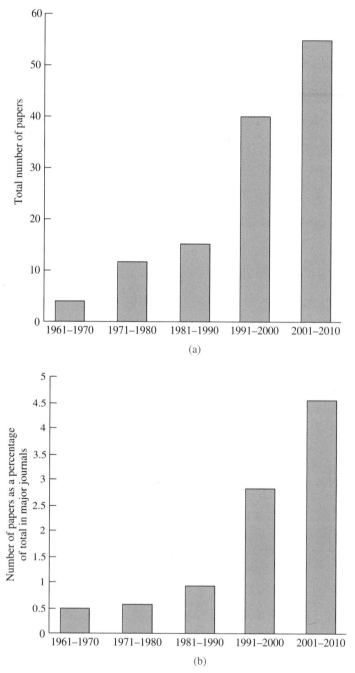

• **FIGURE 1.1** Fifty years of behavioral genetic twin studies. (a) The total number of behavioral genetic papers published per decade from 1961 to 2010, and (b) their percentage of the total number of papers in two of the most highly cited journals in developmental psychology *(Child Development* and *Developmental Psychology)*.

who do better in school, but this **correlation** does not necessarily mean that having more books in the home is an environmental cause for children performing well in school. Genetic factors could affect parental traits that relate both to the number of books parents have in their home and to their children's achievement at school. Genetic involvement has also been found for many other ostensible measures of the environment, including childhood accidents, life events, and social support. To some extent, people create their own experiences for genetic reasons.

These are examples of what you will learn about in this book. The simple message is that genetics plays a major role in behavior. Genetics integrates the behavioral sciences into the life sciences. Although research in behavioral genetics has been conducted for many years, the field-defining text was published only in 1960 (Fuller & Thompson, 1960). Since that date, discoveries in behavioral genetics have grown at a rate that few other fields in the behavioral sciences can match. This growth is accelerating following the sequencing of the human **genome**, that is, identifying each of the more than 3 billion steps in the spiral staircase that is DNA, leading to the identification of the DNA differences among us that are responsible for the heritability of normal and abnormal behavior.

Recognition of the importance of genetics is one of the most dramatic changes in the behavioral sciences during the past three decades. Over 80 years ago, Watson's (1930) behaviorism detached the behavioral sciences from their budding interest in heredity. A preoccupation with the environmental determinants of behavior continued until the 1970s, when a shift began toward the more balanced contemporary view that recognizes genetic as well as environmental influences. This shift toward genetics in the behavioral sciences can be seen in the increasing number of publications in the behavioral sciences that involve genetics. Figure 1.1 illustrates this trend in developmental psychology, a trend that is accelerating with the new advances in DNA research.

CHAPTER TWO

Mendel's Laws of Heredity

Huntington disease (HD) begins with personality changes, forgetfulness, and involuntary movements. It typically strikes in middle adulthood; during the next 15 to 20 years, it leads to complete loss of motor control and intellectual function. No treatment has been found to halt or delay the inexorable decline. This is the disease that killed the famous Depression-era folksinger Woody Guthrie. Although it affects only about 1 in 20,000 individuals, a quarter of a million people in the world today will eventually develop Huntington disease.

When the disease was traced through many generations, it showed a consistent pattern of heredity. Afflicted individuals had one parent who also had the disease, and approximately half the children of an affected parent developed the disease. (See Figure 2.1 for an explanation of symbols traditionally used to describe family trees, called *pedigrees.* Figure 2.2 shows an example of a Huntington disease pedigree.)

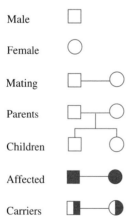

Male

Female

Mating

Parents

Children

Affected

Carriers

• **FIGURE 2.1** Symbols used to describe family pedigrees.

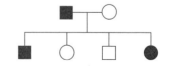

• **FIGURE 2.2** Huntington disease. HD individuals have one HD parent. About 50 percent of the offspring of HD parents will have HD.

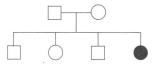

• **FIGURE 2.3** Phenylketonuria. PKU individuals do not typically have parents with PKU. If one child has PKU, the risk for other siblings is 25 percent. As explained later, parents in such cases are carriers for one allele of the PKU gene, but a child must have two alleles in order to be afflicted with recessive disorders such as PKU.

What rules of heredity are at work? Why does this lethal condition persist in the population? We will answer these questions in the next section, but first, consider another inherited disorder.

In the 1930s, a Norwegian biochemist discovered an excess of phenylpyruvic acid in the urine of a pair of mentally retarded siblings and suspected that the condition was due to a disturbance in the metabolism of phenylalanine. Phenylalanine is one of the essential **amino acids,** which are the building blocks of proteins, and is present in many foods in the normal human diet. Other retarded individuals were soon found with this same excess. This type of mental retardation came to be known as phenylketonuria (PKU).

Although the frequency of PKU is only about 1 in 10,000, PKU once accounted for about 1 percent of the mentally retarded institutionalized population. PKU has a pattern of inheritance very different from that of Huntington disease. PKU individuals do not usually have affected parents. Although this might make it seem at first glance as if PKU is not inherited, PKU does in fact "run in families." If one child in a family has PKU, the risk for siblings to develop it is about 25 percent, even though the parents themselves may not be affected (Figure 2.3). One more piece of the puzzle is the observation that when parents are genetically related ("blood" relatives), typically in marriages between cousins, they are more likely to have children with PKU. How does heredity work in this case?

Mendel's First Law of Heredity

Although Huntington disease and phenylketonuria, two examples of hereditary transmission of mental disorders, may seem complicated, they can be explained by a simple set of rules about heredity. The essence of these rules was worked out more than a century ago by Gregor Mendel (Mendel, 1866).

Mendel studied inheritance in pea plants in the garden of his monastery in what is now the Czech Republic (Box 2.1). On the basis of his many experiments, Mendel concluded that there are two "elements" of heredity for each trait in each individual and that these two elements separate, or segregate, during reproduction. Offspring receive one of the two elements from each parent. In addition, Mendel concluded that one of these elements can "dominate" the other, so that an individual with just

BOX 2.1 • Gregor Mendel's Luck

Before Mendel (1822–1884), much of the research on heredity involved crossing plants of different species. But the offspring of these matings were usually sterile, which meant that succeeding generations could not be studied. Another problem with research before Mendel was that features of the plants investigated were complexly determined. Mendel's success can be attributed in large part to the absence of these problems.

Mendel crossed different varieties of pea plants of the same species; thus the offspring were fertile. In addition, he picked simple either-or traits, qualitative traits, that happened to be

• Gregor Johann Mendel. A photograph taken at the time of his research. (Courtesy of V. Orel, Mendel Museum, Brno, Czech Republic.)

due to single genes. He was also lucky that in the traits he chose, one allele completely dominated the expression of the other allele, which is not always the case. However, one feature of Mendel's research was not due to luck. Over seven years, while raising over 28,000 pea plants, he counted all offspring rather than being content, as researchers before him had been, with a verbal summary of the typical results.

Mendel studied seven qualitative traits of the pea plant, such as whether the seed was smooth or wrinkled. He obtained 22 varieties of the pea plant that differed in these seven characteristics. All the varieties were true-breeding plants: those that always yield the same result when crossed with the same kind of plant. Mendel presented the results of eight years of research on the pea plant in his 1866 paper. This paper, "Experiments with Plant Hybrids," now forms the cornerstone of genetics and is one of the most influential publications in the history of science.

In one experiment, Mendel crossed true-breeding plants with smooth seeds and true-breeding plants with wrinkled seeds. Later in the summer, when he opened the pods containing their offspring (called the F_1, or first filial generation), he found that all of them had smooth seeds. This result indicated that the then-traditional view of blending inheritance was not correct. That is, the F_1 did not have seeds that were even moderately wrinkled. These F_1 plants were fertile, which allowed Mendel to take the next step of allowing plants of the F_1 generation

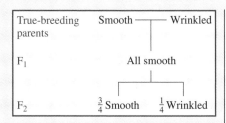

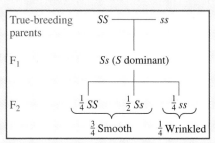

to self-fertilize and then studying their offspring, **F$_2$**. The results were striking: Of the 7324 seeds from the F$_2$, 5474 were smooth and 1850 were wrinkled. That is, ¾ of the offspring had smooth seeds and ¼ had wrinkled seeds. This result indicates that the factor responsible for wrinkled seeds had not been lost in the F$_1$ generation but had merely been dominated by the factor causing smooth seeds. The figure above summarizes Mendel's results.

Given these observations, Mendel deduced a simple explanation involving two hypotheses. First, each individual has two hereditary "elements," now called alleles (alternate forms of a gene). For Mendel's pea plants, these alleles determined whether the seed was wrinkled or smooth. Thus, each parent has two alleles (either the same or different) but transmits only one of the alleles to each offspring. The second hypothesis is that, when an individual's alleles are different, one allele can dominate the other. These two hypotheses neatly explain the data (see the figure above right).

The true-breeding parent plant with smooth seeds has two alleles for smooth seeds (*SS*). The true-breeding parent plant with wrinkled seeds has two alleles for wrinkled seeds (*ss*). First-generation (F$_1$) offspring receive one allele from each

parent and are therefore *Ss*. Because *S* dominates *s*, F$_1$ plants will have smooth seeds. The real test is the F$_2$ population. Mendel's theory predicts that when F$_1$ individuals are self-fertilized or crossed with other F$_1$ individuals, ¼ of the F$_2$ should be *SS*, ½ *Ss*, and ¼ *ss*. Assuming *S* dominates *s*, then *Ss* should have smooth seeds like the *SS*. Thus, ¾ of the F$_2$ should have smooth seeds and ¼ should have wrinkled, which is exactly what Mendel's data indicated. Mendel also discovered that the inheritance of one trait is not affected by the inheritance of another trait. Each trait is inherited in the expected 3:1 ratio.

Mendel was not so lucky in terms of acknowledgment of his work during his lifetime. When Mendel published the paper about his theory of inheritance in 1866, reprints were sent to scientists and libraries in Europe and the United States. However, for 35 years, Mendel's findings on the pea plant were ignored by most biologists, who were more interested in evolutionary processes that could account for change rather than continuity. Mendel died in 1884 without knowing the profound impact that his experiments would have during the twentieth century.

one dominant element will display the trait. A nondominant, or *recessive,* element is expressed only if both elements are recessive. These conclusions are the essence of Mendel's first law, the *law of segregation.*

No one paid any attention to Mendel's law of heredity for over 30 years. Finally, in the early 1900s, several scientists recognized that Mendel's law is a general law of inheritance, not one peculiar to the pea plant. Mendel's "elements" are now known as *genes,* the basic units of heredity. Some genes may possibly have only one form throughout a species, for example, in all pea plants or in all people. Heredity focuses on genes that have different forms: differences that cause some pea seeds to be wrinkled or smooth, or that cause some people to have Huntington disease or PKU. The alternative forms of a gene are called *alleles.* An individual's combination of alleles is its *genotype,* whereas the observed traits are its *phenotype.* The fundamental issue of heredity in the behavioral sciences is the extent to which differences in genotype account for differences in phenotype, observed differences among individuals.

This chapter began with two very different examples of inherited disorders. How can Mendel's law of segregation explain both examples?

Huntington Disease

Figure 2.4 shows how Mendel's law explains the inheritance of Huntington disease. HD is caused by a dominant allele. Affected individuals have one dominant allele (*H*) and one recessive, normal allele (*h*). (It is rare that an HD individual has two *H* alleles, an event that would require both parents to have HD.) Unaffected individuals have two normal alleles.

As shown in Figure 2.4, a parent with HD whose genotype is *Hh* produces **gametes** (egg or sperm) with either the *H* or the *h* allele. The unaffected (*hh*) parent's gametes all have an *h* allele. The four possible combinations of these gametes from the mother and father result in the offspring genotypes shown at the bottom of Figure 2.4. Offspring will always inherit the normal *h* allele from the unaffected parent, but they have a 50 percent chance of inheriting the *H* allele from the HD parent. This pattern

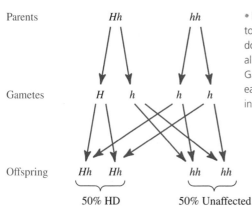

Parents   *Hh*   *hh*

Gametes   *H*   *h*   *h*   *h*

Offspring   *Hh*   *Hh*   *hh*   *hh*

50% HD   50% Unaffected

• **FIGURE 2.4** Huntington disease is due to a single gene, with the allele for HD dominant. *H* represents the dominant HD allele, and *h* is the normal recessive allele. Gametes are sex cells (eggs and sperm); each carries just one allele. The risk of HD in the offspring is 50 percent.

of inheritance explains why HD individuals always have a parent with HD and why 50 percent of the offspring of an HD parent develop the disease.

Why does this lethal condition persist in the population? If HD had its effect early in life, HD individuals would not live to reproduce. In one generation, HD would no longer exist because any individual with the HD allele would not live long enough to reproduce. The dominant allele for HD is maintained from one generation to the next because its lethal effect is not expressed until after the reproductive years.

A particularly traumatic feature of HD is that offspring of HD parents know they have a 50 percent chance of developing the disease and of passing on the HD gene. In 1983, **DNA markers** were used to show that the gene for HD is on chromosome 4, as will be discussed in Chapter 4. In 1993, the HD gene itself was identified. Now it is possible to determine for certain whether a person has the HD gene.

This genetic advance raises its own problems. If one of your parents had HD, you would be able to find out whether or not you have the HD allele. You would have a 50 percent chance of finding that you do not have the HD allele, but you would also have a 50 percent chance of finding that you do have the HD allele and will eventually die from it. In fact, most people at risk for HD decide *not* to take the test. Identifying the gene does, however, make it possible to determine whether a fetus has the HD allele and holds out the promise of future interventions that can correct the HD defect (Chapter 9).

Phenylketonuria

Mendel's law also explains the inheritance of PKU. Unlike HD, PKU is due to the presence of two recessive alleles. For offspring to be affected, they must have two copies of the allele. Those offspring with only one copy of the allele are not afflicted with the disorder. They are called *carriers* because they carry the allele and can pass it on to their offspring. Figure 2.5 illustrates the inheritance of PKU from two unaffected

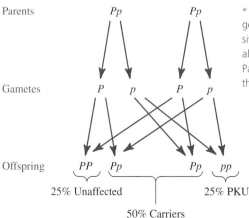

FIGURE 2.5 PKU is inherited as a single gene. The allele that causes PKU is recessive. *P* represents the normal dominant allele, and *p* is the recessive allele for PKU. Parents are carriers; the risk of PKU for their offspring is 25 percent.

carrier parents. Each parent has one PKU allele and one normal allele. Offspring have a 50 percent chance of inheriting the PKU allele from one parent and a 50 percent chance of inheriting the PKU allele from the other parent. The chance of both these things happening is 25 percent. If you flip a coin, the chance of heads is 50 percent. The chance of getting two heads in a row is 25 percent (i.e., 50 percent times 50 percent).

This pattern of inheritance explains why unaffected parents have children with PKU and why the risk of PKU in offspring is 25 percent when both parents are carriers. For PKU and other recessive disorders, identification of the genes makes it possible to determine whether potential parents are carriers. Identification of the PKU gene also makes it possible to determine whether a particular pregnancy involves an affected fetus. In fact, all newborns in most countries are screened for elevated phenylalanine levels in their blood, because early diagnosis of PKU can help parents prevent retardation by serving low-phenylalanine diets to their affected children.

Figure 2.5 also shows that 50 percent of children born of two carrier parents are likely to be carriers, and 25 percent will inherit the normal allele from both parents. If you understand how a recessive trait such as PKU is inherited, you should be able to work out the risk for PKU in offspring if one parent has PKU and the other parent is a carrier. (The risk is 50 percent.)

We have yet to explain why recessive traits like PKU are seen more often in offspring whose parents are genetically related. Although PKU is rare (1 in 10,000), about 1 in 50 individuals are carriers of one PKU allele (Box 2.2). If you are a PKU carrier, your chance of marrying someone who is also a carrier is 2 percent. However, if you marry someone genetically related to you, the PKU allele must be in your family, so the chances are much greater than 2 percent that your spouse will also carry the PKU allele.

It is very likely that we all carry at least one harmful recessive gene of some sort. However, the risk that our spouses are also carriers for the same disorder is small unless we are genetically related to them. In contrast, about half the children born to incestuous relationships between father and daughter show severe genetic abnormalities, often including childhood death or mental retardation. This pattern of inheritance explains why most severe genetic disorders are recessive: Because carriers of recessive alleles do not show the disorder, they escape eradication by **natural selection.**

It should be noted that even single-gene disorders such as PKU are not so simple, because many different **mutations** of the gene occur (more than 500!) and these have different effects (Scriver, 2007). New PKU mutations emerge in individuals with no family history of the disorder. Some single-gene disorders are largely caused by new mutations. In addition, age of onset may vary for single-gene disorders, as it does in the case of HD.

BOX 2.2 • How Do We Know That 1 in 50 People Are Carriers for PKU?

If you randomly mate F_2 plants to obtain an F_3 generation, the frequencies of the S and s alleles will be the same as in the F_2 generation, as will the frequencies of the SS, Ss, and ss genotypes. Shortly after the rediscovery of Mendel's law in the early 1900s, this implication of Mendel's law was formalized and eventually called the **Hardy-Weinberg equilibrium:** The frequencies of alleles and genotypes do not change across generations unless forces such as natural selection or migration change them. This rule is the basis for a discipline called **population genetics,** whose practitioners study forces that change **gene frequencies** (see Chapter 20).

Hardy-Weinberg equilibrium also makes it possible to estimate frequencies of alleles and genotypes. The frequencies of the dominant and recessive alleles are usually referred to as p and q, respectively. Eggs and sperm have just one allele for each gene. The chance that any particular egg or sperm has the dominant allele is p. Because sperm and egg unite at random, the chance that a sperm with the dominant allele fertilizes an egg with the dominant allele is the product of the two frequencies, $p \times p = p^2$. Thus, p^2 is the frequency of offspring with two dominant alleles (called the *homozygous dominant* genotype). In the same way, the *homozygous recessive* genotype has a frequency of q^2. As shown in the diagram, the frequency of offspring with one dominant allele

		Eggs	
Frequencies		p	q
Sperm	p	p^2	pq
	q	pq	q^2

and one recessive allele (called the *heterozygous* genotype) is $2pq$. In other words, if a population is in Hardy-Weinberg equilibrium, the frequency of the offspring genotypes is $p^2 + 2pq + q^2$. In populations with random mating, the expected **genotypic frequencies** are merely the product of $p + q$ for the mothers' alleles and $p + q$ for the fathers' alleles. That is, $(p + q)^2 = p^2 + 2pq + q^2$.

For PKU, q^2, the frequency of PKU individuals (homozygous recessive) is 0.0001. If you know q^2, you can estimate the frequency of the PKU allele and PKU carriers, assuming Hardy-Weinberg equilibrium. The frequency of the PKU allele is q, which is the square root of q^2. The square root of 0.0001 is 0.01, so that 1 in 100 alleles in the population are the recessive PKU alleles. If there are only two alleles at the PKU **locus,** then the frequency of the dominant allele (p) is $1 - 0.01 = 0.99$. What is the frequency of carriers? Because carriers are heterozygous genotypes with one dominant allele and one recessive allele, the frequency of carriers of the PKU allele is 1 in 50 (that is, $2pq = 2 \times 0.99 \times 0.01 = 0.02$).

●KEY CONCEPTS

Gene: Basic unit of heredity. A sequence of DNA bases that codes for a particular product

Allele: Alternative form of a gene.

Genotype: An individual's combination of alleles at a particular locus.

Phenotype: Observed or measured traits.

Dominant allele: An allele that produces the same phenotype in an individual regardless of whether one or two copies are present.

Recessive allele: An allele that produces its phenotype only when two copies are present.

Mendel's Second Law of Heredity

Not only do the alleles for Huntington disease segregate independently during gamete formation, they are also inherited independently from the alleles for PKU. This finding makes sense, because Huntington disease and PKU are caused by different genes; each of the two genes is inherited independently. Mendel experimented systematically with crosses between varieties of pea plants that differed in two or more traits. He found that alleles for the two genes assort independently. In other words, the inheritance of one gene is not affected by the inheritance of another gene. This is Mendel's *law of independent* **assortment.**

Most important about Mendel's second law are its exceptions. We now know that genes are not just floating around in eggs and sperm. They are carried on *chromosomes.* The term *chromosome* literally means "colored body," because in certain laboratory preparations the staining characteristics of these structures are different from those of the rest of the **nucleus** of the cell. Genes are located at places called *loci* (singular, *locus,* from the Latin, meaning "place") on chromosomes. Eggs contain just one chromosome from each pair of the mother's set of chromosomes, and sperm contain just one from each pair of the father's set. An egg fertilized by a sperm thus has the full chromosome complement, which, in humans, is 23 pairs of chromosomes. Chromosomes are discussed in more detail in Chapter 4.

When Mendel studied the inheritance of two traits at the same time (let's call them A and B), he crossed true-breeding parents that showed the dominant trait for both A and B with parents that showed the recessive forms for A and B. He found second-generation (F_2) offspring of all four possible types: dominant for A and B, dominant for A and recessive for B, recessive for A and dominant for B, and recessive for A and B. The frequencies of the four types of offspring were as expected if A and B were inherited independently. Mendel's law is violated, however, when genes for two traits are close together on the same chromosome. If Mendel had studied the joint inheritance of two such traits, the results would have surprised him. The two traits would not have been inherited independently.

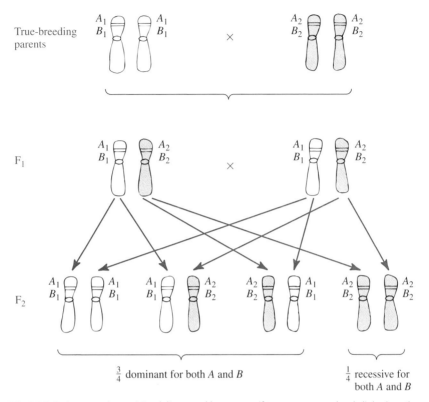

True-breeding parents

F_1

F_2

$\frac{3}{4}$ dominant for both A and B

$\frac{1}{4}$ recessive for both A and B

• **FIGURE 2.6** An exception to Mendel's second law occurs if two genes are closely linked on the same chromosome. The A_1 allele and the B_1 allele are dominant; the A_2 and B_2 alleles are recessive.

Figure 2.6 illustrates what would happen if the genes for traits A and B were very close together on the same chromosome. Instead of finding all four types of F_2 offspring, Mendel would have found only two types: dominant for both A and B and recessive for both A and B.

The reason why such violations of Mendel's second law are important is that they make it possible to map genes to chromosomes. If the inheritance of a particular pair of genes violates Mendel's second law, then it must mean that they tend to be inherited together and thus reside on the same chromosome. This phenomenon is called *linkage.* However, it is actually not sufficient for two linked genes to be on the same chromosome; they must also be very close together on the chromosome. Unless genes are near each other on the same chromosome, they will recombine by a process in which chromosomes exchange parts. **Recombination** occurs during **meiosis** in the ovaries and testes, when gametes are produced.

Figure 2.7 illustrates recombination for three loci (A, C, B) on a single chromosome. The maternal chromosome, carrying the alleles A_1, C_1, and B_2, is represented in white; the paternal chromosome with alleles A_2, C_2, and B_1 is blue. During meiosis,

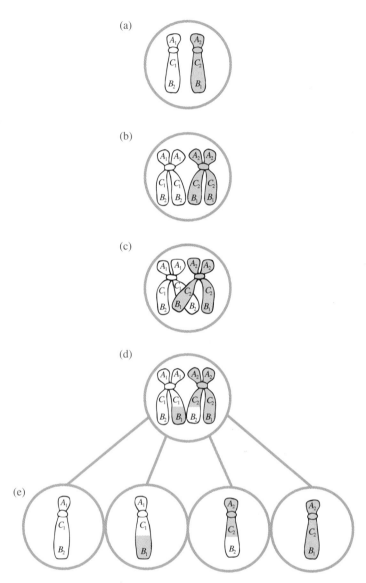

• **FIGURE 2.7** Illustration of recombination. The maternal chromosome, carrying the alleles A_1, C_1, and B_2, is represented in white; the paternal chromosome, with alleles A_2, C_2, and B_1, is blue. The right chromatid (the duplicated chromosome produced during meiosis) of the maternal chromosome crosses over (recombines) with the left chromatid of the paternal chromosome.

each chromosome duplicates to form sister **chromatids** (Figure 2.7b). These sister chromatids may cross over one another, as shown in Figure 2.7c. This overlap happens an average of one time for each chromosome during meiosis. During this stage, the chromatids can break and rejoin (Figure 2.7d). Each of the chromatids will be transmitted to a different gamete (Figure 2.7e). Consider only the A and B loci for the moment. As shown in Figure 2.7e, one gamete will carry the genes A_1 and B_2, as in the mother, and one will carry A_2 and B_1, as in the father. The other two will carry A_1 with B_1 and A_2 with B_2. For the latter two pairs, recombination has taken place—these combinations were not present on the parental chromosomes.

The probability of recombination between two loci on the same chromosome is a function of the distance between them. In Figure 2.7, for example, the *A* and *C* loci have not recombined. All gametes are either A_1C_1 or A_2C_2, as in the parents, because the **crossover** did not occur between these loci. Crossover could occur between the *A* and *C* loci, but it would happen less frequently than between *A* and *B*.

These facts have been used to "map" genes on chromosomes. The distance between two loci can be estimated by the number of recombinations per 100 gametes. This distance is called a map unit or ***centimorgan,*** named after T. H. Morgan, who first identified linkage groups in the fruit fly *Drosophila* (Morgan, Sturtevant, Muller, & Bridges, 1915). If two loci are far apart, like the *A* and *B* loci, recombination will separate the two loci as often as if the loci were on different chromosomes, and they will not appear to be linked.

To identify the location of a gene on a particular chromosome, *linkage analysis* can be used. Linkage analysis refers to techniques that use information about violations of independent assortment to identify the chromosomal location of a gene. DNA markers serve as signposts on the chromosomes, as discussed in Chapter 9. Since 1980, the power of linkage analysis has greatly increased with the discovery of millions of these markers. Linkage analysis looks for a violation of independent assortment between a trait and a DNA marker. In other words, linkage analysis assesses whether the DNA marker and the trait co-assort in a family more often than expected by chance.

In 1983, the gene for Huntington disease was shown to be linked to a DNA marker near the tip of one of the larger chromosomes (chromosome 4; see Chapter 9) (Gusella et al., 1983). This was the first time that the new DNA markers had been used to demonstrate a linkage for a disorder for which no chemical mechanism was known. DNA markers that are closer to the Huntington gene have since been developed and have made it possible to pinpoint the gene. As noted earlier, the gene itself was finally located precisely in 1993.

● KEY CONCEPTS

Chromosome: A threadlike structure that contains DNA and resides in the nucleus of cells. Humans have 23 pairs.

Locus (plural, loci): The site of a specific gene on a chromosome. Latin for "place."

Linkage: Loci that are close together on a chromosome and thus inherited together within families. Linkage is an exception to Mendel's second law of independent assortment.

Recombination: A process that occurs during meiosis in which chromosomes exchange parts.

Once a gene has been found, two things are possible. First, the DNA variation responsible for the disorder can be identified. This identification provides a DNA test that is directly associated with the disorder in individuals and is more than just a risk estimate calculated on the basis of Mendel's laws. That is, the DNA test can be used to diagnose the disorder in individuals regardless of information about other family members. Second, the protein coded by the gene can be studied; this investigation is a major step toward understanding how the gene has its effect and thus can possibly lead to a therapy. In the case of Huntington disease, the gene codes for a previously unknown protein, called huntingtin. This protein interacts with many other proteins, which has hampered efforts to develop drug therapies (Ross & Tabrizi, 2011).

Although the disease process of the Huntington gene is not yet fully understood, Huntington disease, like fragile X mental retardation (mentioned in Chapter 1), also involves a type of genetic defect in which a short sequence of DNA is repeated many times (see Chapter 3). The defective gene product slowly has its effect over the life course by contributing to neural death in the cerebral cortex and basal ganglia. This leads to the motoric and cognitive problems characteristic of Huntington disease.

Finding the PKU gene was easier because its enzyme product was known, as described in Chapter 1. In 1984, the gene for PKU was found and shown to be on chromosome 12 (Lidsky et al., 1984). For decades, PKU infants have been identified by screening for the physiological effect of PKU—high blood phenylalanine levels—but this test is not highly accurate. Developing a DNA test for PKU has been hampered by the discovery that there are hundreds of different mutations at the PKU locus and that these mutations differ in the magnitude of their effects. This diversity contributes to the variation in blood phenylalanine levels among PKU individuals.

Of the several thousand single-gene disorders known (about half of which involve the nervous system), the precise chromosomal location has been identified for most of these genes (Ku, Naidoo, & Pawitan, 2011). The gene sequence and the specific mutation have been found for at least half, and this number is increasing. One of the goals of the Human Genome Project was to identify all genes. Having the essentially complete sequence of the human genome is similar to having all the pages of a manual needed to make the human body. Now the challenge to scientists is to discover the genetic bases of human health and disease by reading and understanding the contents of these pages (National Human Genome Research Institute, 2010). Rapid progress in these challenging areas holds the promise of identifying genes even for complex behaviors influenced by multiple genes as well as environmental factors.

Summary

Huntington disease (HD) and phenylketonuria (PKU) are examples of dominant and recessive disorders, respectively. They follow the basic rules of heredity described by Mendel more than a century ago. A gene may exist in two or more different forms (alleles). One allele can dominate the expression of the other. The two alleles, one from each parent, separate (segregate) during gamete formation. This rule is Mendel's first law, the law of segregation. The law explains many features of inheritance: why 50 percent of the offspring of an HD parent are eventually afflicted, why this lethal gene persists in the population, why PKU children usually do not have PKU parents, and why PKU is more likely when parents are genetically related.

Mendel's second law is the law of independent assortment: The inheritance of one gene is not affected by the inheritance of another gene. However, genes that are closely linked on the same chromosome can co-assort, thus violating Mendel's law of independent assortment. Such violations make it possible to map genes to chromosomes by using linkage analysis. For Huntington disease and PKU, linkage has been established and the genes responsible for the disorders have been identified.

CHAPTER THREE

Beyond Mendel's Laws

C olor blindness shows a pattern of inheritance that does not appear to conform to Mendel's laws. The most common color blindness involves difficulty in distinguishing red and green, a condition caused by a lack of certain color-absorbing pigments in the retina of the eye. It occurs more frequently in males than in females. More interesting, when the mother is color blind and the father is not, all of the sons but none of the daughters are color blind (Figure 3.1a). When the father is color blind and the mother is not, offspring are seldom affected (Figure 3.1b). But something remarkable happens to these apparently normal daughters of a color-blind father: Half of their sons are likely to be color blind. This is the well-known skip-a-generation phenomenon—fathers have it, their daughters do not, but some of the grandsons do. What could be going on here in terms of Mendel's laws of heredity?

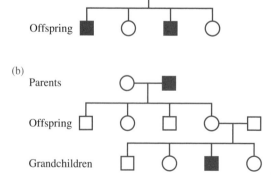

(a)
Parents
Offspring

(b)
Parents
Offspring
Grandchildren

• **FIGURE 3.1** Inheritance of color blindness. (a) A color-blind mother and unaffected father have color-blind sons but unaffected daughters. (b) An unaffected mother and color-blind father have unaffected offspring, but daughters have sons with 50 percent risk for color blindness. (See Figure 2.1 for symbols used to describe family pedigrees).

Genes on the X Chromosome

There are two chromosomes called the **sex chromosomes** because they differ for males and females. Females have two X chromosomes, and males have one X chromosome and a smaller chromosome called Y.

Color blindness is caused by a recessive allele on the X chromosome. But males have only one X chromosome; so, if they have one allele for color blindness (*c*) on their single X chromosome, they are color blind. For females to be color blind, they must inherit the *c* allele on both of their X chromosomes. For this reason, the hallmark of a sex-linked (meaning *X-linked*) recessive gene is a greater incidence in males. For example, if the frequency of an X-linked recessive allele (*q* in Chapter 2) for a disorder is 10 percent, then the expected frequency of the disorder in males would be 10 percent, but the frequency in females (q^2) would be only 1 percent (i.e., $0.10^2 = 0.01$).

Figure 3.2 illustrates the inheritance of the sex chromosomes. Both sons and daughters inherit one X chromosome from their mother. Daughters inherit their father's single X chromosome and sons inherit their father's Y chromosome. Sons cannot inherit an allele on the X chromosome from their father. For this reason, another sign of an **X-linked recessive trait** is that father-son resemblance is negligible. Daughters inherit an X-linked allele from their father, but they do not express a recessive trait unless they receive another such allele on the X chromosome from their mother.

Inheritance of color blindness is further explained in Figure 3.3. In the case of a color-blind mother and unaffected father (Figure 3.3a), the mother has the *c* allele on

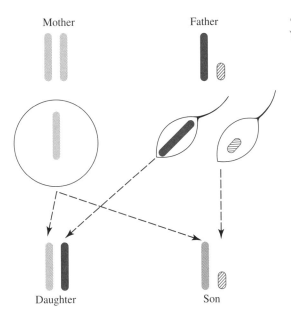

Mother   Father

Daughter   Son

• **FIGURE 3.2** Inheritance of X and Y chromosomes.

(a)

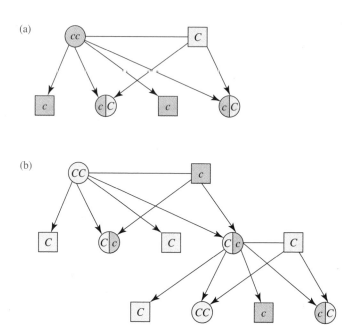

(b)

• **FIGURE 3.3** Color blindness is inherited as a recessive gene on the X chromosome. c refers to the recessive allele for color blindness, and C is the normal allele. (a) Color-blind mothers are homozygous recessive (cc). (b) Color-blind fathers have a c allele on their single X chromosome, which is transmitted to daughters but not to sons.

both of her X chromosomes and the father has the normal allele (*C*) on his single X chromosome. Thus, sons always inherit an X chromosome with the *c* allele from their mother and are color blind. Daughters carry one *c* allele from their mother but are not color blind because they have inherited a normal, dominant *C* allele from their father. They carry the *c* allele without showing the disorder, so they are called *carriers*, a status indicated by the two-toned circles in Figure 3.3.

In the second example (Figure 3.3b), the father is color blind but the mother is neither color blind nor a carrier of the *c* allele. None of the children are color blind, but the daughters are all carriers because they must inherit their father's X chromosome with the recessive *c* allele. You should now be able to predict the risk of color blindness for offspring of these carrier daughters. As shown in the bottom row of Figure 3.3b, when a carrier daughter (*Cc*) has children by an unaffected male (*C*), half of her sons but none of her daughters are likely to be color blind. Half of the daughters are carriers. This pattern of inheritance explains the skip-a-generation phenomenon. Color-blind fathers have no color-blind sons or daughters (assuming normal, noncarrier mothers), but their daughters are carriers of the *c* allele. The daughters' sons have a 50 percent chance of being color blind.

The sex chromosomes are inherited differently for males and females, so detecting X linkage is much easier than identifying a gene's location on other chromosomes.

Color blindness was the first reported human X linkage. Over 1500 genes have been identified on the X chromosome, as well as a disproportionately high number of single-gene diseases (Ross et al., 2005). The Y chromosome has over 200 genes, including those for determining maleness, and the smallest number of genes associated with disease of any chromosome.

● KEY CONCEPTS

Sex-linked (X-linked): A phenotype influenced by a gene on the X chromosome. X-linked recessive diseases occur more frequently in males because they only have one X chromosome.

Carrier: An individual who is heterozygous at a given locus, carrying both a normal allele and a mutant recessive allele, and who appears normal phenotypically.

Other Exceptions to Mendel's Laws

Several other genetic phenomena do not appear to conform to Mendel's laws in the sense that they are not inherited in a simple way through the generations.

New Mutations

The most common type of exceptions to Mendel's laws involve new, or de novo, DNA mutations that do not affect the parent because they occur during the formation of the parent's eggs or sperm. But this situation is not really a violation of Mendel's laws, because the new mutations are then passed on to offspring according to Mendel's laws, even though affected individuals have unaffected parents. Many genetic diseases involve such spontaneous mutations, which are not inherited from the preceding generation. An example is Rett syndrome, an X-linked dominant disorder that has a prevalence of about 1 in 10,000 in girls. Although girls with Rett syndrome develop normally during the first year of life, they later regress and eventually become both mentally and physically disabled. Boys with this mutation on their single X chromosome die either before birth or in the first two years after birth. (See Chapter 11.)

In addition, DNA mutations frequently occur in cells other than those that produce eggs or sperm and are not passed on to the next generation. This mutation type is the cause of many cancers, for example. Although these mutations affect DNA, they are not heritable because they do not occur in the eggs or sperm.

Changes in Chromosomes

Changes in chromosomes are an important source of mental retardation, as discussed in Chapter 11. For example, Down syndrome occurs in about 1 in 1000 births and accounts for more than a quarter of individuals with mild to moderate retardation. It was first described by Langdon Down in 1866, the same year that Mendel published his classic paper. For many years, the origin of Down syndrome defied explanation because it does not "run in families." Another puzzling feature is that it occurs much more often in the offspring of women who gave birth after 40 years of age. This relationship to maternal age suggests environmental explanations.

Instead, in the late 1950s, Down syndrome was shown to be caused by the presence of an entire extra chromosome with its thousands of genes. As explained in Chapter 4, during the formation of eggs and sperm (called *gametes*), each of the 23 pairs of chromosomes separates, and egg and sperm carry just one member of each pair. When the sperm fertilizes the egg, the pairs are reconstituted, with one chromosome of each pair coming from the father and the other coming from the mother. But sometimes the initial division in gamete formation is not even. When this accident happens, one egg or sperm might have both members of a particular chromosome pair and another egg or sperm might have neither. This failure to apportion the chromosomes equally is called *nondisjunction* (Figure 3.4). Nondisjunction is a major reason why so many conceptions abort spontaneously in the first few weeks of prenatal life. However, in the case of certain chromosomes, some fetuses with chromosomal anomalies are able to survive, though with developmental abnormalities. A prominent example is that of Down syndrome, which is caused by the presence of three copies (called *trisomy*) of one of the smallest chromosomes (chromosome 21). No

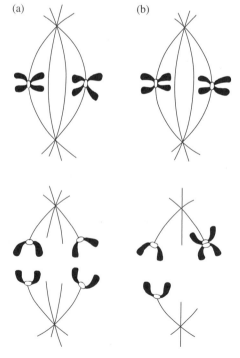

(a)   (b)

• **FIGURE 3.4** An exception to Mendel's laws of heredity: nondisjunction of chromosomes. (a) When eggs and sperm are formed, chromosomes for each pair line up and then split, and each new egg or sperm has just one member of each chromosome pair. (b) Sometimes this division does not occur properly, so one egg or sperm has both members of a chromosome pair and the other egg or sperm has neither.

individuals have been found with only one of these chromosomes (*monosomy*), which might occur when nondisjunction leaves an egg or sperm with no copy of the chromosome and another egg or sperm with two copies. It is assumed that this monosomy is lethal. Too little genetic material is generally more damaging than extra material. Because most cases of Down syndrome are created anew by nondisjunction, Down syndrome generally is not **familial.**

Nondisjunction also explains why the incidence of Down syndrome is higher among the offspring of older mothers. All the immature eggs of a female mammal are present before birth. These eggs have both members of each pair of chromosomes. Each month, one of the immature eggs goes through the final stage of cell division. Nondisjunction is more likely to occur as the female grows older and activates immature eggs that have been dormant for decades. In contrast, fresh sperm are produced all the time. For this reason, the incidence of Down syndrome is not affected by the age of the father.

Many women worry about reproducing later in life because of chromosomal abnormalities such as Down syndrome. Current common tests during pregnancy, such as ultrasound and maternal blood testing, can indicate whether the pregnancy is at greater risk for certain abnormalities. Other tests, such as amniocentesis, can rule out chromosomal abnormalities in the fetus.

Expanded Triplet Repeats

We have known about mutations and chromosomal abnormalities for a long time. Two other exceptions to Mendel's rules were discovered more recently. One is in effect a special form of mutation that involves *repeat sequences* of DNA. Although we do not know why, some very short segments of DNA—two, three, or four nucleotide bases of DNA (Chapter 4)—repeat a few times or up to a few dozen times. Different repeat sequences can be found in more than 50,000 places in the human genome. Each repeat sequence has several, often a dozen or more, alleles that consist of various numbers of the same repeat sequence; these alleles are usually inherited from generation to generation according to Mendel's laws. For this reason, and because there are so many of them, repeat sequences are widely used as DNA markers in linkage studies (see Chapter 9).

Sometimes the number of repeats at a particular locus increases and causes problems (Cooper & Blass, 2011). Several dozen diseases are now known to be associated with such expansions of repeat sequences; all involve the brain and thus lead to behavioral problems. For example, most cases of Huntington disease involve a repeat in the Huntington gene on chromosome 4. It is called a **triplet repeat** because the repeated unit is a certain sequence of three nucleotide bases of DNA. All combinations of the four nucleotide bases of DNA (see Chapter 4) are possible, but certain combinations are more common, such as CGG and CAG. Normal Huntington alleles contain between 11 and 34 copies of the triplet repeat, but Huntington alleles have more than 40 copies. The expanded number of triplet repeats is unstable and can increase in subsequent generations. This phenomenon explains a previously mysterious non-Mendelian process called *genetic anticipation,* in which symptoms appear at an earlier age and with greater severity in successive generations. For Huntington disease, longer expansions lead to earlier onset of the disorder and greater severity. The **expanded triplet repeat** is CAG, which codes for the amino acid glutamine and results in a protein with an expanded number of glutamines in the middle of the protein. The additional glutamines change the conformation of the protein and confer new and toxic properties to the protein. This leads to neural death, especially in the cerebral cortex and basal ganglia. Despite this non-Mendelian twist of genetic anticipation, Huntington disease generally follows Mendel's laws of heredity as a single-gene dominant disorder.

Fragile X mental retardation, the most common cause of mental retardation after Down syndrome, is also caused by an expanded triplet repeat that violates Mendel's laws. Although this type of mental retardation was known to occur almost twice as often in males as in females, its pattern of inheritance did not conform to sex linkage because it is caused by an unstable expanded repeat. As explained in Chapter 11, the expanded triplet repeat makes the X chromosome fragile in a certain laboratory preparation, which is how fragile X received its name. Parents who inherit X chromosomes with a normal number of repeats (5 to 40 repeats) at a particular locus sometimes produce eggs or sperm with an expanded number of repeats (up to 200 repeats), called a *premutation.* This premutation does not cause retardation in the offspring, but it is unstable and often leads to more expansions (200 or more repeats) in the next generation, which do cause retardation (Figure 3.5). Unlike the expanded repeat responsible for Huntington disease, the expanded repeat sequence (CGG) for fragile X mental retardation interferes with **transcription** of the DNA into **messenger RNA** (Bassell & Warren, 2008; see Chapter 11).

Genomic Imprinting

Another example of exceptions to Mendel's laws is called *genomic imprinting* (Reik & Walter, 2001). In genomic imprinting, the expression of a gene depends on whether it is inherited from the mother or from the father, even though, as usual, one allele is inherited from each parent. The precise mechanism by which one parent's allele is imprinted is not known, but it usually involves inactivation of a part of the gene by a process called *methylation,* which is an epigenetic mechanism that silences genes (see Chapter 10). Over 100 such genes have been described in both mice and humans (Morison, Ramsay, & Spencer, 2005; http://igc.otago.ac.nz/home.html). The first discovered and most striking example of genomic imprinting in humans involves deletions of a small part of chromosome 15 that lead to two very different disorders, depending on whether a deletion is inherited from the mother or the father. When it

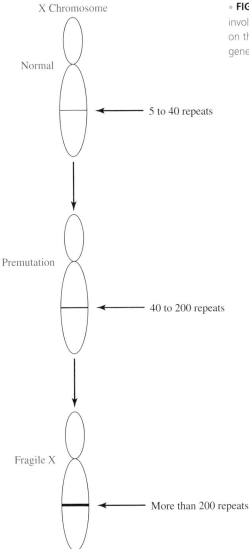

X Chromosome

Normal

5 to 40 repeats

Premutation

40 to 200 repeats

Fragile X

More than 200 repeats

• **FIGURE 3.5** Fragile X mental retardation involves a triplet repeat sequence of DNA on the X chromosome that can expand over generations.

is inherited from the mother, it causes what is known as Angelman syndrome, which causes severe mental retardation and other manifestations, such as an awkward gait and frequent inappropriate laughter. When a deletion is inherited from the father, it causes other behavioral problems, such as overeating, temper outbursts, and depression, as well as physical problems such as obesity and short stature (Prader-Willi syndrome).

Complex Traits

Most psychological traits show patterns of inheritance that are much more complex than those of Huntington disease or PKU. Consider schizophrenia and general cognitive ability.

Schizophrenia

Schizophrenia (Chapter 14) is a severe mental condition characterized by thought disorders. Nearly 1 in 100 people throughout the world are afflicted by this disorder at some point in life, 100 times more than is the case with Huntington disease or PKU. Schizophrenia shows no simple pattern of inheritance like Huntington disease, PKU, or color blindness, but it is familial (Figure 3.6). A special incidence figure used in genetic studies is called a *morbidity risk estimate* (also called the *lifetime expectancy*), which is the chance of being affected during an entire lifetime. The estimate is "age-corrected" for the fact that some as yet unaffected family members have not yet lived through the period of risk. If you have a **second-degree relative** (grandparent or

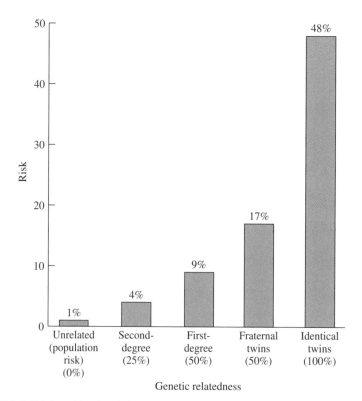

• **FIGURE 3.6** Risk for schizophrenia increases with genetic relatedness. (Data adapted from Gottesman, 1991.)

aunt or uncle) who is schizophrenic, your risk for schizophrenia is about 4 percent, four times greater than the risk in the general population. If a **first-degree relative** (parent or sibling) is schizophrenic, your risk is about 9 percent. If several family members are affected, the risk is greater. If your fraternal twin has schizophrenia, your risk is higher than for other siblings, about 17 percent, even though fraternal twins are no more similar genetically than are other siblings. Most striking, the risk is about 48 percent for an identical twin whose co-twin is schizophrenic. Identical twins develop from one embryo, which in the first few days of life splits into two embryos, each with the same genetic material (Chapter 6).

Clearly, the risk of developing schizophrenia increases systematically as a function of the degree of genetic similarity that an individual has to another who is affected. Heredity appears to be implicated, but the pattern of affected individuals does not conform to Mendelian proportions. Are Mendel's laws of heredity at all applicable to such a complex outcome?

General Cognitive Ability

Many psychological traits are **quantitative dimensions,** as are physical traits such as height and biomedical traits such as blood pressure. Quantitative dimensions are often continuously distributed in the familiar bell-shaped curve, with most people in the middle and fewer people toward the extremes.

For example, an intelligence test score from a general test of intelligence is a composite of diverse tests of cognitive ability and is used to provide an index of general cognitive ability. Intelligence test scores are normally distributed for the most part. (See Chapter 12.)

Because general cognitive ability is a quantitative dimension, it is not possible to count "affected" individuals. Nonetheless, it is clear that general cognitive ability runs in families. For example, parents with high intelligence test scores tend to have children with higher than average scores. As with schizophrenia, transmission of general cognitive ability does not seem to follow simple Mendelian rules of heredity.

The statistics of quantitative traits are needed to describe family resemblance (see Appendix). Over a hundred years ago, Francis Galton, the father of behavioral genetics, tackled this problem of describing family resemblance for quantitative traits. He developed a statistic that he called co-relation and that has become the widely used correlation coefficient. More formally, it is called the Pearson product-moment correlation, named after Karl Pearson, Galton's colleague. The *correlation* is an index of resemblance that ranges from -1.0, indicating an inverse relationship; to 0.0, indicating no resemblance; to 1.0, indicating perfect resemblance.

Correlations for intelligence test scores show that the resemblance of family members depends on the closeness of the genetic relationship (Figure 3.7). The correlation of intelligence test scores for pairs of individuals taken at random from the population is 0.00. The correlation for cousins is about 0.15. For **half siblings,** who

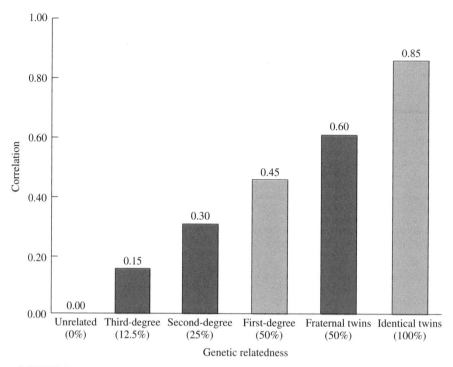

• **FIGURE 3.7** Resemblance for general cognitive ability increases with genetic relatedness. (Data adapted from Bouchard & McGue, 1981, as modified by Loehlin, 1989.)

have just one parent in common, the correlation is about 0.30. For **full siblings,** who have both parents in common, the correlation is about 0.45; this correlation is similar to that between parents and offspring. Scores for fraternal twins correlate about 0.60, which is higher than the correlation of 0.45 for full siblings but lower than the correlation for identical twins, which is about 0.85. In addition, husbands and wives correlate about 0.40, a result that has implications for interpreting sibling and **twin correlations,** as discussed in Chapter 12.

How do Mendel's laws of heredity apply to continuous dimensions such as general cognitive ability?

Pea Size

Although pea plants might not seem relevant to schizophrenia or cognitive ability, they provide a good example of complex traits. A large part of Mendel's success in working out the laws of heredity came from choosing simple traits that are either-or qualitative traits. If Mendel had studied, for instance, the size of the pea seed as indexed by its diameter, he would have found very different results. First, pea seed size, like most traits, is continuously distributed. If he had taken plants with big seeds and crossed them with plants with small seeds, the seed size of their offspring would have

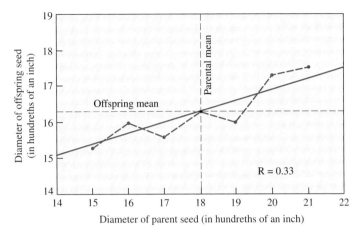

• **FIGURE 3.8** First regression line (solid blue line), drawn by Galton in 1877 to describe the quantitative relationship between pea seed size in parents and offspring. The dashed blue line connects actual data points. (Courtesy of the Galton Laboratory.)

been neither big nor small. In fact, the seeds would have varied in size from small to large, with most offspring seeds of average size.

Only ten years after Mendel's report, Francis Galton studied pea seed size and concluded that it is inherited. For example, parents with large seeds were likely to have offspring with larger than average seeds. In fact, Galton developed the fundamental statistics of regression and correlation mentioned above in order to describe the quantitative relationship between pea seed size in parents and offspring. He plotted parent and offspring seed sizes and drew the regression line that best fits the observed data (Figure 3.8). The slope of the regression line is 0.33. This means that, for the entire population, as parental size increases by one unit, the average offspring size increases one-third of one unit.

Galton also demonstrated that human height shows the same pattern of inheritance. Children's height correlates with the average height of their parents. Tall parents have taller than average children. Children with one tall and one short parent are likely to be of average height. Inheritance of this trait is quantitative rather than qualitative. Quantitative inheritance is the way in which nearly all complex behavioral as well as biological traits are inherited.

Does quantitative inheritance violate Mendel's laws? When Mendel's laws were rediscovered in the early 1900s, many scientists thought this must be the case. They thought that heredity must involve some sort of blending, because offspring resemble the average of their parents. Mendel's laws were dismissed as a peculiarity of pea plants or of abnormal conditions. However, recognizing that quantitative inheritance does *not* violate Mendel's laws is fundamental to an understanding of behavioral genetics, as explained in the following section.

Multiple-Gene Inheritance

The traits that Mendel studied, as well as Huntington disease and PKU, are examples in which a single gene is necessary and sufficient to cause the disorder. That is, you will have Huntington disease only if you have the *H* allele (necessary); if you have the *H* allele, you will have Huntington disease (sufficient). Other genes and environmental factors have little effect on its inheritance. In such cases, a dichotomous (either-or) disorder is found: You either have the specific allele, or not, and thus you have the disorder, or not. More than 3000 (Ku et al., 2011) such single-gene disorders are known definitely and again as many are considered probable.

In contrast, more than just one gene is likely to affect complex disorders such as schizophrenia and continuous dimensions such as general cognitive ability. When Mendel's laws were rediscovered in the early 1900s, a bitter battle was fought between so-called Mendelians and biometricians. Mendelians looked for single-gene effects, and biometricians argued that Mendel's laws could not apply to complex traits because they showed no simple pattern of inheritance. Mendel's laws seemed especially inapplicable to quantitative dimensions.

In fact, both sides were right and both were wrong. The Mendelians were correct in arguing that heredity works the way Mendel said it worked, but they were wrong in assuming that complex traits will show simple Mendelian patterns of inheritance. The biometricians were right in arguing that complex traits are distributed quantitatively, not qualitatively, but they were wrong in arguing that Mendel's laws of inheritance are particular to pea plants and do not apply to higher organisms.

The battle between the Mendelians and biometricians was resolved when biometricians realized that Mendel's laws of inheritance of single genes also apply to complex traits that are influenced by *several* genes. Such a complex trait is called a *polygenic trait*. Each of the influential genes is inherited according to Mendel's laws.

Figure 3.9 illustrates this important point. The top distribution shows the three genotypes of a single gene with two alleles that are equally frequent in the population. As discussed in Box 2.1, 25 percent of the genotypes are homozygous for the A_1 allele (A_1A_1), 50 percent are heterozygous (A_1A_2), and 25 percent are homozygous for the A_2 allele (A_2A_2). If the A_1 allele were dominant, individuals with the A_1A_2 genotype would look just like individuals with the A_1A_1 genotype. In this case, 75 percent of individuals would have the observed trait (phenotype) of the dominant allele. For example, as discussed in Box 2.1, in Mendel's crosses of pea plants with smooth or wrinkled seeds, he found that in the F_2 generation, 75 percent of the plants had smooth seeds and 25 percent had wrinkled seeds.

However, not all alleles operate in a completely dominant or recessive manner. Many alleles are additive in that they each contribute something to the phenotype. In Figure 3.9a, each A_2 allele contributes equally to the phenotype, so if you have two A_2 alleles, you would have a higher score than if you had just one A_2 allele. Figure 3.9b adds a second gene (B) that affects the trait. Again, each B_2 allele makes a contribution. Now there are nine genotypes and five phenotypes. Figure 3.9c adds a third gene (C),

(a)

Number of increasing alleles:

(b)

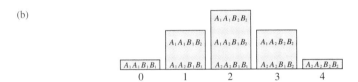

(c)

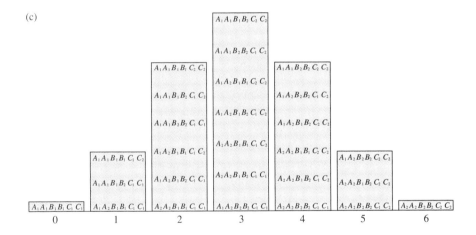

(d)

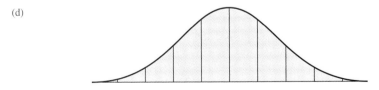

• **FIGURE 3.9** Single-gene and multiple-gene distributions for traits with additive gene effects. (a) A single gene with two alleles yields three genotypes and three phenotypes. (b) Two genes, each with two alleles, yield nine genotypes and five phenotypes. (c) Three genes, each with two alleles, yield twenty-seven genotypes and seven phenotypes. (d) Normal bell-shaped curve of continuous variation.

and there are 27 genotypes. Even if we assume that the alleles of the different genes equally affect the trait and that there is no environmental variation, there are still seven different phenotypes.

So, even with just three genes and two alleles for each gene, the phenotypes begin to approach a normal distribution in the population. When we consider environmental sources of variability and the fact that the effects of alleles are not likely to be equal, it is easy to see that the effects of even a few genes will lead to a quantitative distribution. Moreover, the complex traits that interest behavioral geneticists may be influenced by dozens or even hundreds of genes. Thus, it is not surprising to find continuous variation at the phenotypic level, even though each gene is inherited in accord with Mendel's laws.

Quantitative Genetics

The notion that multiple-gene effects lead to quantitative traits is the cornerstone of a branch of genetics called *quantitative genetics.*

Quantitative genetics was introduced in papers by R. A. Fisher (1918) and by Sewall Wright (1921). Their extension of Mendel's single-gene model to the multiple-gene model of quantitative genetics (Falconer & MacKay, 1996) is described in the Appendix. This multiple-gene model adequately accounts for the resemblance of relatives. If genetic factors affect a quantitative trait, phenotypic resemblance of relatives should increase with increasing degrees of **genetic relatedness.** First-degree relatives (parents/offspring, full siblings) are 50 percent similar genetically. The simplest way to think about this is that offspring inherit half their genetic material from each parent (X linkage aside). If one sibling inherits a particular allele from a parent, the other sibling has a 50 percent chance of inheriting that same allele. Other relatives differ in their degree of genetic relatedness.

Figure 3.10 illustrates degrees of genetic relatedness for the most common types of relatives, using male relatives as examples. Relatives are listed in relation to an individual in the center, the **index case.** The illustration goes back three generations and forward three generations. First-degree relatives (e.g., fathers/sons), who are 50 percent similar genetically, are each one step removed from the index case. Second-degree relatives (e.g., uncles/nephews) are two steps removed and are only half as similar genetically (i.e., 25 percent) as first-degree relatives are. **Third-degree relatives** (e.g., cousins) are three steps removed and half as similar genetically (i.e., 12.5 percent) as second-degree relatives are. Identical twins are a special case, because they are the same person genetically.

For our two examples, schizophrenia and general cognitive ability, phenotypic resemblance of relatives increases with genetic relatedness (see Figures 3.6 and 3.7). How can there be a dichotomous disorder if many genes cause schizophrenia? One possible explanation is that genetic risk is normally distributed but that schizophrenia is not seen until a certain threshold is reached. Another explanation is that disorders

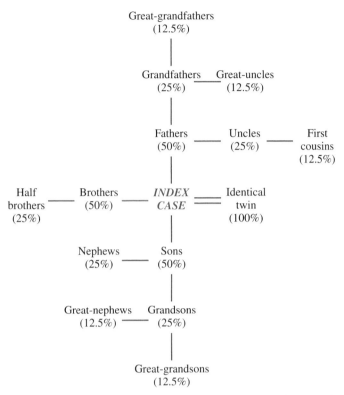

• **FIGURE 3.10** Genetic relatedness: Male relatives of male index case (proband), with degree of genetic relatedness in parentheses.

are actually dimensions artificially established on the basis of a diagnosis. That is, there may be a continuum between what is normal and abnormal. These alternatives are described in Box 3.1.

These data for schizophrenia (Figure 3.6) and general cognitive ability (Figure 3.7) are consistent with the hypothesis of genetic influence, but they do not *prove* that genetic factors are important. It is possible that familial resemblance increases with genetic relatedness for environmental reasons. First-degree relatives might be more similar because they live together. Second-degree and third-degree relatives might be less similar because of less similarity of rearing.

Two experiments of nature are the workhorses of human behavioral genetics that help to disentangle genetic and environmental sources of family resemblance. One is the *twin study*, which compares the resemblance within pairs of identical twins, who are genetically identical, to the resemblance within pairs of fraternal twins, who, like other siblings, are 50 percent similar genetically. The second is the *adoption study*, which separates genetic and environmental influences. For example, when a child is placed for adoption at birth, any resemblance between the adopted child and the

BOX 3.1 • Liability-Threshold Model of Disorders

If complex disorders such as schizophrenia are influenced by many genes, why are they diagnosed as **qualitative disorders** rather than assessed as quantitative dimensions? Theoretically, there should be a continuum of genetic risk, from people having none of the alleles that increase risk for schizophrenia to those having most of the alleles that increase risk. Most people should fall between these extremes, with only a moderate susceptibility to schizophrenia.

One model assumes that risk, or liability, is distributed normally but that the disorder occurs only when a certain threshold of liability is exceeded, as represented in the accompanying figure by the shaded area in (a). Relatives of an affected person have a greater liability, that is, their distribution of liability is shifted to the right, as in (b). For this reason, a greater proportion of the relatives of affected individuals exceed the threshold and manifest the disorder. If there is such a threshold, familial risk can be high only if genetic or shared environmental influence is substantial, because many of an affected individual's relatives will fall just below the threshold and not be affected.

Liability and threshold are hypothetical constructs. However, it is possible to use the **liability-threshold model** to estimate correlations from family risk data (Falconer, 1965; Smith, 1974). For example, the correlation estimated for first-degree relatives for schizophrenia is 0.45, an estimate based on a population base rate of 1 percent and risk to first-degree relatives of 9 percent.

Although correlations estimated from the liability-threshold model are widely reported for psychological disorders, it should be emphasized that this statistic refers to hypothetical constructs of a threshold and an underlying liability derived from diagnoses, not to the risk for the actual diagnosed disorder. In the previous example, the actual risk for schizophrenia for first-degree relatives is 9 percent, even though the liability-threshold correlation is 0.45.

Alternatively, a second model assumes that disorders are actually continuous phenotypically. That is, symptoms might increase continuously from the normal to the abnormal; a diagnosis occurs only when a certain level of symptom severity is reached. The implication is that common disorders are in fact quantitative traits (Plomin, Haworth, & Davis, 2009). A continuum from normal to abnormal seems likely for common disorders such as depression and alcoholism. For example, people vary in the frequency and severity of their depression. Some people rarely get the blues; for others, depression completely disrupts their lives. Individuals diagnosed as depressed might be extreme cases that differ quantitatively, not qualitatively, from the rest of the population. In such cases, it may be possible to assess the continuum directly, rather than assuming a continuum from dichotomous diagnoses using the liability-threshold model. Even for less common disorders like schizophrenia, there is increasing interest in the possibility that there may be no sharp threshold dividing the normal from the abnormal, but rather a continuum from normal to abnormal thought processes. A method called **DF extremes analysis** can be used to investigate the links between the normal and abnormal (see Box 11.1).

The relationship between dimensions and disorders is a key issue, as discussed in later chapters. The best evidence for genetic links between dimensions and disorders will come as specific genes are found for behavior. For example, will a gene associated with diagnosed depression also relate to differences in mood within the normal range?

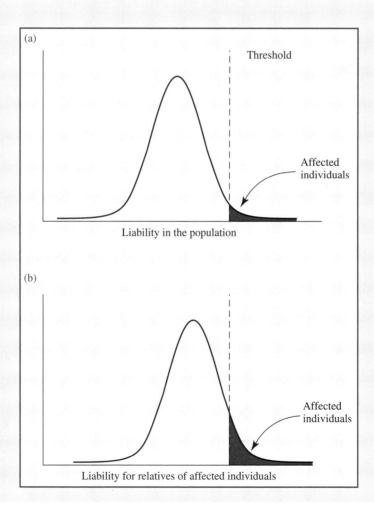

child's birth parents can be attributed to shared heredity rather than to **shared environment** if there is no **selective placement.** In addition, any resemblance between the adoptive parents and their adopted children can be attributed to shared environment rather than to shared heredity. The twin and adoption methods are discussed in Chapter 6.

●KEY CONCEPTS

Polygenic: Influenced by multiple genes.

Genetic relatedness: The extent or degree to which relatives have genes in common. *First-degree relatives* of the **proband** (parents and siblings) are 50 percent similar genetically. *Second-degree relatives* of the proband (grandparents, aunts, and uncles) are 25 percent similar genetically. *Third-degree relatives* of the proband (first cousins) are 12.5 percent similar genetically.

Liability-threshold model: A model which assumes that dichotomous disorders are due to underlying genetic liabilities that are distributed normally. The disorder appears only when a threshold of liability is exceeded.

Summary

Mendel's laws of heredity do not explain all genetic phenomena. For example, genes on the X chromosome, such as the gene for color blindness, require an extension of Mendel's laws. Other exceptions to Mendel's laws include new mutations, changes in chromosomes such as the chromosomal nondisjunction that causes Down syndrome, expanded DNA triplet repeat sequences responsible for Huntington disease and fragile X mental retardation, and genomic imprinting.

Most psychological dimensions and disorders show more complex patterns of inheritance than do single-gene disorders such as Huntington disease, PKU, or color blindness. Complex disorders such as schizophrenia and continuous dimensions such as cognitive ability are likely to be influenced by multiple genes as well as by multiple environmental factors. Quantitative genetic theory extends Mendel's single-gene rules to multiple-gene systems. The essence of the theory is that complex traits can be influenced by many genes, but each gene is inherited according to Mendel's laws. Quantitative genetic methods, especially adoption and twin studies, can detect genetic influence for complex traits.

CHAPTER FOUR

DNA: The Basis of Heredity

M endel was able to deduce the laws of heredity even though he had no idea of how heredity works at the chemical or physiological level. Quantitative genetics, such as twin and adoption studies, depends on Mendel's laws of heredity but does not require knowledge of the biological basis of heredity. However, it is important to understand the biological mechanisms underlying heredity for two reasons. First, understanding the biological basis of heredity makes it clear that the processes by which genes affect behavior are not mystical. Second, this understanding is crucial for appreciating the exciting advances in attempts to identify genes associated with behavior. This chapter describes the biological basis of heredity. There are many excellent genetics texts that provide great detail about this subject (e.g., Hartl & Ruvolo, 2011). The biological basis of heredity includes the fact that genes are contained on structures called chromosomes. The linkage of genes that lie close together on a chromosome has made possible the **mapping** of the human genome. Moreover, abnormalities in chromosomes contribute importantly to behavioral disorders, especially mental retardation.

DNA

Nearly a century after Mendel did his experiments, it became apparent that DNA (deoxyribonucleic acid) is the molecule responsible for heredity. In 1953, James Watson and Francis Crick proposed a molecular structure for DNA that could explain how genes are replicated and how DNA codes for proteins. As shown in Figure 4.1, the DNA molecule consists of two strands that are held apart by pairs of four bases: adenine, thymine, guanine, and cytosine. As a result of the structural properties of these bases, adenine always pairs with thymine and guanine always pairs with cytosine. The backbone of each strand consists of sugar and phosphate molecules. The strands coil around each other to form the famous double helix of DNA (Figure 4.2).

The specific pairing of bases in these two-stranded molecules allows DNA to carry out its two functions: to replicate itself and to direct the synthesis of proteins.

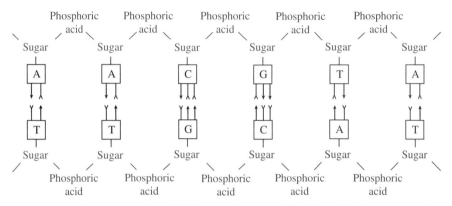

• **FIGURE 4.1** Flat representation of the four DNA bases in which adenine (A) always pairs with thymine (T) and guanine (G) always pairs with cytosine (C). (From *Heredity, Evolution, and Society* by I. M. Lerner. W. H. Freeman and Company. ©1968.)

Replication of DNA occurs during the process of cell division. The double helix of the DNA molecule unzips, separating the paired bases (Figure 4.3). The two strands unwind, and each strand attracts the appropriate bases to construct its complement. In this way, two complete double helices of DNA are created where there was previously only one. This process of replication is the essence of life, which began billions of years ago when the first cells replicated themselves. It is also the essence of each of our lives, beginning with a single cell and faithfully reproducing our DNA in trillions of cells.

The second major function of DNA is to direct the synthesis of proteins according to the genetic information that resides in the particular sequence of bases. DNA encodes the various sequences of the 20 amino acids making up the thousands of specific enzymes and other proteins that are the stuff of living organisms. Box 4.1 describes this process, the so-called central dogma of **molecular genetics.**

What is the genetic code contained in the sequence of DNA bases, which is transcribed to messenger RNA (mRNA; see Box 4.1) and then translated into amino acid sequences? The code consists of various sequences of three bases, which are called *codons*

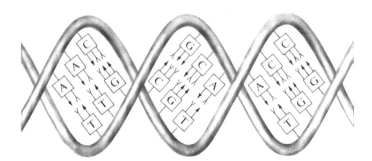

• **FIGURE 4.2** A three-dimensional view of a segment of DNA. (From *Heredity, Evolution, and Society* by I. M. Lerner. W. H. Freeman and Company. ©1968.)

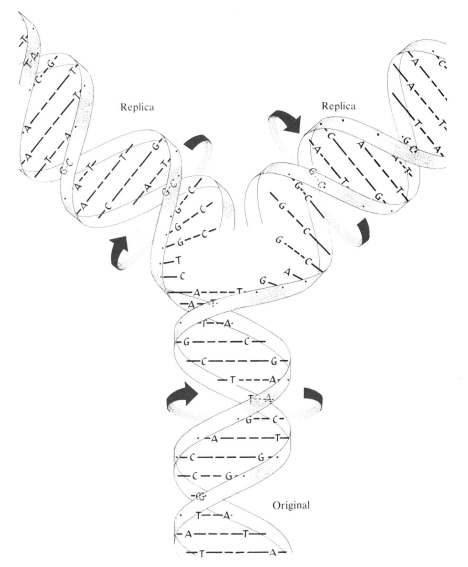

• **FIGURE 4.3** Replication of DNA. (After *Molecular Biology of Bacterial Viruses* by G. S. Stent. W. H. Freeman and Company. ©1963.)

(Table 4.1). For example, three adenines in a row (AAA) in the DNA molecule will be transcribed in mRNA as three uracils (UUU). This mRNA codon codes for the amino acid phenylalanine. Although there are 64 possible **triplet codons** ($4^3 = 64$), there are only 20 amino acids. Some amino acids are coded by as many as six codons. Any one of three particular codons signals the end of a transcribed sequence (stop signals).

This same genetic code applies to all living organisms. Discovering this code was one of the great triumphs of molecular biology. The human set of **DNA sequences**

| TABLE | 4.1 |

The Genetic Code

Amino Acid*	DNA Code
Alanine	CGA, CGG, CGT, CGC
Arginine	GCA, GCG, GCT, GCC, TCT, TCC
Asparagine	TTA, TTG
Aspartic acid	CTA, CTG
Cysteine	ACA, ACG
Glutamic acid	CTT, CTC
Glutamine	GTT, GTC
Glycine	CCA, CCG, CCT, CCC
Histidine	GTA, GTG
Isoleucine	TAA, TAG, TAT
Leucine	AAT, AAC, GAA, GAG, GAT, GAC
Lysine	TTT, TTC
Methionine	TAC
Phenylalanine	AAA, AAG
Proline	GGA, GGG, GGT, GGC
Serine	AGA, AGG, AGT, AGC, TCA, TCG
Threonine	TGA, TGG, TGT, TGC
Tryptophan	ACC
Tyrosine	ATA, ATG
Valine	CAA, CAG, CAT, CAC
(Stop signals)	ATT, ATC, ACT

*The 20 amino acids are organic molecules that are linked together by peptide bonds to form polypeptides, which are the building blocks of enzymes and other proteins. The particular combination of amino acids determines the shape and function of the polypeptide.

(the genome) consists of about 3 billion **base pairs,** counting just one chromosome from each pair of chromosomes. The 3 billion base pairs contain about 25,000 protein-coding genes, which range in size from about 1000 bases to 2 million bases. The chromosomal locations of most genes are known. About a third of our protein-coding genes are expressed only in the brain; these are likely to be most important for behavior. The human genome sequence is like an encyclopedia of genes with 3 billion letters, equivalent in length to about 3000 books of 500 pages each. Continuing with this simile, the encyclopedia of genes is written in an alphabet consisting of 4 letters (A, T, G, C), with 3-letter words (codons) organized into 23 volumes (chromosomes). This simile, however, does not comfortably extend to the fact that each encyclopedia is different; millions of letters (about 1 in 1000) differ for any two people. There is no single human genome; we each have a different genome, except for identical twins. Most of the life sciences focus on the generalities of the genome, but the genetic causes of diseases and disorders lie in these variations in the genome. These variations on the human theme are the focus of behavioral genetics.

The twentieth century has been called the century of the gene. The century began with the re-discovery of Mendel's laws of heredity. The word *genetics* was first coined in 1905. Almost fifty years later, Crick and Watson described the double helix of DNA, the premier icon of science. The pace of discoveries accelerated greatly during the next fifty years, culminating at the turn of the twenty-first century with the sequencing of the human genome. Most of the human genome was sequenced by 2001 (International Human Genome Sequencing Consortium, 2001; Venter et al., 2001). Subsequent publications have presented the finished sequence for all chromosomes (e.g., Gregory et al., 2006).

Sequencing of the human genome and the technologies associated with it have led to an explosion of new findings in genetics. One of many examples was ***alternative splicing,*** in which mRNA is spliced to create different transcripts, which are then translated into different proteins (Brett, Pospisil, Valcárcel, Reich, & Bork, 2002). Alternative splicing has a crucial role in the generation of biological complexity, and its disruption can lead to a wide range of human diseases (Barash et al., 2010). The speed of discovery in genetics is now so great that it would be impossible to predict what will happen in the next five years, let alone the next fifty years. Most geneticists would agree with Francis Collins (2010), the director of the U.S. National Institutes of Health and leader in the Human Genome Project, who expects that the entire genome of all newborns will soon be sequenced to screen for genetic problems and that eventually we will each possess an electronic chip containing our DNA sequence. Individual DNA chips would herald a revolution in personalized medicine in which treatment could be individually tailored rather than dependent on our present one-size-fits-all approach. The greatest value of DNA lies in its ability to predict genetic risk that could lead to preventative interventions. That is, rather than treating problems after they occur, DNA may allow us to predict problems and intervene to prevent them. This could involve genetic engineering that alters DNA. While such efforts with regard to gene therapy in the human species have been historically difficult, even for single-gene disorders (Rubanyi, 2001), recent results in correcting vision loss from a genetically informed standpoint have been promising (e.g., Komaromy et al., 2010; Roy, Stein, & Kaushal, 2010). Importantly, to prevent complex behavioral problems that are affected by many genes as well as many environmental factors, behavioral and environmental engineering will be needed.

We are now in a better position to understand DNA changes in health, behavior, and disease in ways that would not have been thought possible five years ago. There are detailed maps of genetic variation, and efforts are under way to identify parts of the genome that affect the function of genes. Thanks to decreasing costs of new sequencing technologies (see Chapter 9), researchers are examining genome changes that lead to both inherited diseases and common diseases, such as cancer. Another

BOX 4.1 • The "Central Dogma" of Molecular Genetics

enetic information flows from DNA to messenger RNA (mRNA) to protein. These protein-coding genes are DNA segments that are a few thousand to several million DNA base pairs in length. The DNA molecule contains a linear message consisting of four bases (adenine, thymine, guanine, and cytosine); in this two-stranded molecule, A always pairs with T and G always pairs with C. The message is decoded in two basic steps, shown in the figure: (a) transcription of DNA into a different sort of nucleic acid called ribonucleic acid, or RNA, and (b) **translation** of RNA into proteins.

In the transcription process, the sequence of bases in one strand of the

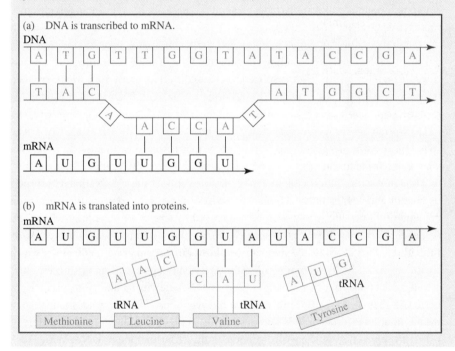

(a) DNA is transcribed to mRNA.

(b) mRNA is translated into proteins.

new direction for research involves efforts to understand the human microbiome (Archie & Theis, 2011; Zhu, Wang, & Li, 2010), the genomes of the microbes that live in and on our bodies, as well as the **epigenome,** chemical marks on our DNA that may play a part in how the human genome functions and contributes to health, behavior, and disease (Rakyan, Down, Balding, & Beck, 2011). For behavioral genetics, the most important thing to understand about the DNA basis of heredity is that the process by which genes affect behavior is not mystical. Genes code for sequences of amino acids that form the thousands of proteins of which organisms are made. Proteins create the skeletal system, muscles, the endocrine system, the immune system, the digestive system, and, most important for behavior, the nervous system. Genes do not code for

DNA double helix is copied to RNA, specifically a type of RNA called messenger RNA (mRNA) because it relays the DNA code. mRNA is single stranded and is formed by a process of base pairing similar to the replication of DNA, except that uracil substitutes for thymine (so that A pairs with U instead of T). In the figure, one DNA strand is being transcribed—the DNA bases ACCA have just been copied as UGGU in mRNA. mRNA leaves the nucleus of the cell and enters the cell body (cytoplasm), where it connects with **ribosomes,** which are the factories where proteins are built.

The second step involves translation of the mRNA into amino acid sequences that form proteins. Another form of RNA, called transfer RNA (tRNA), transfers amino acids to the ribosomes. Each tRNA is specific to 1 of the 20 amino acids. The tRNA molecules, with their attached specific amino acids, pair up with the mRNA in a sequence dictated by the base sequence of the mRNA, as the ribosome moves along the mRNA strand. Each of the 20 amino acids found in proteins is specified by a codon made up of three sequential mRNA bases. In the figure, the mRNA code has

begun to dictate a protein that includes the amino acid sequence methionine-leucine-valine-tyrosine. Valine has just been added to the chain that already includes methionine and leucine. The mRNA triplet code GUA attracts tRNA with the complementary code CAU. This tRNA transfers its attached amino acid valine, which is then bonded to the growing chain of amino acids. The next mRNA codon, UAC, is attracting tRNA with the complementary codon, AUG, for tyrosine. Although this process seems very complicated, amino acids are incorporated into chains at the incredible rate of about 100 per second. Proteins consist of particular sequences of about 100 to 1000 amino acids. The sequence of amino acids determines the shape and function of proteins. Protein shape is subsequently altered in other ways called ***posttranslational changes.*** These changes affect its function and are not controlled by the genetic code.

Surprisingly, DNA that is transcribed and translated like this represents only about 2 percent of the genome. What is the other 98 percent doing? See Chapter 10 for an answer.

behavior directly, but DNA variations that create differences in these physiological systems can affect behavior. We will discuss DNA variations in Chapter 9.

KEY CONCEPTS

Codon: A sequence of three base pairs that codes for a particular amino acid or the end of a transcribed sequence.

Transcription: The synthesis of an RNA molecule from DNA in the cell nucleus.

Translation: Assembly of amino acids into peptide chains on the basis of information encoded in messenger RNA. Occurs on ribosomes in the cell cytoplasm.

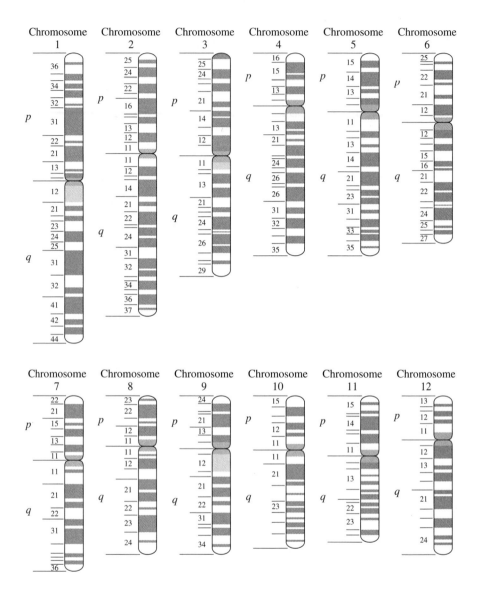

Chromosomes

As discussed in Chapter 2, Mendel did not know that genes are grouped together on chromosomes, so he assumed that all genes are inherited independently. However, Mendel's second law of independent assortment is violated when two genes are close together on the same chromosome. In this case, the two genes are not inherited independently; and, on the basis of this nonindependent assortment, linkages between DNA markers have been identified and used to produce a map of the genome. With

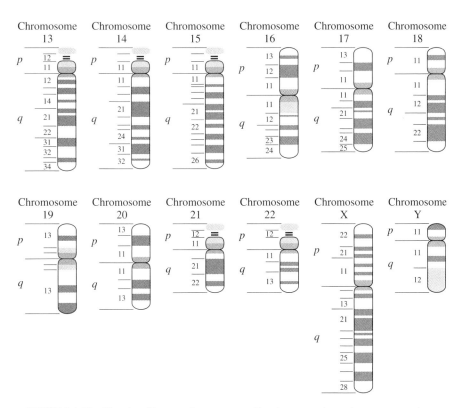

• **FIGURE 4.4** The 23 pairs of human chromosomes. The short arm above the centromere is called *p*, and the long arm below the centromere is called *q*. The bands, created by staining, are used to identify the chromosomes and to describe the location of genes. Chromosomal regions are referred to by chromosome number, arm of chromosome, and band. Thus, 1*p*36 refers to band 6 in region 3 of the *p* arm of chromosome 1. For more details about each chromosome and the locus of major genetic disorders, see http://www.ornl.gov/sci/techresources/Human_Genome/posters/chromosome/chooser.shtml

the same technique, mapped DNA markers are used to identify linkages with disorders and dimensions, including behavior, as described in Chapter 9.

Our species has 23 pairs of chromosomes, for a total of 46 chromosomes. The number of chromosome pairs varies widely from species to species. Fruit flies have 4, mice have 20, dogs have 39, and butterflies have 190. Our chromosomes are very similar to those of the great apes (chimpanzee, gorilla, and orangutan). Although the great apes have 24 pairs, two of their short chromosomes have been fused to form one of our large chromosomes.

As noted above, one pair of our chromosomes is the *sex chromosomes* X and Y. Females are XX and males are XY. All the other chromosomes are called **autosomes.** As shown in Figure 4.4, chromosomes have characteristic banding patterns when stained with a particular chemical. The **bands,** whose function is not known, are used to identify the chromosomes. At some point in each chromosome, there is a *centromere,* a region of the chromosome without genes, where the chromosome is attached to its new copy when cells reproduce. The short arm of the chromosome above the centromere is called *p* and the long arm below the centromere is called *q.* The location of genes is described in relation to the bands. For example, the gene for Huntington disease is at 4*p*16, which means the short arm of chromosome 4 at a particular band, number 6 in region 1.

In addition to providing the basis for **gene mapping,** chromosomes are important in behavioral genetics because mistakes in copying chromosomes during cell division affect behavior. There are two kinds of cell division. Normal cell division, called *mitosis,* occurs in all cells not involved in the production of gametes. These cells are called *somatic cells.* The sex cells produce eggs and sperm, the *gametes.* In mitosis, each chromosome in the somatic cell duplicates and divides to produce two identical cells. A special type of cell division called *meiosis* occurs in the sex cells of the ovaries and testes to produce eggs and sperm, both of which have only one member of each chromosome pair. Each egg and each sperm have 1 of over 8 million (2^{23}) possible combinations of the 23 pairs of chromosomes. Moreover, crossover (recombination) of members of each chromosome pair (see Figure 2.7) occurs about once per meiosis and creates even more genetic variability. When a sperm fertilizes an egg to produce a **zygote,** one chromosome of each pair comes from the mother's egg and the other from the father's sperm, thereby reconstituting the full complement of 23 pairs of chromosomes.

◾KEY CONCEPTS

Centromere: A chromosomal region without genes where the chromatids are held together during cell division.
Mitosis: Cell division that occurs in somatic cells in which a cell duplicates itself and its DNA.
Meiosis: Cell division that occurs during gamete formation and results in halving the number of chromosomes, so that each gamete contains only one member of each chromosome pair.

As indicated in Chapter 3, a common copying error for chromosomes is an uneven split of the pairs of chromosomes during meiosis, called nondisjunction (see Figure 3.4). The most common form of mental retardation, Down syndrome, is caused by nondisjunction of one of the smallest chromosomes, chromosome 21. Many other chromosomal problems occur, such as breaks in chromosomes that lead to inversion, deletion, duplication, and translocation. About half of all fertilized human eggs have a chromosomal abnormality. Most of these abnormalities result in early spontaneous abortions (miscarriages). At birth, about 1 in 250 babies have an obvious chromosomal abnormality. Small abnormalities such as deletions have been difficult to detect but are being made much easier to detect by DNA **microarrays** and sequencing, which are described in Chapter 9. Although chromosomal abnormalities occur for all chromosomes, only fetuses with the least severe abnormalities survive to birth. Some of these babies die soon after they are born. For example, most babies with three chromosomes (trisomy) of chromosome 13 die in the first month, and most of those with trisomy-18 die within the first year. Other chromosomal abnormalities are less lethal but result in behavioral and physical problems. Nearly all major chromosomal abnormalities influence cognitive ability, as expected if cognitive ability is affected by many genes. Because the behavioral effects of chromosomal abnormalities often involve mental retardation, they are discussed in Chapter 11.

Missing a whole chromosome is lethal, except for the X and Y chromosomes. Having an entire extra chromosome is also lethal, except for the smallest chromosomes and the X chromosome, which is one of the largest. The reason why the X chromosome is the exception is also the reason why half of all chromosomal abnormalities that exist in newborns involve the sex chromosomes. In females, one of the two X chromosomes is inactivated, in the sense that most of its genes are not transcribed. In males and females with extra X chromosomes, the extra X chromosomes also are inactivated. For this reason, even though X is a large chromosome with many genes, having an extra X in males or females is not lethal. The most common sex chromosome abnormalities are XXY (males with an extra X), XXX (females with an extra X), and XYY (males with an extra Y), each with an incidence of about 1 in 1000. The incidence of XO (females with just one X) is lower, 1 in 2500 at birth, because 98 percent of such conceptuses abort spontaneously.

Summary

One of the most exciting advances in biology has been understanding Mendel's "elements" of heredity. The double helix structure of DNA relates to its dual functions of self-replication and protein synthesis. The genetic code consists of a sequence of three DNA bases that codes for amino acids. DNA is transcribed to mRNA, which is translated into amino acid sequences

Genes are inherited on chromosomes. Linkage between DNA markers and behavior can be detected by looking for exceptions to Mendel's law of independent assortment, because a DNA marker and a gene for behavior are not inherited independently if they are close together on the same chromosome. Our species has 23 pairs of chromosomes. Mistakes in duplicating chromosomes often affect behavior directly. About 1 in 250 newborns has a major chromosomal abnormality, and about half of these abnormalities involve the sex chromosomes.

CHAPTER FIVE

Animal Models in Behavioral Genetics

B ehavioral genetic research includes both human and animal approaches. In this chapter we will describe the different ways that animal research has been used to help us understand the roles of genes and environments in behavior. The first part of the chapter focuses on quantitative genetic designs, while the second describes how animal studies help to identify genes and clarify their function.

Quantitative Genetic Experiments to Investigate Animal Behavior

Dogs provide a dramatic yet familiar example of genetic variability within species (Figure 5.1). Despite their great variability in size and physical appearance—from a height of six inches for the Chihuahua to three feet for the Irish wolfhound—they are all members of the same species. Molecular genetic research suggests that dogs, which originated from wolves about 30,000 years ago as they were domesticated, may have enriched their supply of genetic variability by repeated intercrossing with wolves (vonHoldt et al., 2010). The genome of the domestic dog has been sequenced (Lindblad-Toh et al., 2005), which makes it possible to identify dog breeds on the basis of DNA alone and suggests that there are four basic genetic clusters of dogs: wolves and Asian dogs (the earliest domesticated dogs, such as Akitas and Lhasa Apsos), mastiff-type dogs (e.g., mastiffs and boxers), working dogs (e.g., collies and sheepdogs), and hunting dogs (e.g., hounds and terriers) (Parker et al., 2004).

Dogs also illustrate genetic effects on behavior. Although physical differences between breeds are most obvious, dogs have been bred for centuries as much for their behavior as for their looks. In 1576, the earliest English-language book on dogs classified breeds primarily on the basis of behavior. For example, terriers (from *terra*, which is Latin for "earth") were bred to creep into burrows to drive out small animals. Another book, published in 1686, described the behavior for which spaniels were originally selected. They were bred to creep up on birds and then spring to frighten the birds

• **FIGURE 5.1** Dog breeds illustrate genetic diversity within species for behavior as well as physical appearance.

into the hunter's net, which is the origin of the *springer spaniel*. With the advent of the shotgun, different spaniels were bred to point rather than to spring. The author of the 1686 work was especially interested in temperament: "Spaniels by Nature are very loveing, surpassing all other Creatures, for in Heat and Cold, Wet and Dry, Day and Night, they will not forsake their Master" (cited by Scott & Fuller, 1965, p. 47). These temperamental characteristics led to the creation of spaniel breeds selected specifically to be pets, such as the King Charles spaniel, which is known for its loving and gentle temperament.

Behavioral classification of dogs continues today. Sheepdogs herd, retrievers retrieve, trackers track, pointers point, and guard dogs guard with minimal training. Breeds also differ strikingly in trainability and in temperamental traits such as emotionality, activity, and aggressiveness, although there is also substantial variation in these traits within each breed (Coren, 2005). The selection process can be quite fine tuned. For example, in France, where dogs are used chiefly for farm work, there are 17 breeds of shepherd and stock dogs specializing in aspects of this work. In England, dogs have been bred primarily for hunting, and there are 26 recognized breeds of hunting dogs. Dogs are unusual in the extent to which different breeds have been intentionally bred to accentuate genetic differences in behavior.

An extensive behavioral genetic research program on breeds of dogs was conducted over two decades by J. Paul Scott and John Fuller (1965). They studied the development of pure breeds and hybrids of the five breeds pictured in Figure 5.2: wire-haired fox terriers, cocker spaniels, basenjis, sheepdogs, and beagles. These breeds are all about the same size, but they differ markedly in behavior. Although considerable genetic variability remains within each breed, average behavioral differences among the breeds reflect their breeding history. For example, as their history would suggest, terriers are aggressive scrappers, while spaniels are nonaggressive and people-oriented. Unlike the other breeds, sheepdogs have been bred, not for hunting, but for performing complex tasks under close supervision from their masters. They are very responsive to training. In short, Scott and Fuller found behavioral breed differences just about everywhere they looked—in the development of social relationships, emotionality, and trainability, as well as many other behaviors. They also found evidence for interactions between breeds and training. For example, scolding that would be brushed off by a terrier could traumatize a sheepdog.

Selection Studies

Laboratory experiments that select for behavior provide the clearest evidence for genetic influence on behavior. As dog breeders and other animal breeders have known for centuries, if a trait is heritable, you can breed selectively for it. Research in Russia aimed to understand how our human ancestors had domesticated dogs from wolves by selecting for tameness in foxes, which are notoriously wary of humans. Foxes that were the tamest when fed or handled were bred for more than 40 generations. The result of this **selection study** is a new breed of foxes that are like dogs in their

• **FIGURE 5.2** J. P. Scott with the five breeds of dogs used in his experiments with J. L. Fuller. Left to right: wire-haired fox terrier, American cocker spaniel, African basenji, Shetland sheepdog, and beagle. (From *Genetics and the Social Behavior of the Dog* by J. P. Scott & J. L. Fuller. ©1965 by The University of Chicago Press. All rights reserved.)

friendliness and eagerness for human contact (Figure 5.3), so much so that these foxes have now become popular house pets in Russia (Kukekova et al., 2011; Trut, Oskina, & Kharlamova, 2009).

Laboratory experiments typically select high and low lines in addition to maintaining an unselected control line. For example, in one of the largest and longest selection studies of behavior (DeFries, Gervais, & Thomas, 1978), mice were selected for activity in a brightly lit box called an open field, a measure of fearfulness that was invented more than 70 years ago (Figure 5.4). In the open field, some animals become immobile, defecate, and urinate, whereas others actively explore it. Lower activity scores are presumed to index fearfulness.

The most active mice were selected and mated with other high-active mice. The least active mice were also mated with each other. From the offspring of the high-active and low-active mice, the most and least active mice were again selected and mated in a similar manner. This selection process was repeated for 30 generations. (In mice, a generation takes only about three months.)

• **FIGURE 5.3** Foxes are normally wary of humans and tend to bite. After selecting for tameness for 40 years, a program involving 45,000 foxes has developed animals that are not only tame but friendly. This one-month-old fox pup not only tolerates being held but is licking the woman's face. (From Trut, 1999. Reprinted with permission.)

• **FIGURE 5.4** Mouse in an open field. The holes near the floor transmit light beams that electronically record the mouse's activity.

The results are shown in Figures 5.5 and 5.6 for replicated high, low, and control lines. Over the generations, selection was successful: The high lines became increasingly more active and the low lines less active (see Figure 5.5). Successful selection can occur only if heredity is important. After 30 generations of such selective breeding, a 30-fold average difference in activity has been achieved. There is no overlap between the activity of the low and high lines (see Figure 5.6). Mice from the high-active line now boldly run the equivalent total distance of the length of a football field during the six-minute test period, whereas the low-active mice quiver in the corners.

Another important finding is that the difference between the high and low lines steadily increases each generation. This outcome is a typical finding from selection studies of behavioral traits and strongly suggests that many genes contribute to variation in behavior. If just one or two genes were responsible for open-field activity, the

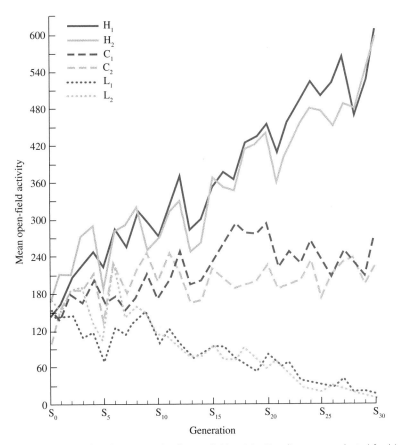

• **FIGURE 5.5** Results of a selection study of open-field activity. Two lines were selected for high open-field activity (H_1 and H_2), two lines were selected for low open-field activity (L_1 and L_2), and two lines were randomly mated within each line to serve as controls (C_1 and C_2). (From "Response to 30 generations of selection for open-field activity in laboratory mice" by J. C. DeFries, M. C. Gervais, & E. A. Thomas. *Behavior Genetics, 8,* 3–13. ©1978 by Plenum Publishing Corporation. All rights reserved.)

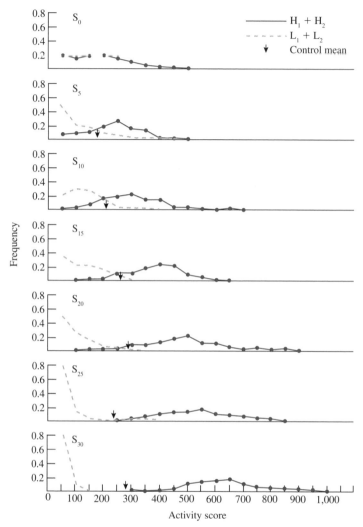

• **FIGURE 5.6** Distributions of activity scores of lines selected for high and low open-field activity for 30 generations (S_0 to S_{30}). Average activity of control lines in each generation is indicated by an arrow. (From "Response to 30 generations of selection for open-field activity in laboratory mice" by J. C. DeFries, M. C. Gervais, & E. A. Thomas. *Behavior Genetics, 8,* 3–13. ©1978 by Plenum Publishing Corporation. All rights reserved.)

two lines would separate after a few generations and would not diverge any further in later generations.

Despite the major investment required to conduct a selection study, the method continues to be used in behavioral genetics, in part because of the convincing evidence it provides for genetic influences on behavior and in part because it produces lines of animals that differ as much as possible genetically for a particular behavior (e.g., Zombeck, DeYoung, Brzezinska, & Rhodes, 2011).

Inbred Strain Studies

The other major quantitative genetic design for animal behavior compares *inbred strains*, in which brothers have been mated with sisters for at least 20 generations. This intensive **inbreeding** makes each animal within the inbred strain virtually a genetic clone of all other members of the strain. Because inbred strains differ genetically from one another, genetically influenced traits will show average differences between inbred strains reared in the same laboratory environment. Differences within strains are due to environmental influences. In animal behavioral genetic research, mice are most often studied; more than 450 inbred strains of mice are available (Beck et al., 2000). Some of the most frequently studied inbred strains are shown in Figure 5.7. A database cataloging differences between inbred mouse strains—including behavioral differences such as anxiety, learning and memory, and stress reactivity—can

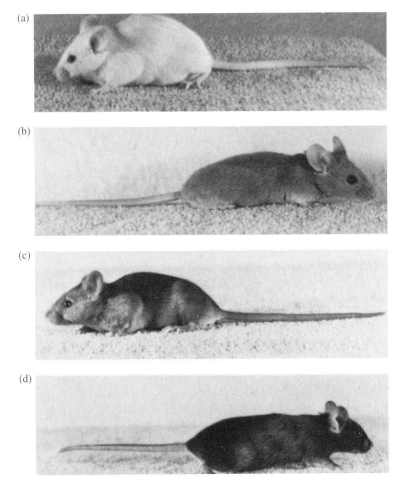

• **FIGURE 5.7** Four common inbred strains of mice: (a) BALB/c; (b) DBA/2; (c) C3H/2; (d) C57BL/6.

be found at: http://phenome.jax.org/, which includes data for over 2000 different measurements for 178 inbred strains (Flint, 2011).

Studies of inbred strains suggest that most mouse behaviors show genetic influence. For example, Figure 5.8 shows the average open-field activity scores of two inbred strains called BALB/c and C57BL/6. The C57BL/6 mice are much more active than the BALB/c mice, an observation suggesting that genetics contributes to open-field activity. The mean activity scores of several crosses are also shown: F_1, F_2, and F_3 crosses (explained in Box 2.1) between the inbred strains, the backcross between the F_1 and the BALB/c strain (B_1 in Figure 5.8), and the backcross between the F_1 and the C57BL/6 strain (B_2 in Figure 5.8). There is a strong relationship between the average open-field scores and the percentage of genes obtained from the C57BL/6 parental strain, which again points to genetic influence.

Rather than just crossing two inbred strains, the ***diallel design*** compares several inbred strains and all possible F_1 crosses between them. Figure 5.9 shows the open-field results of a diallel cross between BALB/c, C57BL/6, and two other inbred strains (C3H/2 and DBA/2). C3H/2 is even less active than BALB/c, and DBA/2 is almost as active as C57BL/6. The F_1 crosses tend to correspond to the average scores of their parents. For example, the F_1 cross between C3H/2 and BALB/c is intermediate to the two parents in open-field activity.

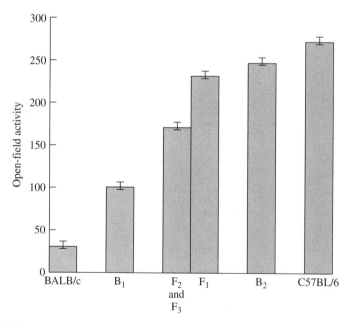

• **FIGURE 5.8** Mean open-field activity ($\pm$ twice the standard error) of BALB/c and C57BL/6 mice and their derived F_1, backcross (B_1 and B_2), F_2, and F_3 generations. (From "Response to 30 generations of selection for open-field activity in laboratory mice" by J. C. DeFries, M. C. Gervais, & E. A. Thomas. *Behavior Genetics, 8,* 3–13. ©1978 by Plenum Publishing Corporation. All rights reserved.)

Studies of inbred strains are also useful for detecting environmental effects. First, because members of an inbred strain are genetically identical, individual differences within a strain must be due to environmental factors. Large differences within inbred strains are found for open-field activity and most other behaviors studied, reminding us of the importance of prenatal and postnatal nurture as well as nature. Second, inbred strains can be used to assess the net effect of mothering by comparing F_1 crosses in which the mother is from either one strain or the other. For example, the F_1 cross between BALB/c mothers and C57BL/6 fathers can be compared to the genetically equivalent F_1 cross between C57BL/6 mothers and BALB/c fathers. In a diallel study like that shown in Figure 5.9, these two hybrids had nearly identical scores, as was the case for comparisons between the other crosses as well. This result suggests that prenatal and postnatal environmental effects of the mother do not importantly affect open-field activity. If maternal effects are found, it is possible to separate prenatal and postnatal effects by cross-fostering pups of one strain with mothers of the other strain. Third, the environments of inbred strains can be manipulated in the laboratory to investigate interactions between genotype and environment, as discussed in Chapter 8. A type of **genotype-environment interaction** was reported in an influential paper in which genetic influences as assessed by inbred strains differed across laboratories for some behaviors, although the results for open-field activity were robust across laboratories (Crabbe, Wahlsten, & Dudek, 1999). Subsequent studies indicated that

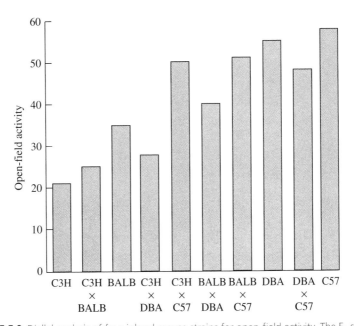

• **FIGURE 5.9** Diallel analysis of four inbred mouse strains for open-field activity. The F_1 strains are ordered according to the average open-field activity score of their parental inbred strains. (After Henderson, 1967.)

the rank order between inbred strains for behaviors showing large strain differences is stable across laboratories (Wahlsten et al., 2003). For example, comparisons over 50 years of research on inbred strains for locomotor activity and ethanol preference yield rank-order correlations of 0.85 to 0.98 across strains (Wahlsten, Bachmanov, Finn, & Crabbe, 2006). Another study of more than 2000 outbred mice also showed few interactions between open-field activity and experimental variables, such as who tests the mice and order of testing (Valdar, Solberg, Gauguier, Cookson, et al., 2006). Nonetheless, there is value in multi-laboratory studies in terms of generalizability of inbred strain results (Kafkafi, Benjamini, Sakov, Elmer, & Golani, 2005).

More than 1000 behavioral investigations involving genetically defined mouse strains were published between 1922 and 1973 (Sprott & Staats, 1975), and the pace accelerated into the 1980s. Studies such as these played an important role in demonstrating that genetics contributes to most behaviors. Although **inbred strain studies** now tend to be overshadowed by more sophisticated genetic analyses, inbred strains still provide a simple and highly efficient test for the presence of genetic influence. For example, inbred strains have recently been used to screen for genetic mediation of associations between genomewide **gene expression** profiles and behavior (Letwin et al., 2006; Nadler et al., 2006), a topic to which we will return in Chapter 10.

● KEY CONCEPTS

Selective breeding: Breeding for a phenotype over several generations by selecting parents with high scores on the phenotype, mating them, and assessing their offspring to determine the response to selection. Bidirectional selection studies also select in the other direction, that is, for low scores.

Inbred strain: A strain of animal (usually mice) that has been mated with siblings for at least 20 generations, resulting in genetically identical individuals. Use of inbred strains allows genetic and environmental influences on behavior to be specified.

Animal Studies for Identifying Genes and Gene Functions

The first part of this chapter described how inbred strain and selection studies with animals provide direct experiments to investigate genetic influence. In contrast, as we will describe in Chapter 6, quantitative genetic research on human behavior is limited to less direct designs, primarily adoption, the experiment of nurture, and twinning, the experiment of nature. Similarly, animal models provide more powerful means to identify genes than are available for our species because genes and genotypes can be manipulated experimentally. Chapter 9 will describe methods for identifying genes in humans.

Long before DNA markers became available in the 1980s (see Box 9.1 for more information on DNA markers), associations were found between single genes and behavior. The first example was discovered in 1915 by A. H. Sturtevant, inventor of the chromosome map. He found that a single-gene mutation that alters eye color in the fruit fly *Drosophila* also affects their mating behavior. Another example involves the single recessive gene that causes albinism and also affects open-field activity in mice. Albino mice are less active in the open field. It turns out that this effect is largely due to the fact that albinos are more sensitive to the bright light of the open field. With a red light that reduces visual stimulation, albino mice are almost as active as pigmented mice. These relationships are examples of what is called ***allelic association,*** the association between a particular allele and a phenotype. Rather than using genes that are known by their phenotypic effect, like those for eye color and albinism, it is now possible to use millions of **polymorphisms** in DNA itself, either naturally occurring DNA polymorphisms, such as those determining eye color or albinism, or artificially created mutations.

Creating Mutations

In addition to studying naturally occurring genetic variation, geneticists have long used chemicals or X-irradiation to create mutations in the DNA itself in order to identify genes affecting complex traits, including behavior. This section focuses on the use of mutational screening to identify genes that affect behavior in animal models.

During the past 40 years, hundreds of behavioral mutants have been created in organisms as diverse as bacteria, worms, fruit flies, zebrafish, and mice (Figure 5.10).

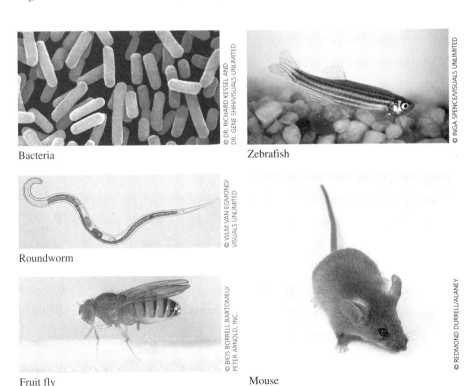

Bacteria

Zebrafish

Roundworm

Fruit fly

Mouse

• **FIGURE 5.10** Behavioral mutants have been created in bacteria (shown magnified 25,000 times), roundworms (about 1 mm in length), fruit flies (about 2–4 mm), zebrafish (about 4 cm), and mice (about 9 cm without the tail).

Information about these and other animal models for genetic research is available from www.nih.gov/science/models. This work illustrates that most normal behavior is influenced by many genes. Although any one of many single-gene mutations can seriously disrupt behavior, normal development is orchestrated by many genes working together. An analogy is an automobile, which requires thousands of parts for its normal functioning. If any one part breaks down, the automobile may not run properly. In the same way, if the function of any gene breaks down through mutation, it is likely to affect many behaviors. In other words, mutations in single genes can drastically affect behavior that is normally influenced by many genes. An important principle is *pleiotropy*, the effect of a single gene on many traits. The corollary is that any complex trait is likely to be *polygenic*, that is, influenced by many genes. Also, there is no necessary relationship between naturally occurring genetic variation and experimentally created genetic variation. That is, creating a mutation that affects a behavior does not imply that naturally occurring variation in that gene is associated with naturally occurring variation in the behavior.

Bacteria Although the behavior of bacteria is by no means attention grabbing, they do behave. They move toward or away from many kinds of chemicals by rotating their propeller-like flagella. Since the first behavioral mutant in bacteria was isolated in 1966, the dozens of mutants that have been created emphasize the genetic complexity of an apparently simple behavior in a simple organism. For example, many genes are involved in rotating the flagella and controlling the duration of the rotation.

Roundworms Among the 20,000 species of nematode (roundworm), *Caenorhabditis elegans* is about 1 mm in length and spends its three-week life span in the soil, especially in rotting vegetation, where it feeds on microbes such as bacteria. Conveniently, it also thrives in laboratory Petri dishes. Once viewed as an uninteresting, featureless tube of cells, *C. elegans* is now studied by thousands of researchers. It has 959 cells, of which 302 are nerve cells, including neurons in a primitive brain system called a nerve ring. A valuable aspect of *C. elegans* is that all its cells are visible with a microscope through its transparent body. The development of its cells can be observed, and it develops quickly because of its short life span.

Its behavior is more complex than that of single-celled organisms like bacteria, and many behavioral mutants have been identified (Hobert, 2003). For example, investigators have identified mutations that affect locomotion, foraging behavior, learning, and memory (Ardiel & Rankin, 2010; Rankin, 2002). *C. elegans* is especially important for functional genetic analysis because the developmental fate of each of its cells and the wiring diagram of its 302 nerve cells are known. In addition, most of its 20,000 genes are known, although we have no idea what half of them do (http://www.wormbase.org/; Harris et al., 2010). About half of the genes are known to match human genes. *C. elegans* has the distinction of being the first animal to have its genome of 100 million base pairs (3 percent of the size of the human genome) completely sequenced (Wilson, 1999). Despite these huge advantages for the experimental analysis of behavior, it has been difficult to connect the dots between genes, brain, and behavior (Schafer, 2005), which is a lesson to which we will return in Chapter 10.

Fruit flies The fruit fly *Drosophila*, with about 2000 species, is the star organism in terms of behavioral mutants, with hundreds identified since the pioneering work of Seymour Benzer (Weiner, 1999). Its advantages include its small size (2–4 mm), the ease of growing it in a laboratory, its short generation time (about two weeks), and its high productivity (females can lay 500 eggs in 10 days). Its genome was sequenced in 1998.

The earliest behavioral research involved responses to light (phototaxis) and to gravity (geotaxis). Normal *Drosophila* move toward light (positive phototaxis) and away from gravity (negative geotaxis). Many mutants that were either negatively phototaxic or positively geotaxic were created. Attempts are continuing to identify the specific genes involved in these behaviors (Toma, White, Hirsch, & Greenspan, 2002).

The hundreds of other behavioral mutants included *sluggish* (generally slow), *hyperkinetic* (generally fast), *easily shocked* (jarring produces a seizure), and *paralyzed* (collapses when the temperature goes above 28°C). A *drop dead* mutant walks and flies normally for a couple of days and then suddenly falls on its back and dies. More complex behaviors have also been studied, especially courtship and learning. Behavioral mutants for various aspects of courtship and copulation have been found. One male mutant, called *fruitless*, courts males as well as females and does not copulate. Another male mutant cannot disengage from the female after copulation and is given the dubious title *stuck*. The first learning behavior mutant was called *dunce* and could not learn to avoid an odor associated with shock even though it had normal sensory and motor behavior.

Drosophila also offer the possibility of creating genetic *mosaics*, individuals in which the mutant allele exists in some cells of the body but not in others (Hotta & Benzer, 1970). As individuals develop, the proportion and distribution of cells with the mutant gene vary across individuals. By comparing individuals with the mutant gene in a particular part of the body—detected by a cell marker gene that is inherited along with the mutant gene—it is possible to localize the site where a mutant gene has its effect on behavior.

The earliest mosaic mutant studies involved sexual behavior and the X chromosome (Benzer, 1973). *Drosophila* were made mosaic for the X chromosome: Some body parts have two X chromosomes and are female, and other body parts have only one X chromosome and are male. As long as a small region toward the back of the brain is male, courtship behavior is male. Of course, sex is not all in the head. Different parts of the nervous system are involved in aspects of courtship behavior such as tapping, "singing," and licking. Successful copulation also requires a male thorax (containing the fly's version of a spinal cord between the head and abdomen) and, of course, male genitals (Greenspan, 1995).

Many other gene mutations in *Drosophila* have been shown to affect behaviors (Sokolowski, 2001). The future importance of *Drosophila* in behavioral research is assured by its unparalleled genomic resources (often called **bioinformatics**) (Matthews, Kaufman, & Gelbart, 2005).

Zebrafish Although invertebrates like *C. elegans* and *Drosophila* are useful in behavioral genetics, many forms and functions are new to vertebrates. The zebrafish, named after its horizontal stripes, is common in many aquaria, grows to about 4 cm, and can live for five years. It has become a key vertebrate for studying early development because the developing embryo can be observed directly—it is not hidden inside the mother as are mammalian embryos. In addition, the embryos themselves are translucent. Nearly 2000 gene mutations that affect embryonic development have been identified. Behavior has recently become a focus of research on the zebrafish (Wright, 2011), including sensory and motor development (Guo, 2004), food and opiate preferences (Lau, Bretaud, Huang, Lin, & Guo, 2006), social behavior (Blaser & Gerlai, 2006; Miller & Gerlai, 2007), and associative learning (Sison & Gerlai, 2010).

Mice and rats The mouse is the main mammalian species used for mutational screening (Kile & Hilton, 2005). Hundreds of lines of mice with mutations that affect behavior have been created (Godinho & Nolan, 2006). Many of these are preserved in frozen embryos that can be "reconstituted" on order. Resources describing the behavioral and biological effects of the mutations are available (e.g., www.informatics. jax.org/). Major initiatives are under way to use chemical mutagenesis to screen mice for mutations on a broad battery of measures of complex traits (Brown, Hancock, & Gates, 2006; Kumar et al., 2011). Behavioral screening is an important part of these initiatives because behavior can be an especially sensitive indicator of the effects of mutations (Crawley, 2003, 2007).

After the human, the mouse was the next mammalian target for sequencing the entire genome, which was accomplished in 2001 (Venter et al., 2001). The rat, whose larger size makes it the favorite rodent for physiological and pharmacological research, is also coming on strong in genomics research (Jacob & Kwitek, 2002; Smits & Cuppen, 2006). The rat genome was sequenced in 2004 (Gibbs et al., 2004). The bioinformatics resources for rodents are growing rapidly (DiPetrillo, Wang, Stylianou, & Paigen, 2005).

Targeted mutations in mice In addition to mutational screening, the mouse is also the main mammalian species used to create **targeted mutations** that knock out the expression of specific genes. A targeted mutation is a process by which a gene is changed in a specific way to alter its function (Capecchi, 1994). Most often, genes are "knocked out" by deleting key DNA sequences that prevent the gene from being transcribed. Many techniques produce more subtle changes that alter the gene's regulation; these changes lead to underexpression or overexpression of the gene rather than knocking it out altogether. In mice, the mutated gene is transferred to embryos (a technique called *transgenics* when the mutated gene is from another species). Once mice homozygous for the **knock-out** gene are bred, the effect of the knock-out gene on behavior can be investigated.

More than 10,000 knock-out mouse lines have been created, many of which affect behavior. For example, since 1996 nearly 100 genes have been genetically engineered for their effect on alcohol responses (Crabbe, Phillips, & Belknap, 2010; Crabbe, Phillips, Harris, Arends, & Koob, 2006). Another example is aggressive behavior in the male mouse, for which 56 genetically engineered genes show effects (Maxson, 2009). An ongoing project, called the Knockout Mouse Project, aims to create knock-outs in 8500 genes (www.nih.gov/science/models/mouse/knockout).

Gene-targeting strategies are not without their limitations (Crusio, 2004). One problem with knock-out mice is that the targeted gene is inactivated throughout the animal's life span. During development, the organism copes with the loss of the gene's function by compensating wherever possible. For example, deletion of a gene coding for a dopamine transporter protein (which is responsible for inactivating dopaminergic neurons by transferring the neurotransmitter back into the presynaptic terminal) results in a mouse that is hyperactive in novel environments (Giros, Jaber, Jones, Wightman, & Caron, 1996). These knock-out mutants exhibit complex compensations throughout the dopaminergic system that are not specifically due to the dopamine transporter itself (Jones et al., 1998). However, in most instances, compensations for the loss of gene function are invisible to the researcher, and caution must be taken to avoid attributing compensatory changes in the animals to the gene itself. These compensatory processes can be overcome by creating conditional knock-outs of regulatory elements; these conditional mutations make it possible to turn expression of the gene on or off at will at any time during the animal's life span, or the mutation can target specific areas of the brain (e.g., Hall, Limaye, & Kulkarni, 2009).

Gene silencing In contrast to knock-out studies, which alter DNA, another method uses double-stranded RNA to "knock down" expression of the gene that shares its sequence (Hannon, 2002). The **gene silencing** technique, which was discovered in 1997 and won the Nobel Prize in 2006 (Bernards, 2006), is called *RNA interference* (**RNAi**) or *small interfering RNA* (**siRNA**), because it degrades complementary RNA transcripts (http://www.ncbi.nlm.nih.gov/projects/genome/probe/doc/ApplSilencing.shtml). siRNA kits are now available commercially that target nearly all the genes in the human and mouse genomes. More than 8000 papers on siRNA were published in 2010 alone, primarily about using cell cultures where delivery of the siRNA to the cells is not a problem. However, in vivo animal model research necessary for behavioral analysis has begun. Although delivery to the brain remains a problem (Thakker, Hoyer, & Cryan, 2006), injecting siRNA in mouse brain has yielded knock-down results on behavior similar to results expected from knock-out studies (Salahpour,

Medvedev, Beaulieu, Gainetdinov, & Caron, 2007). It is hoped that siRNA will soon have therapeutic applications (Kim & Rossi, 2007), for example, for prevention of infection by a respiratory virus (DeVincenzo et al., 2010).

● KEY CONCEPTS

Mutation: A heritable change in DNA base-pair sequences.
Targeted mutation: The changing of a gene in a specific way to alter its function, such as gene knock-outs.
Gene silencing: Suppressing expression of a gene but not altering it and, thus, not heritable.

Quantitative Trait Loci

Creating a mutation that has a major effect on behavior does not mean that this gene is specifically responsible for the behavior. Remember the automobile analogy in which any one of many parts can go wrong and prevent the automobile from running properly. Although the part that goes wrong has a big effect, that part is only one of many parts needed for normal functioning. Moreover, the genes changed by artificially created mutations are not necessarily responsible for the naturally occurring genetic variation detected in quantitative genetic research. Identifying genes responsible for naturally occurring genetic variation that affects behavior has only become possible in recent years. The difficulty is that, instead of looking for a single gene with a major effect, we are looking for many genes, each having a relatively small **effect size**—quantitative trait loci (QTLs).

Animal models have been particularly useful in the quest for QTLs because both genetics and environment can, and may, be manipulated and controlled in the laboratory, whereas for our species neither genetics nor environment may be manipulated. Animal model work on natural genetic variation and behavior has primarily studied the mouse and the fruit fly *Drosophila* (Kendler & Greenspan, 2006). Although this section emphasizes research on mice, similar methods have been used in *Drosophila* (Mackay & Anholt, 2006) and have been applied to many behaviors (Anholt & Mackay, 2004), including mating behavior (Moehring & Mackay, 2004), odor-guided behavior (Sambandan, Yamamoto, Fanara, Mackay, & Anholt, 2006), and locomotor behavior (Jordan, Morgan, & Mackay, 2006). In addition, as mentioned in the previous section, behavioral genetic research on the rat is also increasing rapidly for complex traits, including behavior (Smits & Cuppen, 2006).

In animal models, linkage can be identified by using Mendelian crosses to trace the cotransmission of a marker whose chromosomal location is known and a single-gene trait, as illustrated in Figure 2.6. Linkage, which is also described in Chapter 9, is suggested when the results violate Mendel's second law of independent assortment. However, as emphasized in previous chapters, behavioral dimensions and disorders are likely to be influenced by many genes; consequently, any one gene is likely to have only a small effect. If many genes contribute to behavior, behavioral traits will be distributed quantitatively. The goal is to find some of the many genes (QTLs) that affect these quantitative traits.

F$_2$ crosses Although linkage techniques can be extended to investigate quantitative traits, most QTL analyses with animal models use allelic association, which is more powerful for detecting the small effect sizes expected for QTLs. Allelic association refers to the correlation or association between an allele and a trait. For example, the **allelic frequency** of DNA markers can be compared for groups of animals high or low on a quantitative trait. This approach has been applied to open-field activity in mice (Flint et al., 1995). F$_2$ mice were derived from a cross between high and low lines selected for open-field activity and subsequently inbred by using brother-sister matings for over 30 generations. Each F$_2$ mouse has a unique combination of alleles from the original parental strains because there is an average of one recombination in each chromosome inherited from the F$_1$ strain (see Figure 2.7). The most active and the least active F$_2$ mice were examined for 84 DNA markers spread throughout the mouse chromosomes in an effort to identify chromosomal regions that are associated with open-field activity (Flint et al., 1995). The analysis simply compares the frequencies of marker alleles for the most active and least active groups.

Figure 5.11 shows that regions of chromosomes 1, 12, and 15 harbor QTLs for open-field activity. A QTL on chromosome 15 is related primarily to open-field activity and not to other measures of fearfulness, an observation suggesting the possibility of a gene specific to open-field activity. The QTL regions on chromosomes 1 and 12, on the other hand, are related to other measures of fearfulness, associations suggesting that these QTLs affect diverse measures of fearfulness. QTLs were subsequently mapped in two large ($N = 815$ and 821) F$_2$ crosses from the replicate inbred lines of mice initially selected for open-field activity (Turri, Henderson, DeFries, & Flint, 2001). Results of this study both confirmed and extended the previous findings reported by Flint et al. (1995). QTLs for open-field activity were replicated on chromosomes 1, 4, 12, and 15, and new evidence for additional QTLs on chromosome 7 and the X chromosome was also obtained. An exception is exploration in an enclosed arm of a maze (see Figure 5.11), which was included in the study as a control because other research suggests that this measure is not genetically correlated with measures of fearfulness. Several studies have also reported associations between markers on the distal end of chromosome 1 and quantitative measures of emotional behavior, although it has been difficult to identify the specific gene responsible for the association (Fullerton, 2006).

Heterogeneous stocks and commercial outbred strains Because the chromosomes of F$_2$ mice have an average of only one crossover between maternal and paternal chromosomes, the method has little resolving power to pinpoint a locus,

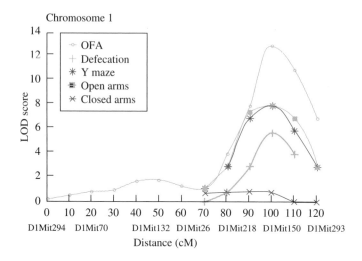

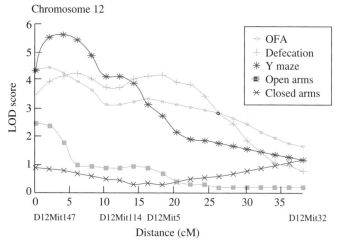

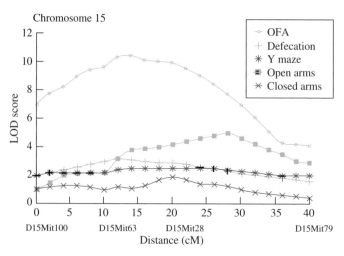

although it has good power to identify the chromosome on which a QTL resides. That is, QTL associations found by using F_2 mice refer only to general "neighborhoods," not specific addresses. The QTL neighborhood is usually very large, about 10 million to 20 million base pairs of DNA, and thousands of genes could reside there. One way to increase the resolving power is to use animals whose chromosomes are recombined to a greater extent, by breeding for many generations animals derived either from two inbred strains (an advanced intercross) (Darvasi, 1998) or from multiple inbred strains (heterogeneous stocks) (Valdar, Solberg, Gauguier, Burnett, et al., 2006). The latter approach was used to increase 30-fold the resolving power of the QTL study of fearfulness (Talbot et al., 1999). Mice in the top and bottom 20 percent of open-field activity scores were selected from 751 heterogeneous stock mice. The results confirmed the association between emotionality and markers on chromosome 1, although the association was closer to the 70-cM region than the 100-cM region of chromosome 1 found in the earlier study (see Figure 5.11). Some supporting evidence for a QTL on chromosome 12 was also found, but none was found for chromosome 15. Even greater mapping resolution is possible using commercially available outbred mice (Yalcin et al., 2010). For example, using commercial outbred strains, the chromosome 1 association with emotionality was mapped to an interval containing a single gene (Yalcin et al., 2004). Commercial outbreds are a resource for **genomewide association** mapping in mice and have the potential to identify multiple genes involved in behavior.

Much QTL research in mice has been in the area of **pharmacogenetics,** a field in which investigators study genetic effects on responses to drugs. Dozens of QTLs have been mapped for drug responses such as alcohol drinking, alcohol-induced loss of righting reflex, acute alcohol and pentobarbital withdrawal, cocaine seizures, and morphine preference and analgesia (Crabbe et al., 2010; Crabbe, Phillips, Buck, Cunningham, & Belknap, 1999). In some instances, the location of a mapped QTL is close enough to a previously mapped gene of known function to make studies of that gene informative for human studies (Ehlers, Walter, Dick, Buck, & Crabbe, 2010). Pharmacogenetics QTL-mapping research also has been extended to rats (Spence et al., 2009).

• **FIGURE 5.11** QTLs for open-field activity and other measures of fearfulness in an F_2 cross between high and low lines selected for open-field activity. The five measures are (1) open-field activity (OFA), (2) defecation in the open field, (3) activity in the Y maze, (4) entry in the open arms of the elevated plus maze, and (5) entry in the closed arms of the elevated plus maze, which is not a measure of fearfulness. LOD (logarithm to the base 10 of the odds) scores indicate the strength of the effect; a LOD score of 3 or greater is generally accepted as significant. Distance in centimorgans (cM) indicates position on the chromosome, with each centimorgan roughly corresponding to 1 million base pairs. Below the distance scale are listed the specific short-sequence repeat markers for which the mice were examined and mapped. (Reprinted with permission from "A simple genetic basis for a complex psychological trait in laboratory mice" by J. Flint et al. *Science, 269,* 1432–1435. ©1995 American Association for the Advancement of Science. All rights reserved.)

Recombinant inbred strains Another method used to identify QTLs for behavior involves special inbred strains called **recombinant inbred (RI) strains.** RI strains are inbred strains derived from an F_2 cross between two inbred strains; this process leads to recombination of parts of chromosomes from the parental strains (Figure 5.12). Thousands of DNA markers have been mapped in RI strains, thus enabling investigators to use these markers to identify QTLs associated with behavior without any additional genotyping (Plomin & McClearn, 1993). The special value of the RI QTL approach is that it enables all investigators to study essentially the same animals because the RI strains are extensively inbred. This feature of RI QTL analysis means that each RI strain needs to be genotyped only once and that genetic correlations can be assessed across measures, across studies, and across laboratories. The QTL analysis itself is much like the F_2 QTL analysis discussed earlier except that, instead of comparing individuals with recombined genotypes, the RI QTL approach compares means of recombinant inbred strains. RI QTL work has also focused on pharmacogenetics. For example, RI QTL research has confirmed some of the associations for responses to alcohol found using F_2 crosses (Buck, Rademacher, Metten, & Crabbe, 2002). Research combining RI and F_2 QTL approaches are also making

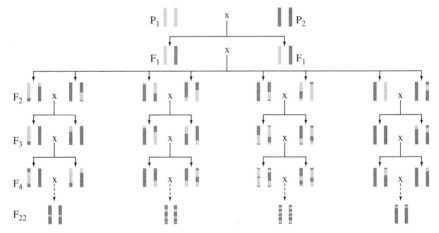

• **FIGURE 5.12** Construction of a set of recombinant inbred strains from the cross of two parental inbred strains. The F_1 is heterozygous at all loci that differ in the parental strains. Crossing F_1 mice produces an F_2 generation in which alleles from the parental strains segregate so that each individual is genetically unique. By inbreeding the F_2 with brother-sister matings for many generations, recombination continues until each RI strain is fixed homozygously at each gene for a single allele inherited from one or the other progenitor inbred strain. Unlike F_2 crosses, RI strains are genetically stable because each strain has been inbred. This means that a set of RI strains needs to be genotyped only once for DNA markers or phenotyped only once for behaviors and the data can be used in any other experiment using that set of RI strains. Similar to the F_2 cross, QTL association can be detected by comparing the quantitative trait scores of RI strains that differ genotypically for a particular DNA marker.

progress toward identifying genes for alcohol-related behaviors (Bennett, Carosone-Link, Zahniser, & Johnson, 2006).

One problem with the RI QTL method has been that only a few dozen RI strains were available, which means that only associations of large effect size could be detected. Also, it has been difficult to locate the specific genes responsible for associations. For example, during the past 15 years, more than 2000 QTL associations have been reported using crosses between inbred strains, but fewer than 1 percent have been localized (Flint, Valdar, Shifman, & Mott, 2005). A major new development is the creation of an RI series that includes as many as 1000 RI strains from crosses between eight inbred strains (Chesler et al., 2008). When eight inbred strains are crossed, the resulting RI strains will show greater recombination than seen in the two-strain RI example shown in Figure 5.12; they will also yield sufficient power to detect QTL associations of modest effect size. The "Collaborative Cross," as the project is known, will provide a valuable resource not only for the identification of genes associated with complex traits but also for integrative analyses of complex systems that include gene expression as well as neural, pharmacological, and behavioral data (Aylor et al., 2011), as described in Chapter 10.

Synteny Homology

QTLs found in mice can be used as candidate QTLs for human research because nearly all mouse genes are similar to human genes. Moreover, chromosomal regions linked to behavior in mice can be used as candidate regions in human studies because parts of mouse chromosomes have the same genes in the same order as parts of human chromosomes, a relationship called *synteny homology.* It is as if about 200 chromosomal regions have been reshuffled onto different chromosomes from mouse to human. (See www.informatics.jax.org/for details about synteny homology.) For example, the region of mouse chromosome 1 shown in Figure 5.11 to be linked with open-field activity has the same order of genes that happen to be part of the long arm of human chromosome 1, although syntenic regions are usually on different chromosomes in mouse and human. As a result of these findings, this region of human chromosome 1 has been considered as a candidate QTL region for human anxiety, and linkage with the syntenic region in human chromosome 1 has been reported in two large studies (Fullerton et al., 2003; Nash et al., 2004). QTLs in syntenic regions for mouse and human chromosomes have also been reported for alcohol use (Ehlers et al., 2010).

Summary

Quantitative genetic studies of animal behavior provide powerful tests of genetic influence. These studies include selection studies and studies of inbred strains; through their use we have learned a great deal about how genes and environments influence behavior. For example, studies of mice have helped to clarify how genes are involved in fearful and aggressive behavior, and there have been many studies of alcohol-related behaviors in mice. Studies of animal behavior have also been used to identify genes. Many behavioral mutants have been identified from studies of chemically induced mutations in organisms as diverse as single-celled organisms, roundworms, fruit flies, and mice. Associations between such single-gene mutations and behavior generally underline the point that disruption of a single gene can drastically affect behavior normally influenced by many genes. Experimental crosses of inbred strains are powerful tools for identifying linkages, even for complex quantitative traits for which many genes are involved. Such quantitative trait loci (QTLs) have been identified for several behaviors in mice, such as fearfulness and responses to drugs.

CHAPTER SIX

Nature, Nurture, and Human Behavior

Most behavioral traits are much more complex than single-gene disorders such as Huntington disease and PKU (see Chapter 2). Complex dimensions and disorders are influenced by heredity, but not by one gene alone. Multiple genes are usually involved, as well as multiple environmental influences. The purpose of this chapter is to describe ways in which we can study genetic effects on complex behavioral traits in humans. Chapter 5 described how complex behavioral traits are examined using animal models. The words *nature* and *nurture* have a rich and contentious history in the field, but they are used here simply as broad categories representing genetic and environmental influences, respectively. They are not distinct categories—Chapter 8 discusses the interplay between them, and the importance of gene-environment interplay is woven throughout this book.

The first question that needs to be asked about behavioral traits is whether heredity is at all important. For single-gene disorders, this is not an issue because it is usually obvious that heredity is important. For example, for dominant genes, such as the gene for Huntington disease, you do not need to be a geneticist to notice that every affected individual has an affected parent. Recessive gene transmission is not as easy to observe, but the expected pattern of inheritance is clear. For complex behavioral traits in the human species, an experiment of nature (twinning) and an experiment of nurture (adoption) are widely used to assess the net effect of genes and environments. The theory underlying these methods is called *quantitative genetics*. Quantitative genetics estimates the extent to which observed differences among individuals are due to genetic differences of any sort and to environmental differences of any sort without specifying what the specific genes or environmental factors are. When heredity is important—and it almost always is for complex traits like behavior—it is now possible to identify specific genes by using the methods of molecular genetics, the topic of Chapter 9. Behavioral genetics uses the methods of both quantitative genetics and molecular genetics to study behavior. Using genetically sensitive designs also facilitates the identification of specific environmental factors, which is the topic of Chapter 8.

Investigating the Genetics of Human Behavior

Quantitative genetic methods to study human behavior are not as powerful or direct as the animal approaches described in Chapter 5. Rather than using genetically defined populations such as inbred strains of mice or manipulating environments experimentally, human research is limited to studying naturally occurring genetic and environmental variation. Nonetheless, adoption and twinning provide experimental situations that can be used to test the relative influence of nature and nurture. As mentioned in Chapter 1, increasing recognition of the importance of genetics during the past three decades is one of the most dramatic shifts in the behavioral sciences. This shift is in large part due to the accumulation of adoption and twin research that consistently points to the important role played by genetics even for complex psychological traits.

Adoption Designs

Many behaviors "run in families," but family resemblance can be due either to nature or to nurture, or to some combination of both. The most direct way to disentangle genetic and environmental sources of family resemblance involves adoption. Adoption creates sets of genetically related individuals who do not share a common family environment because they were adopted apart. Their similarity estimates the contribution of genetics to family resemblance.

Adoption also produces adopted-together family members who share a common family environment but are not genetically related. Their resemblance estimates the contribution of the family environment to family resemblance. In this way, the effects of nature and nurture can be inferred from the adoption design. As mentioned earlier, quantitative genetic research does not in itself identify specific genes or environments. It is possible to incorporate direct measures of genes and environments into quantitative genetic designs, and a few such studies are under way (Chapter 8).

For example, consider parents and offspring. Parents in a **family study** are "genetic-plus-environmental" parents in that they share both heredity and environment with their offspring. The process of adoption results in "genetic" parents and "environmental" parents (Figure 6.1). "Genetic" parents are birth parents who relinquish their child for adoption shortly after birth. Resemblance between birth parents and their adopted offspring directly assesses the genetic contribution to parent-offspring resemblance. "Environmental" parents are adoptive parents who adopt children genetically unrelated to them. When children are placed into adoptive families as infants, resemblance between adoptive parents and their adopted children directly assesses the postnatal environmental contributions to parent-offspring resemblance. Additional environmental influences on the adopted children come from the prenatal environment provided by their birth mothers. Genetic influences can also be

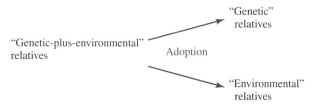

• **FIGURE 6.1** Adoption is an experiment of nurture that creates "genetic" relatives (biological parents and their adopted-away offspring; siblings adopted apart) and "environmental" relatives (adoptive parents and their adopted children; genetically unrelated children adopted into the same adoptive family). Resemblance for these "genetic" and "environmental" relatives can be used to test the extent to which resemblance between the usual "genetic-plus-environmental" relatives is due to either nature or nurture.

assessed by comparing "genetic-plus-environmental" families with adoptive families who share only family environment.

"Genetic" siblings and "environmental" siblings can also be studied. "Genetic" siblings are full siblings adopted apart early in life and reared in different homes. "Environmental" siblings are pairs of genetically unrelated children reared in the same home. This can be due to two children being adopted early in life by the same adoptive parents, to adopted children being reared with children who are biological to the adoptive parents, or to being part of a stepfamily where each parent brings a child from a previous marriage. As described in the Appendix, these adoption designs can be depicted more precisely as path models that are used in **model-fitting** analyses to test the fit of the model, to compare alternative models, and to estimate genetic and environmental influences (see the Appendix; Boker et al., 2011).

Adoption studies often yield evidence for genetic influence on behavioral traits, although results depend on the trait examined and the age of the adopted child. Specifically, studies of young children examining behavioral outcomes find few main effects of genetics (Brooker et al., 2011; Natsuaki et al., 2010), although there is evidence of gene-environment interplay (see Chapter 8). When older adopted children are examined for traits like cognitive ability, genetic factors appear to be important.

Figure 6.2 summarizes adoption results for general cognitive ability (see Chapter 12 for details). "Genetic" parents and offspring and "genetic" siblings significantly resemble each other even though they are adopted apart and do not share family environment. You can see that genetics accounts for about half of the resemblance for "genetic-plus-environmental" parents and siblings. The other half of familial resemblance appears to be explained by shared family environment, assessed directly by the resemblance between adoptive parents and adopted children, and between **adoptive siblings.** Chapter 8 describes a recent important finding that the influence of shared environment on cognitive ability decreases dramatically from childhood to adolescence.

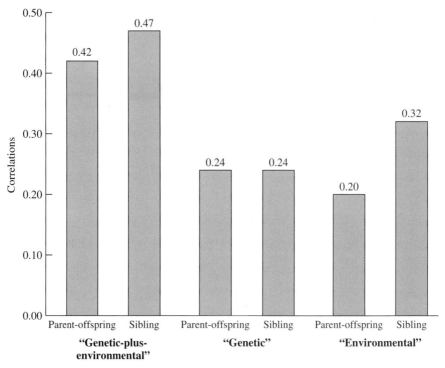

• **FIGURE 6.2** Adoption data indicate that family resemblance for cognitive ability is due both to genetic resemblance and to environmental resemblance. "Genetic" relatives refer to genetically related relatives adopted apart. "Environmental" relatives refer to genetically unrelated individuals adopted together. (Data adapted from Loehlin, 1989.)

One of the most surprising results from genetic research is that, for many psychological traits other than cognitive ability, especially for traits like personality and psychopathology, resemblance between relatives is accounted for by shared heredity rather than by shared environment. For example, the risk of schizophrenia is just as great for offspring of schizophrenic parents whether they are reared by their biological parents or adopted at birth and reared by adoptive parents. This finding implies that sharing a family environment does not contribute importantly to family resemblance for these psychological traits. It does not mean that the environment generally or even the family environment is unimportant. As discussed in Chapter 8, quantitative genetic research, such as adoption studies, provides the best available evidence for the importance of environmental influences. The risk for first-degree relatives of schizophrenic probands who are 50 percent similar genetically is only about 10 percent, not 50 percent. Furthermore, although family environment does not contribute to the resemblance of family members for many traits, such factors could contribute to *differences* among family members, the ***nonshared environment*** (Chapter 8).

The first adoption study of schizophrenia, reported by Leonard Heston in 1966, is a classic study that was highly influential in turning the tide from assuming that schizophrenia was completely caused by early family experiences to recognizing the importance of genetics (Box 6.1). Box 6.2 considers some methodological issues in adoption studies.

BOX 6.1 • The First Adoption Study of Schizophrenia

Environmentalism, which assumes that we are what we learn, dominated the behavioral sciences until the 1960s, when a more balanced view emerged that recognized the importance of nature as well as nurture. One reason for this major shift was an adoption study of schizophrenia reported by Leonard Heston in 1966. Although twin studies had, for decades, suggested genetic influence, schizophrenia was generally assumed to be environmental in origin, caused by early interactions with parents. Heston interviewed 47 adult adopted offspring of hospitalized schizophrenic women. He compared their incidence of schizophrenia with that of matched adoptees whose birth parents had no known mental illness. Of the 47 adoptees whose birth mothers were schizophrenic, 5 had been hospitalized for schizophrenia. Three were chronic schizophrenics hospitalized for several years. None of the adoptees in the control group were schizophrenic.

The incidence of schizophrenia in these adopted offspring of schizophrenic birth mothers was 10 percent. This risk is similar to the risk for schizophrenia found when children are reared by their schizophrenic birth parents. Not only do these findings indicate that heredity makes a major contribution to schizophrenia, they also suggest that rearing environment has little effect. When a birth parent is schizophrenic, the risk for schizophrenia is just as great for the offspring when they are adopted at birth as it is when the offspring are reared by their schizophrenic birth parents.

Several other adoption studies have confirmed the results of Heston's study. His study is an example of what is called the *adoptees' study method* because the incidence of schizophrenia was investigated in the adopted offspring of schizophrenic birth mothers. A second major strategy is called the *adoptees' family method*. Rather than beginning with parents, this method begins with adoptees who are affected (probands) and adoptees who are unaffected. The incidence of the disorder in the biological and adoptive families of the adoptees is assessed. Genetic influence is suggested if the incidence of the disorder is greater for the biological relatives of the affected adoptees than for the biological relatives of the unaffected control adoptees. Environmental influence is indicated if the incidence is greater for the adoptive relatives of the affected adoptees than for the adoptive relatives of the control adoptees.

These adoption methods and their results for schizophrenia are described in Chapter 14.

BOX 6.2 • Issues in Adoption Studies

The adoption design is like an experiment that untangles nature and nurture as causes of family resemblance. The first adoption study, which investigated IQ, was reported in 1924 (Theis, 1924). The first adoption study of schizophrenia was reported in 1966 (see Box 6.1). Adoption studies have become more difficult to conduct as the number of domestic adoptions has declined over the past 50 years. Domestic adoption has become less common as contraception and abortion have increased and as more unmarried mothers have decided to rear their infants. However, there has been an increase in international adoptions, with children typically being adopted at age 1 or older.

One issue about adoption studies is representativeness. If biological parents, adoptive parents, or adopted children are not representative of the rest of the population, the generalizability of adoption results could be affected. However, means are more likely to be affected than variances, and genetic estimates rely primarily on variance. In the population-based Colorado Adoption Project (Petrill, Plomin, DeFries, & Hewitt, 2003), for example, biological and adoptive parents appear to be quite representative of nonadoptive parents, and adopted children seem to be reasonably representative of nonadopted children. Other adoption studies, however, have sometimes shown less representativeness. Restriction of range in the environments of adoptive families can also limit generalizations from adoption studies (Stoolmiller, 1999), although at least one study has found that even though there was some restriction of range, this did not have an impact on the children's development (McGue et al., 2007).

Another issue concerns prenatal environment. Because birth mothers provide the prenatal environment for the children they place for adoption, the resemblance between them might reflect prenatal environmental influences. A strength of adoption studies is that prenatal effects can be tested independently from postnatal environment by comparing correlations for birth mothers and birth fathers. Although it is more difficult to study birth fathers, results for small samples of birth fathers show results similar to those for birth mothers for several behaviors in young children and for IQ and schizophrenia in adult adoptees. Another approach to this issue is to compare adoptees' biological half siblings related through the mother (maternal half siblings) with those related through the father (paternal half siblings). For schizophrenia, paternal half siblings of schizophrenic adoptees

Twin Design

The other major method used to disentangle genetic from environmental sources of resemblance between relatives involves twins (Segal, 1999). Identical twins, also called *monozygotic* (MZ) twins because they derive from one fertilized egg (zygote), are genetically identical (Figure 6.3). If genetic factors are important for a trait, these show the same risk for schizophrenia as maternal half siblings do, an observation suggesting that prenatal factors may not be of great importance (Kety, 1987). Another strategy for disentangling the effects of genetic influences from prenatal environmental influences is to obtain some possible indices of the prenatal environment, such as the birth mother's depressive symptoms during pregnancy. In the only adoption study that has systematically included these effects in the models, prenatal environment was not found to have an influence on child functioning independent of genetic influences; however, the analysis was limited to early childhood (Pemberton et al., 2010).

For the past two decades, most domestic adoptions in the United States have been "open" to some extent. This means that the birth parents and the adoptive families know or share information about each other with the other party and the adopted child. Ongoing studies of domestic adoption have examined the extent to which openness in the adoption influences the functioning of the adoptive parents and the birth parents and found that, in general, more open adoptions were associated with better mental health (Ge et al., 2008). Openness in adoption raises some concerns about the extent to which the adopted child's rearing environment is truly independent of genetic influences from the birth parents. The majority of work in this area indicates that although there may be contact among birth parents, adoptive parents, and adopted children, this contact is relatively infrequent.

Finally, selective placement could cloud the separation of nature and nurture by placing adopted-apart "genetic" relatives into correlated environments. For example, selective placement would occur if the adopted children of the brightest biological parents are placed with the brightest adoptive parents. If selective placement matches biological and adoptive parents, genetic influences could inflate the correlation between adoptive parents and their adopted children, and environmental influences could inflate the correlation between biological parents and their adopted children. If data are available on biological parents as well as adoptive parents, selective placement can be assessed directly. If selective placement is found in an adoption study, its effects need to be considered in interpreting genetic and environmental results. Although some adoption studies show selective placement for IQ, other psychological dimensions and disorders show little evidence for selective placement.

genetically identical pairs of individuals must be more similar than first-degree relatives, who are only 50 percent similar genetically. Rather than comparing identical twins with nontwin siblings or other relatives, nature has provided a better comparison group: fraternal (*dizygotic,* or DZ) twins. Unlike identical twins, fraternal twins develop from separately fertilized eggs. They are first-degree relatives, 50 percent genetically

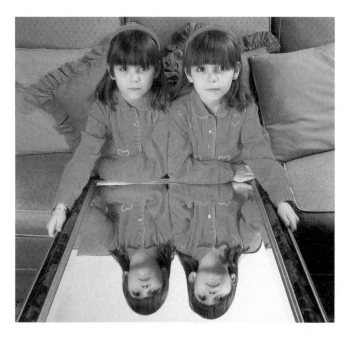

• **FIGURE 6.3** Twinning is an experiment of nature that produces identical twins, who are geneti-cally identical, and fraternal twins, who are only 50 percent similar genetically. If genetic factors are important for a trait, identical twins must be more similar than fraternal twins. DNA markers can be used to test whether twins are identical or fraternal, although for most pairs it is easy to tell because identical twins (top photo) are usually much more similar physically than fraternal twins (bottom photo).

related like other siblings. Half of fraternal twin pairs are same-sex pairs and half are opposite-sex pairs. Twin studies usually focus on same-sex fraternal twin pairs because they are a better comparison group for identical twin pairs, who are always same-sex pairs. If genetic factors are important for a trait, identical twins must be more similar than fraternal twins. (See Box 6.3 for more details about the twin method.)

How can you tell whether same-sex twins are identical or fraternal? DNA markers can tell. If a pair of twins differs for DNA markers (excluding laboratory error or new mutations, called de novo mutations), they must be fraternal because identical twins are nearly identical genetically. If many markers are examined and no differences are found, the twin pair has a high probability of being identical. Physical traits such as eye color, hair color, and hair texture can be used in a similar way to diagnose whether twins are identical or fraternal. Such traits are highly heritable and are affected by many genes. If members of a twin pair differ for one of these traits, they are likely to be fraternal; if they are the same for many such traits, they are probably identical. In most cases, it is not difficult to tell whether twins are identical or fraternal (see Figure 6.3). In fact, a single question works pretty well because it sums up many such physical traits: When the twins were young, how difficult was it to tell them apart? To be mistaken for another person requires that many heritable physical characteristics be identical. Using physical similarity to determine whether twins are identical or fraternal is generally more than 95 percent accurate when compared with the results of DNA markers (e.g., Christiansen et al., 2003; Gao et al., 2006). DNA markers can also be used to determine zygosity prenatally (Levy, Mirlesse, Jacquemard, & Daffos, 2002).

If a trait is influenced genetically, identical twins must be more similar than fraternal twins. However, it is also possible that the greater similarity of MZ twins is caused environmentally rather than genetically because MZ twins are the same sex and age and they look alike. The *equal environments assumption* of the twin method assumes that environmentally caused similarity is roughly the same for both types of twins reared in the same family. If the assumption were violated because identical twins experience more similar environments than fraternal twins, this violation would inflate estimates of genetic influence. The equal environments assumption has been tested in several ways and appears reasonable for most traits (Bouchard & Propping, 1993; Derks, Dolan, & Boomsma, 2006).

Prenatally, identical twins may experience greater environmental *differences* than fraternal twins. For example, identical twins show greater birth weight differences than fraternal twins do. The difference may be due to greater prenatal competition, especially for the majority of identical twins who share the same *chorion* (see Box 6.3). To the extent that identical twins experience less similar environments, the twin method will underestimate heritability. Postnatally, the effect of labeling a twin pair as identical or fraternal has been studied by using twins who were misclassified by their parents or by themselves (e.g., Gunderson et al., 2006; Kendler, Neale, Kessler, Heath, & Eaves, 1993b; Scarr & Carter-Saltzman, 1979). When parents think that twins are fraternal but they really are identical, these mislabeled twins are as similar behaviorally as correctly labeled identical twins.

Another way in which the equal environments assumption has been tested takes advantage of the fact that differences within pairs of identical twins can only be due to environmental influences. The equal environments assumption is supported if

BOX 6.3 • The Twin Method

Francis Galton (1876) studied developmental changes in twins' similarity, but in one of the first real twin studies, conducted in 1924, identical and fraternal twins were compared in an attempt to estimate genetic influence (Merriman, 1924). This twin study assessed IQ and found that identical twins were markedly more similar than fraternal twins, a result suggesting genetic influence. Dozens of subsequent twin studies of IQ confirmed this finding. Twin studies have also been conducted for many other psychological dimensions and disorders; they provide the bulk of the evidence for the widespread influence of genetics on behavioral traits. Although most mammals have large litters, primates, including our species, tend to have single offspring. However, primates occasionally have multiple births. Human twins are more common than people usually realize—about 32 in 1000 births are twins (i.e., 16 pairs of twins). Surprisingly, as many as 20 percent of fetuses are twins, but because of the hazards associated with twin pregnancies, often one member of the pair dies very early in pregnancy. Among live births, the numbers of identical and same-sex fraternal twins are approximately equal. That is, of all twin pairs, about one-third are identical twins, one-third are same-sex fraternal twins, and one-third are opposite-sex fraternal twins.

Identical twins result from a single fertilized egg (called a zygote) that splits for unknown reasons, producing two (or sometimes more) genetically identical individuals. For about a third of identical twins, the zygote splits during the first five days after fertilization as it makes its way down to the womb. In this case, the identical twins have different sacs (called

identical twins who are treated more individually than others do not behave more differently. This is what has been found for most tests of the assumption in research on behavioral disorders and dimensions (e.g., Cronk et al., 2002; Kendler, Neale, Kessler, Heath, & Eaves, 1994; Loehlin & Nichols, 1976; Mazzeo et al., 2010; Morris-Yates, Andrews, Howie, & Henderson, 1990).

A subtle, but important, issue is that identical twins might have more similar experiences than fraternal twins because identical twins are more similar genetically. That is, some experiences may be driven genetically. Such differences between identical and fraternal twins in experience are not a violation of the equal environments assumption because the differences are not caused environmentally (Eaves, Foley, & Silberg, 2003). This topic is discussed in Chapter 8.

As in any experiment, generalizability is an issue for the twin method. Are twins representative of the general population? Two ways in which twins are different are that twins are often born three to four weeks prematurely and intrauterine environments can be adverse when twins share a womb (Phillips, 1993). Newborn twins are also about 30 percent lighter at birth than the average singleton newborn, a difference that disappears chorions) within the placenta. Two-thirds of the time, the zygote splits after it implants in the placenta and the twins share the same chorion. Identical twins who share the same chorion may be more similar for some psychological traits than identical twins who do not share the same chorion, although there is not much support for this in the literature (Fagard, Loos, Beunen, Derom, & Vlietinck, 2003; Gutknecht, Spitz, & Carlier, 1999; Hur & Shin, 2008; Jacobs et al., 2001; Phelps, Davis, & Schwartz, 1997; Riese, 1999; Sokol et al., 1995). When the zygote splits after about two weeks, the twins' bodies may be partially fused—conjoined twins. Fraternal twins occur when two eggs are separately fertilized; they have different chorions. Like other siblings, they are 50 percent similar genetically.

The rate of fraternal twinning differs across countries, increases with maternal age, and may be inherited in some families. Increased use of fertility drugs results in greater numbers of fraternal twins because these drugs make it likely that more than one egg will ovulate. The numbers of fraternal twins have also increased since the early 1980s because of in vitro fertilization, in which several fertilized eggs are implanted and two survive. The rate of identical twinning is not affected by any of these factors.

Identical twins are nearly identical for the sequence of DNA with the exception of de novo mutations. However, identical twins differ for the expression (transcription) of DNA, just as we differ from ourselves for gene expression from minute to minute. These expression differences within pairs of identical twins include epigenetic differences, discussed in Chapter 10.

by middle childhood (MacGillivray, Campbell, & Thompson, 1988). There is also the suggestion that brain development differs in twins vs. singleton children during early infancy (Knickmeyer et al., 2011). In childhood, language develops more slowly in twins, and twins also perform less well on tests of verbal ability and IQ (Deary, Pattie, Wilson, & Whalley, 2005; Ronalds, De Stavola, & Leon, 2005; Voracek & Haubner, 2008). These delays are similar for MZ and DZ twins and appear to be due to the postnatal environment rather than prematurity (Rutter & Redshaw, 1991). Most of this cognitive deficit is recovered in the early school years (Christensen, Petersen, et al., 2006). Twins do not appear to be importantly different from singletons for personality (Johnson, Krueger, Bouchard, & McGue, 2002), for psychopathology (Robbers et al., 2011), or in motor development (Brouwer, van Beijsterveldt, Bartels, Hudziak, & Boomsma, 2006).

In summary, the twin method is a valuable tool for screening behavioral dimensions and disorders for genetic influences (Boomsma, Busjahn, & Peltonen, 2002; Martin, Boomsma, & Machin, 1997). More than 20,000 papers on twins were published during the five years from 2007 to 2011, with more than half of these focused on behavior. The value of the twin method explains why most developed countries have twin registers (Bartels, 2007; Busjahn, 2002). The assumptions underlying the twin method are different from those of the adoption method, yet both methods converge on the conclusion that genetics is important in the behavioral sciences. Recall that for schizophrenia, the risk for a fraternal twin whose co-twin is schizophrenic is about 17 percent; the risk is 48 percent for identical twins (see Figure 3.6). For general cognitive ability, the correlation is about 0.60 for fraternal twins and 0.85 for identical twins (see Figure 3.7). The fact that identical twins are so much more similar than fraternal twins strongly suggests genetic influences. For both schizophrenia and general cognitive ability, fraternal twins are more similar than nontwin siblings, perhaps because twins shared the same uterus at the same time and are exactly the same age (Koeppen-Schomerus, Spinath, & Plomin, 2003).

Combination

During the past two decades, behavioral geneticists have begun to use designs that combine the family, adoption, and twin methods in order to bring more power to bear on these analyses. For example, it is useful to include nontwin siblings in twin studies to test whether twins differ statistically from singletons and whether fraternal twins are more similar than nontwin siblings.

Two major combination designs bring the adoption design together with the family design and with the twin design. The adoption design comparing "genetic" and "environmental" relatives is made much more powerful by including the "genetic-plus-environmental" relatives of a family design. This is the design of one of the largest and longest ongoing genetic studies of behavioral development, the Colorado Adoption Project (Petrill et al., 2003). This project has shown, for example, that genetic influences on general cognitive ability increase during infancy and childhood (Plomin, Fulker, Corley, & DeFries, 1997).

The adoption-twin combination involves twins adopted apart and compares them with twins reared together. Two major studies of this type have been conducted, one in Minnesota (Bouchard, Lykken, McGue, Segal, & Tellegen, 1990; Lykken, 2006) and one in Sweden (Kato & Pedersen, 2005; Pedersen, McClearn, Plomin, & Nesselroade, 1992). These studies have found, for example, that identical twins reared apart from early in life are almost as similar in terms of general cognitive ability as are identical twins reared together, an outcome suggesting strong genetic influence and little environmental influence caused by growing up together in the same family (shared family environmental influence).

An interesting combination of the twin and family methods comes from the study of families of identical twins, which has come to be known as the families-of-twins method (Knopik, Jacob, Haber, Swenson, & Howell, 2009; Schermerhorn et al., 2011; Singh et al., 2011). When identical twins become adults and have their own children, interesting family relationships emerge. For example, in families of male identical twins, nephews are as related genetically to their twin uncle as they are to their own father. That is, in terms of their genetic relatedness, it is as if the first cousins have the same father.

Furthermore, the cousins are as closely related to each other as half siblings are. This design yields similar results in relation to cognitive ability. An extension of the families-of-twins method includes the combination of twins and their children (children-of-twins method) and a sample of children who are twins and their parents (Narusyte et al., 2011; Silberg, Maes, & Eaves, 2010). This extended children-of-twins design allows the effects of parents on children *and* of children on parents to be examined.

Although not as powerful as standard adoption or twin designs, a design that has been used by a few research groups takes advantage of the increasing number of step-families created as a result of divorce and remarriage (Harris et al., 2009; Reiss, Neiderhiser, Hetherington, & Plomin, 2000). Half siblings typically occur in stepfamilies because a woman brings a child from a former marriage to her new marriage and then has another child with her new husband. These children have only one parent (the mother) in common and are 25 percent similar genetically, unlike full siblings, who have both parents in common and are 50 percent similar genetically. Half siblings can be compared with full siblings in stepfamilies to assess genetic influences. Full siblings in stepfamilies occur when the mother brings full siblings from her former marriage or when she and her new husband have more than one child together. A useful test of whether stepfamilies differ from never-divorced families is the comparison between full siblings in the two types of families. This type of design can also include stepsiblings who are genetically unrelated because each parent brought a child from a previous marriage. In the absence of mating (Chapter 12) by the stepparents, the similarity of two stepsiblings tests the importance of shared environmental influences.

Summary

Quantitative genetic methods can detect genetic influence for complex traits. Adoption and twin studies are the workhorses for human quantitative genetics. They capitalize on the quasi-experimental situations caused by adoption and twinning to assess the relative contributions of nature and nurture. For schizophrenia and cognitive ability, for example, resemblance of relatives increases with genetic relatedness, an observation suggesting genetic influence. Adoption studies show family resemblance even when family members are adopted apart. Twin studies show that identical twins are more similar than fraternal twins. Results of such family, adoption, and twin studies converge on the conclusion that genetic factors contribute substantially to complex human behavioral traits, among other traits.

There is a new wave of studies that combine designs, such as including the children of twins or nontwin sibling pairs. These combined and extended designs help to increase our ability to test different questions about the roles of genes and environment in behavior and also increase our confidence that the findings from such studies are generalizable beyond the special populations of twins and adoptees. In Chapters 7 and 8 the importance of such combination designs will be discussed in more detail.

CHAPTER SEVEN

Estimating Genetic and Environmental Influences

U p to this point, we have described different concepts and strategies involved in identifying genetic and environmental influences on behavior. Chapter 5 described animal research and Chapter 6 considered human research in this area. Although it is useful to be able to indicate that environmental and genetic factors contribute to behavior, quantifying those influences allows the relative importance of each to be considered. In this chapter we will describe the techniques used to quantify genetic and environmental influences in human research using the designs presented in Chapter 6. As noted elsewhere in this book, and in more detail in Chapter 8, genes and environments work together to influence behavior, and their influences can and do change over time or depending upon circumstances. Therefore, although it is possible and useful to quantify relative genetic and environmental influences, it is also necessary to recognize that these values can change based on the population studied, the age of the sample, and many other factors.

Heritability

For the complex traits that interest behavioral scientists, it is possible to ask not only whether genetic influences are important but also *how much* genetics contributes to the trait. The question about whether genetic influences are important involves *statistical significance*, the reliability of the effect. For example, we can ask whether the resemblance between "genetic" parents and their adopted offspring is significant or whether identical twins are significantly more similar than fraternal twins. Statistical significance depends on the size of the effect and the size of the sample. For example, a "genetic" parent-offspring correlation of 0.25 will be statistically significant if the adoption study includes at least 45 parent-offspring pairs. Such a result would indicate that it is highly likely (95 percent probability) that the true correlation is greater than zero.

The question about how much genetics contributes to a trait refers to *effect size,* the extent to which individual differences for the trait in the population can be accounted for by genetic differences among individuals. Effect size in this sense refers to individual differences for a trait in the entire population, not to certain individuals. For example, if PKU were left untreated, it would have a huge effect on the cognitive development of individuals homozygous for the recessive allele. However, because such individuals represent only 1 in 10,000 individuals in the population, this huge effect for these few individuals would have little effect overall on the variation in cognitive ability in the entire population. Thus, the size of the effect of PKU in the population is very small.

Many statistically significant environmental effects in the behavioral sciences involve very small effects in the population. For example, birth order is significantly related to intelligence test (IQ) scores (first-born children have higher IQs). This is a small effect in that the mean difference between first- and second-born siblings is less than two IQ points and their IQ distributions almost completely overlap. Birth order accounts for about 1 percent of the variance of IQ scores when other factors are controlled. In other words, if all you know about two siblings is their birth order, then you know practically nothing about their IQs.

In contrast, genetic effect sizes are often very large, among the largest effects found in the behavioral sciences, accounting for as much as half of the variance. The statistic that estimates the genetic effect size is called *heritability.* Heritability is the proportion of phenotypic variance that can be accounted for by genetic differences among individuals. As explained in the Appendix, heritability can be estimated from the correlations for relatives. For example, if the correlation for "genetic" (adopted-apart) relatives is zero, then heritability is zero. For first-degree "genetic" relatives, their correlation reflects half of the effect of genes because they are only 50 percent similar genetically. That is, if heritability is 100 percent, their correlation would be 0.50. In Figure 6.2, the correlation for "genetic" (adopted-apart) siblings is 0.24 for IQ scores. Doubling this correlation yields a heritability estimate of 48 percent, which suggests that about half of the variance in IQ scores can be explained by genetic differences among individuals.

Heritability estimates, like all statistics, include error of estimation, which is a function of the effect size and the sample size. In the case of the IQ correlation of 0.24 for adopted-apart siblings, the number of sibling pairs is 203. There is a 95 percent chance that the true correlation is between 0.10 and 0.38, which means that the true heritability is likely to be between 20 and 76 percent, a very wide range. For this reason, heritability estimates based on a single study need to be taken as very rough estimates surrounded by a large confidence interval unless the study is very large. For example, if the correlation of 0.24 were based on a sample of 2000 instead of 200, there would be a 95 percent chance that the true heritability is between 40 and 56 percent. Replication across studies and across designs also allows more precise estimates.

If identical and fraternal twin correlations are the same, heritability is estimated as zero. If identical twins correlate 1.0 and fraternal twins correlate 0.50, a heritability of 100 percent is implied. In other words, genetic differences among individuals completely account for their phenotypic differences. A rough estimate of heritability in a twin study can be made by doubling the difference between the identical and fraternal twin correlations. As explained in the Appendix, because identical twins are identical genetically and fraternal twins are 50 percent similar genetically, the difference in their correlations reflects half of the genetic effect and is doubled to estimate heritability. For example, in Figure 3.7, IQ correlations for identical and fraternal twins are 0.85 and 0.60, respectively. Doubling the difference between these correlations results in a heritability estimate of 50 percent, which also suggests that about half of the variance of IQ scores can be accounted for by genetic factors. Because these studies include more than 10,000 pairs of twins, the error of estimation is small. There is a 95 percent chance that the true heritability is between 0.48 and 0.52.

Because disorders are diagnosed as either-or dichotomies, familial resemblance for disorders is assessed by **concordances** rather than by correlations. As explained in the Appendix, *concordance* is an index of risk. For example, if sibling concordance is 10 percent for a disorder, we say that siblings of probands have a 10 percent risk for the disorder. The use of concordance to estimate genetic risk for disorders is very common in medical genetics for the study of disorders like heart disease and cancer (Lichtenstein et al., 2000; Wu, Snieder, & de Geus, 2010) and in psychiatric genetics (see Chapters 14 and 15 for more information on behavioral genetic studies of psychiatric disorders).

If identical and fraternal twin concordances are the same, heritability must be zero. To the extent that identical twin concordances are greater than fraternal twin concordances, genetic influences are implied. For schizophrenia (see Figure 3.6), the identical twin concordance of 0.48 is much greater than the fraternal twin concordance of 0.17, a difference suggesting substantial heritability. The fact that in 52 percent of the cases identical twins are *dis*cordant for schizophrenia, even though they are genetically identical, implies that heritability is much less than 100 percent.

One way to estimate heritability for disorders is to use the liability-threshold model (see Box 3.1) to translate concordances into correlations on the assumption that a continuum of genetic risk underlies the dichotomous diagnosis. For schizophrenia, the identical and fraternal twin concordances of 0.48 and 0.17 translate into liability correlations of 0.86 and 0.57, respectively. Doubling the difference between these liability correlations suggests a heritability of about 60 percent. Five of the most recent twin studies yield liability heritability estimates of about 80 percent (Cardno & Gottesman, 2000). As explained in Box 3.1, this statistic refers to a hypothetical construct of continuous liability as derived from a dichotomous diagnosis of schizophrenia rather than to the diagnosis of schizophrenia itself.

For combination designs that compare several groups, and even for simple adoption and twin designs, modern genetic studies are typically analyzed by using an approach called *model fitting*. Model fitting tests the significance of the fit between a model of genetic and environmental relatedness against the observed data. Different models can be compared, and the best-fitting model is used to estimate the effect size of genetic and environmental effects. Model fitting is described in the Appendix.

Quantitative genetic designs estimate heritability indirectly from familial resemblance. Its great strength is that it can estimate genetic influences regardless of the number of genes or magnitude or complexity of the genes' effects. As discussed in Chapter 9, DNA studies to date suggest that the heritability of behavioral disorders and dimensions is highly polygenic, that is, due to the relatively small effects of many genes, which makes it difficult to identify the specific genes responsible for heritability. However, an exciting new approach estimates heritability directly from DNA even though we do not know which genes contribute to heritability. The technique, which is called **genome-wide complex trait analysis (GCTA)**, is described in Box 7.1.

Interpreting Heritability

Heritability refers to the genetic contribution to individual differences (variance), *not* to the phenotype of a single individual. For a single individual, both genotype and environment are indispensable—a person would not exist without both genes and environment. As noted by Theodosius Dobzhansky (1964), the first president of the Behavior Genetics Association:

> The nature-nurture problem is nevertheless far from meaningless. Asking right questions is, in science, often a large step toward obtaining right answers. The question about the roles of genotype and the environment in human development must be posed thus: To what extent are the *differences* observed among people conditioned by the differences of their genotypes and by the differences between the environments in which people were born, grew and were brought up? (p. 55)

This issue is critical for the interpretation of heritability (Sesardic, 2005). You can still read in introductory textbooks that genetic and environmental effects on behavior cannot be disentangled because behavior is the product of genes and environment. An example sometimes given is the area of a rectangle. It is nonsensical to ask about the separate contributions of length and width to the area of a single rectangle because area is the product of length and width. Area does not exist without both length and width. However, if we ask not about a single rectangle but about

BOX 7.1 • **Estimating Quantitative Genetic Parameters Directly from DNA**

An exciting new quantitative genetic technique estimates genetic influences directly from measured genotypes rather than indirectly from comparisons between groups that differ on average genetically, such as MZ and DZ twins. The technique, called *genome-wide complex trait analysis (GCTA;* Yang, Lee, Goddard, & Visscher, 2011), requires thousands of individuals who have been genotyped on hundreds of thousands of DNA markers called **single nucleotide polymorphisms (SNPs),** as described in Chapter 9. As described in Chapter 9, many samples meeting these requirements have been obtained thanks to SNP microarrays that can genotype hundreds of thousands of SNPs quickly and inexpensively.

The GCTA method compares chance genetic similarity across hundreds of thousands of SNPs for each pair of individuals in a matrix of thousands of unrelated individuals. This chance genetic similarity is then used to predict phenotypic similarity for each pair of individuals, as illustrated below. That is, instead of comparing phenotypic resemblance for groups who differ in genetic relatedness such as MZ twins (100 percent) and DZ twins (~50 percent), GCTA uses chance genetic resemblance pair-by-pair for a large sample of individuals even though their overall genetic resemblance varies by only 1 or 2 percent, as shown in the distribution of chance genetic similarity, below right. Despite this minuscule variation in genetic resemblance, the large sample size makes it possible to estimate

	Genetic similarity				Phenotypic similarity			
	S_1	S_2	S_3	S_4	S_1	S_2	S_3	S_4
S_1		+0.1%	−0.5%	+0.1%		+ +	− −	+
S_2			−0.2%	+0.5%			−	+ +
S_3				+0.2%				+
S_4								

• GCTA uses genetic similarity assessed on the basis of hundreds of thousands of SNPs to predict phenotypic resemblance for pairs of individuals in a matrix of thousands of unrelated individuals. This matrix illustrates for just four individuals their genetic similarity, which is used to predict their phenotypic similarity, shown here as minuses and pluses.

heritability directly from DNA markers measured on the microarray. Analogous to quantitative genetic methods for estimating heritability, such as the twin method, GCTA estimates the extent to which phenotypic variance can be explained by genetic variance. The major advance of GCTA estimates of heritability is that they come directly from measured DNA differences between individuals. However, it should be noted that GCTA does not identify which SNPs are responsible for the heritability of a trait.

GCTA will eventually provide direct DNA tests of quantitative genetic results based on twin and adoption studies. One problem is that many thousands of individuals are required to provide reliable estimates. Another problem is that more SNPs are needed than even the million SNPs genotyped on current SNP microar-

rays because there is much DNA variation not captured by these SNPs. As a result, GCTA cannot estimate all heritability, perhaps only about half of the heritability. Indeed, the first reports of GCTA analyses estimate heritability to be about half the heritability estimates from twin and adoption studies for height (Lee, Wray, Goddard, & Visscher, 2011; Yang et al., 2010; Yang, Manolio, et al., 2011), and intelligence (Davies et al., 2011). The value of GCTA is that it does not require special samples such as twins or adoptees: In any large sample with DNA genotyped on microarrays with hundreds of thousands of DNA markers, GCTA can be used to estimate genetic influence for behavioral traits. A multivariate extension of GCTA (Deary et al., 2012) can be used to estimate genetic overlap between traits or across age.

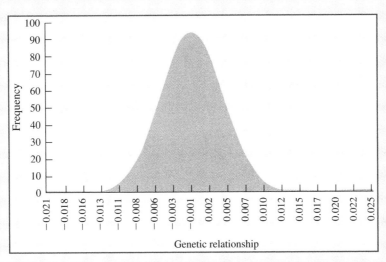

• Distribution of chance genetic similarity for pairs of individuals across hundreds of thousands of SNPs (from Davies et al., 2011, Supplementary Figure 8). The GCTA method estimates genetic influence by predicting phenotypic resemblance from genetic resemblance. (Reprinted by permission from Macmillan Publishers, Ltd: *Molecular Psychiatry, 16*, 996–1005, © 2011.)

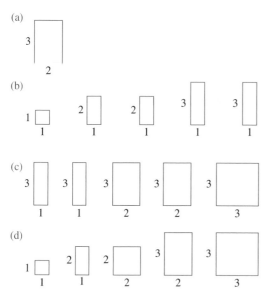

• **FIGURE 7.1** Individuals and individual differences. Genetic and environmental contributions to behavior do not refer to a single individual, just as the area of a single rectangle (a) cannot be attributed to the relative contributions of length and width, because area is the product of length and width. However, in a population of rectangles, the relative contribution of length and width to differences in area can be investigated. It is possible that length alone (b), width alone (c), or both (d) account for differences in area among rectangles.

a population of rectangles (Figure 7.1), the variance in areas could be due entirely to length (b), entirely to width (c), or to both (d). Obviously, there can be no behavior without both an organism and an environment. The scientifically useful question concerns the origins of differences among individuals.

For example, the heritability of height is about 90 percent, but this does not mean that you grew to 90 percent of your height for reasons of heredity and that the other inches were added by the environment. What it means is that most of the height differences among individuals are due to the genetic differences among them. Heritability is a statistic that describes the contribution of genetic differences to observed differences among individuals in a particular population at a particular time. In different populations or at different times, environmental or genetic influences might differ, and heritability estimates in such populations could differ.

A counterintuitive example concerns the effects of equalizing environments. If environments were made the same for everyone in a particular population, heritability would be high in that population because individual differences that remained in the population would be due exclusively to genetic differences. Using education as an example, if a society were able to give all children the same education, the heritability of educational achievement in that society would be high compared to societies in which educational opportunity differed.

It should be emphasized that heritability refers to the contribution of genetic differences to observed differences among individuals for a particular trait in a particular population at a particular time. Most DNA (99.5%) does not vary from person to person. If genes are the same for everyone, they will not contribute to differences among individuals. However, if these genes were disrupted by mutation, they could have a devastating, even lethal, effect on development, even though they may not normally contribute to variation in the population. Similarly, many environmental factors do not vary substantially, for example, the air we breathe and the essential nutrients we eat. Although at this level of analysis such environmental factors do not contribute to differences among individuals, disruption of these essential environments could have devastating effects.

A related issue concerns average differences between groups, such as average differences between males and females, between social classes, or between ethnic groups. It should be emphasized that the causes of individual differences within groups have no implications for the causes of average differences between groups. Specifically, heritability refers to the genetic contribution to differences among individuals within a group. High heritability within a group does not necessarily imply that average differences between groups are due to genetic differences between groups. The average differences between groups could be due solely to environmental differences even when heritability within the groups is very high.

This point extends beyond the politically sensitive issues of gender, social class, and ethnic differences. As discussed in Chapters 14 and 15, a key issue in psychopathology concerns the links between the normal and the abnormal. Finding heritability for individual differences within the normal range of variation does not necessarily imply that the average difference between an extreme group and the rest of the population is also due to genetic factors. For example, if individual differences in depressive symptoms for an unselected sample are heritable, this finding does not necessarily imply that severe depression is also due to genetic factors. This point is worth repeating: The causes of average differences between groups are not necessarily related to the causes of individual differences within groups.

A related point is that heritability describes *what is* in a particular population at a particular time rather than *what could be*. That is, if either genetic influences change (e.g., changes due to migration) or environmental influences change (e.g., changes in educational opportunity), then the relative impact of genes and environment will change. Even for a highly heritable trait such as height, changes in the environment *could* make a big difference, for example, if an epidemic struck or if children's diets were altered. Indeed, the huge increase in children's heights during the past century is likely to be a consequence of improved diet. Conversely, a trait that is largely influenced by environmental factors *could* show a big genetic effect. For example, genetic engineering can knock out a gene or insert a new gene that greatly alters the trait's development, something that can now be done in laboratory animals, as discussed in Chapter 5.

Although it is useful to think about what could be, it is important to begin with what is—the genetic and environmental sources of variance in existing populations. Knowledge about what is can sometimes help guide research concerning what could be, as in the example of PKU, where the effects of this single-gene disorder can be blocked by a diet low in phenylalanine. Most important, heritability has nothing to say about *what should be*. Evidence of genetic influence for a behavior is compatible with a wide range of social and political views, most of which depend on values, not facts. For example, no policies necessarily follow from finding genetic influences or even specific genes for cognitive abilities. It does not mean, for example, that we ought to put all our resources into educating the brightest children. Depending on our values, we might worry more about children falling off the low end of the bell curve in an increasingly technological society and decide to devote more public resources to those who are in danger of being left behind. For example, we might decide that all citizens need to reach basic levels of literacy and numeracy to be empowered to participate in society.

A related point is that heritability does not imply genetic determinism. Just because a trait shows genetic influences does not mean that nothing can be done to change it. Environmental change is possible even for single-gene disorders. For example, when PKU was found to be a single-gene cause of mental retardation, it was not treated by means of eugenic (breeding) intervention or genetic engineering. An environmental intervention was successful in bypassing the genetic problem of high blood levels of phenylalanine: Administer a diet low in phenylalanine. This important environmental intervention was made possible by recognition of the genetic basis for this type of mental retardation.

For behavioral disorders and dimensions, the links between specific genes and behavior are weaker because behavioral traits are generally influenced by multiple genes and environmental factors. For this reason, genetic influences on behavior involve probabilistic propensities rather than predetermined programming. In other words, the complexity of most behavioral systems means that genes are not destiny. Although specific genes that contribute to complex disorders such as late-onset Alzheimer disease are beginning to be identified, these genes only represent genetic risk factors in that they increase the probability of occurrence of the disorder but do not guarantee that the disorder will occur. An important corollary of the point that heritability does not imply genetic determinism is that heritability does not constrain environmental interventions such as psychotherapy.

We hasten to note that finding a gene that is associated with a disorder does not mean that the gene is "bad" and should be eliminated. For example, a gene associated with novelty seeking (Chapter 17) may be a risk factor for antisocial behavior, but it could also predispose individuals to scientific creativity. The gene that causes the flushing response to alcohol in Asian individuals protects them against becoming alcoholics (Chapter 18). The classic evolutionary example is a gene that causes sickle-cell anemia in the recessive condition but protects carriers against malaria in heterozygotes (Chapter 20). As we will see, most complex traits are influenced by multiple genes, so we are all likely to be carrying many genes that contribute to risk for some disorders.

Finally, finding genetic influences on complex traits does not mean that the environment is unimportant. For simple single-gene disorders, environmental factors may have little effect. In contrast, for complex traits, environmental influences are usually as important as, or in some cases more important than, genetic influences. When one member of an identical twin pair is schizophrenic, for example, the other twin is not schizophrenic in about half the cases, even though members of identical twin pairs are identical genetically. Such differences within pairs of identical twins can only be caused by nongenetic factors. Despite its name, behavioral genetics is as useful in the study of environment as it is in the study of genetics. In providing a "bottom line" estimate of all genetic influences on behavior, genetic research also provides a "bottom line" estimate of environmental influences. Indeed, genetic research provides the best available evidence for the importance of the environment. Moreover, genetic research has made some of the most important discoveries in recent years about how the environment works in psychological development (Chapter 8).

●KEY CONCEPTS

Heritability: Proportion of phenotypic variance that can be accounted for by genetic differences among individuals.

Effect size: The size of the estimate or effect in the population.

Twin correlation: Correlation of twin 1 with twin 2. Typically computed separately for MZ and DZ twins. Used to estimate genetic and environmental influences.

Concordance: The presence of the same trait in both members of a twin pair. Used to estimate risk for disorder.

Model fitting: A statistical strategy for testing the significance of the fit between a model of genetic and environmental relatedness against the observed data.

Environmentality

From Freud onward, most theories about how the environment works in behavioral development have implicitly assumed that offspring resemble their parents because parents provide the family environment for their offspring and that siblings resemble each other because they share that family environment. Twin and adoption research during the past two decades has dramatically altered this view. In fact, genetic designs, such as twin and adoption methods, were devised specifically to address the possibility that some of this widespread familial resemblance may be due to shared heredity rather than to shared environment.

As with heritability, we can estimate *how much* environmental influences contribute to individual differences in complex behaviors. The twin, adoption, sibling, and combination designs described in Chapter 6 help to clarify environmental influences as much as they help to estimate genetic influences. We can compute the statistical significance of such **environmentality** in the same way as we compute the significance of genetic influences.

Shared Environment

Shared environmental influences refer to all nongenetic influences that make family members similar to one another. This can include a wide range of factors, including neighborhood, parental education, and family factors such as parenting behavior or the amount of conflict that occurs within the household. These factors will be shared environmental influences only if they result in greater similarity among individuals living in the same household and if they do not vary as a function of genetic relatedness. In other words, if fraternal twins are as similar as identical twins, and this similarity is not negligible, then shared environmental influences are important. Similarly, if the similarity of "environmental" siblings is the same as that of "genetic" siblings, then shared environmental influences are indicated. The Appendix provides more detail about how shared environmental influences are estimated in twin, sibling, and combination designs.

There has been confusion about shared environmental influences. As will be described in Chapters 11 through 19, there is little evidence of shared environmental influences on many commonly studied behaviors such as personality and cognitive abilities; the modest shared environmental influences that have been found are often significant only during childhood and adolescence (Plomin, 2011; Plomin & Daniels, 1987). However, shared environmental influences do have some effect through childhood and adolescence (Burt, 2009b), especially for certain types of behavior problems, although they also become less important for explaining similarity for family members no longer living in the same household. In other words, residing in the same household does increase the similarity of family members, although these effects do not appear to persist once children have moved out of the home.

Nonshared Environment

Nonshared environmental influences are all nongenetic influences that are independent (or uncorrelated) for family members, including error of measurement. Because identical twins living in the same household share all of their genes and share their environment, the only thing that can account for differences within pairs of identical twins is nonshared environmental influences. Sources of nonshared environmental influences include differences in their family experience, such as different treatment by parents, or differential experiences outside the family, such as having different friends.

Quantitative genetic designs provide an essential starting point in the quantification of the net effect of genetic and environmental influences in the populations studied. If the net effect of genetic factors is substantial, there may be value in seeking to identify the specific genes responsible for that genetic effect. Similarly, if environmental influences are largely nonshared rather than shared, this finding should deter researchers from relying solely on family-wide risk factors that pay no attention to the ways in which these influences impinge differentially on different children in the same family. Current research is trying to identify specific sources of nonshared environment and to investigate associations between nonshared environment and behavioral traits, as discussed later.

Estimating Shared and Nonshared Environmental Influences

How do genetic designs estimate the net effect of shared and nonshared environment? Heritability is estimated, for example, by comparing identical and fraternal twin resemblance or by using adoption designs. In quantitative genetics, environmental variance is variance not explained by genetics. Shared environment is estimated as family resemblance not explained by genetics. Nonshared environment is the rest of the variance: variance not explained by genetics or by shared environment. The conclusion that environmental variance in adult behavior is largely nonshared refers to this residual component of variance, usually estimated by model-fitting analyses. However, more direct tests of shared and nonshared environments make it easier to understand how they can be estimated.

A direct test of shared environment is resemblance among adoptive relatives. Why do genetically unrelated adoptive siblings correlate about 0.25 for general cognitive ability in childhood? In the absence of selective placement, the answer must be shared environment because adoptive siblings are unrelated genetically. This result fits with the conclusion in Chapter 12 that about one-quarter of the variance of general cognitive ability in childhood is due to shared environment. By adolescence, the correlation for adoptive siblings plummets to zero and is the basis for the conclusion that shared environment has negligible impact in the long run. For personality and some measures of psychopathology in adults, adoptive siblings correlate near zero, a value implying that shared environment is unimportant and that environmental influences, which are substantial, are of the nonshared variety. For some measures of behavior problems in children and adolescents, adoptive siblings correlate significantly greater than zero, indicating that shared environmental influences are present (Burt, 2009b).

Just as genetically unrelated adoptive siblings provide a direct test of shared environment, identical twins reared together provide a direct test of nonshared environment. Because they are essentially identical genetically, differences within pairs of identical twins can only be due to nonshared environment. For example, for self-report personality questionnaires, identical twins typically correlate about 0.45. This value means that about 55 percent of the variance is due to nonshared environment plus error of measurement. Identical twin resemblance is also only moderate for most mental disorders, an observation implying that nonshared environmental influences play a major role.

Differences within pairs of identical twins provide a conservative estimate of nonshared environment because twins often share special environments that increase their resemblance but do not contribute to similarity among "normal" siblings. For example, for general cognitive ability, identical twins correlate about 0.85, a result that does not seem to leave much room for nonshared environment (i.e., $1 - 0.85 = 0.15$). However, fraternal twins correlate about 0.60 and nontwin siblings correlate about 0.40, implying that twins have a special shared twin environment that accounts for as much as 20 percent of the variance (Koeppen-Schomerus et al., 2003). For this reason, the identical twin correlation of 0.85 may be inflated by 0.20 because of this special shared twin environment. In other words, about a third of the variance of general cognitive ability may be due to nonshared environment, that is, $1 - (0.85 - 0.20) = 0.35$. However, a different study that included twins and nontwin siblings in different families found no systematic indication of a special shared twin environment for a wide range of adolescent adjustment measures (Reiss et al., 2000).

Identifying Specific Nonshared Environment

The next step in research on nonshared environment is to identify specific factors that make children growing up in the same family so different. To identify nonshared environmental factors, it is necessary to begin by assessing aspects of the environment specific to each child, rather than aspects shared by siblings. Many measures of the environment used in studies of behavioral development are general to a family rather than specific to a child. For example, whether or not their parents have been divorced is the same for two children in the family. Assessed in this family-general way, divorce cannot be a source of differences in siblings' outcomes because it does not differ for two children in the same family. However, research on divorce has shown that divorce affects children in a family differently (Hetherington & Clingempeel, 1992). If the divorce is assessed in a child-specific way (e.g., by assessing the children's perceptions about the stress caused by the divorce, which may, in fact, differ among siblings), divorce could well be a source of differential sibling outcome.

Even when environmental measures are specific to a child, they can be shared by two children in a family. Research on siblings' experiences is needed to assess the extent to which aspects of the environment are shared. For example, to what extent are maternal vocalizing and maternal affection toward the children shared by siblings in the same family? Observational research on maternal interactions with siblings, assessed when each child was 1 and 2 years old, indicates that mothers' spontaneous vocalizing correlates substantially across the siblings (Chipuer & Plomin, 1992). This research implies that maternal vocalizing is an experience shared by siblings. In contrast, mothers' affection yields negligible correlations across siblings, a result indicating that maternal affection is not shared and is thus a better candidate for nonshared environmental influence.

Some family structure variables, such as birth order and sibling age spacing, are, by definition, nonshared environmental factors. However, these factors have generally been found to account for only a small portion of variance in behavioral outcomes. Research on more dynamic aspects of nonshared environment has found that children growing up in the same family lead surprisingly separate lives (Dunn & Plomin, 1990). Siblings perceive their parents' treatment of themselves and the other siblings as quite different, although parents generally perceive that they treat their children similarly, depending on the method of assessment. Observational studies tend to back up the children's perspective.

Table 7.1 shows sibling correlations for measures of family environment in a study focused on these issues, called the Nonshared Environment and Adolescent Development (NEAD) project (Reiss et al., 2000). During two 2-hour visits to 720 families with two siblings ranging from 10 to 18 years of age, a large battery of questionnaire and interview measures of the family environment was administered to both parents and both siblings. Parent-child interactions were videotaped during a session when problems in family relationships were discussed. Sibling correlations for children's reports of their family interactions (e.g., children's reports of their parents' negativity) were modest; they were also modest for observational ratings of child-to-parent interactions and parent-to-child interactions. This finding suggests that these experiences are largely nonshared. In contrast, parent reports yielded high sibling correlations, for example, when parents reported on their own negativity toward each of the children. Although this may be due to a "rater" effect, in that the parent rates both children, the high sibling correlations indicate that parent reports of children's environments are not good sources of candidate variables for assessing nonshared environmental factors.

As mentioned earlier, nonshared environment is not limited to measures of the family environment. Indeed, experiences outside the family, as siblings make their own way in the world, are even more likely candidates for nonshared environmental

TABLE 7.1

Sibling Correlations for Measures of Family Environment

Type of Data	Sibling Correlation
Child reports	
Parenting	0.25
Sibling relationship	0.40
Parent reports	
Parenting	0.70
Sibling relationship	0.80
Observational data	
Child to parent	0.20
Parent to child	0.30

SOURCE: Adapted from Reiss et al. (2000).

influences (Harris, 1998). For example, how similarly do siblings experience peers, social support, and life events? The answer is "only to a limited extent"; correlations across siblings for these experiences range from about 0.10 to 0.40 (Plomin, 1994). It is also possible that nonsystematic factors, such as accidents and illnesses, initiate differences between siblings. Compounded over time, small differences in experience might lead to large differences in outcome.

Identifying Specific Nonshared Environment That Predicts Behavioral Outcomes

Once child-specific factors are identified, the next question is whether these nonshared experiences relate to behavioral outcomes. For example, to what extent do differences in parental treatment account for the nonshared environmental variance known to be important for personality and psychopathology? Some success has been achieved in predicting differences in adjustment from sibling differences in their experiences. The NEAD project mentioned earlier provides an example in that negative parental behavior directed specifically to one adolescent sibling (controlling for parental treatment of the other sibling) relates strongly to that child's antisocial behavior and, to a lesser extent, to that child's depression (Reiss et al., 2000). Most of these associations involve negative aspects of parenting, such as conflict, and negative outcomes, such as antisocial behavior. Associations are generally weaker for positive parenting, such as affection.

A meta-analysis of 43 papers that addressed associations between nonshared experiences and siblings' differential outcomes concluded that "measured nonshared environmental variables do not account for a substantial portion of the nonshared variability" (Turkheimer & Waldron, 2000, p. 78). Looking at the same studies, however, an optimist could conclude that this research is off to a good start (Plomin, Asbury, & Dunn, 2001). The proportion of total variance accounted for in adjustment, personality, and cognitive outcomes was 0.01 for family constellation (e.g., birth order), 0.02 for differential parental behavior, 0.02 for differential sibling interaction, and 0.05 for differential peer or teacher interaction. Moreover, these effects are largely independent because they add up in predicting the outcomes—incorporating all of these measures of differential environment accounts for about 13 percent of the total variance of the outcome measures. If nonshared environment accounts for 40 percent of the variance in these domains, we could say the cup is already more than one-quarter full.

When associations are found between nonshared environment and outcome, the question of the direction of effects is raised. That is, is differential parental negativity the cause or the effect of sibling differences in antisocial behavior? Genetic research is beginning to suggest that most differential parental treatment of siblings is in fact the effect rather than the cause of sibling differences. One of the reasons why siblings differ is genetics. Siblings are 50 percent similar genetically, but this statement implies that siblings are also 50 percent different. Research on nonshared environment needs to be embedded in genetically sensitive designs in order to distinguish true nonshared environmental effects from sibling differences due to genetics. For this reason, the

NEAD project included identical and fraternal twins, full siblings, half siblings, and genetically unrelated siblings. Multivariate genetic analysis of associations between parental negativity and adolescent adjustment yielded an unexpected finding: Most of these associations were mediated by genetic factors, although some nonshared environmental influence was also found (Pike, McGuire, Hetherington, Reiss, & Plomin, 1996). This finding and similar research (Burt, McGue, Krueger, & Iacono, 2005; Moberg, Lichtenstein, Forsman, & Larsson, 2011) implies that differential parental treatment of siblings to a substantial extent reflects genetically influenced differences between the siblings, such as differences in personality. The role of genetics in environmental influences is given detailed consideration in the next chapter.

Because MZ twins are identical genetically, they provide an excellent test of nonshared environmental influences. Nonshared environmental influence is implicated if MZ differences in experience correlate with MZ differences in outcome. In the NEAD project, analyses of MZ differences confirmed the results of the full multivariate genetic analysis mentioned above (Pike, McGuire, et al., 1996) in showing that MZ differences in experiences of parental negativity correlated modestly with MZ differences in adjustment outcomes (Pike, Reiss, Hetherington, & Plomin, 1996). Other studies of MZ differences have also identified nonshared environmental factors free of genetic confound (Barclay, Eley, Buysse, Maughan, & Gregory, 2011; Oliver, Pike, & Plomin, 2008; Viding, Fontaine, Oliver, & Plomin, 2009). A longitudinal study of MZ differences that extended from infancy to middle childhood found that MZ differences in birth weight and family environment during infancy related to their differences in behavior problems and academic achievement as assessed by their teachers at age 7 (Asbury, Dunn, & Plomin, 2006b). Another longitudinal study of MZ differences suggested a pernicious downward spiral of the interplay of nonshared environmental influence between negative parenting and children's behavior problems (Burt et al., 2005). A different study also using the MZ twin difference method found that more differences in friends' aggression in kindergarten were linked with increased differences in twin aggression in first grade (Vitaro et al., 2011).

Because such studies have been able to identify specific nonshared environmental factors that account for only a small portion of nonshared environment, the MZ difference method has been used to search for other sources of nonshared environment (Asbury, Dunn, & Plomin, 2006a). From a sample of 1590 MZ pairs rated by their teachers for anxiety at the age of 7, the most discordant pairs were selected and interviewed with their mothers to explore reasons why the twins may have become so different in their level of anxiety. Some of the top reasons reported by the mothers were negative school experiences, peer rejection, illness and accidents, and perinatal life events such as birth weight. Perinatal factors are receiving increased attention as a source of nonshared environmental influence that has a lasting impact on individuals throughout the life span (Salsberry & Reagan, 2010; Stromswold, 2006).

No matter how difficult it may be to find specific nonshared environmental factors within the family, it should be emphasized that nonshared environment is generally the norm in the behavioral sciences. It seems reasonable that experiences outside the family—for example, experiences with peers or life events—might be richer sources of nonshared environment (Harris, 1998). It is also possible that chance contributes to nonshared environment in the sense of random noise, idiosyncratic experiences, or the subtle interplay of a concatenation of events (Davey Smith, 2011; Dunn & Plomin, 1990). Francis Galton, the founder of behavioral genetics, suggested that nonshared environment is largely due to chance: "The whimsical effects of chance in producing stable results are common enough. Tangled strings variously twitched, soon get themselves into tight knots" (Galton, 1889, p. 195).

Support for the hypothesis that chance plays an important role in nonshared environment comes from longitudinal genetic analyses of age-to-age change and continuity. Longitudinal genetic research indicates that nonshared environmental influences are age-specific for psychopathology (Kendler, et al., 1993b; Van den Oord & Rowe, 1997), personality (Loehlin, Horn, & Willerman, 1990; McGue, Bacon, & Lykken, 1993; Pogue-Geile & Rose, 1985), and cognitive abilities (Cherny, Fulker, & Hewitt, 1997). That is, nonshared environmental influences at one age are largely different from nonshared environmental influences at another age. It is difficult to imagine environmental processes, other than chance, that could explain these results. Nonetheless, our view is that chance is the null hypothesis—systematic sources of nonshared environment need to be thoroughly examined before we conclude that chance factors are responsible for nonshared environment.

Multivariate Analysis

The estimation of genetic and environmental influences is not limited to examining the variance of a single behavior. The same model can be applied to investigating genetic and environmental influences on the covariance between two or more traits, which is one of the most important advances in quantitative genetics in the past few decades (Martin & Eaves, 1977). Just as univariate genetic analyses estimate the relative contributions of genetic and environmental factors to the variance of a trait, multivariate genetic analyses estimate the relative contributions of genetic and environmental factors to the covariance between traits. In other words, multivariate genetic analysis estimates the extent to which the same genetic and environmental factors affect different traits. An important developmental application of multivariate genetic analysis is to examine genetic and environmental contributions to stability and change longitudinally in the same individuals from age to age.

As explained in the Appendix, the essence of multivariate genetic analysis is the analysis of *cross-covariance* in relatives. That is, instead of asking whether trait X in one twin covaries with trait X in the co-twin, *cross-covariance* refers to the covariance between trait X in one twin and a different trait, trait Y, in the co-twin. Two new statistical constructs in multivariate genetic analysis are the correlation between genetic influences on X and Y, and the corresponding correlation between environmental influences on the two traits. Focusing on the genetic contribution to the covariance between trait X and trait Y, the *genetic correlation* estimates the extent to which genetic deviations that affect X literally correlate with genetic deviations that affect Y. The genetic correlation is independent of heritability. That is, traits X and Y could be highly heritable but their genetic correlation could be zero. Or traits X and Y could be only slightly heritable yet their genetic correlation could be 1.0. A genetic correlation of zero would indicate that the genetic influences on trait X are not associated with those on trait Y. In contrast, a genetic correlation of 1.0 would mean that all genetic influences on trait X also influence trait Y. Another useful statistic from multivariate genetic analysis is *bivariate heritability*, which weights the genetic correlation between X and Y by the square roots of their heritabilities and estimates the contribution of genetic influences to the phenotypic correlation between the two traits.

Multivariate genetic analysis will be featured in many subsequent chapters. The most interesting result occurs when the genetic structure between traits differs from the phenotypic structure. For example, as explained in Chapter 15, multivariate genetic analysis has shown that the genetic structure of psychopathology differs from phenotypic diagnoses in that many aspects of psychopathology are highly correlated genetically. The same pattern of general effects of genes is found for specific cognitive abilities (Chapter 13). A surprising example is that measures that are ostensibly environmental measures often correlate genetically with behavioral measures (Chapter 8). Another example is that multivariate genetic analyses across age typically find substantial age-to-age genetic correlations, suggesting that genetic factors contribute largely to stability from age to age; environmental factors contribute largely to change.

● KEY CONCEPTS

Environmentality: Proportion of phenotypic variance that can be accounted for by environmental influences.

Shared environmental influences: Nongenetic influences that make family members similar.

Nonshared environmental influences: Nongenetic influences that are uncorrelated for family members.

Genetic correlation: A statistic indexing the extent to which genetic influences on one trait are correlated with genetic influences on another trait independent of the heritabilities of the traits.

Summary

Quantitative genetic methods can detect genetic influences for complex traits. The size of the total genetic effects is quantified by heritability, a statistic that describes the contribution of genetic differences to observed differences in a particular population at a particular time. For most behavioral dimensions and disorders, including cognitive ability and schizophrenia, genetic influences are not only detectable but also substantial, often accounting for as much as half of the variance in the population. Genetic influence in the behavioral sciences has been controversial in part because of misunderstandings about heritability.

Genetic influence on behavior is just that—an influence or contributing factor, not something that is preprogrammed and deterministic. Environmental influences are usually as important as genetic influences; they are quantified as shared environmental influences and nonshared environmental influences. Behavioral genetics focuses on why people differ, that is, the genetic and environmental origins of individual differences that exist at a particular time in a particular population. Behavioral genetic research has helped to increase our understanding of how environmental factors influence behavioral outcomes. A major example is that behavioral genetic research finds only modest evidence for shared environmental influences, a finding that created a new field of research on nonshared environment. Understanding how genetic and environmental influences can make family members similar and different can help to guide work aimed at improving developmental outcomes for individuals. The following chapter continues this discussion about how genes and environments work together.

CHAPTER EIGHT

The Interplay between Genes and Environment

P revious chapters described how genetic and environmental influences can be assessed and the various designs that are typically used in human and animal behavioral genetic research. As described in Chapter 7, behavioral genetic research has helped to advance not just our understanding of how genes influence behavior but also of how environments influence behavior. Although much remains to be learned about the specific mechanisms involved in the pathways between genes and behavior, we know much more about genes than we do about the environment. We know that genes are located on chromosomes in the nucleus of cells, how their information is stored in the four nucleotide bases of DNA, and how they are transcribed and then translated using the triplet code. In contrast, where in the brain are environmental influences expressed, how do they change in development, and how do they cause individual differences in behavior? Given these differences in levels of understanding, genetic influences on behavior may be construed as easier to study than environmental influences.

One thing we know for sure about the environment is that it is important. Quantitative genetic research, reviewed in Chapters 11 to 19, provides the best available evidence that the environment is an important source of individual differences throughout the domain of behavior. Moreover, quantitative genetic research is changing the way we think about the environment. Three of the most important discoveries from genetic research in the behavioral sciences are about nurture rather than nature. The first discovery is that nonshared environmental influences are surprisingly large and important in explaining individual differences. The second discovery is equally surprising: Many environmental measures widely used in the behavioral sciences show genetic influence. This research suggests that people create their own experiences, in part for genetic reasons. This topic has been called the *nature of nurture,* although in genetics it is known as ***genotype-environment correlation*** because it refers to experiences that are correlated with genetic propensities. The third discovery at the interface between nature and nurture is that the effects of the environment can depend on genetics and that the effects of genetics can depend on the environment. This topic is called *genotype-environment interaction,* genetic sensitivity to environments.

Genotype-environment correlation and genotype-environment interaction—often referred to collectively as gene-environment interplay—are the topics of this chapter. The goal of this chapter is to show that some of the most important questions in genetic research involve the environment, and some of the most important questions for environmental research involve genetics. Genetic research will profit if it includes sophisticated measures of the environment, environmental research will benefit from the use of genetic designs, and behavioral science will be advanced by collaboration between geneticists and environmentalists. These are some of the ways in which some behavioral scientists are bringing nature and nurture together in the study of development in their attempt to understand the processes by which genotypes eventuate in phenotypes (Rutter, Moffitt, & Caspi, 2006).

Three reminders about the environment are warranted. First, genetic research provides the best available evidence for the importance of environmental factors. The surprise from genetic research has been the discovery that genetic factors are so important throughout the behavioral sciences, often accounting for as much as half of the variance. However, the excitement about this discovery should not overshadow the fact that environmental factors are at least as important. Heritability rarely exceeds 50 percent and thus environmentality is rarely less than 50 percent.

Second, in quantitative genetic theory, the word *environment* includes all influences other than inheritance, a much broader use of the word than is usual in the behavioral sciences. By this definition, environment includes, for instance, prenatal events and biological events such as nutrition and illness, not just family socialization factors.

Third, as explained in Chapter 7, genetic research describes *what is* rather than predicts *what could be*. For example, high heritability for height means that height differences among individuals are largely due to genetic differences, given the genetic and environmental influences that exist in a particular population at a particular time (*what is*). Even for a highly heritable trait such as height, an environmental intervention such as improving children's diet or preventing illness could affect height (*what could be*). Such environmental factors are thought to be responsible for the average increase in height across generations, for example, even though individual differences in height are highly heritable in each generation.

Beyond Heritability

As mentioned in Chapter 1, one of the most dramatic shifts in the behavioral sciences during the past few decades has been toward a balanced view that recognizes the importance of both nature and nurture in the development of individual differences in behavior. Behavioral genetic research has found genetic influence nearly everywhere it has looked. Indeed, it is difficult to find any behavioral dimension or disorder that reliably shows no genetic influence. On the other hand, behavioral genetic research also provides some of the strongest available evidence for the importance of environmental influences for the simple reason that heritabilities are seldom greater than 50 percent. This means that environmental factors are also important. This message of the importance of both nature and nurture is repeated throughout the following chapters. It is a message that seems to have gotten through to the public as well as academics. For example, a survey of parents and teachers of young children found that over 90 percent believed that genetics is at least as important as environment for mental illness, learning difficulties, intelligence, and personality (Walker & Plomin, 2005).

As a result of the increasing acceptance of genetic influence on behavior, most behavioral genetic research reviewed in the rest of the book goes beyond merely estimating heritability. Estimating whether and how much genetics influences behavior is an important first step in understanding the origins of individual differences. But these are only first steps. As illustrated throughout this book, quantitative genetic research goes beyond heritability in three ways. First, instead of estimating genetic and environmental influence on the variance of one behavior at a time, multivariate genetic analysis investigates the origins of the covariance between behaviors. Some of the most important advances in behavioral genetics have come from multivariate genetic analyses. A second way in which behavioral genetic research goes beyond heritability is to investigate the origins of continuity and change in development. This is why so much recent behavioral genetic research is developmental, as reflected throughout Chapters 11 to 19, most notably in Chapter 16, which addresses developmental psychopathology. Third, behavioral genetics considers the interface between nature and nurture, which is the topic of this chapter. Moreover, the rapid advances in our ability to identify genes (Chapter 9) and to link genes to behaviors via molecular genetics have revolutionized our ability to integrate genetic and social science research. It is possible to address multivariate, developmental, and gene-environment interplay with much greater precision and ease; as described in Chapter 10, we are also making advances in understanding the pathways between genes and behavior. In fact, these many advances have resulted in research cutting across multiple and diverse areas of research, including genetics, sociology, family relations, and prevention science, to name just a few. The rest of this chapter will focus on how genes and environments work together, that is, gene-environment interplay.

Genotype-Environment Correlation

As illustrated in Chapter 7, behavioral genetic research helps to clarify both genetic and environmental influences. Genetic research is also changing the way we think about the environment by showing that we create our experiences in part for genetic reasons. That is, genetic propensities are correlated with individual differences in experiences, an example of a phenomenon known as genotype-environment correlation. In other words, what seem to be environmental effects can reflect genetic influence because these experiences are influenced by genetic differences among individuals.

This genetic influence is just what genetic research during the past decade has found: When environmental measures are examined as phenotypes in twin and adoption studies, the results consistently point to some genetic influence, as discussed later. For this reason, genotype-environment correlation has been described as genetic control of exposure to the environment (Kendler & Eaves, 1986).

Genotype-environment correlation adds to phenotypic variance for a trait (see Appendix), but it is difficult to detect the overall extent to which phenotypic variance is due to the correlation between genetic and environmental effects (Plomin, DeFries, & Loehlin, 1977b). For this reason, these discussions focus on detection of specific genotype-environment correlations rather than on estimating their overall contribution to phenotypic variation.

The Nature of Nurture

The first research on this topic was published over two decades ago, with several dozen studies using various genetic designs and measures converging on the conclusion that measures of the environment show genetic influence (Plomin & Bergeman, 1991). After providing some examples of this research, we will consider how it is possible for measures of the environment to show genetic influences.

A widely used measure of the home environment that combines observations and interviews is the Home Observation for Measurement of the Environment (HOME; Caldwell & Bradley, 1978). HOME assesses aspects of the home environment such as parental responsivity, encouragement of developmental advance, and provision of toys. In an adoption study, HOME correlations for nonadoptive and adoptive siblings were compared when each child was 1 year old and again when each child was 2 years old (Braungart, Fulker, & Plomin, 1992). HOME scores were more similar for nonadoptive siblings than for adoptive siblings at both 1 and 2 years (0.58 versus 0.35 at 1 year and 0.57 versus 0.40 at 2 years), results suggesting genetic influence on HOME scores. Genetic factors were estimated to account for about 40 percent of the variance of HOME scores.

Other observational studies of mother-infant interaction in infancy, using the adoption design (Dunn & Plomin, 1986) and the twin design (Lytton, 1977, 1980), show genetic influences. A study of 8-year-old twins and their mothers found that genetic influences were substantial for observer ratings of maternal control during a mother-child "Etch-a-Sketch" task (Eley, Napolitano, Lau, & Gregory, 2010). The Nonshared Environment and Adolescent Development (NEAD) project, mentioned in Chapter 7, included videotaped observations of each parent interacting with each adolescent child when the parent-child dyad was engaged in ten-minute discussions around problems and conflict relevant to the dyad. Significant heritability was found for all measures (O'Connor, Hetherington, Reiss, & Plomin, 1995).

These observational studies suggest that genetic effects on family interactions are not solely in the eye of the beholder. Most genetic research on the nature of nurture has used questionnaires rather than observations. Questionnaires add another source of possible genetic influence: the subjective processes involved in perceptions of the family environment. The pioneering research in this area included two twin studies of adolescents' perceptions of their family environment (Rowe, 1981, 1983b). Both studies found substantial genetic influence on adolescents' perceptions of their parents' acceptance and no genetic influence on perceptions of parents' control.

The NEAD project was designed in part to investigate genetic contributions to diverse measures of family environment. As shown in Table 8.1, significant genetic influence was found for adolescents' ratings of composite variables of their parents' positivity and negativity (Plomin, Reiss, Hetherington, & Howe, 1994). The highest heritability of the 12 scales that contributed to these composites was for a measure of closeness (e.g., intimacy, supportiveness), which yielded heritabilities of about 50 percent for both mothers' closeness and fathers' closeness as rated by the adolescents. As found in Rowe's original studies and in several other studies (Bulik, Sullivan, Wade, & Kendler, 2000), measures of parental control showed lower heritability than measures of closeness (Kendler & Baker, 2007). The NEAD project also assessed parents' perceptions of their parenting behavior toward the adolescents (lower half of Table 8.1). Parents' ratings of their own behavior yielded heritability estimates similar to those for the adolescents' ratings of their parents' behavior. Because the twins were children in these studies, genetic influence on parenting comes from parents' response to genetically influenced characteristics of their children. In contrast, when the twins are parents, genetic influence on parenting can come from other sources, such as the parents' personality. Nonetheless, studies of twins as parents have generally yielded similar results that show widespread genetic influence (Neiderhiser et al., 2004).

TABLE 8.1

Heritability Estimates for Questionnaire Assessments of Parenting

Rater	Ratee	Measure	Heritability
Adolescent	Mother	Positivity	0.30
		Negativity	0.40
Adolescent	Father	Positivity	0.56
		Negativity	0.23
Mother	Mother	Positivity	0.38
		Negativity	0.53
Father	Father	Positivity	0.22
		Negativity	0.30

SOURCE: Plomin, Reiss, et al. (1994).

More than a dozen other studies of twins and adoptees have reported genetic influence on family environment (Plomin, 1994). For example, for 3-year-olds, observations and ratings of parent-child mutuality (shared positive affect and responsiveness) showed genetic influence in both a twin study and an adoption study (Deater-Deckard & O'Connor, 2000). A longitudinal twin study from ages 11 to 17 found significant genetic influence at both ages but greater genetic influence at age 17 (Elkins, McGue, & Iacono, 1997), a finding replicated in another study (McGue, Elkins, Walden, & Iacono, 2005). Multivariate genetic research suggests that genetic influence on perceptions of family environment is mediated by personality (Horwitz et al., 2011; Krueger, Markon, & Bouchard, 2003) and that genetic influence on personality can also explain covariation among different aspects of family relations, such as marital quality and parenting (Ganiban, Ulbricht, et al., 2009).

Genetic influence on environmental measures also extends beyond the family environment. For example, several studies have found genetic influence on measures of life events and stress, especially life events over which we have some control, such as problems with relationships and financial disruptions (Bolinskey, Neale, Jacobson, Prescott, & Kendler, 2004; Federenko et al., 2006; Kendler, Neale, Kessler, Heath, & Eaves, 1993a; McGuffin, Katz, & Rutherford, 1991; Middeldorp, Cath, Vink, & Boomsma, 2005; Plomin, Lichtenstein, Pedersen, McClearn & Nesselroade, 1990; Thapar & McGuffin, 1996). As is the case for genetic influence on perceptions of family environment, genetic influence on life events and stress is also mediated in part by personality (Kendler, Gardner, & Prescott, 2003; Saudino, Pedersen, Lichtenstein, McClearn, & Plomin, 1997).

Genetic influence has also been found for characteristics of children's friends and peer groups (e.g., Brendgen et al., 2009; Bullock, Deater-Deckard, & Leve, 2006; Guo, 2006; Iervolino et al., 2002; Manke, McGuire, Reiss, Hetherington, & Plomin, 1995) as well as adults' friends (Rushton & Bons, 2005), with genetic influence increasing during adolescence and young adulthood as children leave their homes and create their own social worlds (Kendler, Jacobson, et al., 2007). Several studies have found genetic influences on the tendency to be bullied during middle and late childhood and adolescence (Ball et al., 2008; Beaver, Boutwell, Barnes, & Cooper, 2009; Bowes et al., in press; Brendgen et al., 2008, 2011) and also on the likelihood of repeatedly being victimized (Beaver, Boutwell, et al., 2009). It is important to note that in the studies of bullying and peer victimization, heritabilities were somewhat less when peer nominations were used (Brendgen et al., 2008, 2011) as compared to parent and self-reports (Ball et al., 2008; Beaver et al., 2009; Bowes et al., in press).

There are at least two studies examining genetic influences on the school environment. Specifically, genetic influences have been found in children's perceptions of their classroom environment (Walker & Plomin, 2006) and in the amount of effort teachers report investing in their adolescent students (Houts, Caspi, Pianta, Arseneault, & Moffitt, 2010). Other environmental measures that have shown genetic influence include television viewing (Plomin, Lichtenstein, et al., 1990), school connectedness (Jacobson

& Rowe, 1999), work environments (Hershberger, Lichtenstein, & Knox, 1994), social support (Agrawal, Jacobson, Prescott, & Kendler, 2002; Bergeman, Plomin, Pedersen, McClearn, & Nesselroade, 1990; Kessler, Kendler, Heath, Neale, & Eaves, 1992), accidents in childhood (Phillips & Matheny, 1995), the propensity to marry (Johnson, McGue, Krueger, & Bouchard, 2004), marital quality (Spotts, Prescott, & Kendler, 2006), divorce (McGue & Lykken, 1992), exposure to drugs (Tsuang et al., 1992), and exposure to trauma (Lyons et al., 1993). In fact, there are few measures of experience examined in genetically sensitive designs that do *not* show genetic influence. It has been suggested that other fields, such as demography, also need to consider the impact of genotype-environment correlation (Hobcraft, 2006).

In summary, diverse genetic designs and measures converge on the conclusion that genetic factors contribute to experience. A review of 55 independent genetic studies using environmental measures found an average heritability of 0.27 across 35 different environmental measures (Kendler & Baker, 2007). At least 150 studies of genetic influences on environmental measures have been published since 1991. The large number of different environmental measures that have been found to show genetic influences demonstrates the key role that genetic influences play in the environments that individuals experience. A key direction for research on the interplay between genes and environment is to investigate the causes and consequences of genetic influence on measures of the environment.

Three Types of Genotype-Environment Correlation

What are the processes by which genetic factors contribute to variations in environments that we experience? For example, to what extent are behavioral traits, such as cognitive abilities, personality, and psychopathology, mediators of this genetic contribution? Even more important, does genetic influence on environmental measures contribute to the prediction of behavioral outcomes from environmental measures?

There are three types of genotype-environment correlation: passive, evocative, and active (Plomin et al., 1977b). The passive type occurs when children passively inherit from their parents family environments that are correlated with their genetic propensities. The evocative, or reactive, type occurs when individuals, on the basis of their genetic propensities, evoke reactions from other people on the basis of their genetic propensities. The active type occurs when individuals select, modify, construct, or reconstruct experiences that are correlated with their genetic propensities (Table 8.2).

For example, consider musical ability. If musical ability is heritable, musically gifted children are likely to have musically gifted parents who provide them with both genes and an environment conducive to the development of musical ability (passive genotype-environment correlation). Musically talented children might also be picked out at school and given special opportunities (evocative type). Even if no one does anything about their musical talent, gifted children might seek out their own musical environments by selecting musical friends or otherwise creating musical experiences (active type).

TABLE	8.2

Three Types of Genotype-Environment Correlation

Type	Description	Source of Environmental Influence
Passive	Children receive genotypes correlated with their family environment	Parents and siblings
Evocative	Individuals are reacted to on the basis of their genetic propensities	Anybody
Active	Individuals seek or create environments correlated with their genetic proclivities	Anybody or anything

SOURCE: Plomin et al. (1977b).

Passive genotype-environment correlation requires interactions between genetically related individuals. The evocative type can be induced by anyone who reacts to individuals on the basis of their genetic proclivities. The active type can involve anybody or anything in the environment. We tend to think of positive genotype-environment correlation, such as providing a musical environment, as being positively correlated with children's musical propensities, but genotype-environment correlation can also be negative. As an example of negative genotype-environment correlation, slow learners might be given special attention to boost their performance.

Three Methods to Detect Genotype-Environment Correlation

Three methods are available to investigate the contribution of genetic factors to the correlation between an environmental measure and a behavioral trait. These methods differ in the type of genotype-environment correlation they can detect. The first method is limited to detecting the passive type. The second method detects the evocative and active types. The third method detects all three types. All three methods can also provide evidence for environmental influence free of genotype-environment correlation.

The first method compares correlations between environmental measures and traits in nonadoptive and adoptive families (Figure 8.1). In nonadoptive families, a correlation between a measure of family environment and a behavioral trait of children could be environmental in origin, as is usually assumed. However, genetic factors might also contribute to the correlation. Genetic mediation would occur if genetically influenced traits of parents are correlated with the environmental measure and with the children's trait. For example, a correlation between the Home Observation for Measurement of the Environment and children's cognitive abilities could be mediated by genetic factors that affect both the cognitive abilities of parents and their scores on HOME. In contrast, in adoptive families, this indirect genetic path between

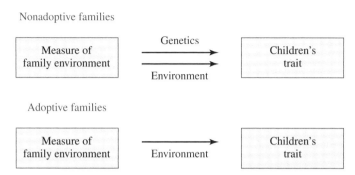

• **FIGURE 8.1** Passive genotype-environment correlation can be detected by comparing correlations between family environment and children's traits in nonadoptive and adoptive families.

family environment and children's traits is not present because adoptive parents are not genetically related to their adopted children. For this reason, a genetic contribution to the covariation between family environment and children's traits is implied if the correlation is greater in nonadoptive families than in adoptive families. The genetic contribution reflects passive genotype-environment correlation because children in nonadoptive families passively inherit from their parents both genes and environment that are correlated with the trait. In both nonadoptive and adoptive families, the environmental measure might be the consequence rather than the cause of the children's traits, which could involve genetic influence of the evocative or active type of genotype-environment correlation. However, this source of genetic influence would contribute equally to environment-outcome correlations in nonadoptive and adoptive families. Increased correlations in nonadoptive families would occur only in the presence of passive genotype-environment correlation. This method uncovered significant genetic contributions to associations between family environment and children's behavioral development in the Colorado Adoption Project. For example, the correlation between HOME scores and the cognitive development of 2-year-olds is higher in nonadoptive families than in adoptive families (Plomin, Loehlin, & DeFries, 1985). The same pattern of results was found for correlations between HOME scores and language development.

The children-of-twins (COT) method can be used to address similar questions (D'Onofrio et al., 2003). As described in Chapter 6, the COT approach provides a powerful pseudo-adoption design that allows for control of genetic risk of parental variables, such as family conflict and parental substance use, in order to examine whether measures of the family environment have a direct effect on child outcomes or are mediated genetically (Knopik et al., 2006). The COT method has shown, for example, that the relationship between parental divorce and early drug use by offspring is genetically mediated, whereas the relationship between parental divorce and emotional difficulties of offspring appears to be a direct environmental effect (D'Onofrio et al., 2006). COT analyses have also suggested that harsh physical punishment has a true environmental effect on children's behavior problems (Lynch et al., 2006), although a twin study of children found that corporal punishment was genetically influenced but more severe physical maltreatment was not (Jaffee et al., 2004). A recent COT analysis looking at a more general measure of family functioning that included family conflict, marital quality, and agreement about parenting found that family conflict had both a direct and genetically mediated association with adolescents' internalizing and externalizing problems (Schermerhorn et al., 2011). Other efforts using the COT design have focused on parental substance use, including drug use during pregnancy, and have found that the association between maternal alcohol use and child attention-deficit hyperactivity disorder (ADHD) is genetically influenced (Knopik et al., 2006), while the association between paternal alcohol use and child ADHD is more likely to be indirect (Knopik, Jacob, et al., 2009). Other maternal variables, such as substance use during pregnancy, appeared to have genetically mediated as well as direct environmental effects on child ADHD (Knopik et al., 2006; Knopik, Jacob, et al., 2009).

Evocative and active genotype-environment correlations are assumed to affect both adopted and nonadopted children and would not be detected using this first method. The second method for finding specific genotype-environment correlations involves correlations between birth parents' traits and adoptive families' environment (Figure 8.2). This method addresses the other two types of genotype-environment correlation, evocative and active. Traits of birth parents can be used as an index of adopted children's genotype, and this index can be correlated with any measure of the adopted children's environment. Although birth parents' traits are a weak index of their adopted children's genotype, the finding that birth parents' traits correlate with the environment of their adopted children suggests that the environmental measure reflects genetically influenced characteristics of the adopted children. That is, adopted children's genetic propensities evoke reactions from adoptive parents. Attempts to use this method in the Colorado Adoption Project yielded only meager evidence for evocative and active genotype-environment correlation. For example, birth mothers' general cognitive ability did not correlate significantly with HOME scores in the adoptive families of their children (Plomin, 1994).

A developmental theory of genetics and experience predicts that the evocative and active forms of genotype-environment correlation become more important as children experience environments outside the family and begin to play a more active role in the selection and construction of their experiences (Scarr & McCartney, 1983). For example, an adoption study found evidence for an evocative genotype-environment

• **FIGURE 8.2** Evocative and active genotype-environment correlation can be detected by the correlation between birth parents' traits (as an index of adopted children's genotype) and the environment of adoptive families.

correlation for antisocial behavior in adolescence (Ge et al., 1996). Genetic risk for the adoptees was indexed by antisocial personality disorder or drug abuse in their birth parents. Adoptees at genetic risk had adoptive parents who were more negative in their parenting than adoptive parents of control adoptees. Moreover, this effect was shown to be mediated by the adolescent adoptees' own antisocial behavior, an observation suggesting evocative genotype-environment correlation. These results were replicated using data from the Colorado Adoption Project (O'Connor, Deater-Deckard, Fulker, Rutter, & Plomin, 1998).

The third method to detect genotype-environment correlation involves multi-variate genetic analysis of the correlation between an environmental measure and a trait (Figure 8.3). This method is the most general in the sense that it detects genotype-environment correlation of any kind—passive, evocative, or active. As explained in the Appendix, multivariate genetic analysis estimates the extent to which genetic effects on one measure overlap with genetic effects on another measure. In this case, genotype-environment correlation is implied if genetic effects on an environmental measure overlap with genetic effects on a trait measure.

Multivariate genetic analysis can be used with any genetic design and with any type of environmental measure, not just measures of the family environment. However, because all genetic analyses are analyses of individual differences, the environmental measure must be specific to each individual. For example, an environmental measure that is the same for all family members, such as the family's socioeconomic status, could not be used in these analyses. However, a child-specific measure, such as children's perceptions of their family's socioeconomic status, could be analyzed in this way. One of the first studies of this type used the sibling adoption design to compare cross-correlations between one sibling's HOME score (a child-specific rather than family-general measure of the environment) and the other sibling's general cognitive ability for nonadoptive and adoptive siblings at 2 years of age in the Colorado Adoption Project (Braungart, Fulker, et al., 1992). Multivariate genetic model fitting indicated that about half of the phenotypic correlation between HOME scores and children's cognitive ability is mediated genetically. A twin study in childhood found that the association between parental negativity and children's prosocial behavior is largely mediated genetically (Knafo & Plomin, 2006a). In adolescence, multivariate genetic analyses have also found substantial genetic mediation of correlations between measures of family environment and adolescents' depression and antisocial behavior in the NEAD project (Reiss et al., 2000) as well as in other

• **FIGURE 8.3** Passive, evocative, and active genotype-environment correlation can be detected by using multivariate genetic analysis of the correlation between environmental measures and traits.

studies (Burt, Krueger, McGue, & Iacono, 2003; Jacobson & Rowe, 1999; Silberg et al., 1999; Thapar, Harold, & McGuffin, 1998). For each of these correlations, more than half of the correlation is mediated genetically. One such report found that adolescent aggressive personality explained the genetic contributions to the association between parenting and adolescent behavior (Narusyte, Andershed, Neiderhiser, & Lichtenstein, 2007). There is also evidence that genetic influences account for the associations among peer characteristics and adolescent drinking (Loehlin, 2010) and young adult smoking (Harakeh et al., 2008).

In adulthood, genetic influence on personality has also been reported to contribute to genetic influence on parenting in several studies (Chipuer, Plomin, Pedersen, McClearn, & Nesselroade, 1993; Ganiban, Ulbricht, et al., 2009; Losoya, Callor, Rowe, & Goldsmith, 1997). In one study, genetic effects on personality traits completely explained genetic influences on life events in a sample of older women (Saudino et al., 1997). Evidence for genetic mediation has also been found in adulthood in correlations between stressful life events and depression (Boardman, Alexander, & Stallings, 2011; Kendler & Karkowski-Shuman, 1997), between social support and depression (Bergeman, Plomin, Pedersen, & McClearn, 1991; Kessler et al., 1992; Spotts et al., 2005), between socioeconomic status and health (Lichtenstein, Harris, Pedersen, & McClearn, 1992), between socioeconomic status and general cognitive ability (Lichtenstein, Pedersen, & McClearn, 1992; Rowe, Vesterdal, & Rodgers, 1999; Tambs, Sundet, Magnus, & Berg, 1989; Taubman, 1976), between education and occupational status (Saudino et al., 1997), and between education and cognitive functioning in elderly individuals (Carmelli, Swan, & Cardon, 1995).

Multivariate genetic analysis can be combined with longitudinal analysis to disentangle cause and effect in the relationship between environmental measures and behavioral measures. For example, if negative parenting at one age is related to children's antisocial behavior at a later age, it would seem reasonable to assume that the negative parenting caused the children's antisocial behavior. However, the first twin study of this type found that this pathway is primarily mediated genetically (Neiderhiser, Reiss, Hetherington, & Plomin, 1999). Similar results were found in other twin studies using a similar approach (Burt et al., 2005; Moberg et al., 2011) and in a systematic examination of parenting and adolescent adjustment constructs in the NEAD project (Reiss et al., 2000). A different longitudinal study of twins, this one concerned with the effects of childhood adversity on antisocial behavior in adolescence and young adulthood, found that although passive genotype-environment correlation was significant, the majority of the variance was due to the direct environmental effects of childhood adversity (Eaves, Prom, & Silberg, 2010).

Recent studies have attempted to clarify whether associations between parenting and child adjustment are due to evocative genotype-environment correlation, passive genotype-environment correlation, or direct environmental effects of parenting on child adjustment. These different mechanisms can be disentangled by combining a multivariate genetic analysis of parenting and child adjustment with a combination of children-of-twins and parents-of-twins designs, referred to as extended children of twins (ECOT; Narusyte et al., 2008). In two studies that have used the ECOT design to examine genotype-environment correlations, mothers' overinvolvement and criticism were related to adolescent internalizing and externalizing behavior, respectively, because of evocative genotype-environment correlation (Narusyte et al., 2008, 2011). In other words, adolescents' behavior evoked a particular type of response from their mothers for genetically influenced reasons. In contrast, fathers' criticism was related to adolescent externalizing behavior through direct environmental influences, with no role for genotype-environment correlation (Narusyte et al., 2011). These findings highlight how multiple strategies can be combined to yield novel information about how genes and environments work together and also help to illustrate the nuances of environmental influences.

Research on the interplay between genes and environment will be greatly facilitated by identifying some of the genes responsible for the heritability of behavior (Jaffee & Price, 2007, in press). The conclusion from research reviewed in this section is that we may be able to identify genes associated with environmental measures because these are heritable. Of course, environments per se are not inherited; genetic influence comes into the picture because these environmental measures involve behavior. For example, many life events and stressors are not things that happen to us passively—to some extent, we contribute to these experiences. The first study to consider the association between DNA and environmental measures used a set of five single nucleotide polymorphisms (SNPs) associated with general cognitive ability in 7-year-old children (Butcher, Meaburn, Dale, et al., 2005; Butcher, Meaburn, Knight, et al., 2005). In a sample of more than 4000 children, this "SNP set" was found to be associated with early proximal measures of the family environment (chaos and discipline) but not with distal measures (maternal education and father's occupational status), suggesting evocative rather than passive genotype-environment correlation (Harlaar, Butcher, et al., 2005). Other studies have reported associations between genes and marital status (Dick, Agrawal, et al., 2006), mothers' behaviors toward their children (Lee, Chronis-Tuscano, et al., 2010), and adults' retrospective reports of how they were parented (Lucht et al., 2006). There have also been reports that parents' genotype was associated with their responsiveness in parenting their infant (Kaitz et al., 2010). A particularly innovative study examined first impression peer rankings of young adults and found that how individuals were ranked by their peers—their "popularity"—was associated with a polymorphism within the serotonergic system (Burt, 2008). In other words, individuals' genotypes influenced (evoked) the way they were viewed by others.

Implications

Research using diverse genetic designs and measures leads to the conclusion that genetic factors often contribute substantially to measures of the environment, especially the family environment. The most important implication of finding genetic contributions to measures of the environment is that the correlation between an environmental measure and a behavioral trait does not necessarily imply exclusively environmental causation. Genetic research often shows that genetic factors are importantly involved in correlations between environmental measures and behavioral traits. In other words, what appears to be an environmental risk might actually reflect genetic factors. Conversely, of course, what appears to be a genetic risk might actually reflect environmental factors.

This research does not mean that experience is entirely driven by genes. Widely used environmental measures show significant genetic influence, but most of the variance in these measures is not genetic. Nonetheless, environmental measures cannot be assumed to be entirely environmental just because they are called environmental. Indeed, research to date suggests that it is safer to assume that measures of the environment include some genetic effects. Especially in families of genetically related individuals, associations between measures of the family environment and children's developmental outcomes cannot be assumed to be purely environmental in origin. Taking this argument to the extreme, two books have concluded that socialization research is fundamentally flawed because it has not considered the role of genetics (Harris, 1998; Rowe, 1994).

These findings support a current shift from thinking about passive models of how the environment affects individuals toward models that recognize the active role we play in selecting, modifying, and creating our own environments. Progress in this field depends on developing measures of the environment that reflect the active role we play in constructing our experience.

●KEY CONCEPTS

Passive genotype-environment correlation: A correlation between genetic and environmental influences that occurs when children inherit genes with effects that covary with their family's environment.

Evocative genotype-environment correlation: A correlation between genetic and environmental influences that occurs when individuals evoke environmental effects that covary with their genetic propensities.

Active genotype-environment correlation: A correlation between genetic and environmental influences that occurs when individuals select or construct environments with effects that covary with their genetic propensities.

Children-of-twins design: A study that includes parents who are twins and the children of each twin.

Extended children-of-twins design: A study that combines a children-of-twins design and a comparable sample of twins who are children and their parents.

Genotype-Environment Interaction

The previous section focused on correlations between genotype and environment. Genotype-environment correlation refers to the role of genetics in exposure to environments. In contrast, genotype-environment interaction involves genetic sensitivity, or susceptibility, to environments. There are many ways of thinking about genotype-environment interaction (Rutter, 2005b, 2006), but in quantitative genetics the term generally means that the effect of the environment on a phenotype depends on genotype or, conversely, that the effect of the genotype on a phenotype depends on the environment (Kendler & Eaves, 1986; Plomin, DeFries, & Loehlin, 1977a). As discussed in Chapter 7, this is quite different from saying that genetic and environmental effects cannot be disentangled because they "interact." When considering the variance of a phenotype, genes can affect the phenotype independent of environmental effects, and environments can affect the phenotype independent of genetic effects. In addition, genes and environments can interact to affect the phenotype beyond the independent prediction of genes and environments.

This point can be seen in Figure 8.4, in which scores on a trait are plotted against low- versus high-risk genotypes for individuals reared in low- versus high-risk environments. Genetic risks can be assessed using animal models, adoption designs, or DNA, as discussed below. The figure shows examples in which (a) genes have an effect with no environmental effect, (b) environment has an effect with no genetic effect, (c) both genes and environment have effects, and (d) both genes and environment have effects *and* there is also an interaction between genetics and environment. In the last case, the interaction involves a greater effect of genetic risk in a high-risk environment. In psychiatric genetics, this type of interaction is called the ***diathesis-stress*** model (Gottesman, 1991; Paris, 1999). That is, individuals at genetic risk for psychopathology (diathesis, or predisposition) are especially sensitive to the effects

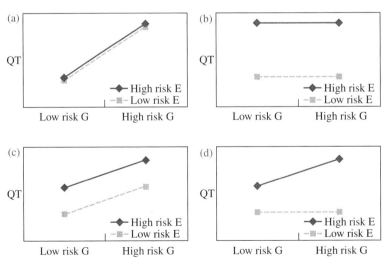

• **FIGURE 8.4** Genetic (G) and environmental (E) effects and their interaction. QT refers to a phenotypic quantitative trait. (a) G can have an effect without an effect of E, (b) E can have an effect without G, (c) both G and E can have an effect, and (d) both G and E can have an effect *and* there can also be an interaction between G and E.

of stressful environments. Although there is evidence for genotype-environment interactions of this sort, some studies show greater genetic influence in permissive, low-risk environments (Kendler, 2001).

As was the case for genotype-environment correlation, genotype-environment interaction adds to phenotypic variance for a trait (see Appendix), but it is difficult to detect the overall extent to which phenotypic variance is due to the interaction between genetic and environmental effects (Jinks & Fulker, 1970; Plomin et al., 1977b; van der Sluis, Dolan, Neale, Boomsma, & Posthuma, 2006). For this reason, the following discussion focuses on detection of specific genotype-environment interactions rather than on estimating their overall contribution to phenotypic variation.

Animal Models

Genotype-environment interaction is easier to study in animals in the laboratory because both genotype and environment can be manipulated. Chapter 12 describes one of the best-known examples of genotype-environment interaction. Maze-bright and maze-dull selected lines of rats responded differently to "enriched" and "restricted" rearing environments (Cooper & Zubek, 1958). The enriched condition had no effect on the maze-bright selected line, but it improved the maze-running performance of the maze-dull rats. The restricted environment was detrimental to the performance of the maze-bright rats but had little effect on the maze-dull rats. This result is an interaction in that the effect of restricted versus enriched environments depends on the genotype of the animals. Other examples from animal research in which environmental effects on behavior differ as a function of genotype have been found (Erlenmeyer-Kimling, 1972; Fuller & Thompson, 1978; Mather & Jinks, 1982), although a series of learning studies in mice failed to find replicable genotype-environment interactions (Henderson, 1972).

As mentioned in Chapter 5, an influential paper reported genotype-environment interaction in which genotype was assessed using inbred strains of mice and environment was indexed by different laboratories (Crabbe, Wahlsten, et al., 1999). However, subsequent studies found much less evidence for genotype-environment interaction of this particular type (Valdar, Solberg, Gauguier, Burnett, et al., 2006; Wahlsten et al., 2003, 2006). Despite the power of animal model research to manipulate genotype and environment, there is surprisingly little systematic research on genotype-environment interaction. (Animal model research in the laboratory is less suited to the study of genotype-environment correlation because such research requires that animals be free to select and modify their environment, which rarely happens in laboratory experiments.)

Adoption Studies

Although genes and environment cannot be manipulated experimentally in the human species as in animal model research, the adoption design can explore genotype-environment interaction, as illustrated in Figure 8.4. Chapter 17 describes an example of genotype-environment interaction for criminal behavior found in two adoption studies (Bohman, 1996; Brennan, Mednick, & Jacobsen, 1996). Adoptees whose birth parents had criminal convictions had an increased risk of criminal behavior, suggesting genetic influence; adoptees whose adoptive parents had criminal convictions also had an increased risk of criminal behavior, suggesting environmental influence. However, genotype-environment interaction was also indicated because criminal convictions of adoptive parents led to increased criminal convictions of their adopted children mainly when the adoptees' birth parents also had criminal convictions.

Another example of a similar type of genotype-environment interaction has been reported for adolescent conduct disorder (Cadoret, Yates, Troughton, Woodworth, & Stewart, 1995b). Genetic risk was indexed by birth parents' antisocial personality diagnosis or drug abuse, and environmental risk was assessed by marital, legal, or psychiatric problems in the adoptive family. Adoptees at high genetic risk were more sensitive to the environmental effects of stress in the adoptive family. Adoptees at low genetic risk were unaffected by stress in the adoptive family. This result confirms previous research that also showed interactions between genetic risk and family environment in the development of adolescent antisocial behavior (Cadoret, Cain, & Crowe, 1983; Crowe, 1974).

A longitudinal adoption study that follows adopted children, their adoptive parents, and their birth mothers and birth fathers is the Early Growth and Development Study (EGDS; Leve, Neiderhiser, Scaramella, & Reiss, 2010). A surprising number of genotype-environment interactions have emerged from EGDS for child behaviors during infancy and toddlerhood. For example, for children whose birth parents had more psychopathology symptoms (depressive and anxiety symptoms, antisocial behaviors, drug and alcohol use), adoptive mothers' use of more structured parenting when the adopted child was 18 months old was associated with significantly fewer child behavior problems than when less structured parenting was used (Leve et al., 2009). Also, elevated depression and anxiety symptoms in adoptive parents increased children's risk for later behavior problems—indexed in toddlers as attention control—only when birth parents' psychopathology was high (Leve, Kerr, et al., 2010). Similarly, elevated adoptive parent depression and anxiety symptoms were related to infant social inhibition (Brooker et al., 2011) and toddler fussiness (Natsuaki et al., 2010) when birth parents were diagnosed with an anxiety disorder. Even the influence of adoptive parents' marital hostility on toddler temperament was moderated by birth parent temperament (Rhoades et al., 2011).

There are, however, many examples in which genotype-environment interaction could not be found. For example, using data from the classic adoption study of Skodak and Skeels (1949), researchers compared general cognitive ability scores for adopted children whose birth parents were high or low in level of education (as an index of genotype) and whose adoptive parents were high or low in level of education (as an index of environment) (Plomin et al., 1977b). Although the level of education of the birth parents showed a significant effect on the adopted children's general cognitive ability, no environmental effect was found for adoptive parents' education and no genotype-environment interaction was found. A similar adoption analysis using more extreme groups found both genetic and environmental effects but, again, no evidence for genotype-environment interaction (Capron & Duyme, 1989, 1996; Duyme, Dumaret, & Tomkiewicz, 1999). Other attempts that used adoption analyses to find genotype-environment interaction for cognitive ability in infancy and childhood have not been successful (Plomin, DeFries, & Fulker, 1988).

Twin Studies

The twin method has also been used to identify genotype-environment interaction. One twin's phenotype can be used as an index of the co-twin's genetic risk in an attempt to explore interactions with measured environments. Using this method, researchers found that the effect of stressful life events on depression was greater for individuals at genetic risk for depression (Kendler et al., 1995). Another study found that the effect of physical maltreatment on conduct problems was greater for children with high genetic risk (Jaffee et al., 2005). The approach is stronger when twins reared apart are studied, an approach that has also yielded some evidence for genotype-environment interaction (Bergeman, Plomin, McClearn, Pedersen, & Friberg, 1988).

The most common use of the twin method in studying genotype-environment interaction simply involves asking whether heritability differs in two environments. Large samples are needed to detect this type of genotype-environment interaction. About 1000 pairs of each type of twin are needed to detect a heritability difference of 60 percent versus 40 percent. For example, Chapter 18 mentions several examples in which the heritability of alcohol use and abuse is greater in more permissive environments. Analyses of differences in heritability as a function of the environment can treat the environment as a continuous variable rather than dichotomizing it (Purcell, 2002; Purcell & Koenen, 2005). In fact, there has been an explosion of studies examining moderation of heritability and environmentality in the past several years (e.g., Brendgen et al., 2009; Feinberg, Button, Neiderhiser, Reiss, & Hetherington, 2007; Tuvblad, Grann, & Lichtenstein, 2006).

Another analysis of this type showed that heritability of general cognitive ability is significantly greater in families with more highly educated parents (74 percent) than in families with less well educated parents (26 percent) (Rowe, Jacobson, & van den Oord, 1999), a finding replicated in four other studies for parental education and socioeconomic status (Harden, Turkheimer, & Loehlin, 2007; Kremen et al., 2005; Tucker-Drob, Rhemtulla, Harden, Turkheimer, & Fask, 2011; Turkheimer, Haley, Waldron, D'Onofrio, & Gottesman, 2003), although opposite results were found in a fifth study (Asbury, Wachs, & Plomin, 2005). A recent report took a longitudinal approach to examining the potential moderating effects of socioeconomic status on children's intelligence assessed eight times from ages 2 to 14 and found no evidence that socioeconomic status moderated heritability (Hanscombe et al., 2012). Life events were found to moderate heritability of cognitive ability in adults, with more life events reducing heritability (Vinkhuyzen, van der Sluis, & Posthuma, 2011). Higher heritability was also found for adolescent antisocial behavior in more economically advantaged families (Tuvblad et al., 2006) and for adolescent externalizing behaviors when environmental adversity was high (Hicks, South, DiRago, Iacono, & McGue, 2009).

In addition, several twin studies have found that aspects of the social environment moderate heritability. For example, more negative and less warm parenting results in higher heritability for adolescent antisocial behavior (Feinberg et al., 2007). Heritability for depressive behavior in children was higher when peer rejection was high (Brendgen et al., 2009) and, using the same sample, heritability for aggressive behavior was lower when children had a positive relationship with a teacher (Brendgen et al., 2011). One of the biggest challenges with twin studies examining genotype-environment interactions is that it is difficult to consider these effects in multivariate models, although at least one published report has done so (Tucker-Drob et al., 2011). As the appropriate data for use in genotype-environment interaction analyses become available within twin and related designs, we will continue to uncover the nuances of how genes and environments work together to influence behavioral outcomes. These processes are also likely to change over time, although longitudinal examinations of genotype-environment interactions are just beginning.

DNA

DNA studies of gene-environment interaction have yielded exciting results in two of the most highly cited papers in behavioral genetics. The first study involved adult antisocial behavior, childhood maltreatment, and a functional polymorphism in the gene for monoamine oxidase A (*MAOA*), which is widely involved in metabolizing a broad range of neurotransmitters (Caspi et al., 2002). As shown in Figure 8.5,

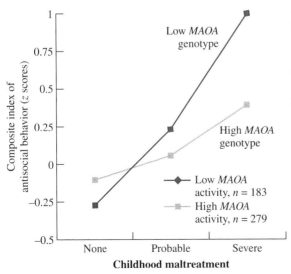

• **FIGURE 8.5** Gene-environment interaction: The effect of a polymorphism in the *MAOA* gene on antisocial behavior depends on childhood maltreatment. (From Caspi et al., 2002. Reprinted with permission from AAAS.)

childhood maltreatment was associated with adult antisocial behavior, as has been known for decades. *MAOA* was not related to antisocial behavior for most individuals who experienced no childhood maltreatment—that is, there was no difference in antisocial behavior between children with low and high *MAOA* genotypes. However, *MAOA* was strongly associated with antisocial behavior in individuals who suffered severe childhood maltreatment, which suggests a genotype-environment interaction of the diathesis-stress type. The rarer form of the gene, which lowers MAOA levels, made individuals especially vulnerable to the effects of childhood maltreatment. Although attempts to replicate this finding have been mixed, it is supported by a meta-analysis of all extant studies (Kim-Cohen et al., 2006), and more recent published reports have also replicated these findings (Aslund et al., 2011).

The second study involved depression, stressful life events, and a functional polymorphism in the promoter region of the serotonin transporter gene (*5-HTTLPR*) (Caspi et al., 2003). As shown in Figure 8.6, there was no association between the gene and depressive symptoms in individuals reporting few stressful life events. An association appeared with increasing number of life events, which is another example of the diathesis-stress model of genotype-environment interaction. This interaction has been replicated in several studies (e.g., Hammen, Brennan, Keenan-Miller, Hazel, & Najman, 2010; Kendler, Kuhn, Vittum, Prescott, & Riley, 2005; Petersen et al., 2012; Zalsman et al., 2006), but not all (e.g., Gillespie, Whitfield, Williams, Heath, & Martin, 2005), and has received support from mouse research, mentioned in Chapter 10, which showed that the serotonin transporter gene was involved in emotional reactions to environmental threats (Hariri & Holmes, 2006). Recently, there have been a series

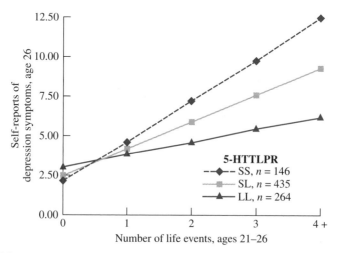

• **FIGURE 8.6** Gene-environment interaction: The effect of a polymorphism in the *5-HTTLPR* gene depends on the number of life events. (From Caspi et al., 2003. Reprinted with permission from AAAS.)

of meta-analyses and debates in the literature about the interaction between stressful life events and the serotonin transporter gene. Two meta-analyses conducted in 2009 found that the evidence for such interactions was due to chance (Munafo, Durrant, Lewis, & Flint, 2009) or simply not present (Risch et al., 2009). The most recent meta-analysis, however, included all published studies through November 2009 in an effort to better represent the state of the field; it found strong evidence for an interaction between stress and the serotonin transporter gene in risk for depression (Karg, Burmeister, Shedden, & Sen, 2011).

Another example of genotype-environment interaction suggests a possible mechanism of effect: Cannabis use was associated with later psychotic symptoms such as hallucinations and delusions only in individuals with a particular allele of the catechol-O-methyltransferase (*COMT*) gene (Caspi et al., 2005).

To date, many studies have reported genotype-environment interactions, most focusing on the genes involved in these first studies. For example, behavioral inhibition was related to *5-HTTLPR* only when maternal overprotectiveness was high (Burkhouse, Gibb, Coles, Knopik, & McGeary, 2011), and children's attentional bias to avoid angry faces was linked with the *5-HTTLPR* genotype only when maternal criticism was high (Gibb et al., 2011). In other studies, the *MAOA* genotype interacted with early life stress to predict hyperactivity (Enoch, Steer, Newman, Gibson, & Goldman, 2010), with physical discipline to predict level of delinquent behavior (Edwards, Dodge, et al., 2010), and with the influence of deviant peer affiliation to predict adolescent boys' antisocial behavior (Lee, 2011). Some studies also found evidence of three-way gene-by-gene-by-environment interactions (Cicchetti, Rogosch, & Oshri, 2011), although the power to detect such effects is limited. Interestingly, there is some evidence that risk may accumulate across genes and environments (e.g., Clasen, Wells, Knopik, McGeary, & Beevers, 2011), suggesting that an approach that considers multiple genes and multiple environmental risk factors will be required to assess genotype-environment interaction more accurately.

There is a need for caution when considering the findings of studies examining candidate gene-by-environment interactions, however. A recent report examined all published studies of candidate gene-by-environment interactions—103 studies published from 2000 to 2009—and found that 96 percent of novel reports were significant, while only 27 percent of replication attempts were significant (Duncan & Keller, 2011). In addition, there appeared to be a publication bias among replication attempts; power analyses suggested that most candidate gene-by-environment interaction studies are underpowered. This report and those described above by Munafo and colleagues (2009) and Risch and colleagues (2009) highlight the critical role that replication has in helping to clarify how genes and environments work together.

Genomewide association approaches have also begun to be applied in the search for genotype-environment interaction (Plomin & Davis, 2006). For example, the set of five SNPs associated with general cognitive ability (*g*) that was mentioned in the previous section yielded significant genotype-environment interaction (Harlaar,

Butcher, et al., 2005). One significant genotype-environment interaction was in line with the quantitative genetic analyses mentioned above: Genetic effects on *g* are stronger for children in families of higher socioeconomic status. Two other significant interactions showed greater associations between the SNP set and general cognitive ability (*g*) at both the low and high ends of the environment. That is, children with a genetic propensity toward high *g* profit disproportionately from a good environment, and children with a genetic propensity toward low *g* suffer disproportionately from a poor environment. Systematic strategies that can be used in mining data from genomewide association studies in examining genotype-environment interaction have been proposed (Thomas, 2010) and include experimental intervention as a way of manipulating the environment (van Ijzendoorn et al., 2011).

■ KEY CONCEPTS

Genotype-environment interaction: Genetic sensitivity or susceptibility to environments. Genotype-environment interaction is usually limited to statistical interactions, such as genetic effects that differ in different environments. The most common use of the twin method in studying genotype-environment interaction involves testing whether heritability differs in different environments.

Diathesis-stress: A type of genotype-environment interaction in which individuals at genetic risk for a disorder (diathesis) are especially sensitive to the effects of risky (stress) environments.

Candidate gene-by-environment interaction: Genotype-environment interaction in which an association between a particular (candidate) gene and a phenotype differs in different environments.

Genomewide gene-by-environment interaction: A method for searching for genotype-environment interaction that assesses DNA variation throughout the genome.

Summary

The interplay between genes and environment has been the focus of a vast amount of research, especially over the past decade. There are two main foci of this work: genotype-environment correlation and genotype-environment interaction. What is clear from all of this research is that genes and environments operate together to influence behavior through genotype-environment correlations and interactions.

One of the most surprising findings in genetic research was that our experiences are influenced in part by genetic factors. This finding is the topic of genotype-environment correlation. Dozens of studies using various genetic designs and measures of the environment converge on the conclusion that genetic factors contribute to the variance of measures of the environment. Genotype-environment correlations are of three types: passive, evocative, and active. Several different methods are available to assess specific genotype-environment correlations between behavioral traits and measures of the environment. These methods have identified several examples of genotype-environment correlation and have helped to clarify how genotype-environment correlations may change over time.

Genotype-environment interaction is the second way that genes and environments work together. Animal studies, in which both genotype and environment can be controlled, have yielded examples in which environmental effects on behavior differ as a function of genotype. A rapidly accumulating number of examples of genotype-environment interaction for human behavior have also been found, especially in molecular genetic studies using functional polymorphisms in **candidate genes.** The general form of these interactions is that stressful environments primarily have their effect on individuals who are genetically at risk, a diathesis-stress type of genotype-environment interaction.

The recognition through behavioral genetic research of genotype-environment correlations and interactions emphasizes the power of genetic research to elucidate environmental risk mechanisms. Understanding how nature and nurture correlate and interact will be greatly facilitated as more genes are identified that are associated with behavior and with experience.

CHAPTER NINE

Identifying Genes

M uch more quantitative genetic research of the kind described in Chapters 6, 7, and 8 is needed to identify the most heritable components and constellations of behavior, to investigate developmental change and continuity, and to explore the interplay between nature and nurture. However, one of the most exciting directions for research in behavioral genetics is the coming together of quantitative genetics and molecular genetics in attempts to identify specific genes responsible for genetic influence on behavior, even for complex behaviors for which many genes as well as many environmental factors are at work.

As illustrated in Figure 9.1, quantitative genetics and molecular genetics both began around the beginning of the twentieth century. The two groups, biometricians (Galtonians) and Mendelians, quickly came into contention, as described in Chapter 3. Their ideas and research grew apart as quantitative geneticists focused on naturally occurring genetic variation and complex quantitative traits, and molecular geneticists analyzed single-gene mutations, often those created artificially by chemicals or X-irradiation (described in Chapter 5). Since the 1980s, however, quantitative genetics and molecular genetics have begun to come together to identify genes for complex, quantitative traits. Such a gene in multiple-gene systems is called a *quantitative trait locus (QTL)*. Unlike single-gene effects that are necessary and sufficient for the development of a disorder, QTLs contribute like probabilistic risk factors, creating quantitative traits rather than qualitative disorders. QTLs are inherited in the same Mendelian manner as single-gene effects; but, if there are many genes that affect a trait, then each gene is likely to have a relatively small effect.

In addition to producing indisputable evidence of genetic influence, the identification of specific genes will revolutionize behavioral genetics by providing measured genotypes for investigating, with greater precision, the multivariate, developmental, and gene-environment interplay issues that have become the focus of quantitative genetic research. In Chapter 5, we briefly presented various ways of identifying genes in animal models. We now turn our attention to identifying genes

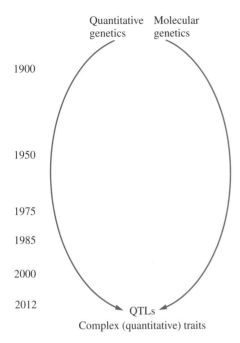

⦁ **FIGURE 9.1** Quantitative genetics and molecular genetics are coming together in the study of complex quantitative traits and quantitative trait loci (QTLs).

associated with human behavior. Once a gene, or cluster of genes, is identified, it is possible to begin to explore the pathways between genes and behavior, which is the topic of Chapter 10.

Mutations

Behavioral genetics asks why people are different behaviorally—for example, why people differ in cognitive abilities and disabilities, psychopathology, and personality. For this reason, it focuses on genetic and environmental differences that can account for these observed differences among people. New DNA differences occur when mistakes, called mutations, are made in copying DNA. These mutations result in different alleles (called polymorphisms), such as the alleles responsible for the variations that Mendel found in pea plants, for Huntington disease and PKU, and for complex behavioral traits such as schizophrenia and cognitive abilities. Mutations that occur in the creation of eggs and sperm will be transmitted faithfully unless natural selection intervenes (Chapter 20). The effects that count in terms of natural selection are effects on survival and reproduction. Because evolution has so finely tuned the genetic system, most new mutations in regions of DNA that are translated into amino acid sequences have deleterious effects. However, sometimes such mutations are neutral overall, and once in a great while a mutation will make the system function a bit better. In evolutionary terms, this outcome means that individuals with the mutation are more likely to survive and reproduce.

A single-base mutation can result in the insertion of a different amino acid into a protein. Such a mutation can alter the function of the protein. For example, in the figure in Box 4.1, if the first DNA codon TAC is miscopied as TCC, the amino acid arginine will be substituted for methionine. (Table 4.1 indicates that TAC codes for methionine and TCC codes for arginine.) This single amino acid substitution in the hundreds of amino acids that make up a protein might have no noticeable effect on the protein's functioning; then again, it might have a small effect or it might have a major, even lethal, effect. A mutation that leads to the loss of a single base is likely to be more damaging than a mutation causing a substitution because the loss of a base shifts the reading frame of the triplet code. For example, if the second base in the box figure were deleted, TAC-AAC-CAT becomes TCA-ACC-AT. Instead of the amino acid chain containing methionine (TAC) and leucine (AAC), the mutation would result in a chain containing serine (TCA) and tryptophan (ACC).

Mutations are often not so simple. For example, a particular gene can have mutations at several locations. As an extreme example, hundreds of different mutations have been found in the gene responsible for PKU, and some of these different mutations have different effects (Scriver, 2007). Another example involves triplet repeats, mentioned in Chapter 3. Most cases of Huntington disease are caused by three repeating bases (CAG). Normal alleles have from 11 to 34 CAG repeats in a gene that codes for a protein found throughout the brain. For individuals with Huntington disease, the number of CAG repeats varies from 37 to more than 100. Because triplet repeats involve three bases, the presence of any number of repeats does not shift the reading frame of transcription. However, the CAG repeat responsible for Huntington disease is transcribed into mRNA and translated into protein, which means that multiple repeats of an amino acid are inserted into the protein. Which amino acid? CAG is the mRNA code, so the DNA code is GTC. Table 4.1 shows that GTC codes for the amino acid glutamine. Having a protein encumbered with many extra copies of glutamine reduces the protein's normal activity; therefore, the lengthened protein would show loss of function. However, although Huntington disease is a dominant disorder, the other allele should be operating normally, producing enough of the normal protein to avoid trouble. This possibility suggests that the Huntington allele, which adds dozens of glutamines to the protein, might confer a new property (gain of function) that creates the problems of Huntington disease.

Many of our 3 billion base pairs differ among individuals, and over 2 million differ for at least 1 percent of the population. As described in the following section, these DNA polymorphisms have made it possible to identify genes responsible for the heritability of traits, including complex behavioral traits.

Detecting Polymorphisms

Much of the success of molecular genetics comes from the availability of millions of DNA polymorphisms. Previously, genetic markers were limited to the products of single genes, such as the red blood cell proteins that define the blood groups. In 1980, new genetic markers that are the actual polymorphisms in the DNA were discovered. Because millions of DNA base sequences are polymorphic, these DNA polymorphisms can be used in genomewide linkage studies to determine the chromosomal location of single-gene disorders, described later in this chapter. In 1983, such DNA markers were first used to localize the gene for Huntington disease at the tip of the short arm of chromosome 4. Technology has advanced to the point where we can now use millions of DNA markers to conduct genomewide association studies to identify genes associated with complex disorders, including behavioral disorders (Hirschhorn & Daly, 2005).

We are also able to detect every single DNA polymorphism by sequencing each individual's entire genome, called **whole-genome sequencing** (Lander, 2011). The race is on to determine how to sequence all 3 billion bases of DNA of an individual for less than $1000 (Kedes & Campany, 2011). The evolution of whole-genome sequencing will allow researchers to focus not just on the 2 percent of DNA involved in coding genes but also on non-coding genes that might also contribute to heritability. The 1000 Genomes Project, launched in 2008, aims to characterize human genetic variation across the world (Altshuler, Durbin, et al., 2010). More recently, in 2010, the 10,000 Genomes Project was started with the goal of identifying even rarer DNA variants (http://www.wellcome.ac.uk/News/Media-office/Press-releases/2010/WTX060061.htm). As mentioned in Chapter 4, with the move toward affordable whole-genome sequencing, there is the very real possibility that the entire genome of all newborns could be sequenced to screen for genetic problems and that eventually we will each have the opportunity to know our own DNA sequence (Collins, 2010). Until whole-genome sequencing becomes affordable, sequencing the 2 percent of the genome that contains protein-coding information has become widely used, especially for discovering rare alleles for unsolved Mendelian disorders (Bamshad et al., 2011).

Although it is possible that rare alleles of large effect explain some of the heritability of complex traits, two types of common DNA polymorphisms can be genotyped affordably in the large samples needed to detect associations of small effect size: *microsatellite markers,* which have many alleles, and *single nucleotide polymorphisms (SNPs),* which have just two alleles (Weir, Anderson, & Hepler, 2006). Box 9.1 describes how microsatellite markers and SNPs are detected and explains the technique of **polymerase chain reaction (PCR).** This is fundamental for detection of all DNA markers because PCR makes millions of copies of a small stretch of DNA. The triplet repeats mentioned in relation to Huntington disease are an example of a microsatellite repeat marker, which can involve two, three, or four base pairs that are repeated up to a hundred times and which have been found at as many as 50,000 loci throughout the genome. The number of repeats at each locus differs among individuals and is inherited in a Mendelian manner. For example, a microsatellite marker might have three alleles, in which the two-base sequence C-G repeats 14, 15, or 16 times.

SNPs (called "snips") are by far the most common type of DNA polymorphisms. As their name suggests, a SNP involves a mutation in a single nucleotide,

BOX 9.1 • DNA Markers

Microsatellite repeats and SNPs are genetic polymorphisms in DNA. They are called DNA markers rather than genetic markers because they can be identified directly in the DNA itself rather than attributed to a gene product, such as the red blood cell proteins responsible for blood types. Investigations of both of these DNA markers are made possible by a technique called *polymerase chain reaction* (PCR). In a few hours, millions of copies of a particular small sequence of DNA a few hundred to two thousand base pairs in length can be created. To do this copying, the sequence of DNA surrounding the DNA marker must be known. From this DNA sequence, 20 bases on both sides of the polymorphism are synthesized. These 20-base DNA sequences, called **primers,** are unique in the genome and identify the precise location of the polymorphism.

Polymerase is an enzyme that begins the process of copying DNA. It begins to do so on each strand of DNA at the point of the primer. One strand is copied from the primer on the left in the right direction and the other strand is copied from the primer on the right in the left direction. In this way, PCR results in a copy of the DNA between the two primers. When this process is repeated many times, even the copies are copied and millions of copies of the double-stranded DNA between the two primers are produced (for an animation, see http://www.dnalc.org/resources/animations/pcr.html). The simplest way to identify a polymor-phism from the PCR-amplified DNA fragment is to sequence the fragment. Sequencing would indicate how many repeats are present for microsatellite markers and which allele is present for SNPs. Because we have two alleles for each locus, we can have two different alleles (heterozygous) or two copies of the same allele (homozygous). For microsatellite markers, a more cost-effective approach that sorts DNA fragments by length is used; this indicates the number of repeats. For SNPs, the DNA fragments can be made single stranded and allowed to find their match (hybridize) to a single-stranded probe for one or the other SNP allele. For example, in the figure in this box, the target probe is ATCATG, with a SNP at the third nucleotide base. The PCR-amplified DNA fragment TAGTAC has hybridized successfully with the probe. In high-throughput approaches, a fluorescent molecule is attached to the DNA fragments so that the fragments light up if they successfully hybridize with the probe. (The TATGAC allele is unable to hybridize with the probe.)

(COURTESY OF AFFYMETRIX.)

for example, a mutation that changes the first codon in Box 4.1 from TAC to TCC, thus substituting arginine for methionine when the gene is transcribed and translated into a protein. SNPs that involve a change in an amino acid sequence are called *nonsynonymous* and are thus likely to be functional: The resulting protein will contain a different amino acid. Most SNPs in **coding regions** are *synonymous:* They do not involve a change in amino acid sequence because the SNP involves one of the alternate DNA codes for the same amino acid (see Table 4.1). Although nonsynonymous SNPs are more likely to be functional because they change the amino acid sequence of the protein, it is possible that synonymous SNPs might have an effect by changing the rate at which mRNA is translated into proteins. The field is just coming to grips with the functional effects of other SNPs throughout the genome, such as SNPs in **non-coding RNA** regions of the genome (see Chapter 10). More than 12 million SNPs have been reported in populations around the world, and most of these have been validated (http://www.ncbi.nlm.nih.gov/SNP/); about 2 million meet the criterion of occurring in at least 1 percent of a population. This work is being systematized by the International HapMap Consortium (http://snp.cshl.org/index.html.en), which has genotyped more than 3 million SNPs for 270 individuals from four ethnic groups (Frazer et al., 2007). The project is called HapMap because its aim is to create a map of correlated SNPs throughout the genome. SNPs close together on a chromosome are unlikely to be separated by recombination, but recombination does not occur evenly throughout the genome. There are blocks of SNPs that are very highly correlated with one another and are separated by so-called *recombinatorial hotspots.* These blocks are called *haplotype blocks.* (In contrast to *genotype,* which refers to a pair of chromosomes, the DNA sequence on one chromosome is called a *haploid genotype,* which has been shortened to *haplotype.*) By identifying a few SNPs that tag a haplotype block, it may be necessary to genotype only half a million SNPs rather than many millions in order to scan the entire genome for associations with phenotypes.

Until recently, only common DNA variants, such as SNPs, occurring at relatively high frequency in the population were well studied. However, rarer variants no doubt also contribute to genetic risk for common diseases (Manolio et al., 2009). These types of polymorphisms have attracted considerable attention. One example is *copy number variants (CNVs),* which involve duplication of long stretches of DNA, often encompassing protein-coding genes as well as non-coding genes (Conrad et al., 2010; Redon et al., 2006). Recent reports suggest a role for rare CNVs in the risk for a range of common diseases, such as autism (Sebat et al., 2007) and schizophrenia (Buizer-Voskamp et al., 2011). Many CNVs, like other mutations, are not inherited and appear uniquely in an individual (de novo). However, a recent project generated a comprehensive map of 11,700 CNVs (Conrad et al., 2010), 80 to 90 percent of which appear at a frequency of at least 5 percent in the population. Recent efforts have expanded our knowledge about common and rare variation across the genome. The International Hapmap 3 Consortium genotyped 1.6 million common SNPs in

1,184 individuals from 11 global populations and sequenced specific regions in 692 of these individuals (Altshuler, Gibbs, et al., 2010). These advances concerning genetic variation in populations will undoubtedly help to answer questions about the role of genetics in human disease and behavior.

● KEY CONCEPTS

Quantitative trait loci (QTLs): Genes of various effect sizes in multiple-gene systems that contribute to quantitative (continuous) variation in a phenotype.
Polymorphism: A locus with two or more alleles; Greek for "multiple forms."
Microsatellite markers: Two, three, or four DNA base pairs that are repeated up to a hundred times. Unlike SNPs, which generally have just two alleles, microsatellite markers often have many alleles that are inherited in a Mendelian manner.
Single nucleotide polymorphism (SNP): The most common type of DNA polymorphism, which involves a mutation in a single nucleotide. SNPs (pronounced "snips") can produce a change in an amino acid sequence (called *nonsynonymous,* i.e., not synonymous).
Polymerase chain reaction (PCR): A method to amplify a particular DNA sequence.
Primer: A short (usually 20-base) DNA sequence that marks the starting point for DNA replication. Primers on either side of a polymorphism mark the boundaries of a DNA sequence that is to be amplified by polymerase chain reaction (PCR).
Recombinatorial hotspot: Chromosomal location subject to much recombination; often marks the boundaries of haplotype blocks.
Haploid genotype (haplotype): The DNA sequence on one chromosome. In contrast to *genotype,* which refers to a pair of chromosomes, the DNA sequence on one chromosome is called a *haploid genotype,* which has been shortened to *haplotype.*
Haplotype block: A series of SNPs that are very highly correlated (i.e., seldom separated by recombination). The HapMap project is systematizing haplotype blocks for several ethnic groups (http://snp.cshl.org/index.html.en).
Copy number variants (CNVs): A polymorphism that involves duplication of long stretches of DNA, often encompassing protein-coding genes as well as non-coding genes. Frequently used more broadly to refer to all structural variations in DNA, including insertions and deletions.

Human Behavior

In studying our species, we cannot manipulate genes or genotypes as in knock-out studies or minimize environmental variation in a laboratory. Although this prohibition makes it more difficult to identify genes associated with behavior, this cloud has the silver lining of forcing us to deal with naturally occurring genetic and environmental variation. The silver lining is that results of human research will generalize to the world outside the laboratory and are more likely to translate to clinically relevant advances in diagnosis and treatment.

As described in Chapter 2, linkage has been extremely successful in locating the chromosomal neighborhood of single-gene disorders. For many decades, the actual residence of a single-gene disorder could be pinpointed when a physical marker for the disorder was available, as was the case for PKU (high phenylalanine levels), which led to identification of the culprit gene in 1984. With the discovery of DNA markers in the 1980s, screening the genome for linkage became possible for any single-gene disorder, which in 1993 led to the identification of the gene that causes Huntington disease (Bates, 2005).

During the past decade, attempts to identify genes responsible for the heritability of complex traits have moved quickly from traditional linkage studies to QTL linkage to candidate gene association to genomewide association studies. Most recently, researchers are using next-generation sequencing to identify all variants in the genome as it became apparent that genetic influence on complex traits is caused by many more genes of much smaller effect size than anticipated. This fast-moving journey is briefly described in this section.

Linkage: Single-Gene Disorders

For single-gene disorders, linkage can be identified by using a few large family pedigrees, in which cotransmission of a DNA marker allele and a disorder can be traced. Because recombination occurs an average of only once per chromosome in the formation of gametes passed from parent to offspring, a marker allele and an allele for a disorder on the same chromosome will usually be inherited together within a family. In 1984, the first DNA marker linkage was found for Huntington disease in a single five-generation pedigree shown in Figure 9.2. In this family, the allele for Huntington disease is linked to the allele labeled C. All but one person with Huntington disease has inherited a chromosome that happens to have the C allele in this family. This marker is not the Huntington gene itself, because a recombination was found between the marker allele and Huntington disease for one individual; the leftmost woman with an arrow in generation IV had Huntington disease but did not inherit the C allele for the marker. That is, this woman received that part of her affected mother's chromosome carrying the gene for Huntington disease, which is normally linked in this family with the C allele but in this woman is recombined with the A allele from the mother's other chromosome. The farther the marker is from the disease gene, the more recombinations will be found within a family. Markers even closer to the Huntington gene were later found. Finally, in 1993, a genetic defect was identified as the CAG repeat sequence associated with most cases of Huntington disease, as described in Chapter 3. A similar approach was used to locate the genes responsible for other single-gene disorders, such as PKU on chromosome 12 and fragile X mental retardation on the X chromosome.

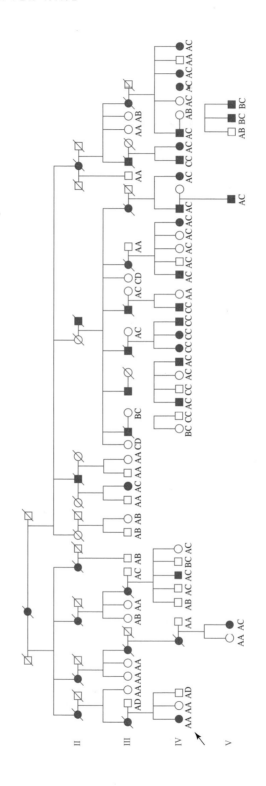

Linkage: Complex Disorders

Although linkage analysis of large pedigrees has been very effective for locating genes for single-gene disorders, it is less powerful when several genes are involved. Another linkage approach has greater power to detect genes of smaller effect size and can be extended to quantitative traits. Rather than studying a few families with many relatives as in traditional linkage, this method studies many families with a small number of relatives, usually siblings. The simplest method examines ***allele sharing*** for pairs of affected siblings in many different families, as explained in Box 9.2. As indicated in later chapters, the **affected sib-pair QTL linkage design** is the most widely used linkage design for studying complex traits such as behavior.

Linkage based on allele sharing can also be investigated for quantitative traits by correlating allele sharing for DNA markers with sibling differences on a quantitative trait. That is, a marker linked to a quantitative trait will show greater than expected allele sharing for siblings who are more similar for the trait. The sib-pair QTL linkage design was first used to identify and replicate a linkage for reading disability on chromosome 6 (6p21; Cardon et al., 1994), a QTL linkage that has been replicated in several other studies (see Chapter 11). As seen in the following chapters, many genomewide linkage studies have been reported. However, replication of linkage results has generally not been as clear as in the case of reading disability, as seen, for example, in a review of 101 linkage studies of 31 human diseases (Altmuller, Palmer, Fischer, Scherb, & Wjst, 2001).

Association: Candidate Genes

A great strength of linkage approaches is that they systematically scan the genome with just a few hundred DNA markers looking for violations of Mendel's law of independent assortment between a disorder and a marker. However, a weakness of linkage approaches is that they cannot detect linkage for genes of small effect size expected for most complex disorders (Risch, 2000; Risch & Merikangas, 1996). Using linkage is like using a telescope to scan the horizon systematically for distant mountains (large QTL effects). However, the telescope goes out of focus when trying to detect nearby hills (small QTL effects).

In contrast to linkage, which is systematic but not powerful, allelic association is powerful but, until recently, not systematic. Association is powerful because, rather than relying on recombination within families as in linkage, it simply compares allelic

• **FIGURE 9.2** Linkage between the Huntington disease gene and a DNA marker at the tip of the short arm of chromosome 4. In this pedigree, Huntington disease occurs in individuals who inherit a chromosome bearing the C allele for the DNA marker. A single individual shows a recombination (marked with an arrow) in which Huntington disease occurred in the absence of the C allele. (From "DNA markers for nervous-system diseases" by J. F. Gusella et al. *Science, 225,* 1320–1326. ©1984. Used with permission of the American Association for the Advancement of Science.)

BOX 9.2 • Affected Sib-Pair Linkage Design

The most widely used linkage design includes families in which two siblings are affected. *Affected* could mean that both siblings meet criteria for a diagnosis or that both siblings have extreme scores on a measure of a quantitative trait. The affected sib-pair linkage design is based on allele sharing—whether affected sibling pairs share 0, 1, or 2 alleles for a DNA marker (see the figure). For simplicity, assume that we can distinguish all four parental alleles for a particular marker. Linkage analyses require the use of markers with many alleles so that, ideally, all four parental alleles can be distinguished. The father is shown as having alleles A and B, and the mother has alleles C and D. There are four possibilities for sib-pair allele sharing: They can share no parental alleles, they can share one allele from the father or one allele from the mother, or they can share two parental alleles. When a marker is not linked to the gene for the disorder, each of these possibilities has a probability of 25 percent. In other words, the probability is 25 percent that sibling pairs share no alleles, 50 percent that they share one allele, and 25 percent that they share two alleles. Deviations from this expected pattern of allele sharing indicate linkage. That is, if a marker is linked to a gene that influences the disorder, more than 25 percent of the affected sibling pairs will share two alleles for the marker. Several examples of affected sib-pair linkage analyses are mentioned in later chapters. A recent example yielded evidence for linkage on chromosome 4 for major depression in a sample of sibling pairs affected with alcohol dependence (Kuo et al., 2010).

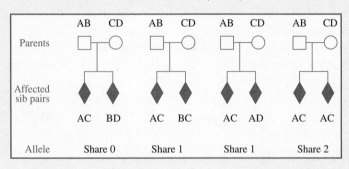

frequencies for groups such as individuals with the disorder (cases) versus controls or low-scoring versus high-scoring individuals on a quantitative trait (Sham, Cherny, Purcell, & Hewitt, 2000). For example, as mentioned in Chapter 1, a particular allele of a gene (for apolipoprotein E on chromosome 19) involved in cholesterol transport is associated with late-onset Alzheimer disease (Corder et al., 1993). In dozens of association studies, the frequency of allele 4 was found to be about 40 percent in individuals with Alzheimer disease and about 15 percent in controls. In recent years, allelic associations have been reported for all domains of behavior, as discussed in later chapters, although none have nearly as large an effect as the association between apolipoprotein E and Alzheimer disease.

The weakness of allelic association is that an association can only be detected if a DNA marker is itself the functional gene (called *direct association*) or very close to it (called *indirect association* or *linkage disequilibrium*). If linkage is a telescope, association is a microscope. As a result, hundreds of thousands of DNA markers need to be genotyped to scan the genome thoroughly. For this reason, until very recently, allelic association has been used primarily to investigate associations with genes thought to be candidates for association. For example, because the drug most commonly used to treat hyperactivity, methylphenidate, acts on the dopamine system, genes related to dopamine, such as the dopamine transporter and dopamine receptors, have been the target of candidate gene association studies of hyperactivity. Evidence for QTL associations with hyperactivity involving the D_4 dopamine receptor (*DRD4*) and other dopamine genes is growing (Banaschewski, Becker, Scherag, Franke, & Coghill, 2010; Sharp, McQuillin, & Gurling, 2009). For example, a meta-analysis of 27 studies found that the *DRD4* 7-repeat (*DRD4-7r*) allele increases the risk for attention-deficit hyperactivity disorder (ADHD; Smith, 2010). Specifically, the frequency of the *DRD4* allele associated with hyperactivity is about 25 percent for children with hyperactivity and about 15 percent in controls. The problem with the candidate gene approach is that we often do not have strong hypotheses as to which genes are candidate genes. Indeed, as discussed in Chapter 5, pleiotropy makes it possible that any of the thousands of genes expressed in the brain could be considered as candidate genes. Moreover, candidate gene studies are limited to the 2 percent of the DNA that lies in coding regions.

The biggest problem is that reports of candidate gene associations have been difficult to replicate (Tabor, Risch, & Myers, 2002). This is a general problem for all complex traits, not just for behavioral traits (Ioannidis, Ntzani, Trikalinos, & Contopoulos-Ioannidis, 2001). For example, in a review of 600 reported associations with common medical diseases, only six have been consistently replicated (Hirschhorn, Lohmueller, Byrne, & Hirschhorn, 2002), although a follow-up meta-analysis indicated greater replication for larger studies (Lohmueller, Pearce, Pike, Lander, & Hirschhorn, 2003). Essentially, as explained in the next section, the failure to replicate is due to the fact that the largest effect sizes are much smaller than expected. In other words, these candidate gene studies were underpowered to detect such effects. Few candidate gene associations have been replicated in genomewide association studies (Siontis, Patsopoulos, & Ioannidis, 2010).

Association: Genomewide

In summary, linkage is systematic but not powerful, and candidate gene allelic association is powerful but not systematic. Allelic association can be made more systematic by using a dense map of markers. Historically, the problem with using a dense map of markers for a genome scan has been the amount of genotyping required and its expense. For example, 750,000 well-chosen SNPs genotyped for 1000 individuals

BOX 9.3 • SNP Microarrays

Microarrays have made it possible to study the entire genome (DNA), the entire **transcriptome** (RNA) (Plomin & Schalkwyk, 2007), and more recently the entire exome (or coding regions), covering variation seen in as little as 1 percent of the population. A microarray is a glass slide the size of a postage stamp dotted with short DNA sequences called probes. Microarrays were first used to assess gene expression, which will be discussed in Chapter 10. In 2000, microarrays were developed to genotype SNPs. Microarrays detect SNPs using the same hybridization method described in Box 9.1. The difference is that microarrays probe for millions of SNPs on a platform the size of a postage stamp. This miniaturization requires little DNA and makes the method fast and inexpensive. This is an advantage in the interim as we wait for whole-genome sequencing to become widely available.

Several types of microarrays are available commercially; the figure shows the microarray manufactured by Affymetrix called GeneChip®. As shown in the figure, many copies of a certain target

nucleotide base sequence surrounding and including a SNP are used to probe reliably for each allele of the SNP. An individual's DNA is cut with **restriction enzymes** into tiny fragments which are then all amplified by *polymerase chain reaction* (*PCR;* see Box 9.1). Using a single PCR to chop up and amplify the entire genome, called **whole-genome amplification,** was the crucial trick that made microarrays possible. The PCR-amplified DNA fragments are made single stranded and washed over the probes on the microarrays so that the individual's DNA fragments will hybridize to the probes if they find exact matches. A fluorescent tag is added to the DNA fragments so that they will fluoresce if they hybridize with a probe, as shown in the figure. The microarray includes probes for both SNP alleles to indicate whether an individual is homozygous or heterozygous.

Microarrays make it possible to conduct genomewide association studies with millions of SNPs. However, any DNA probes can be selected for genotyping on a microarray. As mentioned above, microarrays can include

(500 cases and 500 controls) would require 750 million genotypings. Until recently, such an effort would have cost tens of millions of dollars. This is why, in the past, most association studies have been limited to considering a few candidate genes.

Recently, a revolutionary advance has made genomewide association investigations possible (Hirschhorn & Daly, 2005). Microarrays can be used to genotype millions of SNPs on a "chip" the size of a postage stamp (Box 9.3). With microarrays, the cost of the experiment described above is less than half a million dollars instead of tens of millions. As a result of microarrays, genomewide association analysis has come to dominate attempts to identify genes for complex traits in recent years. Although this is an exciting advance, genomewide studies have found much smaller effects than originally expected,

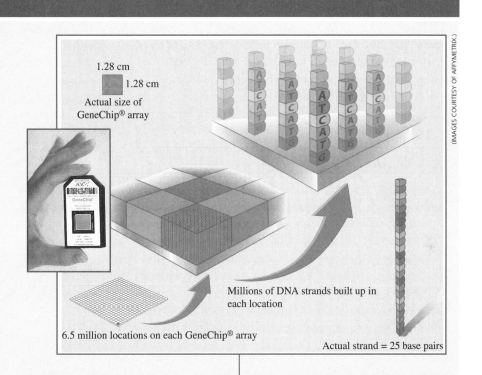

1.28 cm

1.28 cm

Actual size of
GeneChip® array

Millions of DNA strands built up in
each location

6.5 million locations on each GeneChip® array

Actual strand = 25 base pairs

rare SNPs rather than common SNPs or can include probes for CNVs (mentioned earlier in this chapter). Microarrays are also being customized for certain diseases, such as specialized microarrays now available for all DNA variants related to cardiovascular (CardioChip) and immunological (ImmunoChip) function and dysfunction. The cost of microarrays is steadily declining; however, it can still be quite expensive to conduct studies with the hundreds of thousands of subjects needed to detect associations of small effect size.

and evidence suggests that genome scans of 500,000 or more SNPs are needed on very large samples (thousands to tens of thousands of people) to identify replicable associations. As of June 2011, 1449 genomewide association studies had been published for 237 traits (http://www.genome.gov/gwastudies). Recent reports suggest that combining all known SNP associations for any trait explains a small proportion of heritability, ranging from about 5 percent (Manolio et al., 2009) to, at most, 20 percent of the known heritability (Park et al., 2010). This gap between the genomewide identified associations and heritability has become known as the ***missing heritability problem*** (Maher, 2008).

What good will come from identifying genes if they have such small effect sizes? One answer is that we can study pathways between each gene and behavior. Even for genes with a very small effect on behavior, the road signs are clearly marked in a bottom-up analysis that begins with gene expression, although the pathways quickly divide and become more difficult to follow to higher levels of analysis such as the brain and behavior. However, even if there are hundreds or thousands of genes that have small effects on a particular behavior, this set of genes will be useful in top-down analyses that begin with behavior, proceed to investigate multivariate, developmental, and genotype-environment interface issues, and then translate these findings into gene-based diagnosis and treatment as well as prediction and prevention of disorders. These issues about pathways between genes and behavior are the topic of Chapter 10. With DNA microarrays, it would not matter for top-down analyses if there were hundreds or thousands of genes that predict a particular trait. Indeed, for each trait, we can imagine DNA microarrays with thousands of genes that include all the genes relevant to that trait's multivariate heterogeneity and **comorbidity**, its developmental changes, and its interactions and correlations with the environment (Harlaar, Butcher, et al., 2005). Moreover, recent efforts have considered the possibility of aggregating the small effects of many DNA variants associated with a trait (Wray, Goddard, & Visscher, 2008). These composite scores have typically focused on common DNA variants and have been called polygenic susceptibility scores (Pharoah et al., 2002), genomic profiles (Khoury, Yang, Gwinn, Little, & Flanders, 2004), SNP sets (Harlaar, Butcher, et al., 2005), and aggregate risk scores (Purcell et al., 2009). With the advent of rare variant genotyping, new approaches combine the effects of rare and common variants, including variants that are risk-inducing as well as protective (Neale et al., 2011). It is possible that these polygenic composites can aid in explaining more of the genetic variance. Moreover, they could also be useful for identifying groups of individuals at high and low genetic risk in certain areas of research, such as neuroimaging, where large sample sizes are difficult to study. Finally, the inability of association studies to account for most of the reported heritability has also led to a renewed interest in the use of the family design, suggesting that the rare variant approach and next-generation sequencing will improve the power of family-based approaches (Ott, Kamatani, & Lathrop, 2011). Although there is currently no definitive answer to the missing heritability problem, the speed at which the field of behavioral genetics is advancing suggests that it may soon be solved.

● KEY CONCEPTS

Linkage analysis: A technique that detects linkage between DNA markers and traits, used to map genes to chromosomes.

Allelic association: An association between allelic frequencies and a phenotype.

Candidate gene: A gene whose function suggests that it might be associated with a trait. For example, dopamine genes are considered as candidate genes for hyperactivity because the drug most commonly used to treat hyperactivity, methylphenidate, acts on the dopamine system.

Linkage disequilibrium: A violation of Mendel's law of independent assortment in which markers are uncorrelated. It is most frequently used to describe how close together DNA markers are on a chromosome; linkage disequilibrium of 1.0 means that the alleles of the DNA markers are perfectly correlated; 0.0 means that there is no correlation.

Genomewide association study: An association study that assesses DNA variation throughout the genome.

Missing heritability: The difference between the genomewide identified associations and reported heritability estimates from quantitative genetic studies, such as twin and family designs.

Microarray: Commonly known as gene chips, microarrays are slides the size of a postage stamp with hundreds of thousands of DNA sequences that serve as probes to detect gene expression (RNA microarrays) or single nucleotide polymorphisms (DNA microarrays).

Whole-genome amplification: The use of a few restriction enzymes in polymerase chain reactions (PCRs) to chop up and amplify the entire genome; this makes microarrays possible.

Summary

Although much more quantitative genetic research is needed, one of the most exciting directions for genetic research in the behavioral sciences involves harnessing the power of molecular genetics to identify specific genes responsible for the widespread influence of genetics on behavior.

The two major strategies for identifying genes for human behavioral traits are allelic association and linkage. Allelic association is simply a correlation between an allele and a trait for individuals in a population. Linkage is like an association within families, tracing the co-inheritance of a DNA marker and a disorder within families. Linkage is systematic but not powerful for detecting genes of small effect size; association is more powerful but until recently was not systematic and was restricted to candidate genes. SNP microarrays have made possible genomewide association studies using millions of SNPs and incorporating common as well as rare variation.

For complex human behaviors, many associations and linkages have been reported. Ongoing genomewide association studies using SNP microarrays with large samples identify genes of small effect size associated with behavior. The results of genomewide association have yielded genes accounting for much less of the genetic variance than once expected, leaving us with the missing heritability problem. New technologies such as whole-genome sequencing may begin to shed light on this issue; however, in the interim, combining the effects of multiple genes of small effect may aid in accounting for more of the genetic influence on behavior.

As discussed in the next chapter, the goal is not only finding genes associated with behavior but also understanding the pathways between genes and behavior, that is, the mechanisms by which genes affect behavior, sometimes called *functional genomics*.

CHAPTER · TEN

Pathways between Genes and Behavior

Quantitative genetic research consistently shows that genetics contributes importantly to individual differences in nearly all behaviors, such as learning abilities and disabilities, psychopathology, and personality. You will see in later chapters that quantitative genetics and molecular genetics are coming together in the study of complex traits and common disorders. Molecular genetic research, which attempts to identify the specific genes (quantitative trait loci, or QTLs) responsible for the heritability of these behaviors, has begun to identify such genes, although, as noted in Chapter 9, research using genomewide association scans with large samples suggests that the heritabilities of complex traits and common disorders are due to many genes of small effect. Nonetheless, the bottom line for behavioral genetics is this: Heritability means that DNA variation creates behavioral variation, and we need to find these DNA sequences to understand the mechanisms by which genes affect behavior.

The goal is not only finding genes associated with behavior but also understanding the pathways between genes and behavior, that is, the mechanisms by which genes affect behavior, sometimes called *functional genomics* (Figure 10.1). This chapter

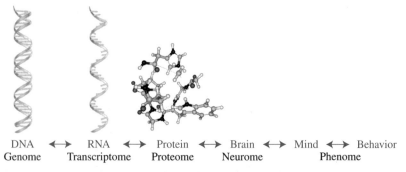

DNA ⟷ RNA ⟷ Protein ⟷ Brain ⟷ Mind ⟷ Behavior
Genome   Transcriptome   Proteome   Neurome   Phenome

• **FIGURE 10.1** Functional genomics includes all levels of analysis from genome (DNA) to phenome (mind and behavior).

BOX 10.1 • Levels of Analysis

The relationship between brain and "mind" (mental constructs) has been a central issue in philosophy for four centuries, since Descartes advocated a mind-body dualism in which the mind was nonphysical. Because this dualism of mind and body is now generally rejected (see Bolton & Hill, 2004; Kendler, 2005), we will simply assert the view that all behavior is biological in the general sense that behavior depends on physical processes. Does this mean that behavior can be reduced to biology (Bickle, 2003)? Because all behavior is biological, it would seem that the answer must logically be "yes." However, saying that all behavior is biological is similar to saying that all behavior is genetic (because without DNA there can be no behavior) or that all behavior is environmental (because without the environment there can be no behavior).

Behavioral genetics' way out of this philosophical conundrum is to focus empirically on individual differences in behavior and to investigate the extent to which genetic and environmental differences can account for these differences in behavior (see Chapter 7). The point of this chapter is to consider some of the levels of analysis that lie between genes and behavior. The ultimate goal of behavioral genetics is to understand the links between genes and behavior at all levels of analysis.

Different levels of analysis are more or less useful for addressing different questions, such as questions about causes and questions about cures (Bolton & Hill, 2004). Functional genomics generally assumes a bottom-up approach that begins at the level of cells and molecular biology. The phrase *behavioral genomics* has been proposed as an antidote emphasizing the value of a top-down approach that attempts to understand how genes work at the level of the behavior of the whole organism (Plomin & Crabbe, 2000). Behavioral genomics may be more fruitful than other levels of analysis in terms of predicting, diagnosing, intervening in, and preventing behavioral disorders.

Finally, relationships between levels of analysis should be considered correlational until proven causal, which is why the connections between levels in Figure 10.1 are double-headed arrows. For example, associations between brain differences and behavioral differences are not necessarily caused by the brain differences: Behavior can cause changes in brain structure and function. A striking example is that the posterior hippocampus, a part of the brain that stores spatial representations of the environment, is significantly larger in London taxi drivers (Maguire et al., 2000); the size is correlated with the number of years spent driving a taxi (Maguire, Woollett, & Spiers, 2006). Similarly, correlations between gene expression and behavior are not necessarily causal because behavior can change gene expression. A crucial point is that the only exception to this rule is DNA: Correlations between differences in DNA sequence and differences in behavior are causal in the sense that behavior does not change the nucleotide sequence of DNA. In this sense, DNA is in a causal class of its own.

considers ways in which researchers are attempting to connect the dots between genes and behavior. (See Box 10.1 for a discussion of some relevant philosophical issues.) We begin with a description of gene expression, including how **epigenetics** relates to expression, and then expand our discussion to consider expression of all the genes in the genome, called the *transcriptome.* The next step along the pathways from genes to behavior is all the proteins coded by the transcriptome, called the ***proteome.*** Next is the brain, which, continuing the *–omics* theme, has been referred to as the ***neurome.*** This chapter stops at the brain level of analysis because the mind (cognition and emotion) and behavior—sometimes called the *phenome*—will be the focus of Chapters 11 to 20.

It should be reiterated that this chapter is about connecting the dots between genes and behavior through the epigenome, the transcriptome, the proteome, and the brain. It is not meant to describe each of these areas per se, four of the most active areas of research in all of the life sciences. Although our focus here is on the links between genes and behavior, it should also be kept in mind that the environment plays a crucial role at each step in the pathways between genes and behavior (Chapter 8).

■ KEY CONCEPTS

Functional genomics: The study of how genes work in the sense of tracing pathways among genes, brain, and behavior. It usually implies a bottom-up approach that begins with molecules in a cell, in contrast to *behavioral genomics.*

Behavioral genomics: The study of how genes in the genome function at the behavioral level of analysis. In contrast to *functional genomics,* behavioral genomics is a top-down approach to understanding how genes work in terms of the behavior of the whole organism.

Genome: All the DNA sequences of an organism. The human genome contains about 3 billion DNA base pairs.

Epigenome: Epigenetic events throughout the genome.

Transcriptome: RNA transcribed from all the DNA in the genome.

Proteome: All the proteins translated from RNA (transcriptome).

Neurome: Effects of the genome throughout the brain.

Gene Expression and the Role of Epigenetics

Genes do not blindly pump out their protein products. As explained in Box 4.1, genetic information flows from DNA to messenger RNA (mRNA) to protein. When the gene product is needed, many copies of its mRNA will be present, but otherwise very few copies of the mRNA are transcribed. In fact, you are changing the rates of transcription of genes for neurotransmitters by reading this sentence. Because mRNA exists for only a few minutes and then is no longer translated into protein, changes in the rate of transcription of mRNA are used to control the rate at which genes produce proteins. This is what is meant by *gene expression.*

RNA is no longer thought of as merely the messenger that translates the DNA code into proteins. In terms of evolution, RNA was the original genetic code, and it still is the genetic code for most viruses. Double-stranded DNA presumably had a selective advantage over RNA because the single strand of RNA left it vulnerable to predatory enzymes. DNA became the faithful genetic code that is the same in all cells, at all ages, and at all times. In contrast, RNA, which degrades quickly, is tissue-specific, age-specific, and state-specific. For these reasons, RNA can respond to environmental changes by regulating the transcription and translation of protein-coding DNA. This is the basis for the process of gene expression.

An area relevant to gene expression that has seen rapid growth over the past few decades is *epigenetics*. Epigenetics is focused on understanding a type of slow-motion, developmentally stable change in certain mechanisms of gene expression that do not alter DNA sequence and can be passed on from one cell to its daughter cells (Bird, 2007). The prefix *epi-* means "above." You can think about the epigenome as the cellular material that sits on top, or outside, of the genome. It is these epigenetic marks that tell your genes to switch on or off, to scream or whisper. It may be through epigenetic marks that environmental factors like diet, stress, and prenatal nutrition can change gene expression from one cell to its daughter cells and, in some cases, from one generation to the next, called imprinting (see Chapter 3).

There are excellent epigenetics texts that provide great detail about these modes of action (e.g., Allis, Jenuwein, & Reinberg, 2007; Tollefsbol, 2011). We will focus briefly on the most widely studied mechanism of epigenetic regulation of gene expression: DNA methylation (Bird, 2007). A methyl group is a basic unit in organic chemistry: one carbon atom attached to three hydrogen atoms. When a methyl group attaches to a specific DNA sequence in a gene's promoter region—a process called DNA methylation—it silences the gene's expression by preventing the gene's transcription. Conversely, when a gene's promoter is not methylated, that gene will not be silenced (Maccani & Marsit, 2009).

Unlike epigenetic marks that effect long-term developmental changes in gene expression, many changes in gene expression are short term, providing quick reactions to changes in the environment. One such recently discovered mechanism of gene regulation is called *non-coding RNA*. As mentioned in Box 4.1, only about 2 percent of the genome involves protein-coding DNA as described by the central dogma. What is the other 98 percent doing? It had been thought that it is "junk" that has just hitched a ride evolutionarily. However, we now know that most human DNA is transcribed into RNA that is not the mRNA translated into amino acid sequences. This so-called non-coding RNA instead plays an important role in regulating the expression of protein-coding DNA, especially in humans.

One type of non-coding RNA has been known for more than 30 years. Embedded in protein-coding genes are DNA sequences, called **introns**, that are transcribed into RNA but are spliced out before the RNA leaves the nucleus. The remaining parts of the RNA are spliced back together, exit the nucleus and are then translated into amino acid sequences. The DNA sequences in protein-coding genes that are transcribed into mRNA and translated into amino acid sequences are called *exons.* Exons usually consist of only a few hundred base pairs, but introns vary widely in length, from 50 to 20,000 base pairs. Only exons are translated into amino acid sequences that make up proteins. However, introns are not "junk." In many cases they regulate the transcription of the gene in which they reside, and in some cases they also regulate other genes.

Introns account for about a quarter of the human genome. An exciting recent finding of great significance is that a further quarter of the human genome produces non-coding RNA anywhere in the genome, not just near protein-coding genes. One class of such non-coding RNA that has attracted much attention is called **microRNA,** small RNAs 21 to 25 nucleotides in length capable of posttranscriptionally regulating genes. Even though they are tiny, microRNAs play a big role in gene regulation and exhibit tissue-specific expression and function. MicroRNAs have also been shown to be responsive to environmental exposures, such as cigarette smoke (Maccani et al., 2010). The human genome is thought to encode over 1000 microRNAs, capable of regulating up to 60 percent of protein-coding genes by binding to (and thus posttranscriptionally silencing) target mRNA (Bentwich et al., 2005; Lim et al., 2005). Moreover, microRNAs are likely to be just the tip of the iceberg of non-coding RNA effects on gene regulation (Mendes Soares & Valcárcel, 2006). The list of novel mechanisms by which non-coding RNA can regulate gene expression is growing rapidly (Costa, 2005; Maccani & Marsit, 2009).

Epigenetics and non-coding RNA are recently discovered mechanisms that regulate gene expression. Figure 10.2 shows how regulation works more generally for

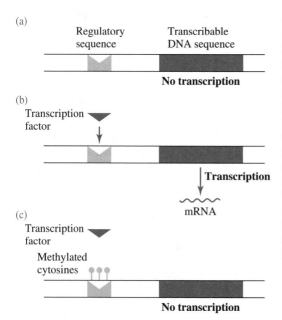

• **FIGURE 10.2** Transcription factors can regulate protein-coding genes by controlling mRNA transcription. (a) A regulatory sequence normally shuts down transcription of its gene; (b) but when a particular transcription factor binds to the regulatory sequence, the gene is freed for transcription. (c) One type of epigenetic regulation involves DNA methylation of cytosine residues in the gene's promoter region; this can regulate transcription by altering the microenvironment so that the transcription factor cannot bind its regulatory sequence, thereby reducing or halting transcription.

classical protein-coding genes. Many of these genes include regulatory sequences that normally block the gene from being transcribed. If a particular molecule binds with the regulatory sequence, it will free the gene for transcription. Figure 10.2 also illustrates epigenetic regulation. Most gene regulation involves several mechanisms that act like a committee voting on increases or decreases in transcription. That is, several transcription factors act together to regulate the rate of specific mRNA transcription. Non-coding RNA transcripts can regulate the expression of other genes without being translated into proteins. This regulation of gene expression by non-coding RNA is primarily affected by altering the rate of transcription, but other factors include changes in the RNA transcript itself and the way the RNA transcript interacts with its regulatory targets, which are often messenger RNA transcripts.

Rather than just looking at the expression of a few genes, researchers can now use microarrays to assess the degree of expression of all genes in the genome simultaneously, including non-coding RNA (the *transcriptome*) and profiles of DNA methylation of all coding genes in the genome (called the *methylome* or *epigenome*), as described in the following section. The importance of microarrays for gene expression and methylome profiling for behavioral genetics lies in the fact that the epigenome and the transcriptome are the first steps in the correlation between genes and behavior. Because gene expression and methylation are sensitive to the environment, the transcriptome and epigenome could be useful as biomarkers of environmental change (Petronis, 2010), including prenatal experiences (Zhang & Meaney, 2010) and mother-infant interaction (Champagne & Curley, 2009; Meaney, 2010).

The Transcriptome: Gene Expression throughout the Genome

As we just outlined, gene expression is the first step on any pathway from genes to behavior: A polymorphism in DNA can have an effect only when the gene is expressed. Some genes, called housekeeping genes, are expressed at a steady rate in most of our cells. Other genes are expressed as their product is needed in response to the environment. For protein-coding genes, expression is most affected by altering the rate of transcription initiation, but other factors that affect expression include alteration of the RNA transcript, passage of the messenger RNA through the nuclear membrane, protection or degradation of the RNA transcript in the cytoplasm, the rate of translation, and posttranslational modification of the protein.

Gene Expression Profiles: RNA Microarrays and Sequence-Based Approaches

For both protein-coding and non-protein-coding DNA, gene expression can be indexed by the number of RNA transcripts, which is the end result of the various processes mentioned, not just the initial transcription process. In contrast to DNA, which faithfully preserves the genetic code in all cells, at all ages, and at all times,

RNA degrades quickly and is tissue-specific, age-specific, and state-specific, as noted above. Two of the key aims of "transcriptomics" are to catalog all types of transcripts, including mRNA, non-coding RNA, and small RNAs, and to quantify their changing expression levels during development and under different conditions (Wang, Gerstein, & Snyder, 2009). Various techniques have been developed to examine the transcriptome or, in other words, to assess the expression of all genes in the genome simultaneously, called *gene expression profiling.*

Specialized gene expression (RNA) microarrays have been designed that are the same as the DNA microarrays described in Box 9.3 except that the probes in RNA microarrays detect a particular sequence of RNA, rather than identifying a particular SNP allele in a DNA sequence. In addition, the goal of RNA microarrays is to detect the quantity of each of the RNA transcripts; for this reason, each probe is represented with many copies. In contrast, SNP probes detect the presence or absence of SNP alleles; multiple probes for each allele are used only in order to increase the accuracy of genotyping. RNA microarrays were originally limited to probes for exons that assessed transcription of the 2 percent of the genome that involves protein-coding genes. One of the most important developments in the recent history of genetics is the ability to sequence an individual's entire genome (Chapter 9). This development has also provided a new method for quantifying the transcriptome by sequencing RNA (Wang et al., 2009). RNA sequencing will undoubtedly drive many exciting discoveries within the next few years. For example, RNA exome sequencing, which involves sequencing only RNA transcribed from exomes, is being widely used to identify rare mutations of large effect in the coding regions of genes (Ng et al., 2010). However, similar to whole-genome sequencing (Chapter 9), the cost of RNA sequencing in individuals remains quite high. Thus, until the costs decrease, a combination of approaches is likely to be used, such as using sequencing data that have detected all possible polymorphisms in DNA and RNA to guide the creation of custom microarrays that are much less expensive than sequencing.

Transcriptomics, including RNA microarrays and sequencing, makes it possible to take snapshots of gene expression throughout the genome at different times (e.g., during development, or before and after interventions) and in different tissues (e.g., in different brain regions). Scores of studies have investigated changes in gene expression profiling in response to drugs (Kreek, Zhou, Butelman, & Levran, 2009; Zhou, Litvin, Piras, Pfaff, & Kreek, 2011) and between groups such as psychiatric cases and controls (Torkamani, Dean, Schork, & Thomas, 2010). Gene expression profiling of the brain is like structural genetic neuroimaging in that it can create an atlas of localized patterns of gene expression throughout the brain. Because genetic neuroimaging requires brain tissue, its use in the human species is limited to postmortem brains and tissue samples removed during surgery, such as tumors (Kleinman et al., 2011; Yamasaki et al., 2005), which raises questions about lack of control concerning gene expression at the time of death (Konradi, 2005). For this reason, structural genetic neuroimaging research has primarily been conducted in mice rather than humans.

Structural brain maps of gene expression are fundamental because genes can only function if they are expressed. Currently, a comprehensive atlas of expression profiles of 20,000 genes in the adult mouse brain is publicly accessible online (Morris et al., 2010; www.brain-map.org). The next goal is functional genetic neuroimaging—studying changes in gene expression in the brain during development or following interventions such as drugs or cognitive tasks. For example, research on mice is under way that aims to create an atlas of profiles of gene expression throughout the brain during learning and memory tasks in the Genes to Cognition research consortium (Manakov, Grant, & Enright, 2009, www.genes2cognition.org, www.brain-map.org). In 2011, BrainCloud was announced as the result of efforts to gain a global molecular perspective on the role of the human genome in brain development, function, and aging. Researchers used an extensive series of postmortem brains from fetal development through aging to examine the timing and genetic control of transcription in the human prefrontal cortex and discovered a wave of gene expression changes occurring during fetal development that are reversed in early postnatal life (Colantuoni et al., 2011, http://braincloud.jhmi.edu/).

Because of the practical and scientific limitations of using postmortem brain tissue, RNA microarrays will be much more widely applicable to human research if easily available tissue such as blood can be used for gene expression profiling. Some similarities between expression in blood and brain have been reported (e.g., Tian et al., 2009). Although gene expression profiling in the blood cannot be used to localize patterns of gene expression in the brain, blood could be used to address some important questions, most notably, gene expression profile differences as a function of development or interventions. Rather than studying the expression of each gene in isolation, researchers can use RNA microarrays and sequencing to study profiles of gene expression across the transcriptome, which leads to understanding the coordination of gene expression throughout the genome (Ghazalpour et al., 2006; Schadt, 2006).

Genetical Genomics

So far, we have discussed gene expression from a normative perspective rather than considering individual differences. The field of gene expression has also considered individual differences as well as their causes and consequences (Cobb et al., 2005; Rockman & Kruglyak, 2006). Much research has been directed toward treating gene expression as a phenotypic trait and finding QTLs (called *expression QTLs* or *eQTLs*) associated with gene expression in mice (Schadt, 2006; Williams, 2006) and humans (Morley et al., 2004). This field has been called *genetical genomics* to emphasize the links between the genome and the transcriptome (Jansen & Nap, 2001; Li & Burmeister, 2005; Petretto et al., 2006). These links have become explicit because recent research using DNA microarrays (see Chapter 9) has begun to scan the genome for SNP associations with genomewide gene expression assessed on RNA microarrays (Skelly, Ronald, & Akey, 2009).

Research on genomewide gene expression in rodents has profited from the availability of inbred lines and especially recombinant inbred lines, which facilitate both quantitative genetic and molecular genetic research (Chesler et al., 2005; Letwin et al., 2006; Peirce et al., 2006) and provide access to brain tissue. However, for rodent research as well as human research, although many eQTL associations have been reported, most suffer from low power and few have been replicated (Skelly et al., 2009). This is a repeat of the story told in Chapter 9 in which genetic effects on complex traits, including individual differences in gene expression, appear to be caused by many QTLs of small effect size. As a result, very large samples will be needed to attain adequate statistical power to detect reliable associations with gene expression traits.

Gene Expression as a Biological Basis for Environmental Influence

Genetical genomics attempts to identify the QTLs responsible for the genetic contribution to individual differences in gene expression, but to what extent are these individual differences genetic in origin? It cannot be assumed that individual differences in gene expression are highly heritable because gene expression has evolved to be responsive to intracellular and extracellular environmental variation. Indeed, quantitative genetic studies of human RNA transcript levels suggest that heritabilities appear to be modest on average across the genome, which implies that most of the variability in transcript levels is due to environmental factors (Cheung et al., 2003; Correa & Cheung, 2004; McRae et al., 2007; Monks et al., 2004; Sharma et al., 2005). Members of identical twin pairs become increasingly different in gene expression profiles throughout the life span (Fraga et al., 2005; Petronis, 2006; Zwijnenburg, Meijers-Heijboer, & Boomsma, 2010). Environmental factors involved in gene expression are part of a rapidly expanding area of research. This was touched on above in the description of epigenetics. It should be reiterated that gene expression is a phenotype; individual differences in expression itself or in epigenetic processes that lead to individual differences in expression may be due to genetic differences (Richards, 2006) or environmental differences. The transcriptome and methylome (or epigenome) could serve as important biomarkers of environmental change because they were designed by evolution to be sensitive to the environment. Examples of such environments include, but are not limited to, prenatal experiences, mother-infant interaction, and exposure to trauma. This perspective could provide a biological foundation upon which to build an understanding of more complex levels of environmental analysis typically studied in behavioral research. It could also have far-reaching impact on translational research by providing biomarkers for differential diagnosis and providing a biological basis for monitoring environmental interventions such as drugs and other therapies (Li, Breitling, & Jansen, 2008).

As noted at the outset of this chapter, we cannot hope to provide a review of all that is known about gene expression or the role of epigenetics in gene expression.

Of special interest in terms of pathways between genes and behavior is the extent to which DNA associations with behavior are mediated by individual differences in gene expression. In the following section, we will continue along the pathways between genes and behavior by considering the next level of analysis, the proteome.

●KEY CONCEPTS

Gene expression: Transcription of DNA into mRNA.

Epigenetics: DNA modifications that affect gene expression without changing the DNA sequence that can be "inherited" when cells divide; can be involved in long-term developmental changes in gene expression.

DNA methylation: An epigenetic process by which gene expression is inactivated by the addition of a methyl group.

Non-coding RNA: RNA that is not translated into amino acid sequences.

Intron: DNA sequence within a gene that is transcribed into messenger RNA but spliced out before the translation into protein. (Compare with *exon.*)

Exon: DNA sequence transcribed into messenger RNA and translated into protein. (Compare with *intron.*)

MicroRNA: A class of non-coding RNA that involves 21 to 25 nucleotides that can degrade or silence gene expression by binding with messenger RNA.

Gene expression profiling: Using microarrays to assess the expression of all genes in the genome simultaneously.

Expression QTL (eQTL): When treating gene expression as a phenotype, QTLs can be identified that account for genetic influence on gene expression.

Genetical genomics: Identifying genes throughout the genome that affect gene expression.

The Proteome: Proteins Coded throughout the Transcriptome

The proteome, which refers to the entire complement of proteins, brings an increase in complexity for three reasons. First, there are many more proteins than genes, in part because alternative splicing of genes can produce different messenger RNA transcripts (Brett et al., 2002). Second, after amino acid sequences are translated from messenger RNA, they undergo modifications, called *posttranslational modifications,* that change their structure and thus change their function. Third, proteins do not work in isolation; their function is affected by their interactions with other proteins as they form protein complexes.

The proteome can be identified using gels in an electrical field (**electrophoresis**) to separate proteins in one dimension on the basis of their charge and in a second dimension on the basis of their molecular weight, called *two-dimensional gel electrophoresis.*

The precision of identifying proteins has been greatly improved by the use of mass spectrometry, which analyzes mass and charge at an atomic level (Aebersold & Mann, 2003). Based on these techniques, a proteome atlas of nearly 5000 proteins and 5000 protein complexes is available for the fruit fly (Giot et al., 2003); similar resources are available for the hippocampus of the mouse (Pollak, John, Hoeger, & Lubec, 2006) and the hippocampus of the rat (Fountoulakis, Tsangaris, Maris, & Lubec, 2005). As the mass spectrometry techniques have been further developed using such techniques as surface-enhanced laser desorption/ionization time-of-flight mass spectrometry (SELDI-TOF MA), high-throughput characterization of proteomic samples has resulted in a wealth of data from numerous biological samples related to medical and psychiatric conditions (Liu, Gong, Cai, & Li, 2011; Xu, Wang, Song, Qiu, & Luo, 2011).

The relative quantity of each protein can also be estimated from these approaches. Individual differences in the quantity of a protein in a particular tissue represent a protein trait that is analogous to the RNA transcript traits discussed in the previous section. As with the transcriptome, the proteome needs to be considered as a phenotype that can be attributed to genetic and environmental factors. Such protein traits can be related to individual differences in behavior. For example, human studies using cerebrospinal fluid have yielded hundreds of differences in protein levels and protein modifications in psychiatric disorders (Fountoulakis & Kossida, 2006). Sophisticated approaches to proteomic characterization of specific brain regions implicated in schizophrenia have also suggested differences that may influence behavior (Matsumoto et al., 2011).

Historically, the transcriptome has been and still is the target of much more genetic research than the proteome; however, the interest in the proteome is gaining momentum. Just as the Human Genome Project revolutionized how biologically driven research is performed, it seems only natural that there is now a systematic effort under way to characterize the protein products of the human genome—the Human Proteome Project (HUPO Views, 2010; Nilsson et al., 2010). The hope is that this project will become a resource to help elucidate biological and molecular function and advance diagnosis and treatment of diseases.

As in research on the transcriptome, the mouse has been the focus of proteomic work because of the availability of brain tissue. A pioneering study that examined 8767 proteins from the mouse brain as well as other tissues found that 1324 of these proteins showed reliable differences in a large backcross (see Chapter 2) (Klose et al., 2002). Of these proteins, 466 were mapped to chromosomal locations. Although such linkages need to be replicated, the genetic results are interesting for two reasons: Most proteins showed linkage to several regions, and the chromosomal positions often differed from those of the genes that code for the proteins. These results suggest that multiple genes affect protein traits. Another study on protein expression in the hippocampus yielded similar results (Pollak, John, Schneider, Hoeger, & Lubec, 2006). As methods have become more efficient, they have been applied to human studies of psychiatric and behavioral phenotypes (Benoit, Rowe, Menard, Sarret, & Quirion, 2011; Filiou, Turck, & Martins-de-Souza, 2011; Focking et al., 2011).

The Brain

Each step along the pathways from genome to transcriptome to proteome involves huge increases in complexity, but these pale in comparison to the complexity of the brain. The brain has trillions of junctions between neurons (*synapses*) instead of billions of DNA base pairs, and hundreds of neurotransmitters, not just the four bases of DNA. Although the three-dimensional structure of proteins and their interaction in protein complexes contribute to the complexity of the proteome, this complexity is nothing compared to the complexity of the three-dimensional structure and interactions among neurons in the brain.

Neuroscience, the study of brain structure and function, is another extremely active area of research. This section provides an overview of neurogenetics as it relates to behavior. Because the brain is so central in the pathways between genes and behavior, brain phenotypes are sometimes referred to as *endophenotypes,* as discussed in Box 10.2. In the remainder of the chapter, we will refer to areas of the brain depicted in Figure 10.3 and to the structure of the neuron shown in Figure 10.4.

As mentioned earlier in this chapter, research on the transcriptome and proteome has begun to build bridges to the brain by creating atlases of gene and protein expression throughout the brain. Most of this research involves animal models because of the access to brain tissue in nonhuman animals. A huge advantage for neurogenetic research in the human species is the availability of neuroimaging, which, as discussed later, makes it possible to assess the structure and function of the human brain. We begin, however, with one major area of neurogenetic research on behavior that focuses on animal models, particularly the fruit fly *Drosophila* and the mouse: learning and memory. The advantage of neurogenetic research with animal models is the ability to use both natural and induced genetic mutations to dissect pathways between neurons and behavior. The second example of neurogenetic research involves emotion in the human species, using neuroimaging.

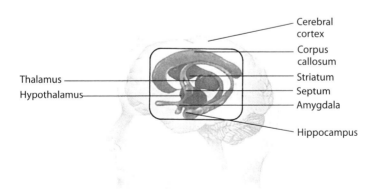

• **FIGURE 10.3** Basic structures of the human brain.

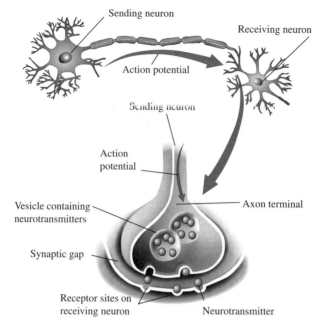

Sending neuron

Receiving neuron

Action potential

Sending neuron

Action potential

Vesicle containing neurotransmitters

Axon terminal

Synaptic gap

Receptor sites on receiving neuron

Neurotransmitter

• **FIGURE 10.4** The neuron. Electrical impulses (action potentials) travel from one neuron to another across a gap at the end of the axon known as a synapse. Action potentials release neurotransmitters into the synaptic gap. The neurotransmitters bind to receptor sites on the receiving neuron.

BOX 10.2 • Endophenotypes

The goal of behavioral genetics is to understand pathways between genes and behavior at all levels of analysis. In addition, each level of analysis warrants attention in its own right (see Box 10.1). Using the brain level of analysis as an example, there is much to learn about the brain itself regardless of the brain's relationship to genes or to behavior. However, the focus of behavioral genetics, and this chapter, is on the brain as a pathway between genes and behavior.

Levels of analysis lower than behavior itself are sometimes called endophenotypes, where *endo* means "inside." The term *intermediate phenotype* has also been used as a synonym for endophenotype. It has been suggested that these lower levels of analysis, such as the brain level, might be more amenable to genetic analysis than behavior (Bearden & Freimer, 2006; Gottesman & Gould, 2003). In addition, lower-level processes, such as neurotransmitter levels in the brain, can be modeled more closely in animals and humans than can behavior itself (Gould & Gottesman, 2006). Specifically, it is hoped that genes will have larger effects on lower levels of analysis and will thus be easier to identify. Recent genetic research on the brain neuroimaging of phenotypes supports this hypothesis (see text), for example, in research on alcoholism (Hill, 2010). However, caution is warranted until these DNA associations are replicated because genetic influences are likely to be pleiotropic and polygenic for brain traits as well as behavioral traits (Kovas & Plomin, 2006). Moreover, a meta-analysis of genetic associations

● KEY CONCEPTS

Posttranslational modification: Chemical change to polypeptides (amino acid sequences) after they have been translated from mRNA.

Electrophoresis: A method used to separate DNA fragments or proteins by size. When an electrical charge is applied to DNA fragments or proteins in a gel, smaller fragments travel farther.

Endophenotype: An "inside" or intermediate phenotype that does not involve overt behavior.

Synapse: A junction between two nerve cells through which impulses pass by diffusion of a neurotransmitter, such as dopamine or serotonin.

Learning and Memory

One important area of neurogenetic research has considered learning and memory, key functions of the brain. Much of this research involves the fruit fly *Drosophila*. *Drosophila* can indeed learn and remember, abilities that have been studied primarily in relation to spatial learning and olfactory learning (Moressis, Friedrich, Pavlopoulos, Davis, & Skoulakis, 2009; Skoulakis & Grammenoudi, 2006). Learning and memory in *Drosophila*

reported for endophenotypes concluded that genetic effect sizes are no greater for endophenotypes than for other phenotypes (Flint & Munafo, 2007). In addition, recent work suggests that careful attention should be paid to claims of causality, measurement error, and environmental factors that can influence both the endophenotype and the final outcome (Kendler & Neale, 2010).

Although less complex than behavioral traits, brain traits are nonetheless very complex, and complex traits are generally influenced by many genes of small effect (see Chapter 9). Indeed, the most basic level of analysis, gene expression, appears to show influence by many genes of small effect as well as substantial influence by the environment. One might think that lower levels of analysis are

more heritable, but this does not seem to be the case. Using gene expression again as an example because it is the most basic level of analysis, individual differences in transcript levels across the genome do not appear to be highly heritable.

Another issue is that the goal of behavioral genetics is to understand pathways among genes, brain, and behavior. Genes found to be associated with brain phenotypes are important in terms of the brain level of analysis, but their usefulness for behavioral genetics depends on their relationship with behavior (Rasetti & Weinberger, 2011). In other words, when genes are found to be associated with brain traits, the extent to which the genes are associated with behavioral traits needs to be assessed rather than assumed.

constitute one of the first areas to connect the dots among genes, brain, and behavior (Davis, 2011; Margulies, Tully, & Dubnau, 2005; McGuire, Deshazer, & Davis, 2005). For example, in studies of chemically created mutations in *Drosophila melanogaster*, investigators have identified dozens of genes that, when mutated, disrupt learning (Waddell & Quinn, 2001). A model of memory has been built by using these mutations to dissect memory processes. Beginning with dozens of mutations that affect overall learning and memory, investigators found, on closer examination, that some mutations (such as *dunce* and *rutabaga*) disrupt early memory processing, called short-term memory (STM). In humans, this is the memory storage system you use when you want to remember a telephone number temporarily. Although STM is diminished in these mutant flies, later phases of memory consolidation, such as long-term memory (LTM), are normal. Other mutations affect LTM but do not affect STM. Genes identified as necessary for learning in *Drosophila* also appear to be important in mammals (Davis, 2005).

Neurogenetic research is now attempting to identify the brain mechanisms by which these genes have their effect. Several of the mutations from mutational screening were found to affect a fundamental signaling pathway in the cell involving cyclic AMP (cAMP). *Dunce*, for example, blocks an early step in the learning process by degrading cAMP prematurely. Normally, cAMP stimulates a cascade of neuronal changes including production of a protein kinase that regulates a gene called *cAMP-responsive element* (*CRE*). CRE is thought to be involved in stabilizing memory by changing the expression of a system of genes that can alter the strength of the synaptic connection between neurons, called *synaptic plasticity*, which has been the focus of research in mice (see below). In terms of brain regions, a major target for research in *Drosophila* has been a type of neuron, called a *mushroom body neuron*, that appears to be the major site of olfactory learning in insects (Busto, Cervantes-Sandoval, & Davis, 2010; Heisenberg, 2003), although many other neurons are also involved (Davis, 2011). Pairing shock with olfactory cues triggers a complex series of signals that results in a cascade of expression of different genes. These changes in gene expression produce long-lasting functional and structural changes in the synapse (Liu & Davis, 2006).

Learning and memory also constitute an intense area of research activity in the mouse. However, rather than relying on randomly created mutations, neurogenetic research on learning and memory in the mouse uses targeted mutations. It also focuses on one area of the brain called the hippocampus (see Figure 10.3), which has been shown in studies of human brain damage to be crucially involved in memory. In 1992, one of the first gene targeting experiments for behavior was reported (Silva, Paylor, Wehner, & Tonegwa, 1992). Investigators knocked out a gene (*a-CaMKII*) that normally codes for the protein $a\text{-}Ca^{2+}$-calmodulin kinase II, which is expressed postnatally in the hippocampus and other forebrain areas critical for learning and memory. Mutant mice homozygous for the knock-out gene learned a spatial task significantly more poorly than control mice did, although otherwise their behavior seemed normal. A spatial memory task used in most of the research of this type is a water maze. In studies using this task, various mutant and control mice are trained to escape from a large pool of opaque water by finding a platform hidden just beneath the water's surface (Figure 10.5).

In the 1990s, there was an explosion of research using targeted mutations in the mouse to study learning and memory (Mayford & Kandel, 1999), with 22 knock-out mutations shown to affect learning and memory in mice (Wahlsten, 1999). Many of these targeted mutations involve changes in the strength of connections across the synapse and have been the topic of more than 10,000 papers, with 500 papers focused on the genetics of synaptic plasticity. Memories are made of long-term synaptic changes, called *long-term potentiation* (Lynch, 2004). The idea that information is stored in neural circuits by changing synaptic links between neurons was first proposed in 1949 (Hebb, 1949).

Although genes drive long-term potentiation, understanding how this occurs is not going to be easy because each synapse is affected by more than a thousand protein components. The *a-CaMKII* gene, mentioned earlier in relation to the first reported knock-out study of learning and memory, activates *CRE*-encoded expression of a protein called CRE-binding protein (CREB), which affects long-term but not short-term memory (Silva, Kogan, Frankland, & Kida, 1998). CREB expression is a critical step in cellular changes in the mouse synapse, as it is in *Drosophila*. In *Drosophila*, another gene that activates CREB was the target of a *conditional* knock-out that can be turned on and off as a function of temperature. These changes in CREB expression were shown to correspond to changes in long-term memory (Yin, Delvecchio, Zhou, & Tully, 1995). A complete knock-out of *CREB* in mice is lethal, but deletions

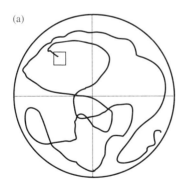
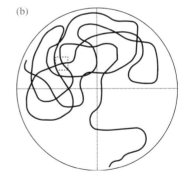

(a)

(b)

• **FIGURE 10.5** The Morris water maze is frequently used in neurogenetic research on spatial memory. A mouse escapes the water by using spatial cues to find a submerged platform. Shown in these diagrams are swim paths to a platform (upper left quadrant) in the Morris water maze. The mouse is trained to know the location of a submerged invisible platform. The animal usually navigates by using distal room clues such as doors and posters on the walls, but it can also be given more proximal cues to control for orientation. (a) The trained animal is tested on its efficiency in finding the platform (time, path length, erroneous entries into the wrong quadrants). (b) The submerged platform is removed, and the time the trained animal spends searching in the correct quadrant is assessed.

that substantially reduce CREB have also been shown to impair long-term memory (Mayford & Kandel, 1999).

A receptor involved in neurotransmission via the basic excitatory neurotransmitter glutamate plays an important role in long-term potentiation and other behaviors in mice as well as humans (Newcomer & Krystal, 2001). The N-methyl-D-aspartate (NMDA) receptor serves as a switch for memory formation by detecting coincident firing of different neurons; it affects the cAMP system among others. Overexpressing one particular *NMDA* gene (*NMDA receptor 2B*) enhanced learning and memory in various tasks in mice (Tang et al., 1999). A conditional knock-out was used to limit the mutation to a particular area of the brain—in this case, the forebrain. Normally, expression of this gene has slowed down by adulthood; this pattern of expression may contribute to decreased memory in adults. In this research, the gene was altered so that it continued to be expressed in adulthood, resulting in enhanced learning and memory. However, this particular *NMDA* gene is part of a protein complex (N-methyl-D-aspartate receptor complex) that involves 185 proteins; mutations in many of the genes responsible for this protein complex are associated with behavior in mice and humans (Grant, Marshall, Page, Cumiskey, & Armstrong, 2005).

Targeted mutations indicate the complexity of brain systems for learning and memory. For example, none of the genes and signaling molecules in flies and mice found to be involved in learning and memory are specific to learning processes. They are involved in many basic cell functions, a finding that raises the question of whether they merely modulate the cellular background in which memories are encoded (Mayford & Kandel, 1999). It seems likely that learning involves a network of interacting brain systems. Another example of complexity can be seen in work on the gene for the *dunce* mutant in *Drosophila*. When it was altered by disabling various combinations of its five DNA start sites for transcription, the investigators found that each combination has different effects on learning and memory processes (Dubnau & Tully, 1998).

Chemical-induced and targeted mutations in *Drosophila* and mice have shown that long-term potentiation of the synapse is a necessary facet of learning and memory, although other processes are also important (Mayford & Kandel, 1999). The number of papers using mutations to study learning and memory has declined recently, in part due to the problems with gene targeting described in Chapter 6 and in part due to the increased use in research of pharmacological and neural interventions rather than genetic interventions. Relatively little neurogenetic research has as yet been conducted on normal variation in learning and memory.

Emotion

In the human species, the structure and function of brain regions can be assessed using noninvasive neuroimaging techniques. There are many ways to scan the brain, each with a different pattern of strengths and weaknesses. As one example, brain structures can be seen clearly using magnetic resonance imaging (MRI) (Figure 10.6). Functional MRI (fMRI) is able to visualize changing blood flow in the brain, which

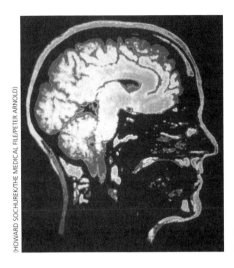

(HOWARD SOCHUREK/THE MEDICAL FILE/PETER ARNOLD)

• **FIGURE 10.6** Magnetic resonance imaging (MRI) scan of the human brain.

is associated with neural activity. The spatial resolution of fMRI is good, about two millimeters, but its temporal resolution is limited to events that take place over several seconds. Electroencephalography (EEG), using electrodes placed on the scalp, measures voltage differences across the brain that index electrical activity. It provides excellent temporal resolution (less than one millisecond), but its spatial resolution is very poor because it averages activity across adjacent regions on the brain's surface. It is possible to combine the spatial strength of fMRI and the temporal strength of EEG (Debener, Ullsperger, Siegel, & Engel, 2006), which can also be accomplished using a different technology, *magnetoencephalography* (MEG; Ioannides, 2006).

Neuroimaging is now often used in genetic research. For example, the IMAGEN study was announced as the first multicenter genetic neuroimaging study aimed at identifying the genetic and neurobiological basis of individual variability in impulsivity, reinforcer sensitivity and emotional reactivity, and how these affect the development of psychiatric disorders (Schumann et al., 2010). Several twin studies using structural neuroimaging have shown that individual differences in the volume of many brain regions are highly heritable and correlated with general cognitive ability (Posthuma et al., 2002; Thompson et al., 2001; Wallace et al., 2006) and may also reflect genetic vulnerability for psychopathic traits (Rijsdijsk et al., 2010). Twin data have recently been used to develop the first brain atlas of human cortical surface area based solely on genetically informative data (Chen et al., 2012). This atlas, shown in Figure 10.7, was created, in part, by using genetic correlations estimated from twin data between different points on the cortical surface. These genetic correlations represent shared genetic influences between cortical regions (Chen et al., 2012).

Candidate gene studies have also begun to report associations with several types of brain function (Mattay & Goldberg, 2004; Winterer, Hariri, Goldman, & Weinberger, 2005). Although much neuroimaging research investigates human learning

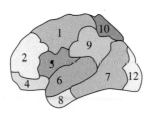 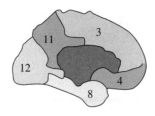

Left Hemisphere   Right Hemisphere

• **FIGURE 10.7** Brain atlas of human cortical surface area (left and right hemispheres) based solely on genetically informative data. (Adapted from Chen et al., 2012.) Map of twelve genetic clusters of the human cortical surface: 1, motor-premotor cortex; 2, dorsolateral prefrontal cortex; 3, dorsomedial frontal cortex; 4, orbitofrontal cortex; 5, pars opercularis and subcentral region; 6, superior temporal cortex; 7, posterolateral temporal cortex; 8, anteromedial temporal cortex; 9, inferior parietal cortex; 10, superior parietal cortex; 11, precuneus; and 12, occipital cortex. These genetic clusters tend to correspond to traditional cortical structures. (Reprinted by permission from Macmillan Publishers Ltd: *Science*, 335, 1634–1636. ©2011.)

and memory, recent attention in neurogenetics has turned to emotion (LeDoux, 2000), especially the role of the amygdala (see Figure 10.3) (Phelps & LeDoux, 2005). For example, a highly cited paper reported that a serotonin transporter gene polymorphism (*5-HTTLPR*) is associated with amygdala neuronal activity in response to threat-related stimuli (looking at angry and fearful faces) as assessed by fMRI (Hariri et al., 2002). This finding has been replicated in several studies, has also received support from mouse knock-out research (Hariri & Holmes, 2006), and is likely to have general behavioral implications in terms of how we react to environmental stress (Hariri et al., 2005). Recent work using various neuroimaging techniques confirms the roles of serotonin, dopamine, norepinephrine, endocannabinoids, and steroid hormones in the responsiveness of the amygdala (Hariri & Whalen, 2011).

Summary

As genes associated with behavior are identified, genetic research will switch from finding genes to using genes to understand the pathways from genes to behavior, that is, the mechanisms by which genes affect behavior. Three general levels of analysis between genes and behavior are the transcriptome (gene expression throughout the genome), the proteome (protein expression throughout the transcriptome), and the brain. RNA sequencing and RNA microarrays make it possible to study the expression of all genes in the genome across the brain, across development, across states, and across individuals. All pathways between genes and behavior travel through the brain, as can be glimpsed in neurogenetic research on learning and memory and on emotion.

CHAPTER · ELEVEN

Cognitive Disabilities

I n an increasingly technological world, cognitive disabilities are important liabilities. More is known about genetic causes of cognitive disabilities than about any other area of behavioral genetics. Many single genes and chromosomal abnormalities that contribute to general cognitive disability are known. Although most of these are rare, together they account for a substantial amount of cognitive disability, especially severe disability, which is often defined as intelligence quotient (IQ) scores below 50. (The average IQ in the population is 100, with a standard deviation of 15, which means that about 95 percent of the population have IQ scores between 70 and 130.) Less is known about mild cognitive disability (IQs from 50 to 70), even though it is much more common. Specific types of cognitive disabilities, especially reading disability and dementia, are the foci of current research because genes linked to these disabilities have been identified.

The American Psychiatric Association's *Diagnostic and Statistical Manual of Mental Disorders-IV* (DSM-IV, due to be revised in 2013 as DSM-5), which is consistent with the *International Classification of Diseases-10* (ICD-10), refers to general cognitive disability as *mental retardation*. For example, DSM-IV defines mental retardation in terms of subaverage intellectual functioning, onset before age 18, and related limitations in adaptive skills. However, the term *mental retardation* is now considered pejorative, as are many other terms such as *developmental delay, developmental disability, intellectual disability,* and *learning disability.* We will use the term *general cognitive disability* when referring to low IQ and *specific cognitive disability* when referring to specific learning disabilities such as those in reading or mathematics. Four levels of general cognitive disability are considered: mild (IQ 50 to 70), moderate (IQ 35 to 50), severe (IQ 20 to 35), and profound (IQ below 20). About 85 percent of all individuals with IQs below 70 are classified as mild, most of whom can live independently and hold a job. Individuals with IQs from 35 to 50 usually have good self-care skills and can carry on simple conversations. Although they generally do not live independently and in the past were usually institutionalized, today they often live in the community in special residences or with their families. People with IQs from 20 to 35 can learn some self-care skills and understand language, but they have trouble speaking and require considerable supervision. Individuals with IQs below 20 may understand a simple communication but usually cannot speak; they remain institutionalized.

General Cognitive Disability: Quantitative Genetics

In the behavioral sciences, it is now widely accepted that genetics substantially influences general cognitive ability; this belief is based on evidence presented in Chapter 12. Although one might expect that low IQ scores are also due to genetic factors, this conclusion does not necessarily follow. For example, cognitive disability can be caused by environmental trauma, such as birth problems, nutritional deficiencies, or head injuries. Given the importance of cognitive disability, it is surprising that no twin or adoption studies of moderate or severe cognitive disability have been reported. Nonetheless, one sibling study suggests that moderate and severe cognitive disability may be due largely to nonheritable factors. In a study of over 17,000 white children, 0.5 percent were moderately to severely disabled (Nichols, 1984). As shown in Figure 11.1, the siblings of these children showed no cognitive disability. The siblings' average IQ was 103, with a range of 85 to 125. In other words, moderate to severe cognitive disability showed no familial resemblance, a finding supported by more recent studies (Collins, Marvelle, & Stevenson, 2011). As discussed later, there are many single-gene causes of cognitive disability that are inherited from generation to generation; however, these are so rare that they may not appear even in large samples that have not been selected for severe disability. Although most moderate and severe cognitive disability may not be inherited from generation to generation, it can be caused by noninherited DNA events, such as new gene mutations and new chromosomal abnormalities, as discussed in the following sections, as well as by environmental events.

In contrast, siblings of mildly disabled children tend to have lower than average IQ scores (see Figure 11.1), as would be expected for an inherited trait. The average IQ for these siblings of mildly disabled children (1.2 percent of the sample were mildly disabled) was only 85. A similar result was found in the largest family study of mild cognitive disability, which considered 80,000 relatives of 289 mentally disabled individuals (Reed & Reed, 1965). This family study showed that mild mental disability is very strongly familial. If one parent is mildly disabled, the risk for cognitive disability in the children is about 20 percent. If both parents are mildly disabled, the risk is nearly 50 percent.

Although mild cognitive disability runs in families, it could do so for reasons of nurture rather than nature. Twin and adoption studies of mild cognitive disability are needed to disentangle the relative roles of nature and nurture. Twin studies of large unselected samples of twins have been used to investigate the origins of low IQ in infancy (Petrill et al., 1997), in childhood (Spinath, Harlaar, Ronald, & Plomin, 2004), and in

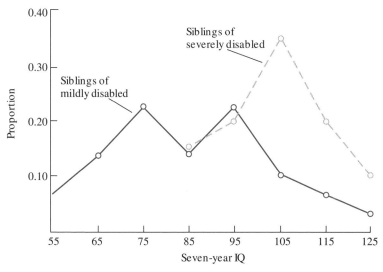

• **FIGURE 11.1** Siblings of children with mild cognitive disability tend to have lower than average IQs. In contrast, siblings of severely disabled children tend to have normal IQs. These trends suggest that mild disability is familial but severe disability is not. (From Nichols, 1984.)

adults (Saudino, Plomin, Pedersen, & McClearn, 1994). These studies found that low IQ is at least as heritable as IQ in the normal range, suggesting that heritable factors might contribute to the familial resemblance found for mild cognitive disability.

General Cognitive Disability: Single-Gene Disorders

The classic single-gene cause of severe cognitive disability is PKU, discussed in Chapter 2; a newer discovery is fragile X, mentioned in Chapter 3. We will first discuss these two single-gene disorders, which are known for their effect on cognitive disability, as well as Rett syndrome, a common cause of cognitive disability in females.

Until recently, much of what was known about these disorders, as well as the chromosomal disorders described in the next section, came from studies of patients in institutions. These earlier studies painted a gloomy picture. But more recent systematic surveys of entire populations show a wide range of individual differences, including individuals whose cognitive functioning is in the normal range. These genetic disorders shift the IQ distribution downward, but a wide range of individual differences remains.

Phenylketonuria

The most well known inherited form of moderate cognitive disability is phenylketonuria (PKU), which occurs in about 1 in 10,000 births, although its frequency varies widely from a high of 1 in 5000 in Ireland to a low of 1 in 100,000 in Finland. In the untreated condition, IQ scores are often below 50, although the range includes some near-normal IQs. As mentioned in Chapter 2, PKU is a single-gene recessive disorder that previously accounted for about 1 percent of mildly disabled individuals in institutions. PKU is the best example of the usefulness of finding genes related to behavior. Knowledge that PKU is caused by a single gene led to an understanding of how the genetic defect causes cognitive disability. Mutations in the gene (*PAH*) that produces the enzyme phenylalanine hydroxylase lead to an enzyme that does not work properly, that is, one that is less efficient in breaking down phenylalanine. Phenylalanine comes from food, especially red meats; if it cannot be broken down properly, it builds up and damages the developing brain.

Although PKU is inherited as a simple single-gene recessive disorder, the molecular genetics of PKU is not so simple (Scriver & Waters, 1999). The *PAH* gene, which is on chromosome 12, shows more than 500 different disease-causing mutations, some of which cause milder forms of cognitive disability (Mitchell, Trakadis, & Scriver, 2011). Similar findings have emerged for many classic single-gene disorders. Different mutations can do different things to the gene's product, and this variability makes understanding the disease process more difficult. It also makes DNA diagnosis more difficult, although DNA sequencing can identify any mutation. A mouse model of a mutation in the *PAH* gene shows similar phenotypic effects and has been widely used to investigate effects on brain and behavioral development (Martynyuk, van Spronsen, & Van der Zee, 2010).

To allay fears about how genetic information will be used, it is important to note that knowledge about the single-gene cause of PKU did not lead to sterilization programs or genetic engineering. Instead, an environmental intervention—a diet low in phenylalanine—successfully prevented the development of cognitive disability. Widespread screening at birth for this genetic effect began in 1961, a program demonstrating that genetic screening can be accepted when a relatively simple intervention is available (Guthrie, 1996). However, despite screening and intervention, PKU individuals still tend to have a slightly lower IQ, especially when the low phenylalanine diet has not been strictly followed (Brumm & Grant, 2010). It is generally recommended that the diet be maintained as long as possible, at least through adolescence. PKU women must return to a strict low-phenylalanine diet before becoming pregnant to prevent their high levels of phenylalanine from damaging the fetus (Mitchell et al., 2011).

Fragile X Syndrome

As mentioned in Chapters 1 and 3, fragile X syndrome is the second most common cause of cognitive disability after Down syndrome and the most common inherited form. It is twice as common in males as in females. The frequency of fragile X is usually given as 1 in 5000 males and 1 in 10,000 females (Rooms & Kooy, 2011). At least 2 percent of the male residents of schools for cognitively disabled persons have fragile X syndrome. Most fragile X males are moderately disabled, but many are only mildly disabled and some have normal intelligence. Only about one-half of girls with fragile X are affected because one of the two X chromosomes for girls inactivates, as mentioned in Chapter 4. Although fragile X syndrome is a major source of the greater incidence of cognitive disability in boys, more than 90 other genes on the X chromosome have been implicated in cognitive disability (Gecz, Shoubridge, & Corbett, 2009).

For fragile X males, IQ declines after childhood. In addition to lowered IQ, about three-quarters of fragile X males show large, often protruding, ears and a long face with a prominent jaw. They also often show unusual behaviors such as odd speech, poor eye contact (gaze aversion), and flapping movements of the hands. Language difficulties range from an absence of speech to mild communication difficulties. Often observed is a speech pattern called "cluttering," in which talk is fast, with occasional garbled, repetitive, and disorganized speech. Spatial ability tends to be affected more than verbal ability. Comprehension of language is often better than expression and better than expected on the basis of an average IQ of about 70. Parents frequently report overactivity, impulsivity, and inattention.

Until the gene for fragile X was found in 1991, the disorder's inheritance was puzzling (Verkerk et al., 1991). It did not conform to a simple X-linkage pattern because its risk increased across generations. Fragile X syndrome is caused by an expanded triplet repeat (CGG) on the X chromosome (Xq27.3). The disorder is called fragile X because the many repeats cause the chromosome to be fragile at that point and to break during laboratory preparation of chromosomes. The disorder is now diagnosed on the basis of DNA sequence. As mentioned in Chapter 3, parents who inherit X chromosomes with a normal number of repeats (6 to 40 repeats) can produce eggs or sperm with an expanded number of repeats (up to 200 repeats), called a *premutation*. This premutation does not cause cognitive disability in their offspring, but it is unstable and often leads to much greater expansions (more than 200 repeats) in later generations, especially when the premutated X chromosome is inherited through the mother. The risk that a premutation will expand to a full mutation increases over four generations from 5 to 50 percent, although it is not yet possible to predict when a premutation will expand to a full mutation. The mechanism by which expansion occurs is not known. The full mutation causes fragile X in almost all males but in only half of the females. Females are mosaics for fragile X in the sense that one X chromosome is inactivated, so some cells will have the full mutation and others will be normal (Willemsen, Levenga, & Oostra, 2011). As a result, females with the full mutation have much more variable symptoms. The triplet repeat is in an untranslated region at the beginning of a gene (*fragile X mental retardation-1, FMR1*) that, when expanded to a full mutation, prevents that gene from being transcribed. The mechanism by which the full mutation prevents transcription is *DNA methylation,* a developmental mechanism for genetic regulation, as discussed in Chapter 10. DNA methylation prevents transcription by binding a methyl group to DNA, usually at CG repeat sites. The full mutation for fragile X, with its hundreds of CGG repeats, causes hypermethylation and thus shuts down transcription of the *FMR1* gene. The gene's protein product (FMRP) binds RNA, which means that the gene product regulates expression of other genes. FMRP facilitates translation of hundreds of neuronal RNAs; thus, the absence of FMRP causes diverse problems. Research on fragile X is moving rapidly from molecular genetics to neurobiology. Researchers hope that, once the functions of FMRP are understood, it can be artificially supplied. In addition, methods for identifying carriers of premutations have improved; these screening tests will help people carrying premutations to avoid producing children who have a larger expansion and therefore suffer from fragile X syndrome (Rooms & Kooy, 2011).

Rett Syndrome

Rett syndrome is the most common single-gene cause of general cognitive disability in females (1 in 10,000) (Neul et al., 2010). The disorder shows few effects in infancy, although the head, hands, and feet are slow to grow. Cognitive development is normal during infancy but, by school age, girls with Rett syndrome are generally unable to talk and about half are unable to walk, with an average IQ of about 55 (Neul et al., 2010). Women with Rett syndrome seldom live beyond age 60, and are prone to seizures and gastrointestinal disorders. This single-gene disorder was mapped to the long arm of the X chromosome (Xq28) and then to a specific gene (*MECP2*, which encodes methyl-CpG-binding protein-2) (Amir et al., 1999). *MECP2* is a gene involved in the methylation process that silences other genes during development and thus has diffuse effects throughout the brain (Bienvenu & Chelly, 2006; Samaco & Neul, 2011). The effects are variable in females because of random X-chromosome inactivation in females (see Chapter 4). Males with *MECP2* mutations usually die before or shortly after birth.

Other Single-Gene Disorders

The average IQ scores of individuals with the most common single-gene causes of general cognitive disability are summarized in Figure 11.2. It should be remembered, however, that the range of cognitive functioning is very wide for these disorders. The defective allele shifts the IQ distribution downward, but a wide range of individual IQs remains. More than 250 other single-gene disorders, whose primary defect is something other than cognitive disability, also show effects on IQ (Inlow & Restifo, 2004; Raymond, 2010). Three of the most common disorders are Duchenne muscular dystrophy, Lesch-Nyhan syndrome, and neurofibromatosis. Duchenne muscular dystrophy is a disorder of muscle tissue caused by a recessive gene on the X chromosome that occurs in 1 in 3500 males and usually leads to death by age 20. The average IQ of males with the disorder is 85, although it is not known how the gene affects the brain (D'Angelo et al., 2011). Lesch-Nyhan syndrome is another rare X-linked recessive disorder, with an incidence of about 1 in 20,000 male births; many medical problems occur that lead to death before age 30. The most striking feature of this disorder is compulsive self-injurious behavior, reported in over 85 percent of cases (Anderson & Ernst, 1994). In terms

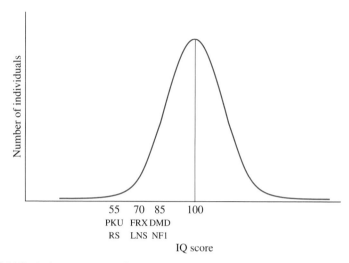

55   70   85   100
PKU   FRX DMD
RS   LNS NF1

IQ score

• **FIGURE 11.2** Single-gene causes of general cognitive disability: phenylketonuria (PKU), Rett syndrome (RS), fragile X syndrome (FRX), Lesch-Nyhan syndrome (LNS), Duchenne muscular dystrophy (DMD), and neurofibromatosis type 1 (NF1). Despite the lower average IQs, a wide range of cognitive functioning is found.

of cognitive disability, most individuals have moderate or severe learning difficulties, and speech is usually impaired, although memory for both recent and past events appears to be unaffected. Neurofibromatosis type 1 is caused by a single dominant allele that is surprisingly common (about 1 in 3000 births) for a dominant allele, which may be related to the fact that most individuals with neurofibromatosis survive until middle age, after the reproductive years. Although the disorder is known for skin tumors and tumors in nerve tissue, the majority of affected individuals also have learning difficulties, with an average IQ of about 85 (Shilyansky, Lee, & Silva, 2010).

Many cases of severe cognitive disability are not familial, as suggested by the sibling study noted earlier (Nichols, 1984). Nonetheless, recent DNA sequencing research is discovering many new noninherited (called *de novo*) dominant mutations responsible for such sporadic cases (Topper, Ober, & Das, 2011). For example, in a sequencing study of 10 children with severe cognitive disability whose parents were unaffected, likely causal mutations were identified for six of the children in six different genes (Vissers et al., 2010). Although more research is needed to prove the causal role of these mutations, sequencing promises to be a powerful approach for identifying de novo mutations for the large number of sporadic cases of severe cognitive ability.

There are hundreds of such rare single-gene disorders; however, together they account for only a small portion of cognitive disability. Most cognitive disability is mild; it represents the low end of the normal distribution of general cognitive ability and is caused by many genes of small effect as well as multiple environmental factors, as discussed in the next chapter.

General Cognitive Disability: Chromosomal Abnormalities

DNA not only affects general cognitive ability at the level of single genes, as described in the previous section. It also has effects at the level of the addition or deletion of an entire chromosome and everything in between, including insertions and deletions of large and small parts of chromosomes. The visual analysis of chromosomes themselves is being replaced by DNA sequencing, which can detect insertions and deletions down to the level of a single nucleotide (Ostrer, 2011). In general, insertions and deletions of DNA, big or small, are detrimental to cognitive development.

This section on chromosomal abnormalities begins with descriptions of the classic whole-chromosome abnormalities that affect cognitive development: Down syndrome and chromosomal abnormalities involving the X chromosome. Chromosomes and chromosomal abnormalities, such as nondisjunction, which causes Down syndrome, and the special case of abnormalities involving the X chromosome, were introduced in Chapters 3 and 4.

Down Syndrome

As described in Chapter 3, Down syndrome is caused by a trisomy of chromosome 21 (Roizen & Patterson, 2003). It was one of the first identified genetic disorders, and its 150-year history parallels the history of genetic research (Patterson & Costa, 2005). It is the single most important cause of general cognitive disability and occurs in about 1 in 1000 births. It is so common that its general features are probably familiar to everyone (Figure 11.3). Although more than 300 abnormal features have been reported for Down syndrome children, a handful of specific physical disorders are diagnostic because they occur so frequently. These features include increased neck tissue, muscle weakness, speckled iris of the eye, open mouth, and protruding tongue. Some symptoms, such as increased neck tissue, become less prominent as the child grows. Other symptoms, such as cognitive disability and short stature, are noted only as the child grows. About two-thirds of affected individuals have hearing deficits, and one-third have heart defects, leading to an average life span of 50 years. As first noted by Langdon Down, who identified the disorder in 1866, children with Down syndrome appear to be obstinate but otherwise generally amiable.

The most striking feature of Down syndrome is general cognitive disability (Lott & Dierssen, 2010). As is the case for all single-gene and chromosomal effects on general cognitive ability, affected individuals show a wide range of IQs. The average IQ among children with Down syndrome is 55, with only the top 10 percent falling within the lower end of the normal range of IQs. By adolescence, language skills are generally at about the level of a 3-year-old child. Most individuals with Down syndrome who reach the age of 45 suffer from the cognitive decline of dementia, which was an early clue suggesting that a gene related to dementia might be on chromosome 21 (see later).

• **FIGURE 11.3** Three-year-old girl with Down syndrome.

In Chapter 3, Down syndrome was used as an example of an exception to Mendel's laws because it does not run in families. Because individuals with Down syndrome do not reproduce, most cases are created anew each generation by non-disjunction of chromosome 21, which is analogous to de novo mutations that are not inherited. Another important feature of Down syndrome is that it occurs much more often in women giving birth later in life, for reasons explained in Chapter 3.

Advances in genetics have stimulated a resurgence of research on Down syndrome with the hope of ameliorating at least some of its symptoms (Lana-Elola, Watson-Scales, Fisher, & Tybulewicz, 2011). The fundamental problem is that because there are three copies of chromosome 21, its several hundred genes are overexpressed. Mouse models have played an important role in understanding and improving cognitive deficits in Down syndrome (Das & Reeves, 2011).

Sex Chromosome Abnormalities

Extra X chromosomes also cause cognitive disabilities, although the effect is highly variable, which is the reason why many cases remain undiagnosed (Lanfranco, Kamischke, Zitzmann, & Nieschlag, 2004). In males, an extra X chromosome causes *XXY male syndrome,* often called Klinefelter syndrome. As indicated in Chapter 4, even though X is a large chromosome with many genes, extra X chromosomes are largely inactivated, as happens with normal females, who have two X chromosomes; however, some genes on the extra X chromosome escape inactivation in XXY males (Tuttelmann & Gromoll, 2010). XXY male syndrome is the most common chromosomal abnormality in males, occurring in about 1 in 500 male births. The major problems involve low testosterone levels after adolescence, leading to infertility, small testes, and breast development. Early detection and hormonal therapy are important to alleviate the condition, although infertility remains (Simm & Zacharin, 2006). Males with XXY male syndrome also have a somewhat lower than average IQ; most have speech and language problems as well as poor school performance (Mandoki, Sumner, Hoffman, & Riconda, 1991).

In females, extra X chromosomes (called *triple X syndrome*) cause the most common whole-chromosome abnormality, occurring in about 1 in 1000 female births. Females with triple X show an average IQ of about 85, lower than for XXY males (Tartaglia, Howell, Sutherland, Wilson, & Wilson, 2010). Unlike XXY males, XXX females have normal sexual development and are able to conceive children; they have so few problems that they are rarely detected clinically. Their scores on verbal tests (such as on vocabulary) are lower than their scores on nonverbal tests (such as puzzles), and many require speech therapy (Bishop et al., 2011). For both XXY and XXX individuals, head circumference at birth is smaller than average, a feature suggesting that the cognitive deficits may be prenatal in origin. As is generally the case for chromosomal abnormalities, structural brain imaging research indicates diffuse effects (Giedd et al., 2007).

In addition to having an extra X chromosome, it is possible for males to have an extra Y chromosome (XYY) and for females to have just one X chromosome (XO, called *Turner syndrome*). There is no equivalent syndrome of males with a Y chromosome but no X because this is fatal. XYY males, about 1 in 1000 male births, are taller than average after adolescence and have normal sexual development. More than 95 percent of XYY males do not even know they have an extra Y chromosome. Although XYY males have fewer cognitive problems than XXY males, about half have speech difficulties as well as language and reading problems (Leggett, Jacobs, Nation, Scerif, & Bishop, 2010). Their average IQ is about 10 points lower than that of their siblings with normal sex chromosomes. Juvenile delinquency is also associated with XYY. The XYY syndrome was the center of a furor in the 1970s, when it was suggested that such males are more violent, a suggestion possibly triggered by the notion of a "super male" with exaggerated masculine characteristics caused by an extra Y chromosome; however, this idea is not supported by research.

Turner syndrome females (XO) occur in about 1 in 2500 female births, although 98 percent of XO fetuses miscarry, accounting for 10 percent of the total number of spontaneous abortions. The main problems are short stature and abnormal sexual development; infertility is common. Puberty rarely occurs without hormone therapy; even with therapy, the individual is infertile because she does not ovulate.

Hormonal treatment is now standard, and many XO women have conceived with in vitro fertilization (Stratakis & Rennert, 2005). Although verbal IQ is about normal, nonverbal IQ is lower, about 90, and social cognition is also impaired (Hong, Dunkin, & Reiss, 2011).

Small chromosomal deletions As noted earlier, chromosomal abnormalities do not just involve a whole chromosome. Three classic small chromosomal deletions that affect cognitive development are Angelman syndrome, Prader-Willi syndrome, and Williams syndrome. After describing these disorders, we will turn to research that uses new DNA techniques to identify even smaller deletions.

A small deletion in chromosome 15 ($15q11$), mentioned in Chapter 3 as an example of genomic imprinting, causes Angelman syndrome (1 in 25,000 births) if the deletion comes from the mother's egg or Prader-Willi syndrome (1 in 15,000 births) if it comes from the father's sperm. In most cases, the deletion occurs spontaneously in the formation of gametes, although in about 10 percent of the cases mutations inherited by the mother or father are responsible (Williams, Driscoll, & Dagli, 2010). This region of chromosome 15, usually millions of base pairs in length, contains several imprinted genes that are differentially silenced by epigenetic methylation of the DNA, depending on whether the deletion comes from the mother's egg or the father's sperm. Angelman syndrome (AS) results in moderate cognitive disability, abnormal gait, speech impairment, seizures, and an inappropriately happy demeanor that includes frequent laughing and excitability. When inherited from the father, the same chromosomal deletion causes Prader-Willi syndrome (PWS), which most noticeably involves overeating and temper outbursts but also leads to multiple learning difficulties and an IQ in the low normal range. New techniques for understanding epigenetic processes are advancing our understanding of how this deletion has its effects on brain development (Mabb, Judson, Zylka, & Philpot, 2011).

Williams syndrome, with an incidence of about 1 in 10,000 births, is caused by a small deletion from chromosome 7 ($7q11.2$), a region that includes about 25 genes. Most cases are spontaneous. Williams syndrome involves disorders of connective tissue that lead to growth retardation and multiple medical problems. General cognitive disability is common (average IQ of 55), and most affected individuals have learning difficulties that require special schooling. Some studies find that language development is less affected than nonverbal abilities (Martens, Wilson, & Reutens, 2008). As adults, most affected individuals are unable to live independently. As is typical of chromosomal abnormalities that include several genes, no consistent brain pathology is found other than a reduction in cerebral volume.

Figure 11.4 summarizes the average effect on IQ of the most common chromosomal causes of general cognitive disability. Again, it should be emphasized that there is a wide range of cognitive functioning around the average IQ scores shown in the figure. In addition to these classic syndromes, new sequencing and microarray research has revealed that as many as 15 percent of cases of severe cognitive disability

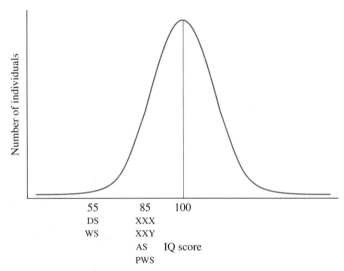

• **FIGURE 11.4** The most common chromosomal causes of general cognitive disability are Down syndrome (DS) and the sex chromosomal abnormalities XXX and XXY. The average IQs of individuals with XYY and XO are only slightly lower than normal and thus are not listed. Deletions of very small parts of chromosomes contribute importantly to general cognitive disability, but most are rare, such as Angelman syndrome (AS), Prader-Willi syndrome (PWS), and Williams syndrome (WS). For all these chromosomal abnormalities, a wide range of cognitive functioning is found.

may be due to smaller deletions or duplications from a thousand to millions of base pairs that can involve a few genes, dozens of genes, or no genes at all (Topper et al., 2011). As mentioned in Chapter 9, these structural variations in chromosomes are called *copy number variants (CNVs)*. Most CNVs arise de novo during meiosis when a DNA segment is deleted on one chromosome and duplicated on the corresponding member of the chromosome pair. As with other chromosomal abnormalities, deletions are generally worse than duplications. There are tens of thousands of CNVs; we all have CNVs peppered throughout our genome without obvious effect, despite all the extra or missing segments of DNA. However, some CNVs, usually rare and de novo (i.e., not seen in either parent), affect neurocognitive development (Morrow, 2010). It appears that unlike inherited single-gene disorders, specific CNVs may not be as important as how many CNVs an individual has.

Specific Cognitive Disabilities

As its name implies, general cognitive disability has general effects on the ability to learn, which is reflected in difficulties at school. We use the term *specific cognitive disabilities* in relation to school-related difficulties such as those affecting reading, communication, and mathematics. Behavioral genetic research brings genetics to the field of educational psychology, which has been slow to recognize the importance of genetic influence (Haworth & Plomin, 2011; Wooldridge, 1994), even though teachers in the classroom do (Walker & Plomin, 2005). This section focuses on low performance in cognitive processes related to academic achievement, whereas Chapter 13 focuses on normal variation in these processes. We begin with reading disability because reading is the primary problem for about 80 percent of children with a diagnosed learning disorder. We then consider communication disorders, mathematics disability, and, finally, the interrelationships of learning disabilities.

Reading Disability

As many as 10 percent of children have difficulty learning to read. Children with reading disability (also known as *dyslexia*) read slowly and often with poor comprehension. When reading aloud, they perform poorly. For some, specific causes can be identified, such as cognitive disability, brain damage, sensory problems, and deprivation. However, many children without such problems find it difficult to read.

Family studies have shown that reading disability runs in families. The largest family study included 1044 individuals in 125 families with a reading-disabled child and 125 matched control families (DeFries, Vogler, & LaBuda, 1986). Siblings and parents of the reading-disabled children performed significantly worse on reading tests than did siblings and parents of control children. The first major twin study indicated that familial resemblance for reading disability involves genetic factors (DeFries, Knopik, & Wadsworth, 1999). For more than 250 twin pairs in which at least one member of the pair was reading disabled, twin concordances were 66 percent for identical twins and 36 percent for fraternal twins, a result suggesting substantial genetic influence. Large twin studies found similar results in the early school years for both reading disability and reading ability in the United Kingdom (Kovas, Haworth, Dale, & Plomin, 2007) and the United States (Hensler, Schatschneider, Taylor, & Wagner, 2010). In all of these studies, shared environmental influence is modest, typically accounting for less than 20 percent of the variance (Willcutt, Pennington, et al., 2010).

As part of DeFries and colleagues' twin study, a new method was developed to estimate the genetic contribution to the mean difference between the reading-disabled probands and the mean reading ability of the population. This type of analysis, called *DF extremes analysis* after its creators (DeFries & Fulker, 1985), is described in Box 11.1. In a meta-analysis of studies of reading disability, DF extremes analysis for reading disability estimates that about 60 percent of the mean difference between the probands and the population is heritable (Plomin & Kovas, 2005). The analysis also suggests genetic links between reading disability and normal variation in reading ability.

As described earlier in this chapter, moderate to severe general cognitive disability is often caused by single-gene mutations and chromosomal abnormalities that do not contribute importantly to variation in the normal range of cognitive ability. In

BOX 11.1 • DF Extremes Analysis

The genetic and environmental causes of individual differences throughout the range of variability in a population can differ from the causes of the average difference between an extreme group and the rest of the population. For example, finding genetic influence on individual differences in reading ability in an unselected sample (Chapter 13) does not mean that the average difference in reading ability between reading-disabled individuals and the rest of the population is also influenced by genetic factors. Alternatively, it is possible that reading disability represents the extreme end of a continuum of reading ability, rather than a distinct disorder. That is, reading disability might be quantitatively rather than qualitatively different from the normal range of reading ability. DF extremes analysis, named after its creators (DeFries & Fulker, 1985, 1988) addresses these important issues concerning the links between the normal and abnormal.

DF extremes analysis takes advantage of the quantitative scores of the relatives of probands rather than just assigning a dichotomous diagnosis to the relatives and assessing concordance for the disorder. The figure below shows hypothetical distributions of reading performance of an unselected sample

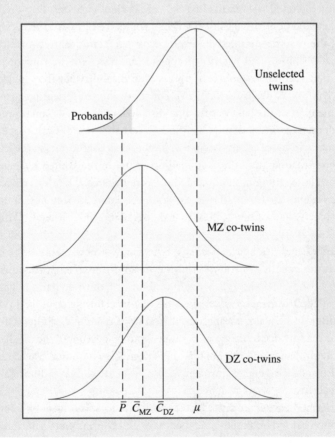

of twins and of the identical (MZ) and fraternal (DZ) co-twins of probands (P) with reading disability (DeFries, Fulker, & LaBuda, 1987). The mean score of the probands is $\bar{P}$. The differential regression of the MZ and the DZ co-twin means ($\bar{C}_{MZ}$ and $\bar{C}_{DZ}$) toward the mean of the unselected population (μ) provides a test of genetic influence. That is, to the extent that reading deficits of probands are heritable, the quantitative reading scores of identical co-twins will be more similar to those of the probands than will the scores of fraternal twins. In other words, the mean reading scores of identical co-twins will regress less far back toward the population mean than will those of fraternal co-twins.

The results for reading disability are similar to those illustrated in the figure. The scores of the identical co-twins regress less far back toward the population mean than do those of the fraternal co-twins. This finding suggests that genetics contributes to the mean difference between the reading-disabled probands and the population. Twin group correlations provide an index of how far the co-twins regress toward the population mean (Plomin, 1991). For reading disability, the twin group correlations are 0.90 for identical twins and 0.65 for fraternal twins. Doubling the difference between these group correlations suggests a group heritability of 50 percent, similar to the results of more sophisticated DF extremes analysis (DeFries & Gillis, 1993; Willcutt, Pennington, et al., 2010). In other words, half of the mean difference between the probands and the population is heritable. This is called "group heritability" to distinguish it from the usual heritability estimate, which refers to differences between individuals rather than to mean differences between groups.

DF extremes analysis is conceptually similar to the liability-threshold model described in Box 3.1. The major difference is that the threshold model assumes a continuous dimension even though it assesses a dichotomous disorder. The liability-threshold analysis converts dichotomous diagnostic data to a hypothetical construct of a threshold with an underlying continuous liability. In contrast, DF extremes analysis assesses rather than assumes a continuum. If all the assumptions of the liability-threshold model are correct for a particular disorder, it will yield results similar to the DF extremes analysis to the extent that the quantitative dimension assessed underlies the qualitative disorder. In the case of reading disability, a liability-threshold analysis of these twin data yields an estimate of group heritability similar to that of the DF extremes analysis (Plomin & Kovas, 2005).

In addition, DF extremes analysis can be used to examine the genetic and environmental origins of the co-occurrence between disorders. For example, language and mathematics problems are often found among reading-disabled children. Multivariate DF extremes analysis suggests that genetic factors are largely responsible for this overlap (Haworth, Kovas, et al., 2009). Genetic overlap is also substantial between reading disability and hyperactivity (Willcutt, Betjemann, et al., 2010). Multivariate DF extremes analysis has also been used to discover that genetic factors account for most of the high stability of reading disability from age 10 to age 15 (Astrom, Wadsworth, Olson, Willcutt, & DeFries, 2011).

contrast, mild cognitive disability appears to be quantitatively, not qualitatively, different from normal variation in cognitive ability. That is, mild cognitive disability is the low end of the same genetic and environmental influences responsible for variation in the normal distribution of cognitive ability. Results for reading disability and other common disorders are similar to those for mild cognitive disability rather than more severe cognitive disability. Phrased more provocatively, these findings from DF extremes analysis suggest that common disorders such as reading disability are not really disorders—they are merely the low end of the normal distribution (Plomin et al., 2009). This view fits with the quantitative trait locus (QTL) hypothesis, which assumes that genetic influence is due to many genes of small effect size that contribute to a normal quantitative trait distribution. What we call disorders and disabilities are the low end of these quantitative trait distributions. The QTL hypothesis predicts that when genes associated with reading disability are identified, the same genes will be associated with normal variation in reading ability.

Early molecular genetic research on reading disability assumed that the target was a single major gene rather than QTLs. Various modes of transmission have been proposed, especially autosomal dominant transmission and X-linked recessive transmission. The autosomal dominant hypothesis takes into account the high rate of familial resemblance but fails to account for the fact that about a fifth of reading-disabled individuals do not have affected relatives. An X-linked recessive hypothesis is suggested when a disorder occurs more often in males than in females, as is the case for reading disability. However, the X-linked recessive hypothesis does not work well as an explanation of reading disability. As described in Chapter 3, one of the hallmarks of X-linked recessive transmission is the absence of father-to-son transmission, since sons inherit their X chromosome only from their mother. Contrary to the X-linked recessive hypothesis, reading disability is transmitted from father to son as often as from mother to son. It is now generally accepted that, like most complex disorders, reading disability is caused by multiple genes as well as by multiple environmental factors (Fisher & DeFries, 2002).

One of the most exciting findings in behavioral genetics in the past two decades is that the first quantitative trait locus for a human behavioral disorder was reported for reading disability, using sib-pair ***QTL linkage analysis*** (Cardon et al., 1994). As explained in Chapter 9, siblings can share zero, one, or two alleles for a particular DNA marker. If siblings who share more alleles are also more similar for a quantitative trait such as reading ability, then QTL linkage is likely. QTL linkage analysis is much more powerful when one sibling is selected because of an extreme score on the quantitative trait. When one sibling was selected for reading disability, the reading ability score of the co-sibling was also lower when the two siblings shared alleles for markers on the short arm of chromosome 6 (6p21). These QTL linkage results for four DNA markers in this region are depicted by the dotted line in Figure 11.5, showing significant linkage for the D6S105 marker. Significant linkage was also found for markers in this region in an independent sample of fraternal twins (see

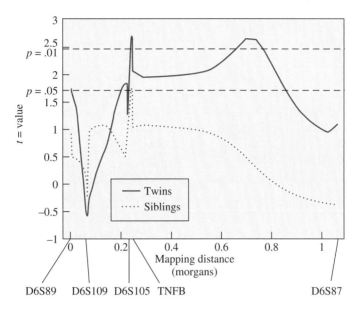

• **FIGURE 11.5** QTL linkage for reading disability in two independent samples in which at least one member of the pair is reading disabled: siblings (dotted line) and fraternal twins (solid line). D6S89, D6S109, D6S105, and TNFB are DNA markers in the 6*p*21 region of chromosome 6. The *t*-values are an index of statistical significance. The marker D6S105 is significant at the $p = 0.05$ level for siblings and at the $p = 0.01$ level for fraternal twins. (After Cardon et al., 1994; modified from DeFries & Alarcón, 1996; courtesy of Javier Gayán.)

solid line in Figure 11.5) and in several replication studies in the broader region of the short arm of chromosome 6 (Fisher & DeFries, 2002). Despite these consistent linkage results, it has been difficult to identify the specific genes responsible for the QTL linkage among the hundreds of genes in this gene-rich region of chromosome 6, but the search has narrowed to two genes very close together at 6*p*22: *KIAA0319* and *DCDC2* (Scerri et al., 2011). Genes in this region and other candidate genes reported to be associated with reading disability provide plausible pathways among genes, brain, and behavior that involve the growth and migration of neurons (Poelmans, Buitelaar, Pauls, & Franke, 2011). Other genomewide linkage analyses have proposed eight other locations in the genome linked to reading disability, although some await independent replication (Willcutt, Pennington, et al., 2010). There have been far fewer candidate gene association studies for reading than for other behaviors, perhaps because there are no obvious candidate genes and also because linkage analysis has dominated reading research. The first genomewide association study of reading disability found no genomewide significant associations, and the largest effect sizes were very small, accounting for less than 0.5% of variance of reading in an unselected population (Meaburn, Harlaar, Craig, Schalkwyk, & Plomin, 2008). The discrepancy between linkage and association results may be that linkage is able to detect multiple causal variants that are closely linked. This suggests what will be a theme for molecular genetic studies: Many genes of small effect are responsible for heritability for complex traits.

Reading disability is generally assumed to be caused by language problems (Hensler et al., 2010); genetic influences on reading disability and on language and speech disorders overlap substantially (Haworth, Kovas, et al., 2009; Pennington & Bishop, 2009). Language and speech disorders are the topic of the following section.

Communication Disorders

DSM-IV includes four types of communication disorders: expressive language (putting thoughts into words) disorder, mixed receptive (understanding the language of others) and expressive language disorder, phonological (articulation) disorder, and stuttering (speech interrupted by prolonged or repeated words, syllables, or sounds). Hearing loss, cognitive disability, and neurological disorders are excluded.

Several family studies, examining communication disorders broadly, indicate that communication disorders are familial (Stromswold, 2001). For children with communication disorders, about a quarter of their first-degree relatives report similar disorders; these communication disorders appear in about 5 percent of the relatives of controls (Felsenfeld, 1994). Twin studies suggest that this familial resemblance is genetic in origin. A review of twin studies of language disability yields twin concordances of 75 percent for MZ twins and 43 percent for DZ twins (Stromswold, 2001). Using DF extremes analysis, the average weighted group heritability was 43 percent for language disabilities (Plomin & Kovas, 2005). A large twin study of language delay in infancy found high heritability, even at 2 years of age (Dale et al., 1998). The only adoption study of communication disorders confirms the twin results, suggesting substantial genetic influence (Felsenfeld & Plomin, 1997).

The high heritability of communication disorders has attracted attention from molecular genetics (Smith et al., 2010). A high-profile paper reported a mutation in a gene (*FOXP2*) that accounted for an unusual type of speech-language impairment that includes deficits in oro-facial motor control in one family (Lai, Fisher, Hurst, Vargha-Khadem, & Monaco, 2001). In the media, this finding was unfortunately trumpeted as "the" gene for language, whereas in fact the mutation has not been found outside the original family (Meaburn, Dale, Craig, & Plomin, 2002; Newbury et al., 2002). Several linkages and candidate gene associations have been reported with communication disorders (Kang & Drayna, 2011).

Stuttering affects about 5 percent of preschool children, but most make a full recovery. Family studies of stuttering over the past 50 years have shown that about a third of stutterers have other stutterers in their families (Kidd, 1983). Twin studies indicate that stuttering is highly heritable (Fagnani, Fibiger, Skytthe, & Hjelmborg, 2011), especially stuttering that persists past early childhood (Dworzynski, Remington, Rijsdijk, Howell, & Plomin, 2007). Results from genomewide linkage studies have not yielded consistent results (Fisher, 2010).

Mathematics Disability

For poor performance on tests of mathematics, the first twin study suggested moderate genetic influence (Alarcón, DeFries, Light, & Pennington, 1997). A study of 7-year-olds using U.K. National Curriculum scores for mathematics reported concordances of about 70 percent for MZ twins and 50 percent for DZ twins (Oliver et al., 2004). Using DF extremes analysis, the average weighted group heritability was 0.61 for twin studies of mathematics disability (Plomin & Kovas, 2005). A more recent twin study using Internet-administered tests of mathematics to select low-performing 10-year-old twins reported a group heritability of 0.47 for low performance in mathematics (Kovas, Haworth, Petrill, & Plomin, 2007). The first genomewide association study of mathematics disability found the usual result of many genes of small effect (Docherty, Davis, et al., 2010).

Comorbidity among Specific Cognitive Disabilities

Learning disabilities are distinguished from cognitive disability because they focus on what is thought to be specific disabilities as distinct from general cognitive disability. Nonetheless, two multivariate genetic analyses suggest that there is substantial genetic overlap between reading and mathematics disabilities (Knopik, Alarcón, & DeFries, 1997; Kovas, Haworth, Harlaar, et al., 2007). Extending DF extremes analysis to bivariate analysis, genetic correlations of 0.53 and 0.67 between reading and mathematics disability were reported. In other words, many of the genes that affect reading disability also affect mathematics disability. The reach of these general effects of genes for cognitive disabilities extends beyond reading and mathematics disability to communication disorders and general cognitive disability (Haworth, Kovas, et al., 2009). Molecular genetic research is beginning to confirm these quantitative genetic results by showing that genes associated with one disability are associated with other disabilities (Docherty, Kovas, Petrill, & Plomin, 2010). Multivariate genetic research has been central to analyses of cognitive abilities; this research also suggests substantial genetic overlap among diverse cognitive abilities, as discussed in Chapter 13.

Dementia

Although aging is a highly variable process, as many as a quarter of individuals over 85 years of age suffer severe cognitive decline known as dementia (Bird, 2008). Prior to age 65, the incidence is less than 1 percent. Among the elderly, dementia accounts for more days of hospitalization than any other psychiatric disorder (Cumings & Benson, 1992). It is the fourth leading cause of death in adults. The number of diagnosed dementia patients is projected to nearly double every 20 years (Alzheimer's Disease International, 2009).

At least half of all cases of dementia involve Alzheimer disease (AD), which has been studied for more than a century (Goedert & Spillantini, 2006). AD occurs very gradually over many years, beginning with loss of memory for recent events. This mild memory loss affects many older individuals but is much more severe in individuals with AD. Irritability and difficulty in concentrating are also often noted. Memory gradually worsens to include simple behaviors, such as forgetting to turn off the stove or bath water and wandering off and getting lost. Eventually—sometimes after 3 years, sometimes after 15 years—individuals with AD become bedridden. Biologically, AD involves extensive changes in brain nerve cells, including plaques and tangles (described later) that build up and result in death of the nerve cells. Although these plaques and tangles occur to some extent in most older people, they are usually restricted to the hippocampus. In individuals with AD, they are much more numerous and widespread.

Another type of dementia is the result of the cumulative effect of multiple small strokes in which blood flow to the brain becomes blocked, thus damaging the brain. This type of dementia is called multiple-infarct dementia (MID). (An infarct is an area damaged as a result of a stroke.) Unlike AD, MID is usually more abrupt and involves focal symptoms such as loss of language rather than general cognitive decline. Co-occurrence of AD and MID is seen in about a third of all cases. DSM-IV recognizes nine other kinds of dementias, such as dementias due to AIDS, to head trauma, and to Huntington disease.

Surprisingly little is known about the quantitative genetics of either AD or MID. Family studies of AD probands estimate risk to first-degree relatives of nearly 50 percent by age 85, when the data are adjusted for age of the relatives (McGuffin, Owen, O'Donovan, Thapar, & Gottesman, 1994). Until recently, the only twin study of dementia was one reported over 50 years ago. That twin study, which did not distinguish AD and MID, found concordances of 43 percent for identical twins and 8 percent for fraternal twins, results suggesting moderate genetic influence (Kallmann & Kaplan, 1955). More recent twin studies of AD also found evidence for genetic influence, with concordances two times greater for identical than for fraternal twins in Finland (Raiha, Kapiro, Koskenvuo, Rajala, & Sourander, 1996), Norway (Bergeman, 1997), Sweden (Gatz et al., 1997), and the United States (Breitner et al., 1995). In the largest twin study to date, liability to AD yielded a heritability estimate of 0.58 (Gatz et al., 2006).

Some of the most important molecular genetic findings for behavioral disorders have come from research on dementia (Bettens, Sleegers, & Van Broeckhoven, 2010). Early research focused on a rare (1 in 10,000) type of Alzheimer disease that appears before 65 years of age and shows evidence for autosomal dominant inheritance. Three genes have been identified that contribute to this rare form of the disorder (Bekris, Yu, Bird, & Tsuang, 2010). The great majority of Alzheimer cases occur after 65 years of age, typically in persons in their seventies and eighties. A major advance toward understanding late-onset Alzheimer disease is the discovery of a strong allelic association with a gene (for apolipoprotein E) on chromosome 19 (Corder et al., 1993). This gene has three alleles (confusingly called alleles 2, 3, and 4). The frequency of allele 4 is about 40 percent in individuals with Alzheimer disease and 15 percent in control samples. This result translates to about a sixfold increased risk for late-onset Alzheimer disease for individuals who have one or two of these alleles.

Apolipoprotein E is a QTL in the sense that allele 4, although a risk factor, is neither necessary nor sufficient for developing dementia. For instance, nearly half of patients with late-onset Alzheimer disease do not have that allele. Assuming a liability-threshold model, allele 4 accounts for about 15 percent of the variance in liability (Owen, Liddle, & McGuffin, 1994). Because apolipoprotein E is known for its role in transporting lipids throughout the body, its association with late-onset AD was puzzling at first. However, the product of allele 4 binds more readily with β-amyloid, leading to amyloid deposits, which in turn lead to plaques and, eventually, to the death of nerve cells (Tanzi & Bertram, 2005). The product of allele 2 may block this buildup of β-amyloid. The product of allele 3 appears to buffer nerve cells against the other characteristic of AD, neurofibrillary tangles. Other roles for the gene product are also known, such as its increased production following injury to the nervous system, as in head injury, and, most important, its role in plaques (Hardy, 1997).

Because the gene for apolipoprotein E does not account for all the genetic influence on AD, the search is on for other QTLs. A meta-analysis of over a thousand reports of associations with over 500 candidate genes finds evidence for significant associations for more than a dozen susceptibility QTLs, although results are often inconsistent (Bertram, McQueen, Mullin, Blacker, & Tanzi, 2007). Genomewide association studies consistently confirm the association with apolipoprotein E, but more than a dozen studies yielded inconsistent results for other associations until three large-scale studies including data from 43,000 individuals provided compelling evidence for small effects of variants in four novel susceptibility genes that might lead to synaptic disintegration (Hollingworth, Harold, Jones, Owen, & Williams, 2011). More than a dozen knock-out mouse models of AD-related genes have been generated, and several of the mutants show β-amyloid deposits and plaques, although no animal model has as yet been shown to have all the expected AD effects, including the critical effects on memory (Bekris et al., 2010).

Summary

Although no twin or adoption studies have been reported for moderate or severe cognitive disability, more than 250 single-gene disorders, most extremely rare, include cognitive disability among their symptoms, and many more are being discovered with new advances in DNA analysis. A classic disorder is PKU, caused by a recessive mutation on chromosome 12. The discovery of fragile X syndrome is especially important because it is the most common cause of inherited cognitive disability (1 in several thousand males, half as common in females). It is caused by a triplet repeat (CGG) on the X chromosome that expands over several generations until it reaches more than 200 repeats, when it causes cognitive disability in males. The most common single-gene cause of severe cognitive disability in females is Rett syndrome. Other single-gene mutations known primarily for other effects also contribute to cognitive disability, such as genes for Duchenne muscular dystrophy, Lesch-Nyhan syndrome, and neurofibromatosis.

For all of the single-gene disorders, the defective allele shifts the IQ distribution downward, but a wide range of individual IQs remains. Also, although there are hundreds of such rare single-gene disorders, together they account for only a tiny portion of cognitive disability. Most cognitive disability is mild and appears to be the low end of the normal distribution of general cognitive ability and caused by many QTLs of small effect as well as multiple environmental factors. Molecular genetic research on general cognitive ability is discussed in the next chapter.

Chromosomal abnormalities play an important role in cognitive disability. The most common cause of cognitive disability is Down syndrome, caused by the presence of three copies of chromosome 21. Down syndrome occurs in about 1 in 1000 births and is responsible for about 10 percent of cognitively disabled individuals in institutions. Risk for cognitive disability is also increased by having an extra X chromosome (XXY males, XXX females). An extra Y chromosome (XYY males) or a missing X chromosome (Turner females) cause less disability. Small deletions of chromosomes can result in cognitive disability, as in Angelman syndrome, Prader-Willi syndrome, and Williams syndrome. XYY males have speech and language problems; Turner females (XO) generally perform less well on nonverbal tasks such as spatial tasks. Similar to single-gene disorders, there is a wide range of cognitive functioning around the lowered average IQ scores found for all these chromosomal causes of cognitive disability. An exciting area of research uses DNA microarrays and sequencing to detect subtle chromosomal abnormalities, especially de novo (noninherited) deletions and duplications called *copy number variants (CNVs)*, that might account for as many as 15 percent of cases of severe cognitive disability.

Twin studies suggest genetic influence for specific cognitive disabilities, including reading disability, communication disorders, and mathematics disability. For these cognitive disabilities, DF extremes analysis suggests that genetic and environmental influences have effects at the low end of the normal distribution of cognitive abilities that are similar to their effects on the rest of the distribution. In addition, multivariate genetic research indicates substantial genetic correlations among learning disabilities. For reading disability, a replicated linkage on chromosome 6 was the first QTL linkage discovered for human behavioral disorders; two genes in this region are the best candidates, although eight other linkage regions have been proposed. Several linkages and candidate gene associations have also been proposed for communication disorders. The first genomewide association study of mathematics disability found the usual result of many genes of small effect. The substantial co-morbidity between specific cognitive disabilities is largely due to genetic factors, meaning that the same genes affect different learning disabilities although there are also disability-specific genes.

For dementia, three genes have been found that account for most cases of early-onset Alzheimer disease, a rare (1 in 10,000) form of the disease that occurs before 65 years of age and often shows pedigrees consistent with autosomal dominant inheritance. Late-onset Alzheimer disease is very common, striking as many as a quarter of individuals over 85 years of age. Its heritability is about 60 percent. The gene for apolipoprotein E is associated with late-onset Alzheimer disease. Although the apolipoprotein E gene association is the largest effect size found for a behavioral disorder, it is a QTL in the sense that it is a probabilistic risk factor, not a single gene necessary and sufficient to develop the disorder. Large-scale genomewide association studies have identified novel susceptibility genes that contribute to the heritability of late-onset Alzheimer disease.

CHAPTER TWELVE

General Cognitive Ability

G eneral cognitive ability (*g*) predicts key social outcomes such as educational and occupational levels far better than any other trait (Gottfredson, 1997; Schmidt & Hunter, 2004). *g* is increasingly important in our knowledge-based society and central to society's intellectual capital (Neisser et al., 1996). *g* is also one of the most well studied domains in behavioral genetics. Nearly all this genetic research is based on a model in which cognitive abilities are organized hierarchically (Carroll, 1993, 1997), from specific tests to broad factors to general cognitive ability (Figure 12.1). There are hundreds of tests of diverse cognitive abilities. These tests measure several broad factors (specific cognitive abilities), such as verbal ability, spatial ability, memory, and speed of processing. Such tests are widely used in schools, industry, the military, and clinical practice.

These broad factors intercorrelate modestly. In general, people who do well on tests of verbal ability tend to do well on tests of spatial ability. *g*, that which is in common among these broad factors, was discovered by Charles Spearman over a century ago, about the same time that Mendel's laws of inheritance were rediscovered

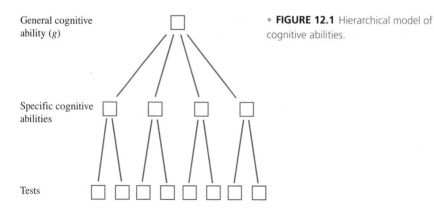

General cognitive ability (*g*)

Specific cognitive abilities

Tests

• **FIGURE 12.1** Hierarchical model of cognitive abilities.

(Spearman, 1904). The phrase *general cognitive ability* is a better choice to describe *g* than the word *intelligence* because the latter has so many different meanings in psychology and in the general language (Jensen, 1998). General texts on *g* are available (Hunt, 2011; see Deary, 2012, for an overview of other books).

Most people are familiar with intelligence tests, often called IQ tests (intelligence quotient tests). These tests typically assess several cognitive abilities and yield total scores that are reasonable indices of *g*. For example, the Wechsler tests of intelligence, widely used clinically, include ten subtests such as vocabulary, picture completion (indicating what is missing in a picture), analogies, and block design (using colored blocks to produce a design that matches a picture). In research contexts, *g* is usually derived by using a technique called *factor analysis* that weights tests differently, according to how much they contribute to *g*. This weight can be thought of as the average of a test's correlations with every other test. This is not merely a statistical abstraction—one can simply look at a matrix of correlations among such measures and see that all the tests intercorrelate positively and that some measures (such as spatial and verbal ability) intercorrelate more highly than do other measures (such as nonverbal memory tests). A test's contribution to *g* is related to the complexity of the cognitive operations it assesses. More complex cognitive processes such as abstract reasoning are better indices of *g* than less complex cognitive processes such as simple sensory discriminations.

Although *g* explains about 40 percent of the variance among such tests, most of the variance of specific tests is independent of *g*. Clearly there is more to cognition than *g*. Specific cognitive abilities assessed in the psychometric tradition are the focus of the next chapter. An important direction for research attempts to understand *g* at more basic levels, especially through information-processing theory and experimental cognitive psychology (Deary, 2000; Duncan, 2010), and, increasingly, through measures of brain structure and function (Blokland et al., 2011; Deary, Penke, & Johnson, 2010; Toga & Thompson, 2005). In addition, just as there is more to cognition than *g*, there is clearly much more to achievement than cognition. Personality, motivation, and creativity all play a part in how well someone does in life. However, it makes little sense to stretch a word like *intelligence* to include all aspects of achievement, such as emotional sensitivity (Goleman, 2005) and musical ability (Gardner, 2006), that do not correlate with tests of cognitive ability (Visser, Ashton, & Vernon, 2006).

Despite the massive amount of data pointing to the reality of *g*, considerable controversy continues to surround *g* and IQ tests, especially in the media. There is a wide gap between what laypeople (including scientists in other fields) believe and what experts believe. Most notably, laypeople often read in the popular press that the assessment of intelligence is circular—intelligence is what intelligence tests assess. On the contrary, *g* is one of the most reliable and valid measures in the behavioral domain. Its long-term stability after childhood is greater than the stability of any other behavioral trait (Deary, Whiteman, Starr, Whalley, & Fox, 2004). Although a few critics remain, *g* is widely accepted as a valuable concept by experts (Carroll, 1997). It is less clear what *g* is and whether *g* is due to a single general process, such as executive function or speed of information processing, or whether it represents a concatenation of more specific cognitive processes (Deary, 2000). The idea of a genetic contribution to *g* has produced controversy in the media, especially follow- ing the publication of *The Bell Curve* by Herrnstein and Murray (1994). In fact, these authors scarcely touched on genetics and did not view genetic evidence as crucial to their arguments. Despite this controversy, there is considerable consensus among scientists—even those who are not geneticists—that *g* is substantially heritable (Brody, 1992; Mackintosh, 1998; Neisser, 1997; Snyderman & Rothman, 1988; Stern- berg & Grigorenko, 1997). The evidence for a genetic contribution to *g* is presented in this chapter.

Historical Highlights

The relative influences of nature and nurture on *g* have been studied since the be- ginning of the behavioral sciences. Indeed, a year before the publication of Gregor Mendel's seminal paper on the laws of heredity, Francis Galton (1865) published a two-article series on high intelligence and other abilities, which he later expanded into the first book on heredity and cognitive ability, *Hereditary Genius: An Enquiry into Its Laws and Consequences* (Galton, 1869; see Box 12.1). The first twin and adoption studies in the 1920s also focused on *g* (Burks, 1928; Freeman, Holzinger, & Mitchell, 1928; Merriman, 1924; Theis, 1924).

Animal Research

Cognitive ability, at least problem-solving behavior and learning, can also be studied in other species. For example, in a well-known experiment in learning psychology, begun in 1924 by the psychologist Edward Tolman and continued by Robert Tryon, rats were selectively bred for their performance in learning to navigate a maze in order to find food. The results of subsequent selective breeding by Tryon for "maze- bright" rats (few errors) and "maze-dull" rats (many errors) are shown in Figure 12.2. Substantial response to selection was achieved after only a few generations of selec- tive breeding. There was practically no overlap between the maze-bright and maze- dull lines; all rats in the maze-bright line were able to learn to run through a maze with fewer errors than any of the rats in the maze-dull line. The difference between the bright and dull lines did not increase after the first half-dozen generations, pos- sibly because brothers and sisters were often mated. Such inbreeding greatly reduces the amount of genetic variability within selected lines, a loss that inhibits progress in a selection study; measurement issues could also contribute to the lack of progress.

Maze-bright and maze-dull selected rats were used in one of the best-known psychological studies of genotype-environment interaction (Cooper & Zubek, 1958).

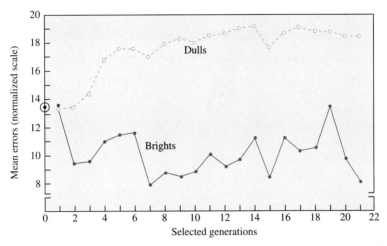

• **FIGURE 12.2** The results of Tryon's selective breeding for maze brightness and maze dullness in rats. (From "The inheritance of behavior" by G. E. McClearn. In L. J. Postman (Ed.), *Psychology in the Making.* © 1963. Used with permission of Alfred A. Knopf, Inc.)

Rats from the two selected lines were reared under one of three conditions. One condition was "enriched," in that the cages were large and contained many movable toys. For the comparison condition, called "restricted," small gray cages without movable objects were used. In the third condition, rats were reared in a standard laboratory environment.

The results of testing the maze-bright and maze-dull rats reared in these conditions are shown in Figure 12.3. Not surprisingly, in the normal environment in which the rats had been selected, there was a large difference between the two selected lines.

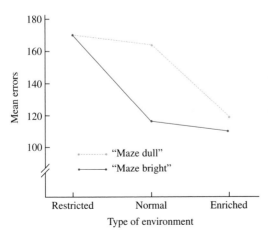

• **FIGURE 12.3** Genotype-environment interaction. The effects of rearing in a restricted, normal, or enriched environment on maze-learning errors differ for maze-bright and maze-dull selected rats. (From Cooper & Zubek, 1958.)

BOX 12.1 • Francis Galton

Francis Galton's life (1822–1911) as an inventor and explorer changed as he read the now-famous book on evolution written by Charles Darwin, his half cousin. Galton understood that evolution depends on heredity, and he began to ask whether heredity affects human behavior. He suggested the major methods of human behavioral genetics—family, twin, and adoption designs—and conducted the first systematic family studies showing that behavioral traits "run in families."

Galton invented correlation, one of the fundamental statistics in all of science, in order to quantify degrees of resemblance among family members (Gillham, 2001).

One of Galton's studies on mental ability was reported in his book *Hereditary Genius: An Enquiry into Its Laws and Consequences* (Galton, 1869). Because there was no satisfactory way at the time to measure mental ability, Galton had to rely on reputation as an index. By "reputation," he did not mean notoriety for a single act, or mere social or official position, but "the reputation of a leader of opinion, or an originator, of a man to whom the world deliberately acknowledges itself largely indebted" (Galton, 1869 p. 37). Galton identified approximately 1000 "eminent" men and found that they belonged to only 300 families, a finding indicating that the tendency toward eminence is familial.

Taking the most eminent man in each family as a reference point, Galton classified the other individuals who attained eminence according to closeness of family relationship. As indicated in the accompanying diagram, eminent status was more likely to appear in close relatives, with the likelihood of eminence

A clear genotype-environment interaction emerged for the enriched and restricted environments. The enriched condition had no effect on the maze-bright rats, but it greatly improved the performance of the maze-dull rats. On the other hand, the restricted environment was very detrimental to the maze-bright rats but had little effect on the maze-dull ones. In other words, there is no simple answer concerning the effect of restricted and enriched environments in this study. It depends on the genotype

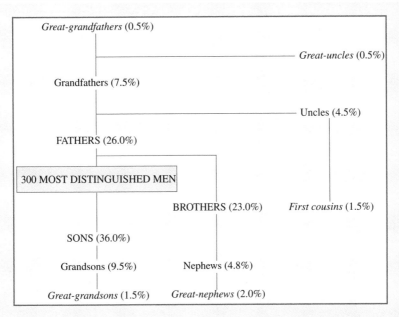

Great-grandfathers (0.5%)

Great-uncles (0.5%)

Grandfathers (7.5%)

Uncles (4.5%)

FATHERS (26.0%)

300 MOST DISTINGUISHED MEN

BROTHERS (23.0%)   First cousins (1.5%)

SONS (36.0%)

Grandsons (9.5%)   Nephews (4.8%)

Great-grandsons (1.5%)   Great-nephews (2.0%)

decreasing as the degree of relationship became more remote.

Galton was aware of the possible objection that relatives of eminent men share social, educational, and financial advantages. One of his counterarguments was that many men had risen to high rank from humble backgrounds. Nonetheless, such counterarguments do not today justify Galton's assertion that genius is solely a matter of nature (heredity) rather than nurture (environment). Family studies by themselves cannot disentangle genetic and environmental influences.

Galton set up a needless battle by pitting nature against nurture, arguing that "there is no escape from the conclusion that nature prevails enormously over nurture" (Galton, 1883, p. 241). Nonetheless, his work was pivotal in documenting the range of variation in human behavior and in suggesting that heredity underlies behavioral variation. For this reason, Galton can be considered the father of behavioral genetics.

of the animals. This example illustrates genotype-environment interaction, the differential response of genotypes to environments, as discussed in Chapter 8. Despite this persuasive example, other systematic research on learning generally failed to find widespread evidence of genotype-environment interaction (Henderson, 1972), although there is evidence for interactions with short-term factors (Crabbe, Wahlsten, et al., 1999).

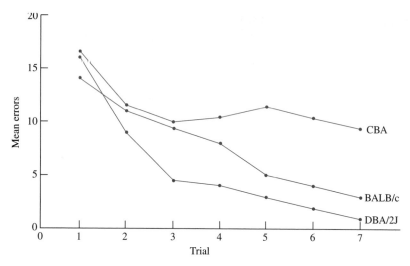

• **FIGURE 12.4** Maze-learning errors (Lashley III maze) for three inbred strains of mice. (From "Genetic aspects of learning and memory in mice" by D. Bovet, F. Bovet-Nitti, & A. Oliverio. *Science, 163,* 139–149. © 1969 by the American Association for the Advancement of Science.)

In the 1950s and 1960s, studies of inbred strains of mice showed the important contribution of genetics to most aspects of learning. Genetic differences have been shown for maze learning as well as for other types of learning, such as active avoidance learning, passive avoidance learning, escape learning, lever pressing for reward, reversal learning, discrimination learning, and heart rate conditioning (Bovet, 1977). For example, differences in maze-learning errors among widely used inbred strains (Figure 12.4) confirm the evidence for genetic influence found in the maze-bright and maze-dull selection experiment. The DBA/2J strain learned quickly, the CBA animals were slow, and the BALB/c strain was intermediate. Similar results were obtained for active avoidance learning, in which mice learn to avoid a shock by moving from one compartment to another whenever a light is flashed on. In this study, however, the CBA strain did not learn at all (Figure 12.5).

A strong *g* factor runs through many learning tasks in mice (Plomin, 2001). In half a dozen studies, intercorrelations among diverse learning tasks indicate that *g* accounts for at least 30 percent of the variance and appears to be moderately heritable (Galsworthy et al., 2005). *g* emerges even when other possible sources of intercorrelations among learning tasks, such as emotional reactivity or sensory and motoric ability, are controlled (Matzel & Kolata, 2010). *g* has also been observed in dogs (Coren, 2005) and in primate species other than ours (Banerjee et al., 2009). Animal models of *g* will be useful for functional genomic investigations of the brain pathways between genes and *g* (see Chapter 10).

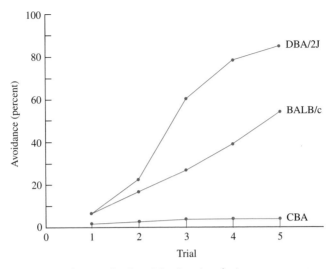

• **FIGURE 12.5** Avoidance learning for three inbred strains of mice. (From "Genetic aspects of learning and memory in mice" by D. Bovet, F. Bovet-Nitti, & A. Oliverio. *Science, 163,* 139–149. © 1969 by the American Association for the Advancement of Science.)

Human Research

Highlights in the history of human research on genetics and *g* include two early adoption studies which found that IQ correlations were greater in nonadoptive than in adoptive families, suggesting genetic influence (Burks, 1928; Leahy, 1935). The first adoption study that included IQ data for birth parents of adopted offspring also showed a significant parent-offspring correlation, again suggesting genetic influence (Skodak & Skeels, 1949). Begun in the early 1960s, the Louisville Twin Study was the first major longitudinal twin study of IQ that charted the developmental course of genetic and environmental influences (Wilson, 1983).

In 1963, a review of genetic research on *g* was influential in showing the convergence of evidence pointing to genetic influence (Erlenmeyer-Kimling & Jarvik, 1963). In 1966, Cyril Burt summarized his decades of research on MZ twins reared apart, which added the dramatic evidence that MZ twins reared apart are nearly as similar as MZ twins reared together. After his death in 1973, Burt's work was attacked, with allegations that some of his data were fraudulent (Hearnshaw, 1979). Two subsequent books reopened the case (Fletcher, 1990; Joynson, 1989). Although the jury is still out on some of the charges (Mackintosh, 1995; Rushton, 2002), it appears that at least some of Burt's data are dubious.

During the 1960s, environmentalism, which had been rampant until then in American psychology, was beginning to wane, and the stage was set for increased acceptance of genetic influence on *g*. Then, in 1969, a monograph on the genetics of intelligence by Arthur Jensen almost brought the field to a halt because a few pages in this lengthy monograph suggested that ethnic differences in IQ might involve genetic differences. Twenty-five years later, this issue was resurrected in *The Bell Curve* (Herrnstein & Murray, 1994) and caused a similar uproar. As we emphasized in Chapter 7, the causes of average differences between groups need not be related to the causes of individual differences within groups. The former question is much more difficult to investigate than the latter, which is the focus of the vast majority of genetic research on IQ. The storm raised by Jensen's monograph led to intense criticism of all behavioral genetic research, especially in the area of cognitive abilities (e.g., Kamin, 1974). These criticisms of older studies had the positive effect of generating bigger and better behavioral genetic studies that used family, adoption, and twin designs. These new projects produced much more data on the genetics of *g* than had been obtained in the previous 50 years. The new data contributed in part to a dramatic shift that occurred in the 1980s in psychology toward acceptance of the conclusion that genetic differences among individuals are significantly associated with differences in *g* (Snyderman & Rothman, 1988).

Overview of Genetic Research

In the early 1980s, a review of genetic research on *g* was published that summarized results from dozens of studies (Bouchard & McGue, 1981). Figure 12.6 is an expanded version of the summary of the review presented earlier in Chapter 3 (see Figure 3.7).

Genetic Influence

First-degree relatives living together are moderately correlated for *g* (about 0.45). As in Galton's original family study on hereditary genius (see Box 12.1), this resemblance could be due to genetic or to environmental influences because such relatives share both. Adoption designs disentangle these genetic and environmental sources of resemblance. Because birth parents and their offspring who are separated by adoption, as well as siblings who are adopted by different families, share heredity but not family environment, their similarity indicates that resemblance among family members is due in part to genetic factors. For *g,* the correlation between adopted children and their genetic parents is 0.24. The correlation between genetically related siblings reared apart is also 0.24. Because first-degree relatives are only 50 percent similar genetically, doubling these correlations gives a rough estimate of heritability of 48 percent. As discussed in Chapter 7, this outcome means that about half of the variance in IQ scores in the populations sampled in these studies can be accounted for by genetic differences among individuals.

The twin method supports this conclusion. Identical twins are nearly as similar as the same person tested twice. (Test-retest correlations for *g* are generally between 0.80 and 0.90.) The average twin correlations are 0.86 for identical twins and 0.60 for fraternal twins. Doubling the difference between MZ and DZ correlations estimates

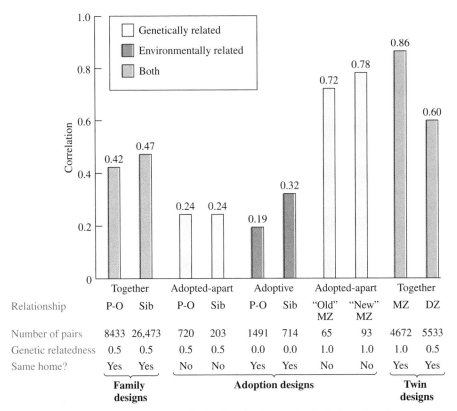

Relationship	Together		Adopted-apart		Adoptive		Adopted-apart		Together	
	P-O	Sib	P-O	Sib	P-O	Sib	"Old" MZ	"New" MZ	MZ	DZ
Number of pairs	8433	26,473	720	203	1491	714	65	93	4672	5533
Genetic relatedness	0.5	0.5	0.5	0.5	0.0	0.0	1.0	1.0	1.0	0.5
Same home?	Yes	Yes	No	No	Yes	Yes	No	No	Yes	Yes
	Family designs		**Adoption designs**						**Twin designs**	

• **FIGURE 12.6** Average IQ correlations for family, adoption, and twin designs. Based on reviews by Bouchard and McGue (1981), as amended by Loehlin (1989). "New" data sources for adopted-apart MZ twins include Bouchard et al. (1990) and Pedersen, McClearn, et al. (1992).

heritability as 52 percent. The most dramatic adoption design involves MZ twins who were reared apart. Their correlation provides a direct estimate of heritability. For obvious reasons, the number of such twin pairs is small. For several small studies published before 1981, the average correlation for MZ twins reared apart is 0.72 (excluding the suspect data of Cyril Burt). This outcome suggests higher heritability (72 percent) than do the other designs. This high heritability estimate has been confirmed in two other studies of twins reared apart. In a report on 45 pairs of MZ twins reared apart, the correlation was 0.78 (Bouchard et al., 1990). A study of Swedish twins that included 48 pairs of MZ twins reared apart reported the same correlation of 0.78 (Pedersen, McClearn, et al., 1992). Although the small sample sizes warrant caution in interpreting this higher heritability estimate for adopted-apart MZ twins, a possible explanation is discussed later.

Model-fitting analyses that simultaneously analyze all the family, adoption, and twin data summarized in Figure 12.6 yield heritability estimates of about 50 percent

(Chipuer, Rovine, & Plomin, 1990; Loehlin, 1989). It is noteworthy that genetics can account for half of the variance of a trait as complex as general cognitive ability. In addition, the total variance includes error of measurement. Corrected for unreliability of measurement, heritability estimates would be higher. Regardless of the precise estimate of heritability, the point is that genetic influence on *g* is not only statistically significant, it is also substantial.

These quantitative genetic estimates of heritability of *g* have been confirmed using genome-wide complex trait analysis (GCTA). As explained in Chapter 7, GCTA uses hundreds of thousands of SNPs genotyped on large samples to estimate heritability directly from DNA. GCTA does not specify which SNPs are associated with a phenotype. Instead, it relates chance genetic similarity on SNPs to phenotypic similarity pair by pair in a large sample of conventionally unrelated individuals. GCTA applied to *g* scores for more than 3000 unrelated older individuals estimated heritability from 40 to 50 percent (Davies et al., 2011), an estimate replicated in other genomewide association studies of *g* in childhood (Benyamin et al., in press; Deary et al., 2012) as well as adulthood (Chabris et al., in press).

Although heritability could differ in different cultures, it appears that the level of heritability of *g* also applies to populations outside North America and Western Europe, where most studies have been conducted. Similar heritabilities have been found in twin studies in Russia (Malykh, Iskoldsky, & Gindina, 2005) and in the former East Germany (Weiss, 1982), as well as in rural India, urban India, and Japan (Jensen, 1998). Another interesting finding is that the more a test relates to *g*, the higher are the heritabilities for cognitive test scores (Jensen, 1998). This result has been found in studies of older twins (Pedersen, McClearn, et al., 1992), in research on individuals with cognitive disability (Spitz, 1988), and in a twin study using information-processing tasks (Vernon, 1989). These results suggest that *g* is the most highly heritable composite of cognitive tests.

What about high *g*? In Chapter 11, we saw that most cognitive disability appears to be the low end of the same genetic and environmental factors that affect individual differences throughout the *g* distribution. The same story appears to apply to high *g*, as indicated by the first large-scale twin study of high *g* (Haworth, Wright, et al., 2009).

Environmental Influence

If half of the variance of *g* can be accounted for by heredity, the other half can be attributed to environment (plus errors of measurement). Some of this environmental influence appears to be shared by family members, making them similar to one another. Direct estimates of the importance of shared environmental influence come from correlations for adoptive parents and children and for adoptive siblings. Particularly impressive is the correlation of 0.32 for adoptive siblings (see Figure 12.6). Because they are unrelated genetically, what makes adoptive siblings similar is shared rearing—having the same parents, the same diet, attending the same schools, and so on. The adoptive sibling correlation of 0.32 suggests that about a third of the total variance can be explained by shared environmental influences. The correlation for adoptive parents and their adopted children is lower ($r = 0.19$) than that for adoptive siblings, a result suggesting that shared environment accounts for less resemblance between parents and offspring than between siblings.

Shared environmental effects are also suggested because correlations for relatives living together are greater than correlations for adopted-apart relatives. Twin studies also suggest shared environmental influence. In addition, shared environmental effects appear to contribute more to the resemblance of twins than to that of nontwin siblings because the correlation of 0.60 for DZ twins exceeds the correlation of 0.47 for nontwin siblings. Twins may be more similar than other siblings because they shared the same womb and are exactly the same age. Because they are the same age, twins also tend to be in the same school, even if not the same class, and share many of the same peers (Koeppen-Schomerus, et al., 2003).

Model-fitting estimates of the role of shared environment for g based on the data in Figure 12.6 are about 20 percent for parents and offspring, about 25 percent for siblings, and about 40 percent for twins (Chipuer et al., 1990). The rest of the environmental variance is attributed to nonshared environment and errors of measurement. However, when these data are examined developmentally, a different picture emerges, as discussed later in this chapter.

Assortative Mating

Several other factors need to be considered to obtain a more refined estimate of genetic influence. One is *assortative mating,* which refers to nonrandom mating. Old adages are sometimes contradictory. Do "birds of a feather flock together" or do "opposites attract"? Research shows that, for some traits, "birds of a feather" do "flock together," in the sense that individuals who mate tend to be similar—although not as similar as you might think. For example, although there is some positive assortative mating for physical characteristics, the correlations between spouses are relatively low—about 0.25 for height and about 0.20 for weight (Spuhler, 1968). Spousal correlations for personality are even lower, in the 0.10 to 0.20 range (Vandenberg, 1972). Assortative mating for g is substantial, with average spousal correlations of about 0.40 (Jensen, 1978). In part, spouses select each other for g on the basis of education. Spouses correlate about 0.60 for education, which correlates about 0.60 with g.

Assortative mating is important for genetic research for two reasons. First, assortative mating increases genetic variance in a population. For example, if spouses mated randomly in relation to height, tall women would be just as likely to mate with short men as with tall men. Offspring of the matings of tall women and short men would generally be of moderate height. However, because there is positive assortative mating for height, children with tall mothers are also likely to have tall fathers, and the offspring themselves are likely to be taller than average. The same thing happens for short parents. In this way, positive assortative mating increases variance in that the offspring differ more from the average than they would if mating were random. Even though spousal correlations are modest, assortative mating can greatly increase genetic variability in a population because its effects accumulate generation after generation.

Assortative mating is also important because it affects estimates of heritability. For example, it increases correlations for first-degree relatives. If assortative mating were not taken into account, it could inflate heritability estimates obtained from studies of parent-offspring (e.g., birth parents and their adopted-apart offspring) or sibling resemblance. For the twin method, however, assortative mating could result in underestimates of heritability. Assortative mating does not affect MZ correlations because MZ twins are identical genetically, but it raises DZ correlations because DZ twins are first-degree relatives. In this way, assortative mating lessens the difference between MZ and DZ correlations; it is this difference that provides estimates of heritability in the twin method. The model-fitting analyses described above took assortative mating into account in estimating the heritability of *g* to be about 50 percent. If assortative mating had not been taken into account, its effects would have been attributed to shared environment.

Nonadditive Genetic Variance

Nonadditive genetic variance also affects heritability estimates. For example, when we double the difference between MZ and DZ correlations to estimate heritability, we assume that genetic effects are largely additive. *Additive genetic effects* occur when alleles at a locus and across loci "add up" to affect behavior. However, sometimes the effects of alleles can be different in the presence of other alleles. These interactive effects are called *nonadditive*.

Dominance is a nonadditive genetic effect in which alleles at a locus interact rather than add up to affect behavior. For example, having one PKU allele is not half as bad as having two PKU alleles. Even though many genes operate with a dominant-recessive mode of inheritance, much of the effect of such genes can nonetheless be attributed to the average effect of the alleles. The reason is that, even though heterozygotes are phenotypically similar to the dominant homozygote, there is a substantial linear relationship between genotype and phenotype.

When several genes affect a behavior, the alleles at different loci can add up to affect behavior, or they can interact. This type of interaction between alleles at different loci is called *epistasis*. (See Appendix for details.)

Additive genetic variance is what makes us resemble our parents, and it is the raw material for natural selection. Our parents' genetic decks of cards are shuffled when our hand is dealt at conception. We and each of our siblings receive a random sampling of half of each parent's genes. We resemble our parents to the extent that each allele that we share with our parents has an average additive effect. Because we do not have exactly the same combination of alleles as our parents (we inherit only one of each of their pairs of alleles), we differ from our parents for nonadditive interactions as a result of dominance or epistasis. The only relatives who will resemble each other for all dominance and epistatic effects are identical twins, who are identical for all combinations of genes. For this reason, the hallmark of nonadditive genetic variation is that first-degree relatives are less than half as similar as MZ twins.

For *g*, the correlations in Figure 12.6 suggest that genetic influence is largely additive. For example, first-degree relatives are just about half as similar as MZ twins. However, there is evidence that assortative mating for *g* masks some nonadditive genetic variance. As indicated in the previous section, assortative mating, which is substantial for *g*, inflates correlations for first-degree relatives but does not affect MZ correlations. When assortative mating is taken into account in model-fitting analyses, some evidence appears for nonadditive genetic variance, although most genetic influence on *g* is additive (Chipuer et al., 1990; Fulker, 1979; Vinkhuyzen, van der Sluis, Maes, & Posthuma, 2012). It is very fortunate for attempts to identify *g* genes that most of the genetic variance is additive. As discussed later, it has been very difficult to identify genes because their effects at the population level are so small. However, if genetic effects were nonadditive, this would mean that instead of looking for the additive effects of genes considered individually, it would be necessary to look for the combined interactive effects of alleles at multiple loci.

The presence of dominance can be seen from studies of inbreeding. (Inbreeding is mating between genetically related individuals.) If inbreeding occurs, offspring are more likely to inherit the same alleles at any locus. Thus, inbreeding makes it more likely that two copies of rare recessive alleles will be inherited, including those for harmful recessive disorders. In this sense, inbreeding reduces **heterozygosity** by "redistributing" heterozygotes as dominant homozygotes and recessive homozygotes. Therefore, inbreeding also alters the average phenotype of a population. Because the frequency of recessive homozygotes for harmful recessive disorders is increased with inbreeding, the average phenotype will be lowered.

Inbreeding data suggest some dominance for *g* because inbreeding lowers IQ (Vandenberg, 1971). Children of marriages between first cousins generally perform worse than controls. The risk of cognitive disability is more than three times greater for children of a marriage between first cousins than for unrelated controls (Böök, 1957). Children of double first cousins (double first cousins are the children of two siblings who are married to another pair of siblings) perform even worse (Agrawal, Sinha, & Jensen, 1984; Bashi, 1977). Nonetheless, inbreeding does not have an appreciable effect in general in the population because it is rare, with the exception of a few societies and small isolated groups.

An extreme version of epistasis called *emergenesis* has been suggested as a model for unusual abilities (Lykken, 2006). Luck of the draw at conception can result in certain unique combinations of alleles that have extraordinary effects not seen in parents or siblings. For example, the great racehorse Secretariat was bred to many fine mares to produce hundreds of offspring. Many of Secretariat's offspring were good horses,

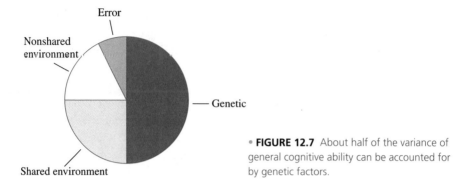

• **FIGURE 12.7** About half of the variance of general cognitive ability can be accounted for by genetic factors.

thanks to additive genetic effects, but none came even close to the unique combination of strengths responsible for Secretariat's greatness. Such genetic luck of the draw might contribute to human genius as well.

Despite the complications caused by assortative mating and nonadditive genetic variance, the general summary of behavioral genetic results for g is surprisingly simple (Figure 12.7). About half of the variance is due to genetic factors. Some, but not much, of this genetic variance might be nonadditive. Of the half of the variance that is due to nongenetic factors, about half of that is accounted for by shared environmental factors. The other half is due to nonshared environment and errors of measurement. However, during the past decade, it has been discovered that these average results are largely based on children; results change dramatically during development, as described in the following section.

Developmental Research

When Francis Galton first studied twins in 1876, he investigated the extent to which the similarity of twins changed during development. Other early twin studies were also developmental (Merriman, 1924), but this developmental perspective faded from genetic research until recent years.

Two types of developmental questions can be addressed in genetic research. Does heritability change during development? Do genetic factors contribute to developmental change?

Does Heritability Change during Development?

Try asking people this question: As you go through life, do you think the effects of heredity become more important or less important? Most people will usually guess "less important" for two reasons. First, it seems obvious that life events such as accidents and illnesses, education and occupation, and other experiences accumulate during a lifetime. This fact implies that environmental differences increasingly contribute to phenotypic differences, so heritability necessarily decreases. Second, most people mistakenly believe that genetic effects never change from the moment of conception.

Because it is so reasonable to assume that genetic differences become less important as experiences accumulate during the course of life, one of the most interesting findings about *g* is that the opposite is closer to the truth. Genetic factors become increasingly important for *g* throughout an individual's life span.

For example, an ongoing longitudinal adoption study called the Colorado Adoption Project (Plomin et al., 1997) provides parent-offspring correlations for general cognitive ability from infancy through adolescence. As illustrated in Figure 12.8, correlations between parents and children for control (nonadoptive) families increase from less than 0.20 in infancy to about 0.20 in middle childhood and to about 0.30 in adolescence. The correlations between birth mothers and their adopted-away children follow a similar pattern, thus indicating that parent-offspring resemblance for *g* is due to genetic factors. Parent-offspring correlations for adoptive parents and their adopted children hover around zero, which suggests that family environment shared by parents and offspring does not contribute importantly to parent-offspring resemblance for *g*. These parent-offspring correlations for adoptive parents and their adopted children are slightly lower than those reported in other adoption studies (see Figure 12.6), possibly because selective placement was negligible in the Colorado Adoption Project (Plomin & DeFries, 1985).

Twin studies also show increases in heritability from childhood to adulthood (McCartney, Harris, & Bernieri, 1990; McGue, Bouchard, Iacono, & Lykken, 1993;

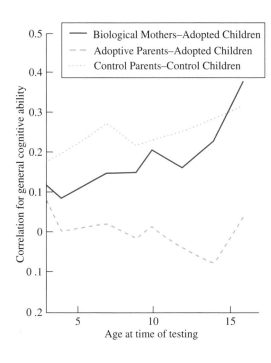

• **FIGURE 12.8** Parent-offspring correlations between parents' *g* scores and children's *g* scores for biological, adoptive, and control parents and their children at ages 3, 4, 7, 9, 10, 12, 14, and 16. Parent-offspring correlations are weighted averages for mothers and fathers to simplify the presentation. (From "Nature, nurture and cognitive development from 1 to 16 years: A parent-offspring adoption study" by R. Plomin, D. W. Fulker, R. Corley, & J. C. DeFries. *Psychological Science, 8,* 442–447. © 1997.)

Plomin, 1986). A recent report on a sample of 11,000 pairs of twins, a larger sample than that in all previous studies combined, showed for the first time that the heritability of general cognitive ability increases significantly from 41 percent in childhood (age 9) to 55 percent in adolescence (age 12) and to 66 percent in young adulthood (age 17) (Haworth et al., 2010), as shown in Figure 12.9. Although the trend of increasing heritability appears to continue throughout adulthood to about 80 percent at age 65, some research suggests that heritability declines in later life, perhaps to about 60 percent after age 80 (Lee, Henry, Trollor, & Sachdev, 2010). The increase in heritability from childhood to adulthood could explain the higher heritability estimate for adopted-apart MZ twins, mentioned earlier: The adopted-apart MZ twins were much older than subjects in the other twin and adoption studies summarized in Figure 12.6.

Why does heritability increase during the life course? Perhaps completely new genes come to affect *g* in adulthood. A more likely possibility is that relatively small genetic effects early in life snowball during development, creating larger and larger phenotypic effects. For the young child, parents and teachers contribute importantly to intellectual experience; but for the adult, intellectual experience is more self-directed. For example, it seems likely that adults with a genetic propensity toward

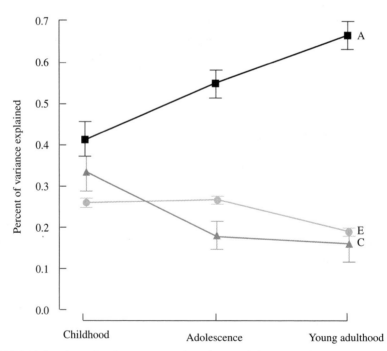

• **FIGURE 12.9** Twin studies show increasing heritability and decreasing shared environmental influence for general cognitive ability from childhood to adulthood. A = additive genetic; C = common or shared environment; E = nonshared environment. (Adapted from Haworth et al., 2010. Reprinted by permission from Macmillan Publishers, Ltd: *Molecular Psychiatry, 15*, 1112–1120, © 2011.)

high *g* keep mentally active by reading, arguing, and simply thinking more than other people do. Such experiences not only reflect but also reinforce genetic differences, creating genotype-environment correlation, as described in Chapter 8.

Another important developmental finding is that the effects of shared environment appear to decrease. Twin study estimates of shared environment are weak because shared environment is estimated indirectly by the twin method; that is, shared environment is estimated as twin resemblance that cannot be explained by genetics. Nonetheless, the twin study illustrated in Figure 12.9 also found that shared environment effects for *g* decline from adolescence to adulthood.

The most direct evidence for the important effect of shared environment on individual differences in *g* comes from the resemblance of adoptive siblings, pairs of genetically unrelated children adopted into the same adoptive families. Figure 12.6 indicates an average IQ correlation of 0.32 for adoptive siblings. However, these studies assessed adoptive siblings when they were children. In 1978, the first study of older adoptive siblings yielded a strikingly different result: The IQ correlation was essentially zero (−0.03) for 84 pairs of adoptive siblings who were 16 to 22 years of age (Scarr & Weinberg, 1978b). Other studies of older adoptive siblings have found similarly low IQ correlations. The most impressive evidence comes from a ten-year longitudinal follow-up study of more than 200 pairs of adoptive siblings. At the average age of 8, the IQ correlation was 0.26. Ten years later, the IQ correlation was near zero (Loehlin, Horn, & Willerman, 1989). Figure 12.10 shows the results of studies of adoptive siblings in childhood and in adulthood (McGue, Bouchard, et al., 1993).

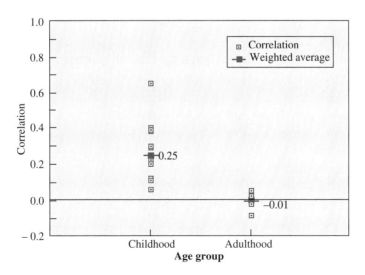

• **FIGURE 12.10** The correlation for adoptive siblings provides a direct estimate of the importance of shared environment. For *g*, the correlation is 0.25 in childhood and −0.01 in adulthood, a difference suggesting that shared environment becomes less important after childhood. (From McGue, Bouchard, et al., 1993, p. 67.)

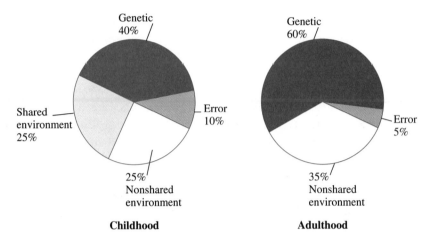

• **FIGURE 12.11** From childhood to adulthood, heritability of *g* increases and shared environment declines in importance.

In childhood, the average adoptive sibling correlation is 0.25; but in adulthood, the correlation for adoptive siblings is near zero.

These results represent a dramatic example of the importance of genetic research for understanding the environment. Shared environment is an important factor for *g* during childhood, when children are living at home. However, its importance fades in adulthood as influences outside the family become more salient.

In summary, from childhood to adulthood, the heritability of *g* increases and the importance of shared environment decreases (Figure 12.11).

Do Genetic Factors Contribute to Developmental Change?

The second type of genetic change in development refers to age-to-age change seen in longitudinal data in which individuals are assessed several times. It is important to recognize that genetic factors can contribute to change as well as to continuity in development. Change in genetic effects does not necessarily mean that genes are turned on and off during development, although this does happen. Genetic change simply means that genetic effects at one age differ from genetic effects at another age. For example, genes that affect cognitive processes involved in language cannot show their effect until language appears in the second year of life.

The issue of genetic contributions to change and continuity can be addressed by using longitudinal genetic data, in which twins or adoptees are tested repeatedly. The simplest way to think about genetic contributions to change is to ask whether changes in scores from age to age show genetic influence. That is, although *g* is quite stable from year to year, some children's scores increase and some decrease. Genetic factors account for part of such changes, especially in childhood (Fulker, DeFries, & Plomin, 1988) and perhaps even in adulthood (Loehlin et al., 1989). Still, not surprisingly, most

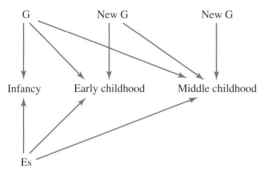

• **FIGURE 12.12** Genetic factors (G) contribute to change as well as continuity in *g* during childhood. Shared environment (Es) contributes only to continuity. (Adapted from Fulker, Cherny, & Cardon, 1993.)

genetic effects on *g* contribute to continuity from one age to the next (Petrill et al., 2004; Rietveld, Dolan, van Baal, & Boomsma, 2003). Model-fitting analysis (see Appendix) is especially useful for longitudinal data because of the complexity of having multiple measurements for each subject. Several types of longitudinal genetic models have been proposed (Loehlin et al., 1989). A longitudinal model applied to twin and adoptive sibling data from infancy to middle childhood found evidence for genetic change at two important developmental transitions (Fulker, Cherny, & Cardon, 1993). The first is the transition from infancy to early childhood, an age when cognitive ability rapidly changes as language develops. The second is the transition from early to middle childhood, at 7 years of age. It is no coincidence that children begin formal schooling at this age—all theories of cognitive development recognize this as a major transition. Figure 12.12 summarizes these findings in childhood. Much genetic influence on *g* involves continuity. That is, genetic factors that affect infancy also affect early childhood and middle childhood. However, some new genetic influence comes into play at the transition from infancy to early childhood. These new genetic factors continue to affect *g* throughout early childhood and into middle childhood. Similarly, new genetic influence also emerges at the transition from early to middle childhood.

Similar results have been reported in analyses from early to middle childhood (Davis, Haworth, & Plomin, 2009a), from childhood to adolescence (van Soelen et al., 2011), from early adulthood to middle adulthood (Lyons et al., 2009), and in old age (Plomin, Pedersen, Lichtenstein, & McClearn, 1994). A surprising amount of genetic influence on general cognitive ability in childhood overlaps with genetic influence even into adulthood, as illustrated in Figure 12.13.

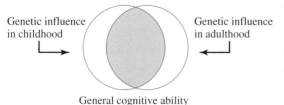

• **FIGURE 12.13** Although genetic influences on *g* in childhood are largely the same as those that affect *g* in adulthood, there is some evidence for genetic change.

As mentioned earlier in this chapter (see also Box 7.1), genome-wide complex trait analysis (GCTA) estimates genetic influence by predicting phenotypic similarity from random SNP similarity for a large sample of unrelated individuals. GCTA has recently been extended to multivariate analysis of g scores from childhood (age 11) to late adulthood (age 65 to 79) (Deary et al., 2012). The essence of the multivariate extension of GCTA is to analyze phenotypic relatedness between each pair of individuals on different traits rather than the same trait; in this example, the different traits are g assessed in childhood and again more than half a century later. Multivariate GCTA indicated substantial age-to-age genetic continuity (a genetic correlation of 0.62) for intelligence from childhood to late adulthood. It also showed significant genetic influence on change: Genetics accounted for nearly a quarter of the variance in cognitive scores in old age independent of scores in childhood. Both of these developmental findings are consistent with results from quantitative genetic research on g, although no longitudinal quantitative genetic studies have extended from childhood to old age.

As discussed earlier, shared environmental influences also affect g in childhood. Unlike genetic effects, which contribute to change as well as to continuity, longitudinal analysis suggests that shared environmental effects contribute only to continuity. That is, the same environmental factors shared by relatives affect g in infancy and in both early and middle childhood (see Figure 12.12). Socioeconomic factors, which remain relatively constant, might account for this shared environmental continuity.

Identifying Genes

Finding genes associated with g will have far-reaching ramifications at all levels from DNA to brain to behavior. Despite its complexity, general cognitive ability is a reasonable candidate for molecular genetic research because it is one of the most heritable dimensions of behavior. As described in Chapter 11, in our species, more than 250 single-gene disorders include cognitive disability among their symptoms (Inlow & Restifo, 2004). The major single-gene effects were described in Chapter 11. The classic example of a single-gene cause of severe cognitive disability is PKU. More recently, researchers have identified a gene causing the fragile X type of cognitive disability. A gene on chromosome 19 that encodes apolipoprotein E contributes substantially to risk for the dementia of late-onset Alzheimer disease.

What about the normal range of general cognitive ability? Some evidence suggests that carriers for PKU show slightly lowered IQ scores (Bessman, Williamson, & Koch, 1978; Propping, 1987). However, differences in the number of fragile X repeats in the normal range do not relate to differences in IQ (Daniels et al., 1994). It is only when the number of repeats expands to more than 200 that cognitive disability occurs, as described in Chapter 11. For apolipoprotein E, a meta-analysis of 77 studies with more than 40,000 healthy subjects shows a weak association with g, primarily in older people, although this effect may be due to as yet undetected dementia in some older individuals (Wisdom, Callahan, & Hawkins, 2011).

Similar to other areas of behavioral genetics, the first attempts to find genes associated with *g* focused on genes involved in brain function (Payton, 2009). One problem with such a candidate gene approach is that we often do not have strong hypotheses as to which genes are true candidate genes. Indeed, the general rule of pleiotropy (each gene has many effects) suggests that most of the thousands of genes expressed in the brain could be considered as candidates. Moreover, many genetic associations are in non-coding regions of DNA rather than in traditional genes, as described in Chapter 10. The major problem for candidate gene association studies is that reports of associations have failed to be replicated, suggesting that published reports of associations are false-positive results caused by the use of samples underpowered to detect the small effect sizes that seem to be the source of heritability for complex traits. Strong support for this conclusion comes from a recent study of nearly 10,000 individuals that was not able to replicate associations for ten of the most frequently reported candidate gene associations (Chabris et al., in press).

Another candidate gene strategy for identifying QTL associations for *g* is to focus on intermediate phenotypes—often called *endophenotypes*—that are presumed to be simpler genetically and thus more likely to yield QTLs of large effect size that can be detected with small samples (Goldberg & Weinberger, 2004; Winterer & Goldman, 2003). As discussed in Chapter 10, although all levels of analysis from genes to *g* are important to study in their own right and in terms of understanding pathways between genes and behavior, it seems unlikely that brain endophenotypes will prove to be simpler genetically or be more useful in identifying genes associated with *g* than studying *g* itself (Kovas & Plomin, 2006). Brain imaging research is discussed in the next chapter.

As discussed in Chapter 9, attempts to find genes associated with complex traits like *g* have gone beyond looking for candidate genes to conducting systematic scans of the genome using linkage and association strategies. Three QTL linkage reports on *g* have suggested several different linkage regions, including linkage near the region of *6p*, which is the region that shows linkage with reading disability, as discussed in Chapter 11 (Dick, Aliev, et al., 2006; Luciano et al., 2006; Posthuma, Luciano, et al., 2005). An early attempt to conduct a systematic association study of *g* before microarrays became available came up empty-handed (Plomin, Hill, et al., 2001). Microarrays have now made it possible to conduct genomewide association (GWA) studies with hundreds of thousands of SNPs. Similar to results from other GWA studies of complex traits in the life sciences, GWA studies of *g* have not yet identified replicable associations (Butcher, Davis, Craig, & Plomin, 2008; Davies et al., 2011; Davis et al., 2010; Need et al., 2009). These GWA studies were powered to detect associations that account for as little as 0.5 percent of the variance, so their results indicate that the largest effect sizes are likely to account for less than this, which is less than 1 IQ point. Even polygenic prediction using many of the strongest associations in discovery samples explained less than 1 percent of the variance of *g* in independent samples (Davies et al., 2011).

As indicated in Chapter 9, one strategy for finding the elusive genes responsible for the heritability of g is to investigate rarer variants than those currently available on SNP microarray platforms, which use the most common SNPs with minor allele frequencies greater than 5 percent because such SNPs are most useful for tagging the entire genome. Weak associations found for common variants could reflect indirect ("synthetic") associations between these common SNP variants and rarer genetic variants of larger effect (Dickson, Wang, Krantz, Hakonarson, & Goldstein, 2010). A recent study reported that individuals with more deletions that are rare had lower g scores, although the sample size was small and the finding requires replication (Yeo, Gangestad, Liu, Calhoun, & Hutchison, 2011). Microarrays with rarer SNPS are being developed, but research is moving toward sequencing the entire genome so that all DNA variation can be detected, not just SNPs but also structural variants (Mills et al., 2011).

Another strategy is to use the common SNPs currently available on microarrays with much larger samples in order to detect smaller effect sizes. A consortium of studies of childhood intelligence with a total sample of nearly 18,000 found no significant associations for individual SNPs even though an association would need to account for only 0.25 percent of the variance to reach statistical significance (Benyamin et al., in press), suggesting that even larger samples will be needed to account for the missing heritability of g. However, SNPs in one gene, *FNBP1L*, were significantly associated with g when SNPs were analyzed in a gene-based rather than SNP-based analysis. *FNBP1L* is interesting for two reasons. First, this gene also emerged as one of the strongest associations in a GWA study of adults (Davies et al., 2011). Second, *FNBP1L* is especially expressed in neurons in developing brains and regulates neuronal morphology. Although GWA studies have not yet identified genes that account for the heritability of g, GCTA results suggest that it should be possible to identify most of the heritability using currently available microarrays consisting of common SNPs—provided the samples are sufficiently large.

Finding genes that account for the heritability of g has important implications for society as well as for science (Plomin, 1999). The grandest implication for science is that g genes will serve as an integrating force across diverse disciplines, with DNA as the common denominator, and will open up new scientific horizons for understanding learning and memory. In terms of implications for society, it should be emphasized that no public policies necessarily follow from finding genes associated with g because policy involves values. For example, finding genes for g does not mean that we ought to put all of our resources into educating the brightest children once we identify them genetically. Depending on our values, we might worry more about the children falling off the low end of the bell curve in an increasingly technological society and decide to devote more public resources to those who are in danger of being left behind. Potential problems related to finding genes associated with g, such as prenatal and postnatal screening, discrimination in education and employment, and group differences, have been considered (Newson & Williamson, 1999; Nuffield

Council on Bioethics, 2002). We need to be cautious and to think carefully about societal implications and ethical issues, but there is also much to celebrate here in terms of increased potential for understanding our species' ability to think and learn.

Summary

Selection and inbred strain studies indicate genetic influence on animal learning, such as the maze-bright and maze-dull selection study of learning in rats. Human studies of general cognitive ability (g) have been conducted for over a century. Family, twin, and adoption studies converge on the conclusion that about half of the total variance of measures of general cognitive ability can be accounted for by genetic factors. For example, twin correlations for general cognitive ability are about 0.85 for identical twins and 0.60 for fraternal twins. Heritability estimates are affected by assortative mating (which is substantial for general cognitive ability) and nonadditive genetic variance (dominance and epistasis). About half of the environmental variance for g appears to be accounted for by shared environmental factors.

The heritability of g increases during the life course, reaching levels in adulthood comparable to the heritability of height. The influence of shared environment diminishes sharply after adolescence. Longitudinal genetic analyses of g suggest that genetic factors primarily contribute to continuity, although some evidence for genetic change has been found, for example, in the transition from early to middle childhood.

Attempts to identify some of the genes responsible for the heritability of g have begun, including candidate gene studies, QTL linkage, and genomewide association studies. This research has demonstrated that many genes of small effect are responsible for the heritability of g. Nonetheless, genome-wide complex trait analysis (GCTA) indicates that common SNPs can explain most of the heritability of g. A multivariate extension of GCTA suggests that the same genes affect g in childhood and in late adulthood.

CHAPTER THIRTEEN

Specific Cognitive Abilities

There is much more to cognitive functioning than general cognitive ability. As discussed in Chapter 12, cognitive abilities are usually considered in a hierarchical model (see Figure 12.1). General cognitive ability (g) is at the top of the hierarchy, representing what all tests of cognitive ability have in common and explaining about 40 percent of the variance of such tests. Below general cognitive ability in the hierarchy are broad factors of specific cognitive abilities, such as verbal ability, spatial ability, memory, and speed of processing. These broad factors are indexed by several tests, such as the assessments of verbal ability and spatial ability in Figure 13.1. The tests are at the bottom of the hierarchical model. Specific cognitive abilities correlate moderately with general cognitive ability, but they are also substantially different. In addition to specific tests, the bottom of the hierarchy can also be considered in terms of the elementary cognitive processes that are thought to be involved in processing information from input to storage and then from retrieval to output. Increasingly, research in this area has employed measures of brain structure and function (Deary et al., 2010).

This chapter presents genetic research on specific cognitive abilities and their relationship to general cognitive ability. It also considers the genetics of a real-world aspect of cognitive abilities, school achievement.

Broad Factors of Specific Cognitive Abilities

More is known about the genetics of the broad factors of specific cognitive abilities than about elementary cognitive processes or brain function (Plomin & DeFries, 1998). The largest family study of specific cognitive abilities, called the Hawaii Family Study of Cognition, included more than a thousand families (DeFries et al., 1979). Like other work in this area, this study used a technique called factor analysis to identify the tightest clusters of intercorrelated tests. Four group factors were derived from 15 tests: verbal (including vocabulary and fluency), spatial (visualizing and rotating

(a) **Tests of verbal ability**

1. Vocabulary: In each row, circle the word that means the same or nearly the same as the underlined word. There is only one correct choice in each line.

a. arid   coarse   clever   modest   dry
b. piquant   fruity   pungent   harmful   upright

2. Word beginnings and endings: For the next three minutes, write as many words as you can that start with F and end with M.

3. Things: For the next three minutes, list all the things you can think of that are flat.

(b) **Tests of spatial ability**

1. Paper form board: Draw a line or lines showing where the figure on the left should be cut to form the pieces on the right. There may be more than one way to draw the lines correctly.

2. Mental rotations: Circle the two objects on the right that are the same as the object on the left.

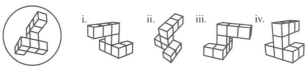

3. Card rotations: Circle the figures on the right that can be rotated (without being lifted off the page) to exactly match the one on the left.

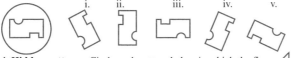

4. Hidden patterns: Circle each pattern below in which the figure appears. The figure must always be in this position, not upside down or on its side.

• **FIGURE 13.1** Tests of specific cognitive abilities, such as those used in the Hawaii Family Study of Cognition, include tasks resembling the ones shown here. (a) The answers for verbal test 1 are (i) dry and (ii) pungent. (b) For spatial test 1, the solution is that, in addition to the rectangle, only one line is needed: The two corners of a short side of the rectangle touch the circle and a single line extends the other short side to bisect the circle. The answers for the other spatial tests are 2. ii, iii; 3. i, iii, iv; 4. i, ii, vi.

objects in two- and three-dimensional space), perceptual speed (simple arithmetic and number comparisons), and visual memory (short-term and longer-term recognition of line drawings). Examples resembling some of the verbal and spatial tests used in the Hawaii Family Study of Cognition are shown in Figure 13.1.

Figure 13.2 summarizes parent-offspring resemblance for the four factors and the 15 cognitive tests for two ethnic groups. The most obvious fact is that familial resemblance differs for the four factors and for tests within each factor. The data were corrected for the unreliability of the tests, so the differences in familial resemblance were not caused by reliability differences among the tests. For both groups, the verbal and spatial factors show more familial resemblance than do the perceptual speed and memory factors. Other family studies also generally indicate that the greatest familial

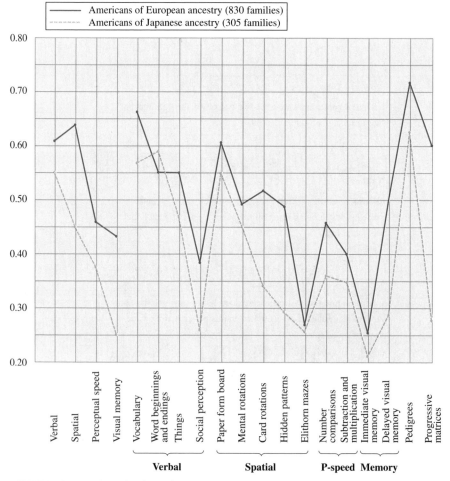

• **FIGURE 13.2** Family study of specific cognitive abilities. Regression of midchild on midparent for four group factors and 15 cognitive tests in two ethnic groups. (Data from DeFries et al., 1979.)

similarity occurs for verbal ability (DeFries, Vandenberg, & McClearn, 1976). It is not known why one group consistently shows greater parent-offspring resemblance than the other. This study is a good reminder of the principle that the results of genetic research can differ in different populations.

Figure 13.2 also makes another important point: Tests within each factor show dramatic differences in familial resemblance. For instance, one spatial test, Paper Form Board, shows high familiality in both groups. The test involves showing how to cut a figure to yield a certain pattern—for example, how to cut a circle to yield a triangle and three crescents (Figure 13.1). Another spatial test, Elithorn Mazes, shows the lowest familial resemblance in both groups. This test involves drawing one line that connects as many dots as possible in a maze of dots. Although these tests correlate with each other and contribute to a broad factor of spatial ability, much remains to be learned about the genetics of the processes involved in each test.

The results of dozens of early twin studies of specific cognitive abilities are summarized in Table 13.1 (Nichols, 1978). When we double the difference between the correlations for identical and fraternal twins to estimate heritability (see Chapter 6), these results suggest that specific cognitive abilities show slightly less genetic influence than general cognitive ability. Memory and verbal fluency show lower heritability, about 30 percent; the other abilities yield heritabilities of 40 to 50 percent. Although the largest twin studies do not consistently find greater heritability for particular cognitive abilities (Bruun, Markkananen, & Partanen, 1966; Schoenfeldt, 1968), it has been suggested that verbal and spatial abilities in general show greater heritability than do perceptual speed and, especially, memory abilities (Plomin, 1988). Estimates of shared

TABLE 13.1

Average Twin Correlations for Tests of Specific Cognitive Abilities

Ability	Number of Studies	Twin Correlations		Parameter Estimates	
		Identical Twins	Fraternal Twins	Heritability	Shared Environment
Verbal comprehension	27	0.78	0.59	0.38	0.40
Verbal fluency	12	0.67	0.52	0.30	0.37
Reasoning	16	0.74	0.50	0.48	0.26
Spatial visualization	31	0.64	0.41	0.46	0.18
Perceptual speed	15	0.70	0.47	0.46	0.24
Memory	16	0.52	0.36	0.32	0.20

SOURCE: Nichols (1978). Estimates of heritability and shared environment were calculated from the twin correlations.

environment vary from 18 to 40 percent. Earlier twin studies of specific cognitive abilities have been reviewed in detail elsewhere (DeFries et al., 1976).

Two studies of identical and fraternal twins reared apart provide additional support for genetic influence on specific cognitive abilities. One is a U.S. study of 72 reared-apart twin pairs of a wide age range in adulthood (McGue & Bouchard, 1989); the other is a Swedish study of older twins (average age of 65), including 133 reared-apart twins and 142 control twin pairs reared together (Pedersen, Plomin, Nesselroade, & McClearn, 1992). Both studies show significant heritability estimates for all four specific cognitive abilities. As shown in Table 13.2, the heritability estimates are generally higher than those implied by the twin results summarized in Table 13.1. This discrepancy may be due to the trend, discussed in Chapter 12, for heritability for cognitive abilities to increase during the life span; the reared-apart twins (Table 13.2) are older than the twins reared together (Table 13.1). In both studies, the lowest heritability is found for memory.

As described in Chapter 12, twin studies of general cognitive ability appear to indicate the influence of shared environment in the sense that twin resemblance cannot be explained entirely by heredity. However, it was noted that both identical and fraternal twins experience more similar environments than do nontwin siblings. For this reason, twin studies inflate estimates of shared environment in studies of general cognitive ability. Adoption designs generally suggest less shared environmental influence, especially after childhood. The twin correlations in Table 13.1 also imply substantial influence of shared environment for specific cognitive abilities. In contrast, the two studies of twins reared apart, which also included control samples of twins reared together, found that shared environment has little influence. That is, twins reared apart were almost as similar as twins reared together.

Studies of adoptive relatives can provide a direct test of shared environment, but only two adoption studies of specific cognitive abilities have been reported. One study found little resemblance for adoptive parents and their adopted children or for adoptive siblings on subtests of an intelligence test, except for vocabulary (Scarr &

TABLE 13.2

Heritability Estimates for Specific Cognitive Abilities in Two Studies of Twins Reared Apart

	Heritability Estimate	
Ability	McGue & Bouchard (1989)	Pedersen, Plomin, et al. (1992)
Verbal	0.57	0.58
Spatial	0.71	0.46
Speed	0.53	0.58
Memory	0.43	0.38

Weinberg, 1978a). Thus, this study supports the results of the two twins-reared-apart adoption studies in suggesting that shared environment has little influence on specific cognitive abilities. Like the twin and twins-reared-apart studies, this adoption study found evidence for genetic influence, in that nonadoptive relatives showed greater resemblance than did adoptive relatives.

Specific cognitive abilities are central to a 30-year longitudinal adoption study called the Colorado Adoption Project (Petrill et al., 2003). Figure 13.3 summarizes parent-offspring results for verbal, spatial, processing speed, and memory abilities from early childhood through adolescence (Plomin et al., 1997). Mother-child and father-child correlations were averaged for both adoptive and control (nonadoptive)

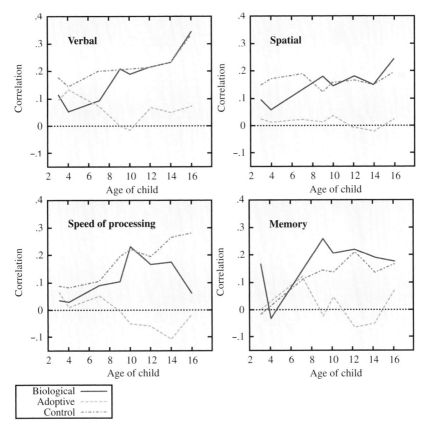

• **FIGURE 13.3** Parent-offspring correlations for factor scores for specific cognitive abilities for adoptive, biological, and control parents and their children at 3, 4, 7, 9, 10, 12, 14, and 16 years of age. Parent-offspring correlations are weighted averages for mothers and fathers. The N's range from 33 to 44 for biological fathers, 159 to 180 for biological mothers, 153 to 197 for adoptive parents, and 136 to 217 for control parents. (From "Nature, nurture and cognitive development from 1 to 16 years: A parent-offspring adoption study" by R. Plomin, D. W. Fulker, R. Corley, & J. C. DeFries. *Psychological Science, 8,* 442–447. © 1997. Used with permission of Psychological Science.)

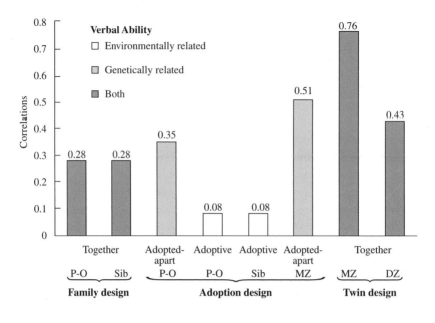

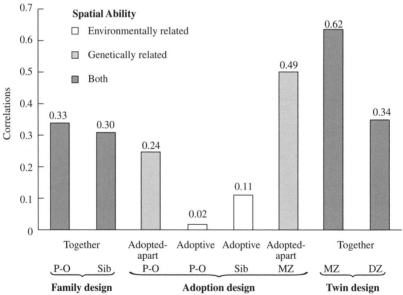

• **FIGURE 13.4** Family, twin, and adoption results for verbal and spatial abilities. The family study results are from the nearly 1000 Caucasian families in the Hawaii Family Study of Cognition, with parent-offspring correlations averaged for mothers and fathers rather than the regression of mid-child on midparent shown in Figure 13.2 (DeFries et al., 1979). The adoption data are from the Colorado Adoption Project, with parent-offspring correlations shown when the adopted children were 16 years old and adoptive sibling correlations averaged across 9 to 12 years (Plomin et al., 1997). The adopted-apart MZ twin data are averaged from the 95 pairs reported by Bouchard et al. (1990) and Pedersen, Plomin, et al. (1992). The twin study correlations are based on more than 1500 pairs of wide age ranges in seven studies from four countries (Plomin, 1988). (From "Human behavioral genetics of cognitive abilities and disabilities" by R. Plomin & I. W. Craig (1997), *BioEssays*, *19*, 1117–1124. Used with permission of BioEssays, ICSU Press.)

families. For each ability, biological parent–adopted child and control parent–control child correlations tend to increase as a function of age. In contrast, adoptive parent–adopted child correlations do not differ substantially from zero at any age. These results indicate increasing heritability and no shared environment.

The results for family, twin, and adoption studies of verbal and spatial ability are summarized in Figure 13.4. The results converge on the conclusion that both verbal and spatial ability show substantial genetic influence but only modest influence of shared environment.

Multivariate Genetic Analysis

Although all specific cognitive abilities are heritable, to what extent are different abilities influenced by the same genes? The hierarchical model of cognitive abilities (see Figure 12.1) is a description of the phenotypic architecture of cognitive abilities. To what extent is the genetic architecture similar? Multivariate genetic analysis can address this question by going beyond the analysis of the variance of a single variable to consider genetic and environmental sources of covariance between traits (see Chapter 7 and Appendix). It yields a key statistic called the *genetic correlation*, which indexes the extent to which genetic influences on one trait also affect another trait. A high genetic correlation implies that if a gene is associated with one trait, there is a good chance that this gene would also be associated with the other trait.

Multivariate genetic analyses of specific cognitive abilities suggest that genetic influences create a hierarchical structure of cognitive abilities that is even stronger than the phenotypic structure (Petrill, 1997). The most surprising finding is how high the genetic correlations are among diverse cognitive abilities such as verbal, spatial, and memory. On average, genetic correlations exceed 0.50 in childhood (Alarcón, Plomin, Fulker, Corley, & DeFries, 1999; Cardon, Fulker, DeFries, & Plomin, 1992; Labuda, DeFries, & Fulker, 1987; Luo, Petrill, & Thompson, 1994; Petrill, Luo, Thompson, & Detterman, 1996; Thompson, Detterman, & Plomin, 1991), adolescence (Luciano et al., 2003; Rijsdijk, Vernon, & Boomsma, 2002), adulthood (Finkel & Pedersen, 2000; Martin & Eaves, 1977; Pedersen, Plomin, & McClearn, 1994; Tambs, Sundet, & Magnus, 1986), and old age (Petrill et al., 1998). These genetic correlations of 0.50 or greater provide strong support for genetic *g*, but they also indicate that there are some genetic effects specific to each of the specific cognitive abilities because the genetic correlations are far less than 1.0. In addition, longitudinal analyses suggest that genetic correlations among specific cognitive abilities increase during development (Price, Dale, & Plomin, 2004; Rietveld, Dolan, et al., 2003). This finding suggests that the developmental rise in heritability of *g* described in Chapter 12 is due to genes that have increasingly general effects across specific cognitive abilities.

A possible exception to the hierarchical model is memory of human faces. Two recent twin studies reported that memory of human faces is heritable and only modestly correlated phenotypically with *g* (Wilmer et al., 2010; Zhu, Song, et al., 2010).

These results have been used to claim that genetic influence on face perception is highly specific, that is, not part of the hierarchical model. However, in these twin studies, genetic correlations between face perception and *g* were not reported, so it remains to be seen whether memory of faces is in fact genetically independent of other cognitive abilities.

Information-Processing Measures

Research on the genetics of specific cognitive abilities has also used laboratory tasks developed by cognitive psychologists to assess how information is processed (Deary, 2000). One early twin study focused on speed-of-processing measures, such as rapid naming of objects and letters (Ho, Baker, & Decker, 1988). These measures are similar to those used to assess the specific cognitive ability factor of perceptual speed. The results of this twin study yield evidence for moderate genetic influence. More traditional reaction-time measures of information processing also show genetic influence in twin studies (Finkel & McGue, 2007) and in a study of twins reared apart (McGue & Bouchard, 1989).

A study of 287 twin pairs aged 6 to 13 (Petrill, Thompson, & Detterman, 1995) used a computerized battery of elementary cognitive tasks designed to test a theory that general cognitive ability is a complex system of independent elementary processes (Detterman, 1986). For example, a speed-of-processing factor was assessed by tasks such as decision time in stimulus discrimination. As shown in Figure 13.5, a probe stimulus is presented above an array of six stimuli, one of which matches the probe. The task is simply to touch as quickly as possible the stimulus that matches the probe. Information-processing tasks can subtract movement time from reaction time to obtain a purer measure of the time required to make the decision. In this study, a measure of decision time based on stimulus discrimination was highly reliable. Despite the simplicity of the task,

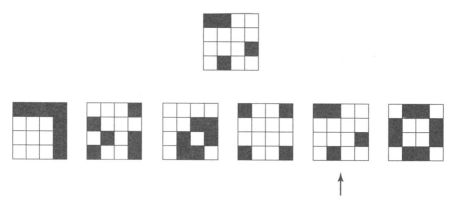

• **FIGURE 13.5** Discrimination task in which the subject simply picks the stimulus from the six below that matches the target above.

it correlates −0.42 with IQ. That is, shorter decision times are associated with higher IQ scores. Twin correlations for this measure of decision time were 0.61 for identical twins and 0.39 for fraternal twins, yielding a heritability of about 45 percent and about 15 percent influence of shared environment. The battery included other measures such as simple reaction time, learning, and memory, most of which showed more modest heritability, ranging down to zero heritability for simple reaction time. Estimates of shared environment also varied widely for the various measures.

Another example is a study of 300 adult twin pairs in which two classic elementary cognitive tasks were assessed: Sternberg's memory scanning and Posner's letter matching (Neubauer, Spinath, Riemann, Borkenau, & Angleitner, 2000). In the Sternberg measure, a random sequence of one, three, or five digits is presented. A target digit is shown, and the task is to indicate as quickly as possible whether the target digit was part of the previously shown set. Reaction time increases linearly from one to three to five digits and is assumed to index the added load for short-term memory. In the Posner task, pairs of letters are shown with the same physical and name identities (A-A), different physical but same name identity (A-a), or different physical and name identities (A-b). The task is to indicate whether the pairs of letters are exactly the same or different in some way. The difference in reaction times for name identity and physical identity is assumed to indicate the time needed for retrieval from long-term memory. These reaction time measures correlate about −0.40 with IQ, that is, individuals with higher IQs respond more quickly. MZ and DZ twin correlations for these five tasks are shown in Figure 13.6. An interesting result is that the more complex tasks such as the five-digit set of the Sternberg measure and the name identity task of the Posner measure showed heritabilities of about 50 percent. In contrast, the simpler tasks showed much lower heritabilities: 6 percent for the one-digit set of the

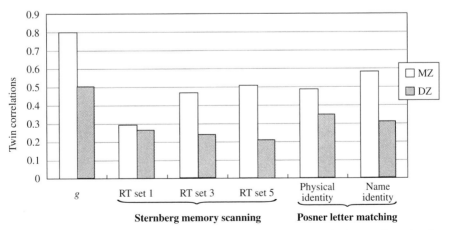

• **FIGURE 13.6** MZ and DZ correlations for two elementary cognitive tasks. See text for a description of the measures. (RT, reaction time.) The *g* measure was an unrotated principal component score derived from standard psychometric tests. (Adapted from Neubauer, Sange, & Pfurtscheller, 1999.)

Sternberg measure and 28 percent for the physical identify task of the Posner measure. A meta-analysis of nine twin studies of reaction-time measures supports the finding that heritability increases as complexity of the task increases (Beaujean, 2005). More recent twin studies using measures of information processing continue to find greater heritability for more complex processing tasks (Singer, MacGregor, Cherkas, & Spector, 2006; Vinkhuyzen, van der Sluis, Boomsma, de Geus, & Posthuma, 2010).

These information-processing measures also suggest a hierarchical structure in that multivariate genetic analyses reveal substantial general as well as some specific genetic effects within the domain of information processing and between information-processing and traditional psychometric measures of cognitive abilities (Plomin & Spinath, 2002). For example, in the information-processing study just described (Figure 13.6; Neubauer et al., 1999), although the two measures were intended to assess very different processes (short-term memory and retrieval from long-term memory), their genetic correlation was 0.84, indicating substantial genetic overlap between these tasks. Moreover, the genetic correlation between a composite based on the two information-processing measures and psychometric g was 0.67 (Plomin & Spinath, 2002). Results supporting the genetic underpinnings of the hierarchical model have also been found for other information-processing measures (Lee et al., 2012; Luciano et al., 2001; Posthuma, de Geus, & Boomsma, 2001; Rijsdijk, Vernon, & Boomsma, 1998).

Working Memory Model

Although cognitive psychology models of information processing have developed separately from the psychometric hierarchical model, they have evolved in a similar direction. The most widely cited model, called the *working memory model,* assumes a central executive system that regulates other subsystems involved in attention, short-term and long-term memory, and other processes (Baddeley, 2007). Twin studies suggest that measures of executive function and working memory are highly heritable (Blokland et al., 2011; Friedman et al., 2008; Panizzon et al., 2011). Although specific tests of these cognitive processes are only moderately correlated with g (Ackerman, Beier, & Boyle, 2005; Friedman et al., 2006), composite measures correlate substantially with g (Colom, Rebollo, Abad, & Shih, 2006). One study reported a genetic correlation of 0.57 between a general executive function factor and IQ (Friedman et al., 2008), but more research is needed to understand the genetic relationships between these cognitive processes and the hierarchical model that has emerged from psychometric tests of cognitive abilities.

Imaging Genetics

Attempts to investigate even more basic processes have led to studies of speed of nerve conduction and brain wave (EEG) measures of event-related potentials. Twin studies of speed of peripheral nerve conduction velocity show high heritability but little correlation with cognitive measures (Rijsdijk & Boomsma, 1997; Rijsdijk, Boomsma, & Vernon, 1995). Twin studies of event-related potentials assessed by EEG yield widely varying heritability estimates across cortical sites, measurement conditions, and age, although much of this inconsistency could be due to the use of small samples (Hansell et al., 2005; van Baal, de Geus, & Boomsma, 1998). An EEG measure called *central coherence*, which assesses the connectivity between cortical regions and is thought to contribute to autism (Happé & Frith, 2006), shows substantial heritability in childhood (van Baal et al., 1998) and adolescence (Van Beijsterveldt, Molenaar, de Geus, & Boomsma, 1998), as does an EEG measure of brain oscillations (Anokhin, Muller, Lindenberger, Heath, & Myers, 2006). However, the genetic as well as phenotypic correlations are low between cognitive abilities and peripheral nerve conduction (Rijsdijk & Boomsma, 1997) and these EEG measures (Posthuma, Neale, Boomsma, & de Geus, 2001; van Baal, Boomsma, & de Geus, 2001).

Magnetic resonance imaging (MRI) and other brain imaging techniques provide greater resolution of brain regions and stronger correlations with cognitive abilities. Combining such brain imaging techniques with genetics has led to a new field called *imaging genetics* (Thompson, Martin, & Wright, 2010). Imaging genetics research began with brain structure, which can be assessed more reliably than brain function. One of the most robust findings is that total brain volume, as well as the volume of most brain regions, correlate moderately (~0.40) with cognitive abilities (Deary et al., 2010). Twin studies have found strong genetic influences on individual differences in the size of many brain regions (Pennington et al., 2000; Thompson et al., 2001). Multivariate genetic twin analyses indicate that the correlation between these measures of brain structure and cognitive ability is largely genetic in origin (Betjemann et al., 2010; Hulshoff Pol et al., 2006; Peper, Brouwer, Boomsma, Kahn, & Hulshoff Pol, 2007; Posthuma et al., 2002) and that most of these genetic effects are explained by total brain volume rather than by the volume of specific brain regions (Schmitt et al., 2010). Twin studies have recently mapped the surface and thickness of areas of cortical brain regions in terms of the genetic correlations among the regions (Chen et al., 2011; Eyler et al., 2011; Rimol et al., 2010). Other more specific measures of brain structure are beginning to be explored, such as asymmetries between the two hemispheres of the brain (Jahanshad et al., 2010). For example, individual differences in the degree of thinning of the cerebral cortex during adolescence are highly heritable (Joshi et al., 2011; van Soelen et al., 2012) and are related to cognitive abilities (Shaw et al., 2006). New structural measures of connectivity also show high heritability and strong correlations with cognitive abilities (Chiang et al., 2009).

Functional imaging studies identify regions of brain activation in response to tasks. A surprising finding is that high cognitive ability is associated with *less* brain activation, presumably because these brains are more efficient (Neubauer & Fink, 2009). Similar to structural imaging results, functional imaging research suggests that activation occurs across diverse brain regions rather than being restricted to a single brain region (Deary et al., 2010). Twin studies are beginning to untangle genetic and environmental sources of these effects. For example, twin studies using functional MRI (fMRI) have found moderate heritability for individual differences in activation of several brain regions during cognitive tasks (Blokland et al., 2011; Koten et al., 2009). fMRI twin studies of functional connectivity between regions of the brain also indicate moderate heritability (Posthuma, de Geus, et al., 2005). A collaborative Human Connectome Project has begun that will provide a comprehensive map of the structural and functional connections between parts of the brain and cognitive abilities in a sample of 1200 twins and their siblings (Schlaggar, 2011). Multivariate genetic analysis is beginning to be used to map genetically driven patterns of activity across brain regions (Park, Shedden, & Polk, 2012). The goal is to understand the genetic and environmental etiologies of individual differences in brain structure and function as they relate to cognitive abilities (Karlsgodt, Bachman, Winkler, Bearden, & Glahn, 2011).

School Achievement

At first glance, tests of school achievement seem quite different from tests of specific cognitive abilities. School achievement tests focus on performance in specific subjects taught at school, such as literacy (reading), numeracy (mathematics), and science. However, although some subjects, such as history, largely require learning facts, others, such as reading, mathematics, and science, are more similar to cognitive abilities because they also involve more general cognitive processes beyond specific content. In the case of reading, most children quickly progress in the early school years from learning to read to reading to learn, that is, to using reading as a domain-general cognitive process. One difference is that the fundamentals of reading are taught in school, whereas the specific cognitive abilities discussed earlier—such as verbal, spatial, memory, and perceptual speed abilities—are not taught explicitly. Nonetheless, as we shall see, multivariate genetic research finds considerable genetic overlap between domains of school achievement and specific cognitive abilities.

The word *achievement* itself implies that school achievement is due to dint of effort, assumed to be an environmental influence, in contrast to *ability,* for which genetic influence seems more reasonable. For the past half-century, the focus of educational research has been on environmental factors, such as characteristics of schools, neighborhoods, and parents. Hardly any attention has been given to the possibility that genetic influences on the characteristics of children affect learning in school (Haworth & Plomin, 2011; Wooldridge, 1994). However, given the strong evidence for genetic influence on general cognitive ability, described in the previous chapter, and on specific cognitive abilities, described earlier in this chapter, it seems reasonable to expect that genetics contributes to individual differences in learning in schools. Moreover, behavioral genetics can go beyond the rudimentary nature-nurture question to ask questions about "how" rather than "how much." For example, we can explore the genetic and environmental etiology of links with cognitive abilities, links between the normal (learning abilities) and abnormal (learning disabilities), and links with developmental changes. Such questions about school achievement have been the target of much behavioral genetic research in the past decade. Reading and mathematics disabilities were discussed in Chapter 11, but the present discussion considers the normal range of individual differences in these and other aspects of school achievement.

The most well studied area by far is reading ability (Olson, 2007). As shown in Figure 13.7, a meta-analysis of a dozen twin studies indicates that reading-related processes such as word recognition, spelling, and reading comprehension show substantial genetic influence, with all average heritability estimates within the narrow range of 0.54 to 0.63 (Harlaar, 2006). General reading composites from such tests yield an average heritability estimate of 0.64. (See Figure 13.7.) Similar results have been reported recently for a twin study in China, despite the different orthography of Chinese (Chow, Ho, Wong, Waye, & Bishop, 2011).

Although it would be reasonable to expect that learning to read (e.g., word recognition) might be less heritable than reading to learn (e.g., reading comprehension), reading ability in the early school years is also highly heritable (Harlaar, Hayiou-Thomas, & Plomin, 2005; Petrill et al., 2007). Even pre-reading skills such as phonological awareness, rapid naming, and verbal memory show substantial genetic influence (Hensler et al., 2010; Samuelsson et al., 2007). Twin studies of genotype-environment interaction reported lower heritability of reading ability for families in low-income neighborhoods (Taylor & Schatschneider, 2010) and greater heritability of reading ability for students with better teachers, which was assessed as the average improvement in reading by each twin's class (Taylor, Roehrig, Hensler, Connor, & Schatschneider, 2010). Although, as in the latter study, improvement in reading and other aspects of academic performance have been used as indices of the quality

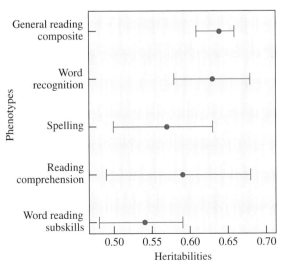

• **FIGURE 13.7** Meta-analysis of heritabilities of reading-related processes. The circles indicate the average heritability, and the lines around the circles indicate the 95 percent confidence intervals. (Adapted from Harlaar, 2006.)

TABLE 13.3

Twin Correlations for Report Card Grades for 13-Year-Olds

Subject Graded	Twin Correlation		Parameter Estimates	
	Identical Twins	Fraternal Twins	Heritability	Shared Environment
History	0.80	0.51	0.58	0.22
Reading	0.72	0.57	0.30	0.42
Writing	0.76	0.50	0.52	0.24
Arithmetic	0.81	0.48	0.66	0.15

SOURCE: Husén (1959). Estimates of heritability and shared environment were calculated from the twin correlations.

or "added value" of teachers and schools, another twin study showed that improvement in academic performance is just as heritable as initial performance, meaning that "added value" cannot be considered a purely environmental measure (Haworth, Asbury, Dale, & Plomin, 2011). An interesting analysis across countries suggests that heritability of reading ability in first grade is similar in Australia, Scandinavian countries, and the United States (Samuelsson et al., 2008).

What about other academic subjects? One of the earliest studies used report card grades in an analysis of data from more than a thousand 13-year-old twins in Sweden (Husén, 1959). Twin correlations for history, reading, writing, and arithmetic (Table 13.3) indicate heritability estimates from 30 to 66 percent and shared environment estimates from 15 to 42 percent. Another early twin study of high school–age twins

TABLE 13.4

Twin Correlations for School Achievement Tests in High School

Test Subject	Twin Correlation		Parameter Estimates	
	Identical Twins	Fraternal Twins	Heritability	Shared Environment
Social studies	0.69	0.52	0.34	0.35
Natural sciences	0.64	0.45	0.38	0.26
English usage	0.72	0.52	0.40	0.32
Mathematics	0.71	0.51	0.40	0.31

SOURCE: Loehlin & Nichols (1976). Estimates of heritability and shared environment were calculated from the twin correlations.

in the United States obtained data from the National Merit Scholarship Qualifying Test for 1300 identical and 864 fraternal twin pairs (Loehlin & Nichols, 1976). The twin correlations shown in Table 13.4 yield heritabilities of about 0.40 and shared environment estimates of about 0.30. Similar results for adolescents have been obtained in the Netherlands (Bartels, Rietveld, van Baal, & Boomsma, 2002b) and in Australia (Wainwright, Wright, Luciano, Geffen, & Martin, 2005).

As with reading, other aspects of early school achievement are also substantially heritable. In a longitudinal study of more than 2000 twin pairs in the United Kingdom, teachers assessed second-graders, using criteria based on the UK National Curriculum for English, mathematics, and science, at 7, 9, and 10 years of age (Kovas, Haworth, Dale, et al., 2007). As shown in Table 13.5, twin correlations are remarkably consistent across subjects and across ages, suggesting heritabilities of about 0.60 and shared environment of only about 0.20, despite the fact that the twins grew up in the same family, attended the same school, and were often taught by the same teacher in the same classroom.

As mentioned earlier, behavioral genetics can go beyond the nature-nurture question of "how much." The first example concerns the genetic links between the normal (learning abilities) and abnormal (learning disabilities). This topic was addressed in relation to cognitive disability in Chapter 11, where DF extremes analysis was introduced (Box 11.1) and we noted that research using this method has led to the conclusion that what we call abnormal may be part of the normal distribution. That is, mild cognitive disability is the low end of the same genetic and environmental influences

TABLE 13.5

Twin Correlations for UK National Curriculum Ratings at 7, 9, and 10 Years

Subject		Twin Correlation		Parameter Estimates	
		Identical Twins	Fraternal Twins	Heritability	Shared Environment
English	7 years	0.82	0.50	0.64	0.18
	9 years	0.78	0.46	0.64	0.14
	10 years	0.80	0.49	0.62	0.18
Math	7 years	0.78	0.47	0.62	0.16
	9 years	0.76	0.41	0.70	0.06
	10 years	0.76	0.48	0.56	0.20
Science	9 years	0.76	0.44	0.64	0.12
	10 years	0.76	0.57	0.38	0.38

SOURCE: Kovas, Haworth, Dale, et al. (2007).

responsible for variation in the normal distribution of general cognitive ability. In other words, mild cognitive disability is not really a disorder—it is the low end of the normal distribution. Similar results have been found for abilities and disabilities in reading, language, and mathematics (Plomin & Kovas, 2005). DF extremes analyses of the data presented in Table 13.5 support the conclusion that the abnormal is normal across domains and ages (Kovas, Haworth, Dale, et al., 2007).

A second example involves developmental change and continuity in genetic and environmental influences. As discussed in Chapter 12 in relation to general cognitive ability, two types of developmental questions can be asked: Does heritability change during development? Do genetic factors contribute to developmental change? In the case of general cognitive ability, the answer to the first question is yes, but for school achievement the answer appears to be no, as suggested, for example, by the results in Table 13.5. However, studies with a larger age range for school achievement measures would be necessary to answer this question more definitely. For general cognitive ability, the answer to the second question is that genetic factors largely contribute to continuity even from childhood to adulthood, although some evidence for genetic change exists, especially during the transition to school. Results appear to be similar for school achievement: Genetics appears to contribute largely to continuity, with some evidence for genetic change (Bartels, Rietveld, van Baal, & Boomsma, 2002a; Byrne et al., 2007; Petrill et al., 2007), especially during the transition to school (Byrne et al., 2005, 2009). For example, longitudinal analyses of the data in Table 13.5 yielded age-to-age genetic correlations from 7 to 10 years of age of about 0.70 (Kovas, Haworth, Dale, et al., 2007). Even preschool speech and language at 4 years of age is related to reading at 10 years of age largely for genetic reasons (Hayiou-Thomas, Harlaar, Dale, & Plomin, 2010).

A third example is multivariate genetic analysis among learning abilities and between learning abilities and general cognitive ability, which has been a major focus of research in recent years. Earlier in this chapter, multivariate genetic research on specific cognitive abilities was presented which suggested that most genetic effects are general although some effects are specific, in line with the hierarchical model. An even stronger hierarchical model is emerging from multivariate genetic research on learning abilities: Genetic correlations are very high within and between learning abilities. For example, the many processes related to reading (see Figure 13.7) show substantial genetic overlap (Harlaar et al., 2010), as do different mathematics abilities (Kovas, Haworth, Petrill, et al., 2007). In addition, these general effects within domains extend across domains. In a review of such studies, genetic correlations varied from 0.67 to 1.0 between reading and language (five studies), 0.47 to 0.98 between reading and mathematics (three studies), and 0.59 to 0.98 between language and mathematics (two studies) (Plomin & Kovas, 2005). The average genetic correlation across all of these studies was about 0.70. Genetic correlations among the measures shown in Table 13.5 are 0.79 on average (Kovas, Haworth, Dale, et al., 2007). A recent study of more than 5000 pairs of 12-year-old twins tested online on a Web-based battery of measures of reading, mathematics, and language found high genetic correlations of 0.75, 0.78, and 0.91 between latent factors of these learning abilities, as shown in Figure 13.8 (Davis, Haworth, & Plomin, 2009b). Other recent twin studies continue to find that genetic overlap across learning abilities is substantial (Hart, Petrill, Thompson, & Plomin, 2009).

Could this general genetic factor that affects scores on diverse tests of school achievement be general cognitive ability? The results in Figure 13.8 suggest that the strong genetic correlations among learning abilities extend to general cognitive ability, with genetic correlations of 0.86, 0.88, and 0.91. Other multivariate genetic research has found similar results, although genetic correlations are somewhat lower because the results shown in Figure 13.8 are based on latent factors, which are more reliable. Multivariate genetic analyses between tests of school achievement and general cognitive ability suggest that genetic effects on school achievement test scores show moderate genetic correlations with general cognitive ability but that the genetic correlations

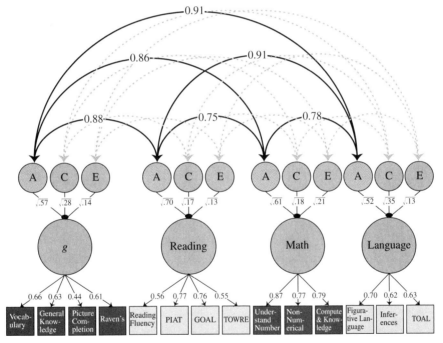

• **FIGURE 13.8** Genetic correlations among learning abilities and *g*. A = additive genetic effects; C = shared (common) environmental effects; E = nonshared environmental effects. Squares represent measured traits; circles represent latent factors. Multiple tests are used to index latent factors of *g*, reading, mathematics, and language. The lower tier of arrows represents factor loadings of the tests on the latent factor. The second tier of coefficients represents the genetic and environmental components of the variance of the latent variables—the path coefficients in this path diagram are the square roots of these coefficients. The curved arrows at the top represent correlations between genetic influences. (From "Learning abilities and disabilities: Generalist genes in early adolescence" by O. S. P. Davis, C. M. A. Haworth & R. Plomin (2009b), *Cognitive Neuropsychiatry, 14,* 312–331. Reprinted by permission of the publisher (Taylor & Francis Ltd., http://www.tandf.co.uk/journals).)

are higher among the school achievement measures. For example, based on the school achievement measures listed in Table 13.5, the average genetic correlation is 0.61 between the school achievement measures and general cognitive ability, in contrast to the average genetic correlation of 0.79 between the school achievement measures (Kovas, Haworth, Dale, et al., 2007). A multivariate genetic analysis of several reading-related processes yielded an average genetic correlation of 0.51 with general cognitive ability, considerably lower than the average genetic correlation of 0.76 among the reading-related processes (Gayán & Olson, 2003). A review of a dozen such studies reached a similar conclusion, with average genetic correlations of about 0.70 between school achievement measures and about 0.60 between these measures and general cognitive ability (Plomin & Kovas, 2005). These high genetic correlations suggest a "top-heavy" hierarchical model in the sense that most genetic effects are general across learning and cognitive abilities but some genetic variance is specific to group factors such as reading and mathematics and some genetic variance is specific to the individual tests. It has been suggested that information-processing speed, discussed earlier, may be key to these general genetic effects (Willcutt, Pennington, et al., 2010).

Identifying Genes

As mentioned in earlier chapters, research has begun to identify specific genes associated with cognitive disabilities such as dementia and reading disability (Chapter 11) and with general cognitive ability (Chapter 12). DF extremes results mentioned in the previous section suggest that QTLs associated with learning disabilities such as reading disability are also likely to be associated with learning abilities such as reading ability, that is, with variation throughout the distribution.

QTL linkage analyses of specific cognitive abilities and school achievement in the normal range have reported weak linkages for a memory task (Singer, Falchi, MacGregor, Cherkas, & Spector, 2006), reading ability and spelling (Bates et al., 2007), and academic achievement (Wainwright et al., 2006). Genomewide association studies have also begun to be reported for specific cognitive abilities, including reading ability (Luciano, Montgomery, Martin, Wright, & Bates, 2011; Meaburn et al., 2008), mathematics ability (Docherty, Davis, et al., 2010), memory tasks (Papassotiropoulos et al., 2006), and information-processing measures (Cirulli et al., 2010; Luciano, Hansell, et al., 2011; Need et al., 2009). These first genomewide association studies suggest a familiar refrain: No associations of sufficiently large effect size have emerged that reach genomewide significance, suggesting that heritability is caused by many genes of small effect. As in other domains, the major strategy for identifying these genes of small effect is to increase the sample sizes by conducting meta-analyses across studies. A recent remarkable example is a meta-analysis of nearly 20,000 individuals with structural MRI data from 17 studies, which identified with genomewide significance a SNP associated with hippocampal volume and another SNP associated with intracranial volume; the latter SNP was also associated with general intelligence (Stein et al., 2012).

Summary

Family studies of specific cognitive abilities, most notably the Hawaii Family Study of Cognition, show greater familial resemblance for verbal and spatial abilities than for perceptual speed and memory. Tests within each ability vary in their degree of familial resemblance. Twin studies indicate that most of this familial resemblance is genetic in origin, as do studies of identical twins reared apart. Developmental analyses of adoption data indicate that heritability increases during childhood. The results for family, twin, and adoption studies of verbal and spatial ability, summarized in Figure 13.4, converge on the conclusion that both verbal and spatial ability show substantial genetic influence but only modest influence of shared environment. Other cognitive abilities, especially certain types of memory, may be less heritable.

Multivariate genetic research suggests that the hierarchical model of cognitive abilities is largely genetic in origin. That is, genetic correlations among diverse cognitive abilities generally exceed 0.50, suggesting strong general effects of genes, especially for more complex processing tasks. These same genetic correlations indicate that some genetic effects are specific to each ability. Genetic research on measures of information processing, including executive function and working memory, also indicates substantial genetic influence and supports a hierarchical structure. In addition, these measures of cognitive processes are genetically related to psychometric tests of cognitive abilities. Measures of brain structure and function show similar results: substantial heritability and substantial genetic correlations among brain structures and functions and between these brain measures and cognitive abilities.

In the field of education, individual differences in school achievement have been assumed to be due primarily to environmental influences; however, twin studies consistently show substantial genetic influence, not just for reading but also for other subject areas such as mathematics and science. In each of these domains, the abnormal is normal; that is, learning disabilities are the low end of the same genetic and environmental influences responsible for variation in the normal distributions of learning abilities. Similar to general cognitive ability, genetic effects on learning abilities largely contribute to continuity during childhood, although some significant change is observed. Multivariate genetic analyses of diverse measures of school achievement provide strong support for a hierarchical model in that genetic correlations are high between domains of school performance (about 0.70). Multivariate genetic analyses between tests of school achievement and general cognitive ability suggest that genetic influence on school achievement overlaps substantially with genetic influence on general cognitive ability, although some genetic influences are specific to achievement.

Research is under way to identify associations between DNA markers and cognitive abilities, not just disabilities. As with other complex traits, the first genomewide association studies have not found genomewide significant associations, suggesting that heritability is caused by many genes of small effect.

CHAPTER FOURTEEN

Schizophrenia

P sychopathology has been, and continues to be, one of the most active areas of behavioral genetic research, largely because of the social importance of mental illness. One out of two persons in the United States has some form of disorder during their lifetime, and one out of three persons suffered from a disorder within the last year (Kessler et al., 2007). The costs in terms of suffering to patients and their friends and relatives, as well as the economic costs, make psychopathology one of the most pressing problems today.

The genetics of psychopathology led the way toward the acceptance of genetic influence in psychology and psychiatry. The history of psychiatric genetics is described in Box 14.1. This chapter and the next two provide an overview of what is known about the genetics of several major categories of psychopathology: schizophrenia, mood disorders, and anxiety disorders. Other disorders, such as posttraumatic stress disorder, somatoform disorders, and eating disorders, are also briefly reviewed, as are disorders usually first diagnosed in childhood: autism, attention-deficit hyperactivity, and tic disorders. Other major categories in the *Diagnostic and Statistical Manual of Mental Disorders-IV* (DSM-IV) include cognitive disorders such as dementia (Chapter 11), personality disorders (Chapter 17), and drug-related disorders (Chapter 18). The DSM-IV includes several other disorders for which no genetic research is as yet available (e.g., dissociative disorders such as amnesia and fugue states). Much has been written about the genetics of psychopathology, including recent texts (Faraone, Tsuang, & Tsuang, 2002; Jang, 2005; Kendler & Prescott, 2007) and several edited books (e.g., Dodge & Rutter, 2011; Hudziak, 2008; Kendler & Eaves, 2005; Ritsner, 2009). Many questions remain concerning diagnosis, most notably the extent of comorbidity and heterogeneity (Cardno et al., 2012). Diagnoses to date depend on symptoms, and it is possible that the same symptoms have different causes and that different symptoms could have the same causes (Ritsner & Gottesman, 2011). One of the hopes for genetic research is that it can begin to provide diagnoses based on causes rather than symptoms. We will return to this issue in Chapter 15.

This chapter focuses on schizophrenia, the most highly studied area in behavioral genetic research on psychopathology. Schizophrenia involves persistent abnormal beliefs (delusions), hallucinations (especially hearing voices), disorganized speech (odd associations and rapid changes of subject), grossly disorganized behavior, and so-called negative symptoms, such as flat affect (lack of emotional response) and avolition (lack of motivation). A diagnosis of schizophrenia requires that such symptoms occur for at least six months. It usually strikes in late adolescence or early adulthood. Early onset in adolescence tends to be gradual but has a worse prognosis. Although it derives from Greek words meaning "split mind," schizophrenia has nothing to do with the notion of a "split personality."

More genetic research has focused on schizophrenia than on other areas of psychopathology for three reasons. First, it is the most severe form of psychopathology and one of the most debilitating of all disorders (Üstün et al., 1999). Second, it is so common, with a lifetime risk in nearly 1 percent of the population (Saha, Chant, Welham, & McGrath, 2005). Third, it generally lasts a lifetime, although a few people recover, especially if they have had just one episode (Robinson, Woerner, McMeniman, Mendelowitz, & Bilder, 2004); there are signs, however, that recovery rates are improving (Bellack, 2006). Unlike patients of two decades ago, most people with schizophrenia are no longer institutionalized because drugs can control some of their worst symptoms. Nonetheless, schizophrenics still occupy half the beds in mental hospitals, and those discharged make up about 10 percent of the homeless population (Fischer & Breakey, 1991). It has been estimated that the cost to our society of schizophrenia alone is greater than that of cancer (McEvoy, 2007).

Family Studies

The basic genetic results for schizophrenia were described in Chapter 3 to illustrate genetic influence on complex disorders. Family studies consistently show that schizophrenia is familial (Ritsner & Gottesman, 2011). In contrast to the base rate of about 1 percent lifetime risk in the population, the risk for relatives increases with genetic relatedness to the schizophrenic proband: 4 percent for second-degree relatives and 9 percent for first-degree relatives.

The average risk of 9 percent for first-degree relatives differs for parents, siblings, and offspring of schizophrenics. In 14 family studies of over 8000 schizophrenics, the median risk was 6 percent for parents, 9 percent for siblings, and 13 percent for offspring (Gottesman, 1991; Ritsner & Gottesman, 2011). The low risk for parents of schizophrenics (6 percent) is probably due to the fact that schizophrenics are less likely to marry and those who do marry have relatively few children. For this reason, parents of schizophrenics are less likely than expected to be schizophrenic. When schizophrenics do become parents, the rate of schizophrenia in their offspring is high (13 percent). The risk is the same regardless of whether the mother or the father is schizophrenic. When both parents are schizophrenic, the risk for their offspring

BOX 14.1 • The Beginnings of Psychiatric Genetics: Bethlem Royal and Maudsley Hospitals

Founded in London, in 1247, Bethlem Hospital is one of the oldest institutions in the world caring for people with mental disorders. However, there have been times in Bethlem's long history when it was associated with some of the worst images of mental illness, and it gave us the origin of the word *bedlam*. Perhaps the most famous portrayal is in the final scene of Hogarth's series of paintings *A Rake's Progress,* which shows the Rake's decline into madness at Bethlem (see figure). Hogarth's portrayal assumes that madness is the consequence

A Rake's Progress. (William Hogarth, *A Rake's Progress,* 1735. Plate 8. The British Museum.)

shoots up to 46 percent. Siblings provide the least biased risk estimate, and their risk (9 percent) is in between the estimates for parents and for offspring. Although the risk of 9 percent is high, nine times the population risk of 1 percent, it should be remembered that the majority of schizophrenics do not have a schizophrenic first-degree relative.

Eliot Slater

of high living and therefore, it is implied, a wholly environmental affliction.

The observation that mental disorders have a tendency to run in families is ancient, but among the first efforts to record this association systematically were those at Bethlem Hospital. Records from the 1820s show that one of the routine questions that doctors had to attempt to answer about the illness of a patient they were admitting was "whether hereditary?" This, of course, predated the development of genetics as a science, and it was not until a hundred years later that the first research group on psychiatric genetics was established in Munich, Germany, under the leadership of Emil Kraepelin. The Munich department attracted many visitors and scholars, including a mathematically gifted young psychiatrist from Maudsley Hospital, Eliot Slater, who obtained a fellowship to study psychiatric genetics there. In 1935, Slater returned to London and started his own research group, which led to the creation in 1959 of the Medical Research Council's (MRC) Psychiatric Genetics Unit. The Bethlem and Maudsley Twin Register, set up by Slater in 1948, was among the important resources that underpinned a number of influential studies, and he introduced sophisticated statistical approaches to data evaluation. The MRC Psychiatric Genetics Unit became one of the key centers for training and played a major role in the career development of many overseas postdoctoral students, including Irving Gottesman, Leonard Heston, and Ming Tsuang.

In 1971, Slater published the first psychiatric genetics textbook in English, the *Genetics of Mental Disorders* (Slater & Cowie, 1971). Later in the 1970s, following Slater's retirement, psychiatric genetics became temporarily unfashionable in the United Kingdom but was continued as a scientific discipline in North America and mainland Europe by researchers trained by Slater or influenced by his work.

The family design provides the basis for genetic high-risk studies of the development of children whose mothers were schizophrenic. In one of the first such studies, begun in the early 1960s in Denmark, 200 such offspring were followed until their forties (Parnas et al., 1993). In the high-risk group whose mothers were schizophrenic, 16 percent were diagnosed as schizophrenic (whereas 2 percent in the low-risk group were schizophrenic), and the children who eventually became schizophrenic had mothers whose schizophrenia was more severe. These children experienced a less stable home life and more institutionalization, reminding us that family studies do not disentangle nature and nurture in the way an adoption study does. The children who became schizophrenic were more likely to have had birth complications, particularly prenatal viral infection (Cannon et al., 1993). They also showed attention problems in childhood, especially problems in "tuning out" incidental stimuli like the ticking of a clock (Hollister, Mednick, Brennan, & Cannon, 1994). Similar results were found in childhood in one of the best U.S. genetic high-risk studies, which also found more personality disorders in the offspring of schizophrenic parents when the offspring were young adults (Erlenmeyer-Kimling et al., 1995). Recent studies also suggest that schizophrenia and bipolar disorder frequently co-occur (Laursen, Agerbo, & Pedersen, 2009) and that such comorbidity is due primarily to genetic influences (Lichtenstein et al., 2009) (see below).

Twin Studies

Twin studies show that genetics contributes importantly to familial resemblance for schizophrenia. As was shown in Figure 3.6, the probandwise concordance for MZ twins is 48 percent and the concordance for DZ twins is 17 percent. In a meta-analysis of 14 twin studies of schizophrenia using a liability-threshold model (see Chapter 3), these concordances suggest a heritability of liability of about 80 percent (Sullivan, Kendler, & Neale, 2003). Studies continue to confirm these earlier findings, yielding probandwise concordances of 41 to 65 percent in MZ and 0 to 28 percent in DZ pairs (Cardno et al., 2012).

A dramatic case study involved identical quadruplets, called the Genain quadruplets, all of whom were schizophrenic, although they varied considerably in severity of the disorder (DeLisi et al., 1984) (Figure 14.1). For 14 pairs of reared-apart identical twins in which at least one member of each pair became schizophrenic, 9 pairs (64 percent) were concordant (Gottesman, 1991).

Despite the strong and consistent evidence for genetic influence provided by the twin studies, it should be remembered that the average concordance for identical twins is only about 50 percent. In other words, half of the time these genetically identical pairs of individuals are discordant for schizophrenia, an outcome that provides strong evidence for the importance of nongenetic factors, which could include epigenetic (see Chapter 10) or nonshared environmental factors (see Chapter 7), despite the heritability of the hypothetical construct of liability being 80 percent.

Because differences within pairs of identical twins cannot be genetic in origin, the co-twin control method can be used to study nongenetic reasons why one identical twin is schizophrenic and the other is not. One early study of discordant identical twins found few life history differences except that the schizophrenic co-twins were more likely to have had birth complications and some neurological abnormalities

• **FIGURE 14.1** Identical quadruplets (known under the fictitious surname Genain), each of whom developed symptoms of schizophrenia between the ages of 22 and 24. (Courtesy of Miss Edna Morlok.)

(Mosher, Polling, & Stabenau, 1971). Follow-up studies also found differences in brain structures and more frequent birth complications for the schizophrenic co-twin in discordant identical twin pairs (Torrey, Bowler, Taylor, & Gottesman, 1994). Recent research has found epigenetic (DNA methylation) differences within pairs of identical twins discordant for schizophrenia (Dempster et al., 2011).

An interesting finding has emerged from another use of discordant twins: studying their offspring or other first-degree relatives. Discordant identical twins provide direct proof of nongenetic influences because the twins are identical genetically yet discordant for schizophrenia. Even though one twin in discordant pairs is spared from schizophrenia for environmental reasons, that twin still carries the same high genetic risk as the twin who is schizophrenic. That is why nearly all studies find rates of schizophrenia as high in the families of discordant as in concordant pairs (Gottesman & Bertelsen, 1989; McGuffin, Farmer, & Gottesman, 1987).

For the offspring of discordant fraternal twins, the children of the schizophrenic twin are at much greater risk than are the children of the nonschizophrenic twin.

Members of discordant fraternal twin pairs, unlike identical twins, differ genetically as well as environmentally. However, sample sizes have been small, and one such small study did not support earlier conclusions (Kringlen & Cramer, 1989; see also, Torrey, 1990). Nonetheless, these data provide food for thought about the complex interactions between nature and nurture in schizophrenia and schizophrenia-related disorders.

Adoption Studies

Results of adoption studies agree with those of family and twin studies in pointing to genetic influence in schizophrenia. As described in Chapter 6, the first adoption study of schizophrenia by Leonard Heston in 1966 is a classic study. The results (see Box 6.1) showed that the risk of schizophrenia in adopted offspring of schizophrenic birth mothers was 11 percent (5 of 47), much greater than the 0 percent risk for 50 adoptees whose birth parents had no known mental illness. The risk of 11 percent is similar to the risk for offspring reared by their schizophrenic biological parents. This finding not only indicates that family resemblance for schizophrenia is largely genetic in origin, but it also implies that growing up in a family with schizophrenics does not increase the risk for schizophrenia beyond the risk due to heredity.

Box 6.1 also mentioned that Heston's results have been confirmed and extended by other adoption studies. Two Danish studies began in the 1960s with 5500 children adopted between 1924 and 1947 as well as 10,000 of their 11,000 biological parents. One of the studies (Rosenthal, Wender, Kety, & Schulsinger, 1971; Rosenthal et al., 1968) used the adoptees' study method. This method is the same as that used in Heston's study, but important experimental controls were added. At the time of these studies, birth parents were typically teenagers when they placed children for adoption. Consequently, because schizophrenia does not usually occur until later in life, often neither the adoption agencies nor the adoptive parents were aware of the diagnosis. In addition, both schizophrenic fathers and mothers were studied to assess whether Heston's results, which involved only mothers, were influenced by prenatal maternal factors.

The first Danish study began by identifying biological parents who had been admitted to a psychiatric hospital. Biological mothers or fathers who were diagnosed as schizophrenic and whose children had been placed in adoptive homes were selected. This procedure yielded 44 birth parents (32 mothers and 12 fathers) who were diagnosed as chronic schizophrenics. Their 44 adopted children were matched to 67 control adoptees whose birth parents had no psychiatric history, as indicated by the records of psychiatric hospitals. The adoptees, with an average age of 33, were interviewed for three to five hours by an interviewer blind to the status of their birth parents.

Three (7 percent) of the 44 proband adoptees were chronic schizophrenics, whereas none of the 67 control adoptees were (Figure 14.2). Moreover, 27 percent of the probands showed schizophrenic-like symptoms, whereas 18 percent of the controls had similar symptoms. Results were similar for 69 proband adoptees whose parents were selected by using broader criteria for schizophrenia. Results were also similar

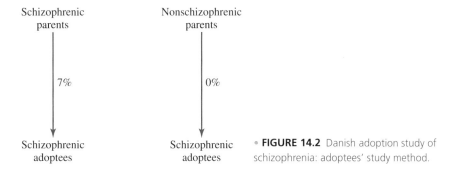

Schizophrenic parents

7%

Schizophrenic adoptees

Nonschizophrenic parents

0%

Schizophrenic adoptees

• **FIGURE 14.2** Danish adoption study of schizophrenia: adoptees' study method.

regardless of whether the mother or the father was schizophrenic. The unusually high rates of psychopathology in the Danish control adoptees may have occurred because the study relied on hospital records to assess the psychiatric status of the birth parents. For this reason, the study may have overlooked psychiatric problems of control parents that had not come to the attention of psychiatric hospitals. To follow up this possibility, the researchers interviewed the birth parents of the control adoptees and found that one-third fell in the schizophrenic spectrum. Thus, the researchers concluded that "our controls are a poor control group and our technique of selection has minimized the differences between the control and index groups" (Wender, Rosenthal, Kety, Schulsinger, & Welner, 1974, p. 127). This bias is conservative in terms of demonstrating genetic influence.

An adoptees study in Finland confirmed these results (Tienari et al., 2004). About 10 percent of adoptees who had a schizophrenic biological parent showed some form of psychosis, whereas 1 percent of control adoptees had similar disorders. This study also suggested genotype-environment interaction, because adoptees whose biological parents were schizophrenic were more likely to have schizophrenia-related disorders when the adoptive families functioned poorly.

The second Danish study (Kety et al., 1994) used the adoptees' family method, focusing on 47 of the 5500 adoptees diagnosed as chronically schizophrenic. A matched control group of 47 nonschizophrenic adoptees was also selected. The biological and adoptive parents and siblings of the index and control adoptees were interviewed. The rate of chronic schizophrenia was 5 percent (14 of 279) for the first-degree biological relatives of schizophrenic adoptees and 0 percent (1 of 234) for the biological relatives of the control adoptees. The adoptees' family method also provides a direct test of the influence of the environmental effect of having a schizophrenic relative. If familial resemblance for schizophrenia is caused by the family environment created by schizophrenic parents, schizophrenic adoptees should be more likely to come from adoptive families with schizophrenia, relative to the control adoptees. To the contrary, 0 percent (0 of 111) of the schizophrenic adoptees had adoptive parents or siblings who were schizophrenic—like the 0 percent incidence (0 of 117) for the control adoptees (Figure 14.3).

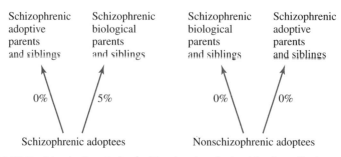

• **FIGURE 14.3** Danish adoption study of schizophrenia: adoptees' family method.

This study also included many biological half siblings of the adoptees (Kety, 1987). Such a situation arises when biological parents place a child for adoption and then later have another child with a different partner. The comparison of biological half siblings who have the same father (paternal half siblings) with those who have the same mother (maternal half siblings) is particularly useful for examining the possibility that the results of adoption studies may be affected by prenatal factors rather than by heredity. Paternal half siblings are less likely to be influenced by prenatal factors because they were born to different mothers. Among half siblings of schizophrenic adoptees, 16 percent (16 of 101) were schizophrenic; among half siblings of control adoptees, only 3 percent (3 of 104) were schizophrenic. The results were the same for maternal and paternal half siblings, an outcome suggesting that prenatal factors are not likely to be of major importance in the origin of schizophrenia.

In summary, the adoption studies clearly point to genetic influence. Moreover, adoptive relatives of schizophrenic probands do not show increased risk for schizophrenia. These results imply that familial resemblance for schizophrenia is due to heredity rather than to shared family environment.

Schizophrenia or Schizophrenias?

Is schizophrenia one disorder or is it a heterogeneous collection of disorders? When the disorder was named in 1908, it was called "the schizophrenias." Multivariate genetic analysis can address this fundamental issue of heterogeneity. The classic subtypes of schizophrenia—such as catatonic (disturbance in motor behavior), paranoid (persecution delusions), and disorganized (both thought disorder and flat affect are present)—are not supported by genetic research. That is, although schizophrenia runs in families, the particular subtype does not. This result is seen most dramatically in a follow-up of the Genain quadruplets (DeLisi et al., 1984). Although they were all diagnosed as schizophrenic, their symptoms varied considerably.

There is evidence that more severe schizophrenia is more heritable than milder forms (Gottesman, 1991). Furthermore, the evidence from both early studies and more recent work, using multivariate statististical methods such as cluster analysis, suggests that the classic "disorganized" subtype of schizophrenia, even if it does not "breed true," shows an especially high rate of affected family members (Cardno et al., 1999; Farmer, McGuffin, & Gottesman, 1987). An alternative to the classic subtypes is a distinction largely based on severity (Crow, 1985). Type I schizophrenia, which has a better prognosis and a better response to drugs, involves active symptoms, such as hallucinations. Type II schizophrenia, which is more severe and has a poorer prognosis, involves passive symptoms, such as withdrawal and lack of emotion. Type II schizophrenia appears to be more heritable than type I (Dworkin & Lenzenweger, 1984).

Another approach to the problem of heterogeneity divides schizophrenia on the basis of family history (Murray, Lewis, & Reveley, 1985), although there are problems with this approach (Eaves, Kendler, & Schulz, 1986) and there is clearly no simple dichotomy (Jones & Murray, 1991). These typologies seem more likely to represent a continuum from less to more severe forms of the same disorder rather than genetically distinct disorders (McGuffin et al., 1987).

A related strategy is the search for behavioral or biological markers of genetic liability, called endophenotypes (Gottesman & Gould, 2003), discussed in Chapter 10. Many potential endophenotypes have been suggested for schizophrenia, including various structural and functional markers in the brain, olfactory deficits, and attention and memory deficits (Ritsner & Gottesman, 2011). One additional example of a behavioral endophenotype in schizophrenia research is called smooth-pursuit eye tracking. This term refers to the ability to follow a moving object smoothly with one's eyes without moving the head (Levy, Holzman, Matthysse, & Mendell, 1993). Some studies have shown that schizophrenics whose eye tracking is jerky tend to have more negative symptoms and that their relatives with poor eye tracking are more likely to show schizophrenic-like behaviors (Clementz, McDowell, & Zisook, 1994). However, other research does not support this hypothesis (Torrey et al., 1994). The hope is that such endophenotypes will clarify the inheritance of schizophrenia and assist attempts to find specific genes responsible for schizophrenia.

Although some researchers assume that schizophrenia is heterogeneous and needs to be split into subtypes, others argue in favor of the opposite approach, lumping schizophrenia-like disorders into a broader spectrum of schizoid disorders (Farmer et al., 1987; McGue & Gottesman, 1989). Because schizophrenia co-occurs with various other disorders, including depression, anxiety, and substance abuse disorders, future analyses of such comorbidity may shed new light on the genetic factors that underlie schizophrenia and related disorders (Ritsner & Gottesman, 2011).

Identifying Genes

Before the new DNA markers were available, attempts were made to associate classic genetic markers, such as blood groups, with schizophrenia. For example, several early studies suggested a weak association of schizophrenia marked by paranoid delusions with the major genes encoding human leukocyte antigens (HLAs) of the immune response, a gene cluster associated with many diseases (McGuffin & Sturt, 1986).

Although schizophrenia was one of the first behavioral domains put under the spotlight of molecular genetic analysis, it has been slow to reveal evidence for specific genes. During the euphoria of the 1980s, when the new DNA markers were first being used to find genes for complex traits, some claims were made for linkage, but they could not be replicated. The first was a claim for linkage with an autosomal dominant gene on chromosome 5 for Icelandic and British families (Sherrington et al., 1988). However, combined data from five other studies in other countries failed to confirm the linkage (McGuffin et al., 1990).

More than 20 genomewide linkage scans (with more than 350 genetic markers) have been published, but none have suggested a gene of major effect for schizophrenia (Riley & Kendler, 2006). Hundreds of reports of linkage for schizophrenia in the 1990s led to a confusing picture because few studies were replicated. However, greater clarity has emerged since around 2000. For example, a meta-analysis of 20 genomewide linkage scans of schizophrenia in diverse populations indicated greater consistency of linkage results than previously recognized (Lewis et al., 2003). Significant linkage was found on the long arm of chromosome 2 (2q); linkage was suggested for ten other regions, including 6p and 8p. It has been difficult to detect linkage signals because linkage analysis requires very large samples to discern small effects.

Association studies of schizophrenia have also provided their own challenges. Over 1000 genes have been tested for association with schizophrenia, making it one of the most studied disorders through a candidate gene approach (Gejman, Sanders, & Kendler, 2011). Despite this fact, there is considerable inconsistency in the results. Over the past 10 years, multiple genes have been suggested, such as *neuregulin 1* on chromosome 8 (Stefansson et al., 2002) and *dysbindin* at 6p22.3 (Straub et al., 2002), as well as other genes related to neurotransmitters expressed in the brain, such as dopamine. However, many of these findings do not replicate across individual studies, possibly due in part to small effect sizes, small sample sizes, or the selective reporting of positive results.

Over the past few years, efforts have been made to try to resolve some of these issues. Larger samples obtained by combining studies are showing greater power to detect genes that increase risk for schizophrenia. Genomewide association studies (GWAS), which systematically look at the whole genome, have detected new possible loci, such as the **major histocompatibility complex** (Purcell et al., 2009; Ripke et al., 2011; Shi et al., 2009; Stefansson et al., 2009), *TCF4* (Stefansson et al., 2009), and at least another half dozen genes (Bergen & Petryshen, 2012). Moreover, as mentioned in Chapter 9, success has also been found when looking at the risk across a set of genes. For example, the International Schizophrenia Consortium has found that hundreds of genes, each with small individual effects, contribute to the risk for developing the disorder (Purcell et al., 2009). There is also growing evidence that copy number variants (CNVs, see Chapter 9) are also seen more frequently in individuals with schizophrenia. Rare and large CNVs associated with schizophrenia have been found on several chromosomes (see Gejman et al., 2011, for a review). Many more common smaller CNVs associated with schizophrenia also appear to be associated with a broad range of neurodevelopmental problems (Sahoo et al., 2011). These results collectively support a very complex genetic architecture underlying this disorder.

Summary

Psychopathology is the most active area of research in behavioral genetics. For schizophrenia, lifetime risk is about 1 percent in the general population, 10 percent in first-degree relatives whether reared together or adopted apart, 17 percent for fraternal twins, and 48 percent for identical twins. This pattern of results indicates substantial genetic influence as well as nonshared environmental influence. Genetic high-risk studies and co-twin control studies suggest that birth complications and attention problems in childhood are weak predictors of schizophrenia, which usually strikes in early adulthood. Genetic influence has been found for both the adoptees' study method, like that used in the first adoption study by Heston, and the adoptees' family method. More severe schizophrenia may be more heritable than less severe forms.

Linkage studies with schizophrenia have begun to yield consistent results and, combined with results from association studies, have led to the identification of several genes or regions that have significant but small associations with schizophrenia. Overall, it seems likely that genetic liability to schizophrenia results from multiple genes of small effect.

CHAPTER FIFTEEN

Other Adult Psychopathology

Although schizophrenia has been the most highly studied disorder in behavioral genetics, in recent years the spotlight has turned to mood disorders. In this chapter, we provide an overview of genetic research on mood disorders as well as other adult psychopathology. The chapter ends with a discussion of the extent to which genes that affect one disorder also affect other disorders.

Mood Disorders

Mood disorders involve severe swings in mood, not just the "blues" that all people feel on occasion. For example, the lifetime risk for suicide for people diagnosed as having mood disorders has been estimated as 19 percent (Goodwin & Jamison, 1990). There are two major categories of mood disorders: major depressive disorder, consisting of episodes of depression, and bipolar disorder, in which there are episodes of both depression and mania.

Major depressive disorder usually has a slow onset over weeks or even months. Each episode typically lasts several months and ends gradually. Characteristic features include depressed mood, loss of interest in usual activities, disturbance of appetite and sleep, loss of energy, and thoughts of death or suicide. Major depressive disorder affects an astounding number of people. In a U.S. survey, the lifetime risk is about 17 percent, with about half of these in a severe or very severe category; risk is two times greater for women than for men after adolescence (National Comorbidity Study: http://www.hcp.med.harvard.edu/ncs/). Moreover, the problem is getting worse: Each successive generation born since World War II has higher rates of depression (Burke, Burke, Roe, & Regier, 1991), and prevalence rates more than doubled from the early 1990s to the early 2000s (Compton, Conway, Stinson, & Grant, 2006). These temporal trends could possibly be due to changes in environmental influences, diagnostic criteria, or clinical referral rates. Major depressive disorder is sometimes called unipolar depression because it involves only depression. In contrast, bipolar disorder, also known as manic-depressive illness, is a disorder in which the mood of the affected individual alternates between the depressive pole and the other pole of mood, called mania. Mania involves euphoria, inflated self-esteem, sleeplessness, talkativeness, racing thoughts, distractibility, hyperactivity, and reckless behavior. Mania typically begins and ends suddenly, and it lasts from several days to several months. Mania is sometimes difficult to diagnose; for this reason, DSM-IV (the American Psychiatric Association's *Diagnostic and Statistical Manual of Mental Disorders-IV*) has distinguished bipolar I disorder, with a clear manic episode, from bipolar II disorder, with a less clearly defined manic episode. Bipolar disorder is much less common than major depression, with an incidence of about 4 percent in the adult population and no gender difference (National Comorbidity Study: http://www.hcp.med.harvard.edu/ncs/), although this estimate is based on a broader concept than has traditionally been applied.

Family Studies

For more than 70 years, family studies have shown increased risk for first-degree relatives of individuals with mood disorders (Slater & Cowie, 1971). Since the 1960s, researchers have considered major depression and bipolar disorder separately. In seven family studies of major depression, the family risk was 9 percent on average, whereas risk in control samples was about 3 percent (McGuffin & Katz, 1986). Age-corrected morbidity risk estimates that take into account lifetime risk (see Chapter 3) are about twice as high (Sullivan, Neale, & Kendler, 2000). A review of 18 family studies of bipolar I and II disorder yielded an average risk of 9 percent, as compared to less than 1 percent in control individuals (Smoller & Finn, 2003). (See Figure 15.1.) The risks in these studies are low relative to the frequency of the disorder mentioned earlier because these studies focused on severe depression and bipolar disorder, often requiring hospitalization.

It has been hypothesized that the distinction between unipolar major depression and bipolar disorder is primarily a matter of severity; bipolar disorder may be a more severe form of depression (McGuffin & Katz, 1986). The basic multivariate finding from family studies is that relatives of unipolar probands are not at increased risk for bipolar disorder (less than 1 percent), but relatives of bipolar probands are at increased risk (14 percent) for unipolar depression (Smoller & Finn, 2003). If we postulate that bipolar disorder is a more severe form of depression, this model would explain why familial risk is greater for bipolar disorder, why bipolar probands have an excess of unipolar relatives, and why unipolar probands do not have many relatives with bipolar disorder. However, a twin study discussed in the next section does not provide much support for the hypothesis that bipolar disorder is a more severe form of unipolar depression (McGuffin et al., 2003). Identifying genes associated with these disorders will provide crucial evidence for resolving such issues, although to date the findings are mixed. A recent meta-analysis of gene variants in the methylenetetrahydrofolate reductase (*MTHFR*) gene and schizophrenia, bipolar disorder, and unipolar major depression found an association with the combined disorders for

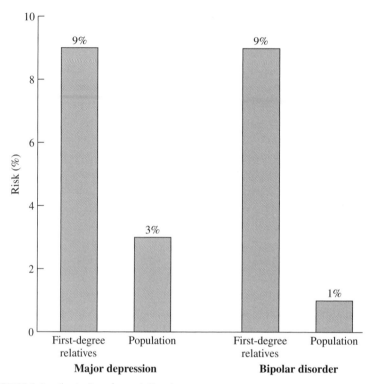

FIGURE 15.1 Family studies of mood disorders.

one *MTHFR* variant, suggesting a shared genetic influence on the three disorders (Peerbooms et al., 2011). Other genes have also been associated with both bipolar disorder and unipolar depression, further supporting the likelihood of a common genetic liability to these disorders (e.g., Weber et al., 2011).

Are some forms of depression more familial? For example, there is a long history of trying to subdivide depression into reactive (triggered by an event) and endogenous (coming from within) subtypes, but family studies provide little support for this distinction (Rush & Weissenburger, 1994). However, severity and especially recurrence show increased familiality for major depressive disorder (Janzing et al., 2009; Milne et al., 2009; Sullivan et al., 2000). Early onset appears to increase familial risk for bipolar disorder (Smoller & Finn, 2003). Drug use and suicide attempts are also familial features of bipolar disorder (Schulze, Hedeker, Zandi, Rietschel, & McMahon, 2006). Another potentially promising direction for subdividing depression is in terms of response to drugs (Binder & Holsboer, 2006). For example, there is some evidence that the therapeutic response to specific antidepressants tends to run in families (Tsuang & Faraone, 1990). The main drug treatment for bipolar disorder is lithium; responsiveness to lithium appears to be strongly familial (Grof et al., 2002).

Twin Studies

Twin studies yield evidence for substantial genetic influence for mood disorders. For major depressive disorder, six twin studies yielded average twin probandwise concordances of 0.43 for MZ twins and 0.28 for DZ twins (Sullivan et al., 2000). Liability-threshold model fitting of these data estimated heritability of liability as 0.37, with no shared environmental influence. The largest twin study to date yielded highly similar results: 0.38 heritability and no shared environmental influence (Kendler, Gatz, Gardner, & Pedersen, 2006b). However, family studies suggest that more severe depression might be more heritable. In line with this suggestion, the only clinically ascertained major depressive disorder twin sample large enough to perform model-fitting analyses estimated heritability of liability as 70 percent (McGuffin, Katz, Watkins, & Rutherford, 1996). However, it is also possible that the higher heritability of depression in the clinical sample represents higher reliability of clinical assessment.

For bipolar disorder, average twin concordances were 72 percent for MZ twins and 40 percent for DZ twins in early studies (Allen, 1976); three more recent twin studies yield average twin concordances of 65 and 7 percent, respectively (Smoller & Finn, 2003). Two twin studies of bipolar disorder using different samples from different countries yield strikingly similar results: MZ and DZ twin concordances were 40 and 5 percent in a U.K. study (McGuffin et al., 2003) and 43 and 6 percent in a Finnish study (Kieseppa, Partonen, Haukka, Kaprio, & Lonnqvist, 2004). Model-fitting liability-threshold analyses suggest extremely high heritabilities of liability (0.89 and 0.93, respectively) and no shared environmental influence. The average MZ and DZ twin concordances for the five more recent studies described above are 55 and 7 percent, respectively (Figure 15.2).

As mentioned earlier, one of the most important goals of genetic research is to provide diagnostic classifications based on etiology rather than symptoms. For example, are unipolar depression and bipolar disorder genetically distinct? One twin study investigated the model described earlier that bipolar disorder is a more extreme version of major depressive disorder (McGuffin et al., 2003). Part of the problem in addressing this issue is that conventional diagnostic rules assume that an individual has either unipolar or bipolar disorder and that bipolar disorder trumps unipolar disorder. However, in this twin study, this diagnostic assumption was relaxed and a genetic correlation of 0.65 was found between depression and mania, a finding that supports the model that bipolar disorder is a more extreme version of unipolar depression. However, 70 percent of the genetic variance on mania was independent of depression, a finding that does not support the model. A model that explicitly tested the assumption that bipolar disorder is a more extreme form of unipolar depression was rejected, but so was a model in which the two disorders were assumed to be genetically distinct. This lack of resolution is probably due to a lack of power: Although this was the largest clinically ascertained twin study, there were only 67 pairs in which at least one twin was diagnosed with bipolar disorder and 244 pairs in which at least one twin was diagnosed with unipolar depression. Resolution of this important diagnostic issue can be addressed definitively when

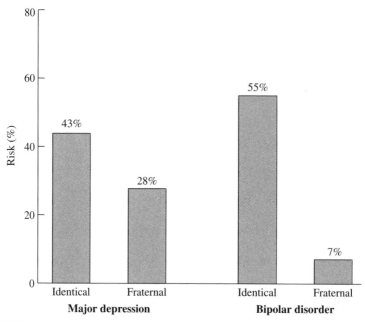

• **FIGURE 15.2** Approximate twin results for mood disorders.

genes are identified for the two disorders: To what extent will the same genes be associated with depression and mania? There is some emerging evidence (discussed below) for overlapping linkages and associations (Farmer, Elkin, & McGuffin, 2007).

As in the research on schizophrenia (Chapter 14), a study of offspring of identical twins discordant for bipolar disorder has been reported (Bertelsen, 1985). Similar to the results for schizophrenia, the same 10 percent risk for mood disorder was found in the offspring of the unaffected twin and in the offspring of the affected twin. This outcome implies that the identical twin who does not succumb to bipolar disorder nonetheless transmits a liability for the illness to offspring to the same extent as does the ill twin.

Adoption Studies

Results of adoption research on mood disorders are mixed. The largest study began with 71 adoptees with a broad range of mood disorders (Wender et al., 1986). Mood disorders were found in 8 percent of the 387 biological relatives of the probands, a risk only slightly greater than the risk of 5 percent for the 344 biological relatives of control adoptees. The biological relatives of the probands showed somewhat greater rates of alcoholism (5 percent versus 2 percent) and of attempted or actual suicide (7 percent versus 1 percent). Two other adoption studies relying on medical records of depression found little evidence for genetic influence (Cadoret, O'Gorman, Heywood, & Troughton, 1985; von Knorring, Cloninger, Bohman, & Sigvardsson, 1983). Although the sample size is necessarily small, 12 pairs of identical twins reared apart have been identified in which at least one member of each pair had suffered from major depression (Bertelsen, 1985). Eight of the 12 pairs (67 percent) were concordant for major depression, which is consistent with a hypothesis of at least some genetic influence on depression.

An adoption study that focused on adoptees with bipolar disorder found stronger evidence for genetic influence (Mendlewicz & Rainer, 1977). The rate of bipolar disorder in the birth parents of the bipolar adoptees was 7 percent, but it was 0 percent for the parents of control adoptees. As in the family studies, birth parents of these bipolar adoptees also showed elevated rates of unipolar depression (21 percent) relative to the rate for birth parents of control adoptees (2 percent), a result suggesting that the two disorders are not distinct genetically. Adoptive parents of the bipolar and control adoptees differed little in their rates of mood disorders.

Identifying Genes

For decades, the greater risk of major depression for females led to the hypothesis that a dominant gene on the X chromosome might be involved. As explained in Chapter 3, females can inherit the gene on either of their two X chromosomes, whereas males can only inherit the gene on the X chromosome they receive from their mother. Although initially linkage was reported between depression and color blindness, which is caused by genes on the X chromosome (Chapter 3), studies of DNA markers on the X chromosome failed to confirm linkage (Baron, Freimer, Risch, Lerer, & Alexander, 1993). Father-to-son inheritance is common for both major depression and bipolar disorder, which argues against X-linkage inheritance. Moreover, as mentioned earlier, bipolar disorder shows little sex difference. For these reasons, X linkage seems unlikely (Hebebrand, 1992).

In 1987, researchers reported linkage between bipolar disorder and markers on chromosome 11 in a genetically isolated community of Old Order Amish in Pennsylvania (Egeland et al., 1987). Unfortunately, this highly publicized finding was not replicated in other studies. The original report was withdrawn when follow-up research on the original pedigree with additional data showed that the evidence for linkage disappeared (Kelsoe et al., 1989).

These false starts led to greater caution in the search for genes for mood disorders. Linkage studies of major depressive disorder have lagged behind those for schizophrenia and bipolar disorder because, as discussed above, major depressive disorder appears to be less heritable, at least in community-based samples (McGuffin, Cohen, & Knight, 2007). Three genomewide linkage studies of major depressive disorder converge on linkage at $15q$ (Camp et al., 2005; Holmans et al., 2007; McGuffin et al., 2005). Follow-up fine mapping showed modestly positive evidence for linkage at $15q25$-$q26$ (Levinson et al., 2007). These studies focused on early-onset depression and recurrent depression because quantitative genetic results, mentioned above, suggest that early-onset depression and recurrent depression are more heritable. A recent review of molecular genetic studies of major depressive disorder found five genes replicated in three large genomewide association studies, but these gene regions are not in $15q$ (Hettema, 2010).

Genomewide linkage scans of bipolar disorder led to a surprising discovery. A meta-analysis of 11 linkage studies with more than 1200 individuals diagnosed as having bipolar disorder found strong evidence for linkage at 13*q* and 22*q* (Badner & Gershon, 2002). The same study also conducted a meta-analysis of 18 linkage studies of schizophrenia and found the strongest evidence for linkage in the same two regions, 13*q* and 22*q*, in addition to other regions. Moreover, although many candidate genes have been reported to be associated with bipolar disorder (Craddock & Forty, 2006; Farmer et al., 2007), the genes that appear to replicate most are those that were identified initially for their association with schizophrenia: *neuregulin* and *dysbindin* (see Chapter 14) (Farmer et al., 2007; Kato, 2007). More molecular genetic research is needed to resolve this critical issue of genetic overlap between schizophrenia and bipolar disorder. For example, later meta-analyses of linkage studies of bipolar disorder do not support the linkages to 13*q* and 22*q* (McQueen et al., 2005; Segurado et al., 2003); however, these meta-analyses used different analytic techniques.

Nonetheless, the possibility of genetic overlap between schizophrenia and bipolar disorder is important because the distinction between them is fundamental to diagnostic classifications, such as described in the DSM-IV. Bipolar disorder can only be diagnosed if the individual is *not* schizophrenic, which precludes comorbidity, although a mixed category called schizoaffective disorder has been acknowledged. For this reason, relatives of schizophrenics have not shown an increased risk for bipolar disorder in family studies. However, the assumption that schizophrenia and bipolar disorder are distinct has been brought into question by the molecular genetic evidence (Craddock & Owen, 2005). Based on symptoms rather than acceptance of the traditional diagnostic categories, family studies are beginning to show overlap between schizophrenic symptoms and bipolar symptoms (Craddock, O'Donovan, & Owen, 2005). Similarly, a twin study has shown genetic overlap between symptoms of schizophrenia and bipolar disorder (Cardno, Rijsdijk, Sham, Murray, & McGuffin, 2002).

Anxiety Disorders

A wide range of disorders involve anxiety (panic disorder, generalized anxiety disorder, phobias) or attempts to ward off anxiety (obsessive-compulsive disorder). In panic disorder, recurrent panic attacks come on suddenly and unexpectedly, usually lasting for several minutes. Panic attacks often lead to a fear of being in a situation that might bring on more panic attacks (e.g., agoraphobia, which literally means "fear of the marketplace"). Generalized anxiety refers to a more chronic state of diffuse anxiety marked by excessive and uncontrollable worrying. In a phobia, the fear is attached to a specific stimulus, such as fear of heights (acrophobia), enclosed places (claustrophobia), or social situations (social phobia). In obsessive-compulsive disorder, anxiety occurs when the person does not perform some compulsive act driven by an obsession—for example, repeated hand-washing in response to an obsession with hygiene.

Anxiety disorders are usually not as crippling as schizophrenia or severe depressive disorders. However, they are the most common form of mental illness, with a lifetime prevalence of 29 percent (Kessler, Berglund, et al., 2005), and can lead to other disorders, notably depression and alcoholism. Median age of onset is much earlier for anxiety (age 11) than for mood disorders (age 30). The lifetime risks for anxiety disorders are 5 percent for panic disorder, 6 percent for generalized anxiety disorder, 13 percent for specific phobias, 12 percent for social phobia, and 2 percent for obsessive-compulsive disorder. Panic disorder, generalized anxiety disorder, and specific phobias are twice as common in women as in men.

There has been much less genetic research on anxiety disorders than on schizophrenia and mood disorders. In general, results for anxiety disorders appear to be similar to those for depression in suggesting moderate genetic influence, as compared to the more substantial genetic influence seen for schizophrenia and bipolar disorder. As discussed later, the similarity in results for anxiety and depression may be caused by genetic overlap between them. Nonetheless, we will briefly review evidence for genetic influence for panic disorder, generalized anxiety disorder, phobias, and obsessive-compulsive disorder.

A review of eight family studies of panic disorder yielded an average morbidity risk of 13 percent in first-degree relatives of cases and 2 percent in controls (Shih, Belmonte, & Zandi, 2004). In an early twin study of panic disorder, the concordance rates for identical and fraternal twins were 31 and 10 percent, respectively (Torgersen, 1983). In two large twin studies with nonclinical samples, the heritability of liability was about 40 percent, with no evidence of shared environmental influence (Kendler, Gardner, & Prescott, 2001; Mosing, Gordon, et al., 2009); in a third large twin study, heritability was approximately 30 percent, with no shared environmental influence (Tambs et al., 2009). A meta-analysis of five twin studies yielded a similar liability heritability (43 percent), with no shared environmental influence (Hettema, Neale, & Kendler, 2001). No adoption data are available for panic disorder or any other anxiety disorders.

Generalized anxiety disorder appears to be as familial as panic disorder, but the evidence for heritability is weaker. A review of family studies indicates an average risk of about 10 percent among first-degree relatives as compared to a risk of 2 percent in controls (Eley, Collier, & McGuffin, 2002). However, two twin studies found no evidence for genetic influence (Andrews, Stewart, Allen, & Henderson, 1990; Torgersen, 1983); three other twin studies suggested modest genetic influence of about 20 percent and little shared environmental influence (Hettema, Prescott, & Kendler, 2001; Kendler, Neale, Kessler, Heath, & Eaves, 1992; Scherrer et al., 2000). A recent study of Norwegian twins found a somewhat higher heritability for generalized anxiety disorder of 27 percent, although nearly all of this genetic variance was shared with other anxiety disorders (Tambs et al., 2009).

Phobias show familial resemblance: 30 percent familial risk versus 10 percent in controls for specific phobias excluding agoraphobia (Fyer, Mannuzza, Chapman, Martin, & Klein, 1995), 5 percent versus 3 percent for agoraphobia (Eley et al., 2002), and 20 percent versus 5 percent for social phobia (Stein et al., 1998). One twin study found heritability of liability of about 30 percent for these phobias (Kendler, Myers, Prescott, & Neale, 2001). A more recent study of Norwegian twins found that about

40 to 60 percent of the variance could be explained by genetic influences, with no significant shared environmental influences (Czajkowski, Kendler, Tambs, Roysamb, & Reichborn-Kjennerud, 2011). Although there is little evidence of shared environmental influence, phobias are learned, even fears of evolutionarily fear-relevant stimuli such as snakes and spiders. An interesting twin study of fear conditioning showed moderate genetic influence on individual differences in learning and extinguishing fears (Hettema, Annas, Neale, Kendler, & Fredrikson, 2003).

For obsessive-compulsive disorder (OCD), family studies yield a wide range of results because of differences in diagnostic criteria. However, nine family studies that used criteria from the DSM and had more than 100 cases yielded more consistent results, with an average risk of 7 percent for family members and 3 percent for controls (Shih et al., 2004). Family studies also suggest that early-onset OCD is more familial. Only three small twin studies of OCD have been reported, and two of these found no heritability (Shih et al., 2004). A meta-analysis of 14 reports of twin studies of obsessive-compulsive symptoms found that genetic influences accounted for around 40 percent of the variance, while shared environmental effects accounted for less than 10 percent of the variance (Taylor, 2011).

DSM-IV also includes posttraumatic stress disorder (PTSD) as an anxiety disorder, even though its diagnosis depends on a prior traumatic event that threatens death or serious injury, such as war, assault, or natural disaster. PTSD symptoms include reexperiencing the trauma (intrusive memories and nightmares) and denying the trauma (emotional numbing). One survey estimated that the lifetime risk for one PTSD episode is about 1 percent (Davidson, Hughes, Blazer, & George, 1991). The risk is much higher, of course, in those who have experienced trauma. For example, after a plane crash, as many as one-half of the survivors develop PTSD (Smith, North, McColl, & Shea, 1990). About 10 percent of U.S. veterans of the Vietnam War still suffered from PTSD many years later (Weiss et al., 1992). Response to trauma appears to show familial resemblance (Eley et al., 2002). The Vietnam War provided an opportunity to conduct a twin study of PTSD because more than 4000 twin pairs were veterans of the war. A series of studies of these twins began by dividing the sample into those who served in Southeast Asia (who were much more likely to experience trauma) and those who did not (True et al., 1993). The results were similar for both groups regardless of the type of trauma experienced: Heritabilities of 15 PTSD symptoms were all about 40 percent, and there was no evidence of shared environmental influence.

Other Disorders

As mentioned earlier, DSM-IV includes many other categories of disorders, but next to nothing is known about their genetics. Interesting results are emerging, however, from the early stages of genetic research on four of these categories of disorders: seasonal affective disorder, somatoform disorders, chronic fatigue, and eating disorders. Other disorders are discussed in later chapters: impulse-control disorders such as hyperactivity in Chapter 16, antisocial personality disorder in Chapter 17, and substance abuse disorders in Chapter 18.

Seasonal affective disorder (SAD) is a type of major depression that occurs seasonally, typically in the fall or winter (Rosenthal et al., 1984). Family and twin studies suggest results similar to those for depression, with modest heritability (about 30 percent) and little shared environmental influence (Sher, Goldman, Ozaki, & Rosenthal, 1999). However, one twin study reported heritability twice as high (Jang, Lam, Livesley, & Vernon, 1997). It is noteworthy that this study was conducted in British Columbia (Canada) and yielded very high rates of SAD compared to the other studies, which suggests the possibility that the higher heritability and prevalence in the Canadian samples might be due to the northern latitude and more severe winters of Canada (Jang, 2005).

In somatoform disorders, psychological conflicts lead to physical (somatic) symptoms such as stomach pains. Somatoform disorders include somatization disorder, hypochondriasis, and conversion disorder. Somatization disorder involves multiple symptoms with no apparent physical cause. Hypochondriacs worry that a specific disease is about to appear. Conversion disorder, which was formerly called hysteria, involves a specific disability, such as paralysis, with no physical cause. Somatoform disorders show some genetic influence in family, twin, and adoption studies (Guze, 1993). Somatization disorder, which is much more common in women than in men, shows strong familial resemblance for women, but for men it is related to increased family risk for antisocial personality (Guze, Cloninger, Martin, & Clayton, 1986; Lilienfeld, 1992). An adoption study suggests that this link between somatization disorder in women and antisocial behavior in men may be genetic in origin (Bohman, Cloninger, von Knorring, & Sigvardsson, 1984). Biological fathers of adopted women with somatization disorder showed increased rates of antisocial behavior and alcoholism. A twin study of somatic distress symptoms in an unselected sample showed genetic as well as shared environmental influence; it also suggested that some of the genetic influence is independent of depression and phobia (Gillespie, Zhu, Heath, Hickie, & Martin, 2000).

Chronic fatigue refers to fatigue of more than six months' duration that cannot be explained by a physical or other psychiatric disorder. Family studies suggest that chronic fatigue is moderately familial (Albright, Light, Light, Bateman, & Cannon-Albright, 2011; Walsh, Zainal, Middleton, & Paykel, 2001). A twin study of diagnosed chronic fatigue found concordance rates of 55 percent in MZ twins and 19 percent in DZ twins (Buchwald et al., 2001). Twin studies of chronic fatigue symptoms in unselected samples found modest genetic and shared environmental influence (Sullivan, Evengard, Jacks, & Pedersen, 2005; Sullivan, Kovalenko, York, Prescott, & Kendler, 2003), even in childhood (Farmer, Scourfield, Martin, Cardno, & McGuffin, 1999). Fatigue-related symptoms were found to be due mostly to shared environmental influences in women and to genetic and nonshared environmental influences in men (Schur, Afari, Goldberg, Buchwald, & Sullivan, 2007). A recent set of studies that examined chronic fatigue symptoms and other somatic symptoms found that the symptoms could be explained by genetic and nonshared environmental influences (Kato, Sullivan, Evengard, & Pedersen, 2009; Kato, Sullivan, & Pedersen, 2010).

Eating disorders include anorexia nervosa (extreme dieting and avoidance of food) and bulimia nervosa (binge eating followed by vomiting), both of which occur mostly in adolescent girls and young women. Both types of eating disorders appear to run in families (Eley et al., 2002); in twin studies, both appear to be moderately heritable, with little influence of shared environment (Baker et al., 2009; Mitchell et al., 2010; Root et al., 2010). For example, the largest twin study of anorexia found a heritability of liability estimate of 56 percent and no shared environmental influence (Bulik et al., 2006). A sibling adoption study of disordered eating yielded a similar pattern of findings, with genetic influences accounting for more than half of the variance and no shared environmental influences (Klump, Suisman, Burt, McGue, & Iacono, 2009). Eating disorders is an area that is especially promising for studies of the interplay between genes and environment (Bulik, 2005), and there are at least two reports of biological factors such as puberty moderating genetic and environmental influences on eating disorders (e.g., Culbert, Racine, & Klump, 2011; Klump, Keel, Sisk, & Burt, 2010).

Co-Occurrence of Disorders

The co-occurrence, or comorbidity, of psychiatric disorders is striking. People with one disorder have almost a 50 percent chance of having more than one disorder during a 12-month period (Kessler, Chiu, Demler, Merikangas, & Walters, 2005). In addition, more serious disorders are much more likely to involve comorbidity. Are these really different disorders that co-occur, or does the co-occurrence call into question current diagnostic systems? Diagnostic systems are based on phenotypic descriptions of symptoms rather than on causes. Genetic research offers the hope of systems of diagnosis that take into account evidence on causation. As explained in Chapter 7 and the Appendix, multivariate genetic analysis of twin and adoption data can be used to ask whether genes that affect one trait also affect another trait.

More than a hundred genetic studies have addressed this key question of comorbidity in psychopathology. Earlier in this chapter, we considered the surprising finding of genetic overlap between major depressive disorder and bipolar disorder as well as the even more surprising possibility of genetic overlap between bipolar disorder and schizophrenia. Scores of multivariate family and twin studies have examined comorbidity across the many anxiety disorders as well as between anxiety disorders and other disorders such as depression and alcoholism. Rather than describe studies that compare two or three disorders (see, for example, Jang, 2005; McGuffin, Gottesman, & Owen, 2002), we will provide an overview of multivariate genetic results that point to a surprising degree of genetic comorbidity.

For example, consider the diverse anxiety disorders. A multivariate genetic analysis of lifetime diagnoses of major anxiety disorders indicated substantial genetic overlap among generalized anxiety disorder, panic disorder, agoraphobia, and social phobia

(Hettema, Prescott, Myers, Neale, & Kendler, 2005). The only specific genetic effects were found for specific phobias such as fear of animals. Differences between the disorders are largely caused by nonshared environmental factors. Results were similar for men and women despite the much greater frequency of anxiety disorders in women. A subsequent twin study examined panic disorder, generalized anxiety disorder, phobias, obsessive-compulsive disorder, and posttraumatic stress disorder (Tambs et al., 2009). Again, all of the anxiety disorders were influenced by a common genetic factor, with only phobias and obsessive-compulsive disorder showing some specific genetic influences; no shared environmental influences were significant for any of the disorders.

Broadening this multivariate genetic approach beyond anxiety disorders to include major depression yields the most surprising finding in this area: Anxiety (especially generalized anxiety disorder) and depression are largely the same thing genetically. This finding was initially reported in a paper in 1992 for lifetime estimates (Kendler et al., 1992), with results summarized in Figure 15.3. Heritability of liability in this study was 42 percent for major depression and 69 percent for generalized anxiety disorder. There was no significant shared environmental influence; nonshared environment accounted for the remainder of the liability of the two disorders. The amazing finding was the genetic correlation of 1.0 between the two disorders, indicating that the same genes affect depression and anxiety. Nonshared environmental influences correlated 0.51, suggesting that nonshared environmental factors differentiate the disorders to some extent. These findings for lifetime estimates of depression and anxiety were replicated using one-year prevalences obtained from follow-up interviews (Kendler, 1996). A review of 23 twin studies and 12 family studies confirms that anxiety and depression are largely the same disorder genetically and that the disorders are differentiated by nonshared environmental factors (Middeldorp, Cath, Van Dyck, & Boomsma, 2005).

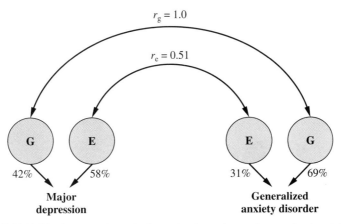

• **FIGURE 15.3** Multivariate genetic results for major depression and generalized anxiety disorder. (Adapted from Kendler, Neale, Kessler, Heath, & Eaves, 1992 [Figure 2]. Copyright 1992 by the American Medical Association. Used with permission.)

Going beyond depression and anxiety disorders to include drug abuse and antisocial behavior suggests a genetic structure of common psychiatric disorders (not including schizophrenia and bipolar disorder) that differs substantially from current diagnostic classifications based on symptoms (Kendler, Prescott, Myers, & Neale, 2003). As summarized in Figure 15.4, genetic research suggests two broad categories

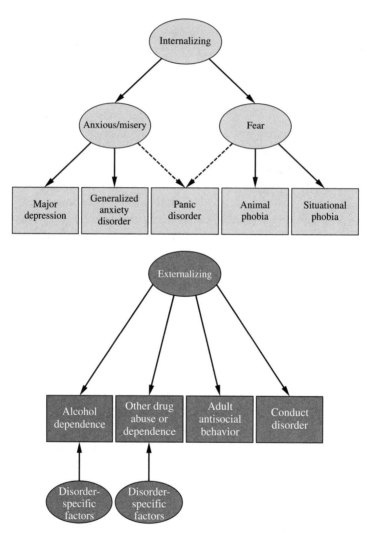

• **FIGURE 15.4** A structure for the genetic risk factors for common psychiatric and substance use disorders. Strong relationships are depicted by solid lines and, in the case of panic disorder only, weaker relationships by dashed lines. (From "The Structure of Genetic and Environmental Risk Factors for Common Psychiatric and Substance Use Disorders in Men and Women," by K. S. Kendler, C. A. Prescott, J. Myers, & M. C. Neale, 2003, *Archives of General Psychiatry, 60,* p. 936. Used with permission.)

of disorders, called *internalizing* and *externalizing*. Internalizing disorders include depression and anxiety disorders; externalizing disorders include alcohol and other drug abuse as well as antisocial behavior in adulthood (and conduct disorder in childhood). Internalizing disorders can be separated into an anxious/misery factor, which includes depression and anxiety disorders, and a fear factor, which includes phobias. Both internalizing factors are involved in panic disorder. As discussed in Chapter 17, the internalizing disorders might represent the extreme of the broad personality trait called *neuroticism*.

The disparate externalizing disorders (Chapters 16 and 17) are part of a general genetic factor, and both alcohol dependence and other drug abuse include some disorder-specific genetic effects. The genetic structure of internalizing and externalizing disorders applies equally to men and women despite the much greater risk of internalizing disorders for women and externalizing disorders for men. Because few disorders show shared environmental influence, it does not affect the structure. Nonshared environment largely contributes to heterogeneity rather than comorbidity. Thus, the phenotypic structure of comorbidity is largely driven by the genetic structure shown in Figure 15.4 (Krueger, 1999).

These multivariate genetic results predict that when genes are found that are associated with any of the internalizing disorders, the same genes are highly likely to be associated with other internalizing disorders. Similarly, genes associated with any of the externalizing disorders will likely be associated with the other externalizing disorders but not with the internalizing disorders. This result suggests that genetic influences are broad in their effect in psychopathology. It mirrors a similar finding concerning "generalist genes" in the area of cognitive abilities (see Chapters 12 and 13).

Identifying Genes

Although multivariate genetic research suggests that the genetic action lies at the level of broad categories of internalizing and externalizing disorders, molecular genetic research on anxiety disorders has focused on traditional diagnoses. Moreover, not nearly as much molecular genetic research has been conducted on these disorders as compared to the mood disorders. As a result, linkage studies have not yet converged, and the "usual suspect" candidate gene studies have not yet revealed replicable results (e.g., Eley et al., 2002; Jang, 2005; Smoller, Block, & Young, 2009).

Panic disorder has been studied most, in part because it appears to be more heritable than the other anxiety disorders and in part because it can be so debilitating. Five earlier linkage studies of panic disorder did not yield consistent results (Villafuerte & Burmeister, 2003), suggesting that genetic effects may be relatively small. However, recent reports have yielded more promising results, suggesting linkage on 15q and possibly 2q (Fyer et al., 2006) as well as for regions that are specific for different forms of panic or phobic disorder (Smoller et al., 2009). As with other complex traits, candidate gene associations have largely failed to replicate (e.g., Maron et al., 2007). The strongest case so far can be made for an association between panic disorder in females and a functional polymorphism (Val158Met) in the catechol-O-methyltransferase (*COMT*) gene (McGrath et al., 2004; Rothe et al., 2006), a polymorphism that has been reported to be associated with many other common disorders and complex traits (Craddock, Owen, & O'Donovan, 2006). Similar mixed stories are beginning to emerge for candidate gene studies of obsessive-compulsive disorders (Hemmings & Stein, 2006; Stewart et al., 2007) and for linkage and candidate gene association studies of eating disorders (Slof-Op 't Landt et al., 2005). A genomewide association study of depression using sets of SNPs selected from GWA studies for their association with a particular phenotype found evidence for significant associations for many genetic loci of small effect that influence both depression and anxiety (Demirkan et al., 2011). Thus, although there is some evidence for genes related to specific disorders, there is also emerging evidence of substantial overlap in the genes across multiple disorders. As more studies consider multiple diagnoses, it is likely that, consistent with findings from twin studies, specific genes will be found to be related to categories of disorders.

Summary

Moderate genetic influence has been found for major depressive disorder, and substantial genetic influence has been found for bipolar disorder. More severe and recurrent forms of these mood disorders are more heritable. Bipolar disorder may be a more severe form of depression. Surprisingly, molecular genetic studies for bipolar disorder suggest linkages and associations similar to those found for schizophrenia.

Anxiety disorders yield quantitative genetic results that are similar to depression—moderate genetic influence with little evidence of shared environmental influence. Some evidence for genetic influence has also been found for seasonal affective disorder, somatoform disorders, chronic fatigue, and eating disorders.

Some of the most far-reaching genetic findings in psychopathology concern genetic comorbidity. Genetic research has begun to call into question the fundamental diagnostic distinction between schizophrenia and bipolar disorder, including the molecular genetic findings of similar linkages and associations for the two disorders. The most striking finding regarding the mood disorders is that major depressive disorder and generalized anxiety disorder are the same disorder from a genetic perspective. Multivariate genetic research suggests a genetic structure of common psychiatric disorders that includes just two broad categories, internalizing and externalizing disorders.

CHAPTER SIXTEEN

Developmental Psychopathology

S chizophrenia is typically diagnosed in adulthood. Other disorders emerge in childhood. General cognitive disability, learning disorders, and communication disorders were discussed in Chapter 11. Other DSM-IV diagnostic categories that first appear in childhood include pervasive developmental disorders (e.g., autistic disorder), attention-deficit and disruptive behavior disorders (e.g., attention-deficit hyperactivity disorder, conduct disorder), anxiety disorders, tic disorders (e.g., Tourette disorder), elimination disorders (e.g., enuresis), and, most recently, mood disorders. In a recent nationwide sample of unselected households with children from 8 to 15 years of age, 12 percent of children met 12-month criteria for diagnosis of disruptive disorders (attention-deficit hyperactivity disorder or conduct disorder), mood or anxiety disorder (depression, dysthymia, anxiety or panic), or eating disorder (Merikangas et al., 2010). Even more surprising is that approximately 14 percent of those children met the criteria for two or more of the disorders.

Only in the past two decades has genetic research begun to focus on disorders of childhood (Rutter, Silberg, O'Connor, & Simonoff, 1999). Developmental psychopathology is not limited to childhood: It considers change and continuity throughout the life course, including disorders such as dementia, which develops later in life (see Chapters 11 and 19). However, genetic research on childhood disorders has blossomed recently, as is reflected in this chapter. One reason to consider childhood disorders is that some adult disorders emerge in childhood. Median age of onset is much earlier for anxiety disorders (age 11) and impulse-control disorders (age 11) than for mood disorders (age 30). Half of all lifetime cases of diagnosed disorders start by age 14, which suggests that interventions aimed at prevention or early treatment need to focus on childhood and adolescence (Kessler, Berglund, et al., 2005), especially because only half of the children aged 8 to 15 who met criteria for diagnosis for a mental health disorder had sought treatment with a mental health professional (Merikangas et al., 2010). However, the main reason for the increased interest in the genetics of childhood disorders is that the two major childhood disorders—autism and attention-deficit hyperactivity disorder—have been shown to be among the most heritable of all mental disorders, as described in the following sections.

Autism

Autism was once thought to be a childhood version of schizophrenia, but it is now known to be a distinct disorder marked by abnormalities in social relationships, communication deficits, and restricted interests. As traditionally diagnosed, it is relatively uncommon; however, a 2008 survey by the U.S. Centers for Disease Control and Prevention found a higher rate than previously reported, about 1 in 110 children, with rates four to five times higher for boys than girls (http://www.cdc.gov/ncbddd/autism/addm.html). Even higher rates, 1 in 38 children, have been reported in a study that screened over 55,000 children in a South Korean community (Kim et al., 2011). During the 1990s, there was a fivefold increase in the diagnosis of autism, in part because of heightened awareness and changing diagnostic criteria (Muhle, Trentacoste, & Rapin, 2004), and the rates have continued to increase in the 2000s. The diagnosis of autism has been broadened to *autism spectrum disorders* (ASDs), which include autism, Asperger syndrome, and other pervasive developmental disorders. Traditionally, a diagnosis of autism was limited to children who showed impairments in all three areas (social, communication, interests) before 3 years of age. In contrast, Asperger syndrome was diagnosed if children were impaired in the social and interests domains but appeared to have normal language and cognitive development before 3 years of age. The "other" diagnosis was used for children who showed severe impairment in just one or two of the domains. Most researchers now consider these three disorders as part of a single continuum or spectrum of disorder. In the early 2000s, great concern among parents was driven by media reports that the supposed increase in ASDs was caused environmentally by the measles-mumps-rubella (MMR) vaccine. However, the evidence on this putative environmental cause of ASDs has been consistently negative (Mrozek-Budzyn, Kieltyka, & Majewska, 2010; Rutter, 2005a; Smeeth et al., 2004).

Family and Twin Studies

When Kanner (1943) first characterized autism in 1943, he assumed it was caused "constitutionally." However, in subsequent decades, autism was thought to be environmentally caused, either by cold and rejecting parents or by brain damage (Hanson & Gottesman, 1976). Genetics did not seem to be important because there were no reported cases of an autistic child having an autistic parent and because the risk to siblings was only about 5 percent (Bailey, Phillips, & Rutter, 1996; Smalley, Asarnow, & Spence, 1988). However, this rate of 5 percent was 100 times greater than the population rate of autism as diagnosed in those original studies, a difference implying strong familial resemblance. The reason why autistic children do not have autistic parents is that few autistic individuals marry and have children.

In 1977, the first systematic twin study of autism began to change the view that autism was environmental in origin (Folstein & Rutter, 1977). Four of 11 pairs of identical twins were concordant for autism, whereas none of 10 pairs of fraternal twins were. These pairwise concordance rates of 36 and 0 percent rose to 92 and 10 percent when the diagnosis was broadened to include communication and social problems. Co-twins of autistic children are more likely to have communication problems as well as social difficulties. In a follow-up of the twin sample into adult life, problems with social relationships were prominent (Le Couteur et al., 1996). These findings were replicated in other twin studies (Ronald & Hoekstra, 2011). A conservative estimate of the concordance in monozygotic pairs is 60 percent. A review of four independent twin studies suggests a heritability of liability for autism greater than 90 percent (Freitag, 2007). Recent twin studies of autism spectrum disorders find similar results, suggesting substantial heritability (Lichtenstein, Carlstrom, Rastam, Gillberg, & Anckarsater, 2010; Ronald & Hoekstra, 2011; Rosenberg et al., 2009; Taniai, Nishiyama, Miyachi, Imaeda, & Sumi, 2008; but see Hallmayer et al., 2011, for a contrasting view of the role of shared environment).

On the basis of these twin and family findings, views regarding autism have changed radically. Instead of being seen as an environmentally caused disorder, it is now considered to be one of the most heritable mental disorders (Freitag, 2007; Ronald & Hoekstra, 2011). One unusual aspect of genetic research on autism is that, as traditionally diagnosed, autism is so severe that it nearly always results in affected children being seen by clinical services rather than remaining undetected in the community (Thapar & Scourfield, 2002). As a result, nearly all twin studies have been based on clinical cases rather than community samples. However, recent research has considered ASDs as continua that extend well into common behavioral problems seen in undiagnosed children in the community. This trend was driven in part by the results of early family studies in which relatives of autistic individuals were found to have some communication and social difficulties (Bailey, Palferman, Heavey, & Le Couteur, 1998). Twin studies have also supported the hypothesis that the genetic and environmental causes of ASD symptoms are distributed continuously throughout the population and that the etiology of autistic traits does not differ across the full range of severity (Constantino & Todd, 2003; Lundstrom et al., 2012; Robinson et al., 2011). This is an emerging rule in behavioral genetics—that disorders are actually the quantitative extreme of a continuum of normal variation (see Chapters 14 and 15).

In contrast to the assumption that autism involves a triad of impairments—poor social interaction, language and communication problems, and restricted range of interests and activities—twin studies of ASD symptoms in community samples have found evidence for genetic heterogeneity, especially between social impairments (interaction and communication) and nonsocial impairments (interests and activities). Several multivariate genetic analyses of the triad of symptoms have found high heritability (about 80 percent) for each of the three types of symptoms but surprisingly low genetic correlations among them (Happé & Ronald, 2008; Kolevzon, Smith, Schmeidler, Buxbaum, & Silverman, 2004; Ronald, Happé, Price, Baron-Cohen, & Plomin, 2006; Ronald, Larsson, Anckarsater, & Lichtenstein, 2011). These findings suggest that, although some children by chance have all three types of symptoms, the ASD triad of symptoms are different genetically. This surprising conclusion, which contradicts the traditional diagnosis of autism, is supported by cognitive and brain data (Happé, Ronald, & Plomin, 2006).

Identifying Genes

Quantitative genetic evidence suggesting substantial genetic influence on autism led to autism being the early target of affected sib-pair linkage analysis after the success of QTL linkage in the area of reading disability in 1994 (see Chapter 11). In 1998, an international collaborative linkage study reported evidence of a locus on chromosome 7 (7q31-q33) in a study of 87 affected sibling pairs (International Molecular Genetic Study of Autism Consortium, 1998). This 7q linkage was replicated in other studies, although several studies did not replicate the linkage (Trikalinos et al., 2006). No specific gene has been implicated reliably (Freitag, 2007). Many other linkage regions have been reported in several genomewide linkage studies, but none has been replicated in more than two studies (Ma et al., 2007). Despite the sex difference in ASDs, no consistent evidence for linkage to the X chromosome has emerged.

As with other common disorders, these linkage results could be viewed as demonstrating that there are no genes of sufficiently large effect size to be detected by sib-pair linkage analyses with samples of fewer than many hundreds of affected sibling pairs. The most straightforward way to address the issue of power to detect smaller QTL effect sizes is to increase the sample size, although it is difficult to obtain such samples because only about 5 percent of the siblings of autistic children are also autistic. One large-scale collaborative project conducted a sib-pair linkage analysis of more than 1000 families across 19 countries, involving 120 scientists from more than 50 institutions (Szatmari et al., 2007). Although previously reported linkages were not replicated, including the linkage on 7q, linkage was suggested for 11p12-q13. Linkage results appeared stronger when families with copy number variants (see Chapter 9) were removed from the analysis.

Similar to other disorders, more than 100 candidate gene associations have been reported but no consistent associations have as yet been found (Geschwind, 2011; Xu et al., 2012). Although many genomewide association studies have also been conducted, the results of such studies have been similarly inconclusive for finding a particular gene or set of genes associated with ASD. Both genomewide linkage and association studies examine common variants. There is accumulating evidence that as many as 10 percent of ASD cases can be accounted for by rare mutations due to copy number variants (CNVs) (Levy et al., 2011). Because autism does run in families, rare CNVs, which are usually not heritable but are de novo mutations, cannot be the only explanation for ASD. Instead, both common variants and CNVs are likely to play a role. Because so many common variants and CNVs appear to operate in the development of ASD, there has been some question as to whether ASD should be viewed as a single disorder (Geschwind, 2011). It may be more appropriate to consider the specific symptoms of ASD or to examine subtypes rather than the spectrum as traditionally diagnosed (Liu, Georgiades, et al., 2011; Ronald et al., 2010). This view is consistent with findings from twin studies that have found evidence for distinct genetic influences on the three types of symptoms (Happé & Ronald, 2008), although there is also evidence, as described above, of some genetic overlap among the symptoms (Ronald et al., 2011).

In an effort to organize the vast number of genes identified for autism and to provide a resource for researchers, a recent review and analysis of existing data identified more than 2000 genes, 4500 CNVs, and 158 linkage regions reported to be associated with ASD (Xu et al., 2012). This information is in an online searchable database (http://autismkb.cbi.pku.edu.cn/). As this work moves forward, the multivariate genetic research described above indicating genetic heterogeneity for the three types of symptoms suggests that molecular genetic studies might profit by focusing more on the three types of symptoms separately rather than beginning with diagnoses of autism, which requires the presence of all three impairments.

Attention-Deficit and Disruptive Behavior Disorders

The DSM-IV grouping of attention-deficit and disruptive behavior disorders is interesting because it includes a disorder that appears to be substantially heritable, attention-deficit hyperactivity disorder, and a disorder that shows only modest genetic influence, conduct disorder, when it occurs in the absence of overactivity/inattention. Although all children have trouble learning self-control, most have made considerable progress by the time they enter school. Those who have not learned self-control are often disruptive, impulsive, and aggressive, and they have problems adjusting to school.

Attention-deficit hyperactivity disorder (ADHD), as defined by DSM-IV, refers to children who exhibit very high activity, have a poor attention span, and act impulsively. Estimates of the prevalence of ADHD in North America are about 9 percent of children, with boys greatly outnumbering girls (Faraone, Sergeant, Gillberg, & Biederman, 2003; Merikangas et al., 2010; Polanczyk, de Lima, Horta, Biederman, & Rohde, 2007). European psychiatrists have tended to take a more restricted approach to diagnosis, with an emphasis on hyperactivity that not only is severe and pervasive across situations but also is of early onset and unaccompanied by high anxiety (Polanczyk et al., 2007; Taylor, 1995). There is continuing uncertainty about the merits of these narrower and broader approaches to diagnosis. However conceptualized, ADHD usually continues into adolescence and, depending on the criteria used, may persist into adulthood (Faraone, Biederman, & Mick, 2006; Klein & Mannuzza, 1991).

Twin Studies

ADHD runs in families, with first-degree relatives five times more likely to be diagnosed as compared to controls (Biederman et al., 1992) and with greater familial risk when ADHD persists into adulthood (Faraone, Biederman, & Monuteaux, 2000). Twin studies have consistently shown a strong genetic effect on hyperactivity regardless of whether it is measured by questionnaire (Goodman & Stevenson, 1989; Silberg et al., 1996) or by standardized and detailed interviewing (Eaves et al., 1997), regardless of whether it is rated by parents or teachers (Saudino, Ronald, & Plomin, 2005), and regardless of whether it is treated as a continuously distributed dimension (Thapar, Langley, O'Donovan, & Owen, 2006) or as a clinical diagnosis (Gillis, Gilger, Pennington, & DeFries, 1992; Larsson, Anckarsater, Rastam, Chang, & Lichtenstein, 2012). A heritability estimate of 76 percent was computed for pooled findings across 20 twin studies (Faraone et al., 2005), and a more recent meta-analysis of 21 studies confirmed these findings with a heritability estimate of about 70 percent (Burt, 2009b). These results suggest that heritability is greater for ADHD than for other childhood disorders with the exception of autism.

As is almost always the case in behavioral genetics, stability of ADHD symptoms is largely driven by genetics (Kuntsi, Rijsdijk, Ronald, Asherson, & Plomin, 2005; Larsson, Dilshad, Lichtenstein, & Barker, 2011; Price et al., 2005; Rietveld, Hudziak, Bartels, Van Beijsterveldt, & Boomsma, 2004). As is usually the case for psychopathology, heritability appears to be greater for persistent ADHD that extends into adulthood (Faraone, 2004). An unusual aspect of ADHD results is that DZ correlations are lower than expected relative to MZ correlations, especially for parental ratings. This could be due to a contrast effect in which parents inflate differences between their DZ twins, but this pattern of twin results is also consistent with nonadditive genetic variance (Eaves et al., 1997; Hudziak, Derks, Althoff, Rettew, & Boomsma, 2005; Nikolas & Burt, 2010; Rietveld, Posthuma, Dolan, & Boomsma, 2003), as discussed in Chapter 12. Although adoption studies to date have been few and quite limited methodologically (McMahon, 1980), they lend some support to the hypothesis of genetic influence for ADHD (e.g., Cantwell, 1975). Two children-of-twins studies (Chapter 6) attempted to clarify the joint roles of genetic and environmental influences in the development of ADHD in children of alcoholics and found that maternal alcohol use disorder and ADHD relate to child ADHD largely via genetic effects (Knopik et al., 2006; Knopik, Jacob, et al., 2009).

The activity and attention components of ADHD are both highly heritable (Greven, Asherson, Rijsdijk, & Plomin, 2011; Nikolas & Burt, 2010). Multivariate genetic twin analyses of the inattention and hyperactivity components of ADHD indicate substantial genetic overlap between the two components, providing genetic justification for the syndrome of ADHD (Eaves et al., 2000; Greven, Asherson, et al.,

2011; Greven, Rijsdijk, & Plomin, 2011; Knopik, Heath, Bucholz, Madden, & Waldron, 2009; Larsson, Lichtenstein, & Larsson, 2006; McLoughlin, Ronald, Kuntsi, Asherson, & Plomin, 2007; Rasmussen et al., 2004). Another multivariate issue concerns the genetic overlap between parental and teacher ratings of ADHD, both of which are highly heritable. Multivariate genetic analyses suggest some genetic overlap but also some genetic effects specific to parents and teachers (McLoughlin, Rijsdijk, Asherson, & Kuntsi, 2011; Thapar et al., 2006). In other words, these results predict that to some extent different genes will be associated with ADHD viewed by parents in the home and ADHD viewed by teachers in school. In addition, pervasive ADHD that is seen both at home and in school is more heritable than ADHD specific to just one situation (Thapar et al., 2006), and hyperactivity-impulsivity and inattention are seen, in part, as distinct by both parents and teachers (McLoughlin et al., 2011).

Genetic studies of conduct disorder yield results quite different from those for ADHD. DSM-IV criteria for conduct disorder include aggression, destruction of property, deceitfulness or theft, and other serious violations of rules such as running away from home. Some 5 to 10 percent of children and adolescents meet these diagnostic criteria, with boys again greatly outnumbering girls (Cohen et al., 1993; Rutter et al., 1997). In contrast to ADHD, the combined data from several early twin studies of juvenile delinquency yield concordance rates of 87 percent for identical twins and 72 percent for fraternal twins, rates that suggest only modest genetic influence and substantial shared environmental influence (McGuffin & Gottesman, 1985). This pattern is broadly supported by the results of a twin study of self-reported teenage antisocial behavior in U.S. Army Vietnam-era veterans (Lyons et al., 1995). However, most recent twin studies of delinquent acts and conduct disorder symptoms in normal samples of adolescents have shown greater genetic influence (Thapar et al., 2006) as well as substantial shared environmental influences (e.g., Bornovalova, Hicks, Iacono, & McGue, 2010; Burt, 2009a).

A twin study of adolescent males used a technique called *latent class analysis*, which attempts to account for patterning among symptoms by hypothesizing underlying (latent) classes (Eaves et al., 1993). One class involves symptoms from both ADHD and conduct disorder, for which strong genetic influence was found (Nadder, Rutter, Silberg, Maes, & Eaves, 2002; Silberg et al., 1996). In sharp contrast, there was almost no significant genetic influence for a "pure" class of conduct disorder without hyperactivity, for which there was a strong shared environmental influence (Silberg et al., 1996). This finding fits with multivariate twin studies that found substantial genetic overlap between ADHD and conduct problems (Thapar et al., 2006; Tuvblad, Zheng, Raine, & Baker, 2009), suggesting that what ADHD and conduct problems have in common is largely genetic and that what conduct problems do not share with ADHD is largely environmental.

Heterogeneity in antisocial behavior also contributes to some of the inconsistencies in the published research findings on conduct problems. For example, there is evidence from several twin studies that aggressive antisocial behavior is more heritable than nonaggressive antisocial behavior (e.g., Burt & Neiderhiser, 2009; Eley,

Lichtenstein, & Stevenson, 1999) (Figure 16.1). Moreover, different genetic factors affect aggressive and nonaggressive conduct problems (Burt, 2009a; Gelhorn et al., 2006). Genetic effects are probably greatest with respect to early-onset aggressive antisocial behavior that is accompanied by hyperactivity and that shows a strong tendency to persist into adulthood as antisocial personality disorder (DiLalla & Gottesman, 1989; Lyons et al., 1995; Moffitt, 1993; Robins & Price, 1991; Rutter et al., 1999). (See Chapter 17 for a discussion of personality disorders, including antisocial personality disorder.) In addition, antisocial behavior that is persistent across situations (home, school, laboratory) is more heritable (Arseneault et al., 2003; Baker, Jacobson, Raine, Lozano, & Bezdjian, 2007). In contrast, environmentally mediated risks are probably strongest with respect to nonaggressive juvenile delinquency that has an onset in the adolescent years and does not persist into adult life. The development of conduct disorder and antisocial behavior is a rich vein for studies of gene-environment interplay (Jaffee, Strait, & Odgers, 2012; Moffitt, 2005), as discussed in Chapter 8.

Another aspect of genetic heterogeneity in childhood antisocial behavior is callous-unemotional personality, which involves psychopathic tendencies such as

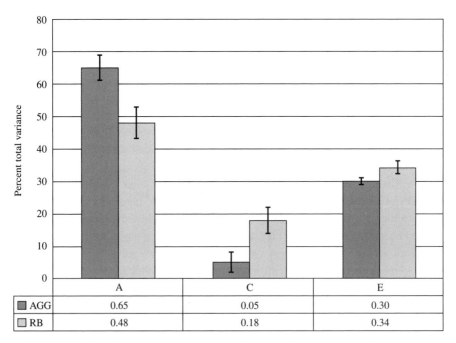

	A	C	E
■ AGG	0.65	0.05	0.30
□ RB	0.48	0.18	0.34

• **FIGURE 16.1** Genetic, shared environmental, and nonshared environmental influences on aggressive (AGG) and nonaggressive rule-breaking (RB) behaviors. A = additive genetic variance; C = common (shared) environmental variance; E = nonshared environmental variance. (Adapted from *Child Psychology Review, 29,* Burt, S. A., Are there meaningful etiological differences within antisocial behavior? Results of a meta-analysis, 163–178, Copyright © 2009, with permission from Elsevier.)

lack of empathy and guilt. In a large twin study of 7-year-old children rated by their teachers, antisocial behavior accompanied by callous-unemotional tendencies was highly heritable (80 percent), with no shared environmental influence, whereas antisocial behavior without callous-unemotional tendencies was only modestly heritable (30 percent) and showed moderate shared environmental influence (35 percent) (Viding, Blair, Moffitt, & Plomin, 2005). These findings persisted longitudinally; moreover, children who showed high or increasing levels of callous-unemotional traits during middle childhood and high levels of conduct problems had the most problematic outcomes at age 12 (Fontaine, McCrory, Boivin, Moffitt, & Viding, 2011).

Identifying Genes

As was the case for autism, the consistent evidence of a large genetic contribution to ADHD attracted the attention of molecular geneticists. However, this recognition came later for ADHD than for autism and at a time when molecular genetic studies had moved on from linkage to association studies in an attempt to identify QTLs of small effect size. Because genomewide association was not available at that time, these early studies were limited to candidate genes. Interest has centered on genes involved in the dopamine pathway because many children with ADHD improve when given psychostimulants, such as methylphenidate, which affect dopamine pathways. The dopamine transporter gene *DAT1* was an obvious candidate because methylphenidate inhibits the dopamine transporter mechanism and *DAT1* knock-out mice are hyperactive (Caron, 1996). An exciting initial finding of an association for *DAT1* (Cook et al., 1998) was replicated in three studies but failed to replicate in three other studies (Thapar & Scourfield, 2002). Somewhat stronger results were found for two other dopamine genes that code for dopamine receptors called *DRD4* and *DRD5*. A meta-analysis found small (**odds ratios** of about 1.3) but significant associations for *DRD4* and *DRD5*, although no association was found for *DAT1* (Li, Sham, Owen, & He, 2006). As expected from the multivariate genetic results indicating substantial genetic overlap between ADHD symptoms, these patterns of associations are similar across symptoms (Thapar et al., 2006).

Associations have been reported for more than 30 other candidate genes, but none have been consistently replicated (Bobb, Castellanos, Addington, & Rapoport, 2006; Thapar, O'Donovan, & Owen, 2005; Waldman & Gizer, 2006). Although candidate gene association studies have dominated genetic research on ADHD, genomewide linkage screens have been reported, including a meta-analysis of seven independent linkage scans (Zhou et al., 2008), a bivariate linkage scan for ADHD and reading disability (Gayán et al., 2005), and a follow-up fine-mapping study of nine candidate linkage regions (Ogdie et al., 2004). No consistent linkage regions have been identified.

Similar to autism, many genomewide association studies have been conducted for ADHD with no clear and consistent findings. A recent report took a systematic approach to search for common variants using both a standard SNP genomewide association analysis and a more focused, hypothesis-driven approach guided by findings from studies of CNVs (Stergiakouli et al., 2012). This study reported convergence between the SNP and CNV-guided analyses for *CHRNA7* and some overlap in regions across the two approaches for cholesterol-related and central nervous system pathways. Another study focusing on CNVs also found evidence for involvement of the *CHRNA7* gene, which is also associated with comorbid conduct disorder (Williams et al., 2012). A different study examining the possible role of CNVs in ADHD found associations with genes in a different system, glutamatergic neurotransmission (*GRM*) (Elia et al., 2012). A database of ADHD genes (*ADHDgene*) has recently been created that includes SNPs, CNVs and other variants, genes, and chromosomal regions gleaned from published genetic studies of ADHD (Zhang et al., 2012).

Anxiety Disorders

The median age of onset for anxiety disorders is 11 years; for this reason, some genetic research has considered anxiety in childhood (Rutter et al., 1999), with recent work identifying relatively stable anxiety symptoms in preschool-aged children (Edwards, Rapee, & Kennedy, 2010). A twin study in the United Kingdom, with more than 4500 4-year-old twin pairs rated by their mothers, examined five components of anxiety (Eley et al., 2003). Three components are comparable to adult anxiety disorders (see Chapter 15): generalized anxiety, fears, and obsessive-compulsive behaviors. Two are specific to childhood: separation anxiety and shyness/inhibition. Heritability was greatest for obsessive-compulsive behaviors (65 percent) and shyness/inhibition (75 percent), with no evidence of shared environmental influence. Another report examining the same sample of twins during middle childhood found moderate stability in parent reports of anxiety symptoms from age 7 to 9, with genetic influences accounting for approximately half of the variance in symptoms (Trzaskowski, Zavos, Haworth, Plomin, & Eley, 2011). The stability in each type of anxiety symptom was due primarily to genetic influences, whereas change from one type of symptom to another over time was due mostly to shared environmental influences. These findings highlight the need for longitudinal research and for examining multiple symptoms.

A study of obsessive-compulsive symptoms in the United States and the Netherlands also found high heritability (55 percent) in both countries in twins aged 7, 10, and 12 (Hudziak et al., 2004). Heritabilities of generalized anxiety and fears were about 40 percent. For fears, there was some evidence of shared environmental influence, which is similar to results for specific fears in adults (Chapter 15). A similar pattern of findings was found in a longitudinal study of fears and phobias for 2490 Swedish twins followed from middle childhood (age 8–9) to early adulthood (age 19–20) (Kendler, Gardner, Annas, et al. 2008). For three categories of fears—animal, blood/injury, and situational—this study showed relatively stable genetic influence over time, decreasing shared environmental influences, and increasing nonshared environmental influences. An interesting developmental pattern of genetic effects also emerged, with only a modest amount of genetic influence from middle childhood persisting into young adulthood and with new genetic influences (or innovations) emerging at each age/time of assessment, especially during early adolescence (age 13–14) (Figure 16.2).

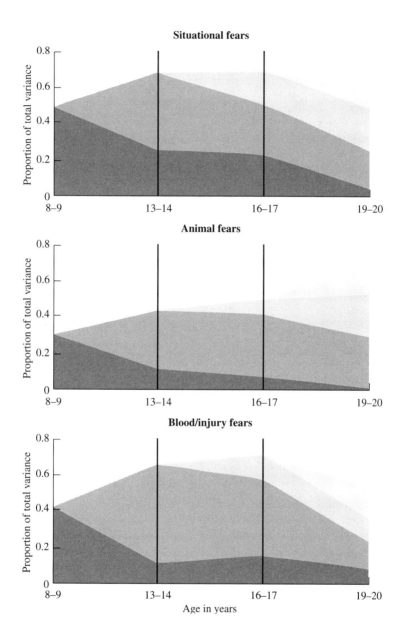

• **FIGURE 16.2** Graphs representing the proportion of total variance in fears accounted for by genetic influences from middle childhood to early adulthood. The three types of fears all show substantial genetic influence at age 8–9 (dark blue shading) but substantial new genetic influences emerge by age 13–14 (lighter blue shading). Less new genetic influence emerges at ages 16–17 (light blue shading) or 19–20 (gray portion). (Adapted, with permission, from Kendler, K. S., Gardner, C. O., Annas, P., Neale, M. C., Eaves, L. J., & Lichtenstein, P. (2008). A longitudinal twin study of fears from middle childhood to early adulthood: Evidence for a developmentally dynamic genome. *Archives of General Psychiatry, 65,* 421–429. Personal use of this material is permitted. However, permission to reuse this material for any other purpose must be obtained from the American Medical Association.)

Separation anxiety is interesting because in addition to showing moderate heritability (about 40 percent), substantial shared environmental influence was also found (35 percent) (Feigon, Waldman, Levy, & Hay, 2001). It is noteworthy that studies of maternal attachment of young children, which is indexed in part by separation anxiety, have also found evidence for shared environmental influence (Fearon et al., 2006; O'Connor & Croft, 2001; Roisman & Fraley, 2006). However, a follow-up of 4-year-old UK twins at 6 years of age using DSM-IV diagnoses of separation anxiety disorder found high heritability of liability (73 percent) and no shared environmental influence (Bolton et al., 2006), although there were significant shared environmental influences on the covariation between specific phobia and separation anxiety symptoms in a subset of the same sample (Eley, Rijsdijk, Perrin, O'Connor, & Bolton, 2008). These results are not necessarily contradictory because the studies that found shared environmental influence and modest heritability analyzed individual differences throughout the distribution, whereas the Bolton et al. study focused on the diagnosable extreme of separation anxiety, that is, on covariation of diagnoses.

Multivariate genetic analysis of the study of 4-year-old twins indicated that the five components of anxiety were moderately correlated genetically, although obsessive-compulsive behaviors were least related genetically to the others (Eley et al., 2003). A subsequent analysis of 7- and 9-year-old twins yielded a similar pattern of findings, with anxiety-related behaviors showing common variance due to genetic and shared environmental influences and specific genetic and nonshared environmental influences on each subtype (Hallett, Ronald, Rijsdijk, & Eley, 2009). These findings were replicated in a study examining 378 pairs of Italian twin children (Ogliari et al., 2010). Specifically, genetic and nonshared environmental influences explained the covariation among generalized anxiety, panic, social phobia, and separation anxiety. The strong genetic overlap between anxiety and depression in adulthood (Chapter 15) suggests that depressive symptoms might also be profitably studied in childhood (Thapar & Rice, 2006). One twin study found differences before and after puberty in the etiology of the association between anxiety and depression (Silberg, Rutter, & Eaves, 2001). Several studies have examined genetic and environmental influences on depressive symptoms and on the covariation among depressive and anxiety symptoms during adolescence and during childhood (e.g., Brendgen et al., 2009; Kendler, Gardner, & Lichtenstein, 2008; Lamb et al., 2010). A review of this work indicates that genes substantially influence stability in both anxiety and depression from age 7 to 12, but not from age 3 to 7, and that the high degree of comorbidity between these disorders can be explained primarily by genetic covariation (Franic, Middeldorp, Dolan, Ligthart, & Boomsma, 2010).

Other Disorders

Although schizophrenia and bipolar disorder do not generally appear until early adulthood, genetic research on possible childhood forms of these disorders has been motivated by the principle that more severe forms of disorders are likely to have an earlier onset (Nicolson & Rapoport, 1999). In relation to childhood-onset schizophrenia, relatives of affected individuals are at increased risk of schizophrenia, suggesting a link between the child and adult forms of the disorder (Nicolson et al., 2003). The only twin study of childhood schizophrenia yielded high heritability, although the sample size was small (Kallmann & Roth, 1956). More recent work has examined child and adolescent deficits in social adjustment and schizotypal personality as precursors of the development of schizophrenia using a twin-family design, finding that schizophrenia was associated with these deficits for primarily genetic reasons (Picchioni et al., 2010). Interesting results concerning links with adult schizophrenia are emerging from molecular genetic research incorporating brain endophenotypes (Addington et al., 2005; Gornick et al., 2005).

Childhood bipolar disorder appears to be more likely in families with adult bipolar disorder (Pavuluri, Birmaher, & Naylor, 2005). Linkage and candidate gene studies of childhood bipolar disorder have been reported, but no consistent results have emerged (Althoff, Faraone, Rettew, Morley, & Hudziak, 2005; Doyle et al., 2010; McGough et al., 2008). When genes are identified that are responsible for the high heritabilities of adult schizophrenia and bipolar disorder, one of the next research questions will be whether these genes are also associated with juvenile forms of these disorders.

Other childhood disorders for which some genetic data are available include enuresis (bedwetting) and tics. Enuresis in children after age 4 is common, about 7 percent for boys and 3 percent for girls. An early family study found substantial familial resemblance (Hallgren, 1957). Strong genetic influence was found in three small twin studies (Bakwin, 1971; Hallgren, 1957; McGuffin et al., 1994). A large study of adult twins reporting retrospectively on enuresis in childhood yielded substantial heritability (about 70 percent) for both males and females (Hublin, Kaprio, Partinen, & Koskenvuo, 1998). However, an equally large study of 3-year-old twins found only moderate genetic influence on nocturnal bladder control as reported by parents for boys (about 30 percent) and an even smaller effect in girls (about 10 percent) (Butler, Galsworthy, Rijsdijk, & Plomin, 2001). Candidate gene studies have not yielded replicable results (von Gontard, Schaumburg, Hollmann, Eiberg, & Rittig, 2001). A large epidemiological family study found that risk for severe childhood nocturnal enuresis was greater when mothers or fathers experienced nocturnal enuresis, and urinary incontinence was nearly 10 times higher in children when fathers were incontinent (with a lower risk from mothers of about 3 times higher), indicating a strong familial influence (von Gontard, Heron, & Joinson, 2011).

Tic disorders involve involuntary twitching of certain muscles, especially of the face, that typically begins in childhood. A twin study indicated that heritability of tics in children and adolescents was modest (about 30 percent) (Ooki, 2005). The same study showed that stuttering was highly heritable (about 80 percent) but that tics and stuttering are genetically different. Genetic research has focused on the most severe form, called Tourette disorder. Tourette disorder is rare (about 0.4

percent), whereas simple tics are much more common. Although family studies show little familial resemblance for simple tics, relatives of probands with chronic, severe tics characteristic of Tourette disorder are at increased risk for tics of all kinds (Pauls, 1990), for obsessive-compulsive disorder (Pauls, Towbin, Leckman, Zahner, & Cohen, 1986), and for ADHD (Pauls, Leckman, & Cohen, 1993). A twin study of Tourette disorder found concordances of 53 percent for identical twins and 8 percent for fraternal twins (Price, Kidd, Cohen, Pauls, & Leckman, 1985). Molecular genetic studies have so far not yielded replicable results. Linkage studies of large family pedigrees have been reported (e.g., Verkerk et al., 2006), but no clear major-gene linkages have been detected. The largest genomewide QTL linkage study of Tourette disorder suggested linkage on chromosome $2p$ (The Tourette Syndrome Association International Consortium for Genetics, 2007). Of the many candidate gene associations that have been reported (Pauls, 2003), only one gene has as yet yielded consistent associations: rare variants of a gene (*SLITRK1*) involved in dendritic growth (Abelson et al., 2005; Grados & Walkup, 2006; O'Roak et al., 2010). At least two association studies, however, have failed to replicate this finding (Keen-Kim et al., 2006; Scharf et al., 2008).

Overview of Twin Studies of Childhood Disorders

Genetic research on childhood disorders has increased dramatically, in part fueled by the finding of high heritabilities for autism and ADHD. A general summary of twin results for the major domains of childhood psychopathology is presented in Figure 16.3. In addition to the high heritabilities of autism and its components and of ADHD and its components, heritability is also exceptionally high for aggressive conduct disorder, obsessive-compulsive symptoms, and shyness. Just as interesting, however, are the moderate heritabilities for nonaggressive conduct disorder, generalized anxiety, fears, and separation anxiety. Especially noteworthy is the evidence for shared environmental influence for nonaggressive conduct disorder and separation anxiety. Nearly all of these results in childhood are based on parent or teacher reports of children's behavior. A twin study of psychopathology in adolescence using interviews with the twins themselves yielded quite different results (Ehringer, Rhee, Young, Corley, & Hewitt, 2006).

Summary

Two decades ago, autism was thought to be an environmental disorder. Now, twin studies suggest that it is one of the most heritable disorders. Although the results of linkage studies and candidate gene studies have not as yet been successful, there is accumulating evidence that rare variants (e.g., CNVs) play an important role. This lack of success in finding common variants might be due in part to the possibility that the

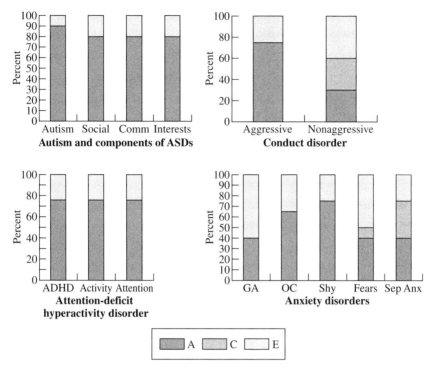

• **FIGURE 16.3** Summary of twin study estimates of genetic and environmental variances for major domains of childhood psychopathology. The components of autism spectrum disorders (ASDs) are social relationships, communication deficits, and restricted interests. The five aspects of anxiety disorders are generalized anxiety (GA), obsessive-compulsive behaviors (OC), shyness/inhibition (Shy), fears, and separation anxiety (Sep Anx). A = additive genetic variance; C = common (shared) environmental variance; E = nonshared environmental variance.

components of the autistic triad—abnormalities in social relationships, communication deficits, and restricted interests—are different genetically, even though each is highly heritable.

The DSM-IV category of attention-deficit and disruptive behavior disorders includes attention-deficit hyperactivity disorder (ADHD), which is highly heritable and shows no shared environmental influence. Multivariate genetic research suggests that its components of activity and attention overlap genetically, providing support for the construct of ADHD. Candidate gene studies of ADHD have yielded two dopamine receptor genes that show small but significant associations. This DSM-IV category also includes conduct disorder. Genetic research suggests that conduct disorder is heterogeneous, with aggressive conduct disorder showing substantial genetic influence and no shared environmental influence, in contrast to nonaggressive conduct disorder, which shows only modest genetic influence and moderate shared environmental influence.

Twin studies of parental ratings of anxiety in childhood suggest an interestingly diverse pattern of results. The highest heritability emerges for shyness, which is one of the most highly heritable personality traits (Chapter 17). Heritability is also very high for obsessive-compulsive symptoms, although results in adulthood are more mixed (Chapter 17). Heritability is more modest for generalized anxiety, which is comparable to results in adulthood (Chapter 15). These three aspects of anxiety show no evidence for shared environmental influence, which is similar to results for adult psychopathology but is even more surprising in childhood because children are living with their families. In contrast, fears and especially separation anxiety are notable for the evidence they show of shared environmental influence.

Some genetic influence has also been reported for childhood schizophrenia, childhood bipolar disorder, enuresis, and chronic tics, although much less genetic research has targeted these disorders.

CHAPTER SEVENTEEN

Personality and Personality Disorders

I f you were asked what someone is like, you would probably describe various personality traits, especially those depicting extremes of behavior. "Jennifer is full of energy, very sociable, and unflappable." "Steve is conscientious, quiet, but quick tempered." Genetic researchers have been drawn to the study of personality because, within psychology, personality has always been the major domain for studying the normal range of individual differences, with the abnormal range being the provenance of psychopathology. A general rule emerging from behavioral genetic research is that common disorders are the quantitative extreme of the same genetic and environmental factors that contribute to the normal range of variation. In other words, some psychopathology may be the extreme of normal variation in personality. We will return to the links between personality and psychopathology later in this chapter, after we have described basic research on personality.

Personality traits are relatively enduring individual differences in behavior that are stable across time and across situations (John, Robins, & Pervin, 2008). In the 1970s, there was an academic debate about whether personality exists, a debate reminiscent of the nature-nurture debate. Some psychologists argued that behavior is more a matter of the situation than of the person, but it is now generally accepted that both are important and can interact (Kenrick & Funder, 1988; Rowe, 1987). Cognitive abilities (Chapters 12 and 13) also fit the definition of enduring individual differences, but they are usually considered separately from personality. Another definitional issue concerns temperament, personality traits that emerge early in life and, according to some researchers (e.g., Buss & Plomin, 1984), may be more heritable. However, there are many different definitions of temperament (Goldsmith et al., 1987), and the supposed distinction between temperament and personality will not be emphasized here.

Genetic research on personality is extensive and is described in several books (Benjamin, Ebstein, & Belmaker, 2002; Cattell, 1982; Eaves, Eysenck, & Martin, 1989; Loehlin, 1992; Loehlin & Nichols, 1976) and hundreds of research papers (Bezdjian, Baker, & Tuvblad, 2011; Krueger, South, Johnson, & Iacono, 2008; Saudino, 2005;

Turkheimer, 2013). We will provide only an overview of this huge literature, in part because its basic message is quite simple: Genes make a major contribution to individual differences in personality whereas shared environment does not, especially when assessed by a self-report questionnaire; environmental influence on personality is almost entirely of the nonshared variety.

Self-Report Questionnaires

The vast majority of genetic research on personality involves self-report questionnaires administered to adolescents and adults. Such questionnaires include a range of dozens to hundreds of items, such as "I am usually shy when meeting people I don't know well" or "I am easily angered." People's responses to these questionnaires are remarkably stable, even over several decades (Costa & McCrae, 1994).

Nearly 40 years ago, a landmark study involving 750 pairs of adolescent twins and dozens of personality traits reached two major conclusions that have stood the test of time (Loehlin & Nichols, 1976). First, nearly all personality traits show moderate heritability. This conclusion might seem surprising because you would expect some traits to be highly heritable and other traits not to be heritable at all. Second, although environmental variance is also important, virtually all the environmental variance makes children growing up in the same family no more similar than children in different families. This category of environmental effects is called nonshared environment. The second conclusion is also surprising because theories of personality from Freud onward assumed that parenting played a critical role in personality development. This important finding is discussed in Chapter 8.

Genetic research on personality has focused on five broad dimensions of personality, called the *Five-Factor Model (FFM)*, that encompass many aspects of personality (Goldberg, 1990). The best studied of these are extraversion and neuroticism. Extraversion includes sociability, impulsiveness, and liveliness. Neuroticism (emotional instability) involves moodiness, anxiousness, and irritability. These two traits plus the three others included in the FFM create the acronym OCEAN: openness to experience (culture), conscientiousness (conformity, will to achieve), extraversion, agreeableness (likability, friendliness), and neuroticism.

Genetic results for extraversion and neuroticism are summarized in Table 17.1 (Loehlin, 1992). In five large twin studies in five different countries, with a total sample size of 24,000 pairs of twins, results indicate moderate genetic influence. Correlations are about 0.50 for identical twins and about 0.20 for fraternal twins. Studies of twins reared apart also indicate genetic influence, as do adoption studies of extraversion. For neuroticism, adoption results point to less genetic influence than do the twin studies—indeed, the sibling data indicate no genetic influence at all. Lower heritability in adoption than in twin studies could be due to nonadditive genetic variance, which makes identical twins more than twice as similar as first-degree relatives (Eaves, Heath, et al., 1999; Eaves, Heath, Neale, Hewitt, & Martin, 1998; Keller, Coventry, Heath, & Martin, 2005; Loehlin, Neiderhiser, & Reiss, 2003; Plomin, Corley,

Twin, Family, and Adoption Results for Extraversion and Neuroticism

Type of Relative	Correlation	
	Extraversion	Neuroticism
Identical twins reared together	0.51	0.46
Fraternal twins reared together	0.18	0.20
Identical twins reared apart	0.38	0.38
Fraternal twins reared apart	0.05	0.23
Nonadoptive parents and offspring	0.16	0.13
Adoptive parents and offspring	0.01	0.05
Nonadoptive siblings	0.20	0.09
Adoptive siblings	−0.07	0.11

SOURCE: Loehlin (1992).

Caspi, Fulker, & DeFries, 1998). It could also be due to a special environmental effect that boosts identical twin similarity (Plomin & Caspi, 1999). Model-fitting analyses across these twin and adoption designs produce heritability estimates of about 50 percent for extraversion and about 40 percent for neuroticism (Loehlin, 1992). The fact that the heritability estimates are much less than 100 percent implies that environmental factors are important, but, as mentioned earlier, this environmental influence is almost entirely due to nonshared environmental effects, although there is some evidence that shared environmental influence may be more important at the extremes of personality (Pergadia, Madden, et al., 2006).

Heritabilities in the 30 to 50 percent range are typical of personality results (Figure 17.1), although much less genetic research has been done on the other three traits

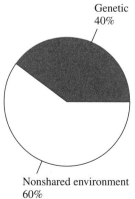

Genetic
40%

Nonshared environment
60%

• **FIGURE 17.1** Genetic results for personality traits assessed by self-report questionnaires are remarkably similar, suggesting that 30 to 50 percent of the variance is due to genetic factors. Environmental variance is also important, but hardly any environmental variance is due to shared environmental influence.

of the FFM. Also, openness to experience, conscientiousness, and agreeableness have been measured differently in different studies because, until recently, no standard measures were available. A model-fitting summary of family, twin, and adoption data for scales of personality thought to be related to these three traits yielded heritability estimates of 45 percent for openness to experience, 38 percent for conscientiousness, and 35 percent for agreeableness, with no evidence of shared environmental influence (Loehlin, 1992). The first genetic study to use a measure specifically designed to assess the FFM factors found similar estimates in an analysis of twins reared together and twins reared apart, except that agreeableness showed lower heritability (12 percent) (Bergeman et al., 1993). Another twin study yielded similar moderate heritabilities for all of the FFM factors (Jang, Livesley, & Vernon, 1996).

Do these broad FFM factors represent the best level of analysis for genetic research? Multivariate genetic research supports the FFM structure but also suggests a hierarchical model in which subtraits within each FFM factor show significant unique genetic variance not shared with other traits in the factor (Jang et al., 2006; Jang, McCrae, Angleitner, Riemann, & Livesley, 1998; Loehlin, 1992). For example, extraversion includes diverse traits such as sociability, impulsiveness, and liveliness, as well as activity, dominance, and sensation seeking. Each of these traits has received some attention in genetic research but not nearly as much as the more global traits of extraversion and neuroticism.

Several theories of personality development have been proposed about other ways in which personality should be sliced, and similar results have been found for the different traits highlighted in these theories (Kohnstamm, Bates, & Rothbart, 1989). For example, a neurobiologically oriented theory organizes personality into four different domains: novelty seeking, harm avoidance, reward dependence, and persistence (Cloninger, 1987). Similar twin study results have been found for these dimensions (Heiman, Stallings, Young, & Hewitt, 2004; Stallings, Hewitt, Cloninger, Heath, & Eaves, 1996). Sensation seeking, which is related to conscientiousness as well as to extraversion (Zuckerman, 1994), is especially interesting because it is the domain of the first association reported between a specific gene and normal personality, as described later. In two large twin studies, heritabilities of about 60 percent were found for a measure of general sensation seeking (Fulker, Eysenck, & Zuckerman, 1980; Koopmans, Boomsma, Heath, & van Doornen, 1995). This evidence for substantial genetic influence is supported by results from a study of identical twins reared apart, which yielded a correlation of 0.54 (Tellegen et al., 1988). Sensation seeking itself can be broken down into components, such as disinhibition (seeking sensation through social activities such as parties), thrill seeking (desire to engage in physically risky activities), experience seeking (seeking novel experiences through the mind and senses), and boredom susceptibility (intolerance for repetitive experience). Each of these subscales also shows moderate heritability. A study that combined multiple personality scales from different measures using latent factors found three dimensions that showed heritabilities of about 50 to 65 percent and no evidence of shared environmental influences (Ganiban, Chou, et al., 2009). The heritabilities are somewhat higher than usual for personality because they are estimated from reliable variance from latent personality constructs rather than from estimates of total variance from single-reporter indices.

One of the most surprising findings from genetic research on personality questionnaires is that the many traits that have been studied all show moderate genetic influence and no influence of shared environment. It is also surprising that studies have not found any personality traits assessed by self-report questionnaire that consistently show low or no heritability in twin studies. This is in contrast to childhood psychopathology (Chapter 16), where some disorders are more heritable than others and some disorders yield more shared environmental influence than others. But can this be true for personality? One way to explore this issue is to use measures of personality other than self-report questionnaires to investigate whether this result is somehow due to self-report measures.

Other Measures of Personality

A study of nearly 1000 adult twin pairs in Germany and Poland compared results from self-report questionnaires and from ratings by peers for measures of the FFM personality factors (Riemann, Angleitner, & Strelau, 1997). Each twin's personality was rated by two different peers. The average correlation between the two peer ratings was 0.61, a result indicating substantial agreement concerning each twin's personality. The averaged peer ratings correlated 0.55 with the twins' self-report ratings, a result indicating moderate validity of self-report ratings. Figure 17.2 shows the results of twin analyses for self-report data and peer ratings averaged across two peers. The results for self-report ratings are similar to those in other studies. The exciting result is that peer ratings also show significant genetic influence, although somewhat less than

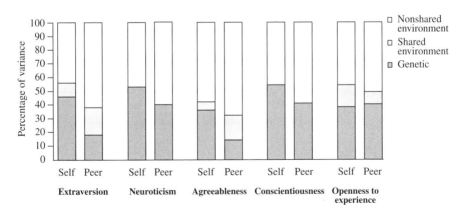

• **FIGURE 17.2** Genetic (dark blue), shared environment (light blue), and nonshared environment (white) components of variance for self-report ratings and peer ratings for the FFM personality traits. Components of variance were calculated from identical twin (660 pairs) and same-sex fraternal twin (200 pairs) correlations presented by Riemann and colleagues (1997). (Used with permission from Plomin & Caspi, 1999.)

self-report ratings. For two of the five traits (extraversion and agreeableness), peer ratings suggest greater influence of shared environment than do self-report ratings, although these differences are not statistically significant. Importantly, multivariate genetic analysis indicates that the same genetic factors are largely involved in self-report and peer ratings, a result providing strong evidence for the genetic validity of self-report ratings. An earlier study used twin reports about each other, and it also found similar evidence for genetic influence on personality traits, whether assessed by self-report or by the co-twin (Heath, Neale, Kessler, Eaves, & Kendler, 1992).

Genetic researchers interested in personality in childhood were forced to use measures other than self-report questionnaires. For the past 30 years, this research has relied primarily on ratings by parents, but twin studies using parent ratings have yielded odd results. Correlations for identical twins are high and correlations for fraternal twins are very low, sometimes even negative. It is likely that these results are due to contrast effects, which result when parents of fraternal twins contrast the twins (Plomin, Chipuer, & Loehlin, 1990). For example, parents might report that one twin is the active twin and the other is the inactive twin, even though, relative to other children that age, the twins are not really very different from each other (Carey, 1986; Eaves, 1976; Neale & Stevenson, 1989).

Furthermore, adoption studies using parent ratings in childhood find little evidence for genetic influence (Loehlin, Willerman, & Horn, 1982; Plomin, Coon, Carey, DeFries, & Fulker, 1991; Scarr & Weinberg, 1981; Schmitz, 1994). A combined twin study and stepfamily study of parent ratings of adolescents found significantly greater heritability estimates for twins than for nontwins and confirmed that parent ratings are subject to contrast effects (Saudino, McGuire, Reiss, Hetherington, & Plomin, 1995). Interestingly, using the same sample, sibling contrast effects also were significant during later adolescence but did not impact genetic and shared environmental estimates (Ganiban, Saudino, Ulbricht, Neiderhiser, & Reiss, 2008). As mentioned in relation to self-report questionnaires, such findings might also be due to nonadditive genetic variance. However, the weight of evidence indicates that genetic results for parent ratings of personality are due in part to contrast effects (Mullineaux, Deater-Deckard, Petrill, Thompson, & DeThorne, 2009; Saudino, Wertz, Gagne, & Chawla, 2004).

Other measures of children's personality, such as behavioral ratings by observers, show more reasonable patterns of results in both twin and adoption studies (Braungart, Plomin, DeFries, & Fulker, 1992; Cherny et al., 1994; Goldsmith & Campos, 1986; Lemery-Chalfant, Doelger, & Goldsmith, 2008; Matheny, 1980; Plomin et al., 1993; Plomin & Foch, 1980; Plomin, Foch, & Rowe, 1981; Saudino, 2012; Saudino, Plomin, & DeFries, 1996; Wilson & Matheny, 1986). For example, genetic influence has been found in observational studies of young twins for a dimension of fearfulness called behavioral inhibition (Matheny, 1989; Robinson, Kagan, Reznick, & Corley, 1992), for shyness observed in the home and the laboratory (Cherny et al., 1994), for effortful control during middle childhood (Lemery-Chalfant et al., 2008), and for activity level measured by actometers that record movement (Saudino, 2012; Saudino & Eaton,

1991). Because evidence for genetic influence is so widespread, even for observational measures, it is interesting that observer ratings of personality in the first few days of life have found no evidence for genetic influence (Riese, 1990) and that individual differences in smiling in infancy also show no genetic influence (Plomin, 1987).

Other Findings

There is a renaissance of genetic research on personality, which will be accelerated by research showing the association between personality and psychopathology and molecular genetic studies of personality (Bouchard & Loehlin, 2001). Both of these trends will be discussed later in this chapter. As just described, another example of new directions for personality research is increasing interest in measures other than self-report questionnaires. Three other examples include research on personality in different situations, studies of developmental change and continuity, and the role of personality in the interplay between nature and nurture.

Situations

It is interesting, in relation to the person-situation debate mentioned earlier, that some evidence suggests that genetic factors are involved in situational change as well as in stability of personality across situations (Phillips & Matheny, 1997). For example, in one study, observers rated the adaptability of infant twins in two laboratory settings: unstructured free play and test taking (Matheny & Dolan, 1975). Adaptability differed to some extent across these situations, but identical twins changed in more similar ways than fraternal twins did, an observation implying that genetics contributes to change as well as to continuity across situations for this personality trait. Similar results were found in a more recent study of person-situation interaction (Borkenau, Riemann, Spinath, & Angleitner, 2006). Such results might differ for other personality traits. For example, a twin study of shyness found that genetic factors largely contribute to stability across observations in the home and in the laboratory; environmental factors account for shyness differences between these situations (Cherny et al., 1994). A twin study using a questionnaire to assess personality in different situations found that genetic factors contribute to personality changes across situations (Dworkin, 1979). Even patterns of responding across items of personality questionnaires show genetic influence (Eaves & Eysenck, 1976; Hershberger, Plomin, & Pedersen, 1995).

Development

Does heritability change during development? Unlike general cognitive ability, which shows increases in heritability throughout the life span (Chapter 12), it is more difficult to draw general conclusions concerning personality development, in part because there are so many personality traits. In general, heritability appears to increase during infancy (Goldsmith, 1983; Loehlin, 1992), starting with zero heritability for personality during the first days of life (Riese, 1990). Of course, what is assessed as personality during the first few days of life is quite different from what is assessed later in development, and the sources of individual differences might also be quite different in neonates. Throughout the rest of the life span, it is clear that twins become less similar as time goes by, but this decreasing similarity occurs for identical twins as much as for fraternal twins for most personality traits, an observation suggesting that heritability does not change in childhood (McCartney et al., 1990), adolescence (Rettew et al., 2006), or adulthood (Loehlin & Martin, 2001).

A second important question about development concerns the genetic contribution to either continuity or change from age to age. For cognitive ability, genetic factors largely contribute to stability from age to age rather than to change, although some evidence can be found, especially in childhood, for genetically influenced change (Chapter 12). Although less well studied than cognitive ability, developmental findings for personality appear to be similar (Bratko & Butkovic, 2007; Ganiban et al., 2008; Hopwood et al., 2011; Kupper, Boomsma, de Geus, Denollet, & Willemsen, 2011; Loehlin, 1992; Saudino, 2012), even in late adulthood (Johnson, McGue, & Krueger, 2005; Read, Vogler, Pedersen, & Johansson, 2006).

Nature-Nurture Interplay

Another new direction for genetic research on personality involves the role of personality in explaining a fascinating finding: Environmental measures widely used in psychological research show genetic influence (Kendler & Baker, 2007). As discussed in Chapter 8, genetic research consistently shows that family environment, peer groups, social support, and life events often show as much genetic influence as measures of personality. The finding is not as paradoxical as it might seem at first. Measures of social environments, in part, assess genetically influenced characteristics of the individual. Personality is a good candidate to explain this genetic influence because personality can affect how people select, modify, construct, or perceive their environments. For example, in adulthood, genetic influence on personality has been reported to contribute to genetic influence on parenting in two studies (Chipuer & Plomin, 1992; Losoya et al., 1997), although not in another (Vernon, Jang, Harris, & McCarthy, 1997).

Genetic influence on perceptions of life events can be entirely accounted for by the FFM personality factors (Saudino et al., 1997). These findings are not limited to self-report questionnaires. For example, genetic influence found on an observational measure of home environments can be explained entirely by genetic influence on a tester-rated measure of attention called task orientation (Saudino & Plomin, 1997).

Personality and Social Psychology

Social psychology focuses on the behavior of groups, whereas individual differences are in the spotlight for personality research. For this reason, there is not nearly as much genetic research relevant to social psychology as there is for personality.

However, some areas of social psychology border on personality, and genetic research has begun at these borders. Four examples are relationships, self-esteem, attitudes, and behavioral economics.

Relationships

Genetic research has addressed parent-offspring relationships, peer relationships, romantic relationships, and sexual orientation.

Parent-offspring relationships Relationships between parents and offspring vary widely in their warmth (such as affection and support) and control (such as monitoring and organization). To what extent do genetic influences on parents and on offspring contribute to relationships? If identical twins are more similar in the qualities of their relationships than fraternal twins, this difference indicates genetic influence on relationships. For example, the first research of this sort involved adolescent twins' perceptions of their relationships with their parents. In two studies with different samples and different measures, genetic influence was found for twins' perceptions of their mothers' and fathers' warmth toward them (Rowe, 1981, 1983a). In contrast, adolescents' perceptions of their parents' control did not show genetic influence. One possible explanation is that parental warmth reflects genetically influenced characteristics of their children, but parental control does not (Lytton, 1991). Dozens of subsequent twin and adoption studies have found similar results that point to substantial genetic influences in most aspects of relationships, not just between parents and offspring but also between siblings, friends, and spouses (Plomin, 1994; Ulbricht & Neiderhiser, 2009).

A major area of developmental research on parent-offspring relationships involves attachment between infant and caregiver, as assessed in the so-called Strange Situation, a laboratory-based assessment in which mothers briefly leave their child with an experimenter and then return (Ainsworth, Blehar, Waters, & Wall, 1978). Sibling concordance of about 60 percent has been reported for attachment classification (van Ijzendoorn et al., 2000; Ward, Vaughn, & Robb, 1988). The first systematic twin study of attachment using the Strange Situation found only modest genetic influence and substantial influence of shared environment (O'Connor & Croft, 2001). For 110 twin pairs, MZ and DZ concordances for attachment type were 70 and 64 percent, respectively; for a continuous measure of attachment security, MZ and DZ correlations were 0.48 and 0.38, respectively. Although another twin study based on observations rather than the Strange Situation found evidence for greater genetic influence (Finkel, Wille, & Matheny, 1998), three other studies using the Strange Situation also found modest heritability and substantial shared environmental influence (Bokhorst et al., 2003; Fearon et al., 2006; Roisman & Fraley, 2006). In addition, a small twin study using a different measure of attachment found similar results for infant-father attachment (Bakermans-Kranenburg, van Uzendoorn, Bokhorst, & Schuengel, 2004). As described in Chapter 16, twin studies of separation anxiety disorder, which is related to attachment, also generally show modest heritability and substantial shared environmental influence.

Another component of relationships is empathy. One twin study of infants used videotaped observations of the empathic responding of infant twins following simulations of distress in the home and in the laboratory (Zahn-Waxler, Robinson, & Emde, 1992). Evidence was found for genetic influence for some aspects of the infants' empathic responses. A twin study of parent and teacher ratings of children's prosocial behavior during infancy and childhood found increasing genetic influence and decreasing shared environmental influence (Knafo & Plomin, 2006b), a finding replicated in another study (Knafo, Zahn-Waxler, Van Hulle, Robinson, & Rhee, 2008). Twin studies of empathic emotional responses also yielded evidence for genetic influence in adolescence (Davis, Luce, & Kraus, 1994) and in adulthood (Rushton, 2004).

Peer relationships The types of friends we select (or who select us) have been the focus of an enormous amount of research in developmental psychology, with many studies finding that children and adolescents with delinquent peers are more likely to engage in delinquent behaviors themselves. Most of these studies, however, have not considered whether something about the child may explain the types of friends the child has. Studies of adolescent twins and siblings that examined peer group characteristics found some evidence of genetic influence, especially for parent reports (Beaver, Shutt, et al., 2009; Iervolino et al., 2002). In addition, a twin study of the quality of young children's friendships found evidence for mostly nonshared environmental influences (Pike & Atzaba-Poria, 2003).

Romantic relationships Like parent-offspring and peer relationships, romantic relationships differ widely in various aspects, such as closeness and passion. The first genetic study of styles of romantic love is interesting because it showed no genetic influence (Waller & Shaver, 1994). The average twin correlations for six scales (for example, companionship and passion) were 0.26 for identical twins and 0.25 for fraternal twins, results implying some shared environmental influence but no genetic influence. Similar results have been found for initial attraction in mate selection (Lykken & Tellegen, 1993). In other words, genetics may play no role in the type of romantic relationships we choose. Perhaps love is blind, at least from the DNA point of view, although as we will see in Chapter 20, research in **evolutionary psychology** suggests that genetic factors may be involved in mate preference.

Although more research is needed to pin down the role of genetics in initial attraction, research suggests that genetic factors are important when the *quality* of romantic relationships is considered. There are now a handful of studies that have examined self-report, partner report, and observational ratings of relationship quality in married and long-term cohabiting twins; these studies have yielded heritability estimates ranging from about 15 to 35 percent, depending on the construct, and no shared environmental influence (Spotts et al., 2004, 2006). There is also some evidence that personality accounts for nearly half of the genetic variance in relationship quality (Spotts et al., 2005). Therefore, although genetic factors may not influence the type of romantic relationships we choose, they may affect our satisfaction with those relationships.

Sexual orientation An early twin study of male homosexuality reported remarkable concordance rates of 100 percent for identical twins and 15 percent for fraternal twins (Kallmann, 1952). However, a later twin study found less extreme concordances of 52 and 22 percent, respectively, and a concordance of 22 percent for genetically unrelated adoptive brothers (Bailey & Pillard, 1991); other twin studies found even less genetic influence and more influence of shared environment (Bailey, Dunne, & Martin, 2000; Kendler, Thornton, Gilman, & Kessler, 2000). A small twin study of lesbians also yielded evidence for moderate genetic influence (Bailey, Pillard, Neale, & Agyei, 1993). A population-based study of nearly 4000 Swedish twins found heritabilities ranging from 34 to 39 percent and no shared environmental influences for the total number of same-sex partners for men and much lower heritability, of around 20 percent, and modest shared environmental influence (~16 percent) for women (Langstrom, Rahman, Carlstrom, & Lichtenstein, 2010). This area of research received considerable attention because of reports of linkage between homosexuality and a region at the tip of the long arm of the X chromosome (Hamer, Hu, Magnuson, Hu, & Pattatucci, 1993; Hu et al., 1995). The X chromosome was targeted because it was thought that male homosexuality is more likely to be transmitted from the mother's side of the family, but later studies did not find an excess of maternal transmission (Bailey et al., 1999). The X linkage was not replicated in a subsequent study (Rice, Anderson, Risch, & Ebers, 1999). When genetic research touches on especially sensitive issues such as sexual orientation, it is important to keep in mind earlier discussions (see Chapter 7) about what it does and does not mean to show genetic influence (Pillard & Bailey, 1998).

Self-Esteem

A key variable for adjustment is self-esteem, which is also referred to as a sense of self-worth. Research on the etiology of individual differences in self-esteem has focused on the family environment (Harter, 1983). It is surprising that the possibility of genetic influence had not been considered previously, because it seems likely that genetic influence on personality and psychopathology (especially depression, for which low self-esteem is a core feature) could also affect self-esteem. Twin and adoption studies of self-esteem have been reported for teacher and parent ratings in middle childhood (Neiderhiser & McGuire, 1994), for self-ratings in adolescence (Kamakura, Ando, & Ono, 2007; Neiss, Sedikides, & Stevenson, 2006; Neiss, Stevenson, Legrand, Iacono, & Sedikides, 2009), for teacher, parent, and self-ratings in adolescence (McGuire, Neiderhiser, Reiss, Hetherington, & Plomin, 1994), and for self-ratings in adulthood (Roy, Neale, & Kendler, 1995). These studies point to modest genetic influence on self-esteem but no influence of shared family environment.

Attitudes and Interests

Social scientists have long been interested in the impact of group processes on change and continuity in attitudes and beliefs. Although it is recognized that social factors are not solely responsible for attitudes, it has been a surprise to find that genetics makes a major contribution to individual differences in attitudes. A core dimension of attitudes is traditionalism, which involves conservative versus liberal views on a wide range of issues. A measure of this attitudinal dimension was included in an adoption study of personality as a control variable because it was not expected to be heritable (Scarr & Weinberg, 1981). However, the results indicated that this measure was as heritable as the personality measures. In several twin studies (Eaves et al., 1989), including a study of twins reared apart (McCourt, Bouchard, Lykken, Tellegen, & Keyes, 1999; Tellegen et al., 1988), identical twin correlations are typically about 0.65 and fraternal twin correlations are about 0.50.

This pattern of twin correlations suggests heritability of about 30 percent and shared environmental influence of about 35 percent. However, assortative mating is higher for traditionalism than for any other psychological trait, with spouse correlations of about 0.50, unlike personality, which shows little assortative mating. Assortative mating inflates the fraternal twin correlation for interests, thereby lowering estimates of heritability and raising estimates of shared environment (Chapter 12). When assortative mating is taken into account, heritability is estimated to be about 50 percent and shared environmental influence is about 15 percent (Eaves et al., 1989; Olson, Vernon, Harris, & Jang, 2001). A twin-family analysis confirmed heritabilities of about 50 percent for traditionalism as well as showing similarly high heritabilities for sexual and religious attitudes but lower heritabilities for attitudes about taxes, the military, and politics (15 to 30 percent) (Eaves, Heath, et al., 1999). Religious attitudes were the focus of a special issue of the journal *Twin Research* (Eaves, D'Onofrio, & Russell, 1999). A recent study suggests that the heritability of religiousness increases from adolescence to adulthood (Koenig, McGue, Krueger, & Bouchard, 2005).

Sometimes these results are held up for ridicule: How can attitudes about royalty or nudist camps be heritable? We hope that by now you can answer this question (see Chapter 8), but it has been put particularly well in the context of social attitudes:

> We may view this as a kind of cafeteria model of the acquisition of social attitudes. The individual does not inherit his ideas about fluoridation, royalty, women judges, and nudist camps; he learns them from his culture. But his genes may influence which ones he elects to put on his tray. Different cultural institutions—family, church, school, books, television—like different cafeterias, serve up somewhat different menus, and the choices a person makes will reflect those offered him as well as his own biases. (Loehlin, 1997, p. 48)

This theme of nature operating via nurture was discussed in Chapter 8.

Social psychology traditionally uses the experimental approach rather than investigating naturally occurring variation. There is a need to bring together these two research traditions. For example, Tesser (1993), a social psychologist, separated attitudes into those that were more heritable (such as attitudes about the death penalty) and those that were less heritable (such as attitudes about coeducation and the truth of the Bible). In standard social psychology experimental situations, the more heritable items were found to be less susceptible to social influence and more important in interpersonal attraction (Tesser, Whitaker, Martin, & Ward, 1998).

A rapidly growing area of research has focused on the role of genetic influences on political psychology, the study of political attitudes and behaviors (e.g., Fowler & Schreiber, 2008; Hatemi & McDermot, 2011). One study in this area with a large American twin sample found that political party identification was due mostly to shared environmental influences, while the intensity of party identification was equally split between genetic and nonshared environmental influences (Hatemi, Alford, Hibbing, Martin, & Eaves, 2009). At least two studies have reported that political participation is heritable (Baker, Barton, Lozano, Raine, & Fowler, 2006; Fowler, Baker, & Dawes, 2008). Political participation also shows some evidence of increased heritability for those residing in counties that supported a third-party U.S. presidential candidate (G × E interaction; see Chapter 8) (Boardman, 2011). A recent report examined the direction of effects in the association between personality and political attitudes and found that although there was substantial genetic correlation between the two, personality traits did not seem to cause specific political attitudes (Verhulst, Eaves, & Hatemi, 2012).

Behavioral Economics

A fast-growing area of genetic research is behavioral economics. For example, results obtained from twin and adoption studies of vocational interests are similar to those that have been reported for personality questionnaires (Betsworth et al., 1994; Roberts & Johansson, 1974; Scarr & Weinberg, 1978a). Evidence for genetic influence was also found in twin studies of work values (Keller, Bouchard, Segal, & Dawes, 1992) and job satisfaction (Arvey, Bouchard, Segal, & Abraham, 1989). Recent research in this area links behavioral economics, genetics, and addictive behaviors (MacKillop et al., 2011) with the ability to delay reward, an important construct in behavioral economics.

Recent genetic research in behavioral economics has also begun to focus on other behaviors central to economics, such as investor behavior (Barnea, Cronqvist, & Siegel, 2010), financial decision making (Cesarini, Johannesson, Lichtenstein, Sandewall, & Wallace, 2010), philanthropy (Cesarini, Dawes, Johannesson, Lichtenstein, & Wallace, 2009), and economic risk-taking (Le, Miller, Slutske, & Martin, 2010; Zhong et al., 2009; Zyphur, Narayanan, Arvey, & Alexander, 2009). The field is moving quickly toward molecular genetic research (Beauchamp et al., 2011; Koellinger et al., 2010).

Personality Disorders

To what extent is psychopathology the extreme manifestation of normal dimensions of personality? It has long been suggested that this is the case for some psychiatric disorders (e.g., Cloninger, 2002; Eysenck, 1952; Livesley, Jang, & Vernon, 1998). As noted earlier, an important general lesson from behavioral genetic research on psychopathology (Chapters 14 and 15) as well as cognitive disabilities (Chapter 11) is that common disorders are the quantitative extreme of the same genetic and environmental factors that contribute to the normal range of variation. With cognitive disabilities such as reading disability, it is easy to see what normal variation is—variation in reading ability is normally distributed, and reading disability is the low end of that distribution. However, what are the dimensions of normal variation associated with depression or other types of psychopathology?

Chapter 15 ended with a multivariate genetic model that proposes two broad categories of psychopathology. The internalizing category includes depression and the anxiety disorders, and the externalizing category includes antisocial behavior and drug abuse. One of the most important findings from genetic research on personality is the extent of genetic overlap between the internalizing category of psychopathology and the personality factor of neuroticism. As mentioned earlier, neuroticism does not mean *neurotic* in the sense of being nervous; neuroticism refers to a general dimension of emotional instability, which includes moodiness, anxiousness, and irritability. Recent twin studies found that genetic factors shared between neuroticism and internalizing disorders accounted for between one-third and one-half of the genetic risk (Hettema et al., 2006; Kendler & Gardner, 2011; Mackintosh, Gatz, Wetherell, & Pedersen, 2006). Another study reported genetic correlations of about 0.50 between neuroticism and major depression (Kendler, Gatz, Gardner, & Pedersen, 2006a). Similar findings had emerged from earlier multivariate genetic studies (Eaves et al., 1989).

In summary, the internalizing category of psychopathology is similar genetically to the personality factor of neuroticism. What about the externalizing category of psychopathology? Although it would be wonderfully symmetrical if extraversion predicted externalizing psychopathology, this is not the case (Khan, Jacobson, Gardner, Prescott, & Kendler, 2005). However, several studies have shown that aspects of extraversion—especially novelty seeking, impulsivity, and disinhibition—predict externalizing psychopathology (Krueger, Caspi, Moffitt, Silva, & McGee, 1996). Two different twin studies have addressed the causes of overlap between disinhibitory dimensions of personality and externalizing psychopathology; both found that some of the overlap is genetic in origin, although most of the genetic influence on disinhibitory personality is independent of externalizing psychopathology (Krueger et al., 2002; Young et al., 2009).

Genetic research on the overlap between personality and psychopathology has focused on an area of psychopathology called *personality disorders*. Unlike psychopathology, described in Chapters 14, 15, and 18, personality disorders are personality traits that cause significant impairment or distress. People with personality disorders regard their disorder as part of who they are, their personality, rather than as a condition that can be treated. That is, they do not feel that they were once well and are now ill. For this reason, DSM-IV separates personality disorders from clinical syndromes. This category of disorders (called Axis II), which also includes general cognitive disability (called *mental retardation* in DSM-IV), refers to long-term disorders that date from childhood. Although the reliability, validity, and utility of diagnosing personality disorders have long been questioned, research has addressed the genetics of personality disorders and their links to normal personality and to other psychopathology (Jang, 2005; Nigg & Goldsmith, 1994; Torgersen, 2009). Increasingly, personality disorders are being considered as dimensions rather than categories, a distinction that will be reflected in the DSM-5 and will increase genetic research on their links with personality (Widiger & Trull, 2007).

DSM-IV recognizes ten personality disorders, but only three have been investigated systematically in genetic research: schizotypal, obsessive-compulsive, and antisocial personality disorders.

Schizotypal Personality Disorder

Schizotypal personality disorder involves less intense schizophrenic-like symptoms and, like schizophrenia, clearly runs in families (e.g., Baron, Gruen, Asnis, & Lord, 1985; Siever et al., 1990). The results of a small twin study suggested genetic influence, yielding 33 percent concordance for identical twins and 4 percent for fraternal twins (Torgersen et al., 2000). Twin studies using dimensional measures of schizotypal symptoms in unselected samples of twins also found evidence for genetic influence, with heritability estimates ranging widely from about 20 to 80 percent (Claridge & Hewitt, 1987; Coolidge, Thede, & Jang, 2001; Kendler, Aggen, et al., 2008; Kendler, Czajkowski, et al., 2006; Torgersen, 2009).

Genetic research on schizotypal personality disorder focuses on its relationship to schizophrenia and has consistently found an excess of the disorder among first-degree relatives of schizophrenic probands. A summary of such studies found that the risks of schizotypal personality disorder are 11 percent for the first-degree relatives of schizophrenic probands and 2 percent for control families (Nigg & Goldsmith, 1994).

Adoption studies have played an important role in showing that the disorder is part of the genetic spectrum of schizophrenia. For example, in a Danish adoption study (see Chapter 14), the rate of schizophrenia was 5 percent in the biological first-degree relatives of schizophrenic adoptees but 0 percent in their adoptive relatives and relatives of control adoptees (Kety et al., 1994). When schizotypal personality disorder was included in the diagnosis, the rates rose to 24 and 3 percent, respectively, implying greater genetic influence for the spectrum of schizophrenia that includes schizotypal personality disorder (Kendler, Gruenberg, & Kinney, 1994). Twin studies also suggest that schizotypal personality disorder is genetically related to schizophrenia (Farmer et al., 1987), especially for the negative (anhedonia) rather than the positive (delusions) aspects of schizotypy (Torgersen et al., 2002). Studies using community samples of twins suggest that the negative and positive aspects of schizotypy differ genetically (Linney et al., 2003) and that schizotypy is genetically related to the schizophrenia spectrum (Jang, Woodward, Lang, Honer, & Livesley,

2005). Recent genetic research considers subclinical psychotic experiences more generally as a personality trait that is normally distributed in the population and that predicts genetic liability for psychosis (Lataster, Myin-Germeys, Derom, Thiery, & van Os, 2009; Wigman et al., 2011).

Obsessive-Compulsive Personality Disorder

Obsessive-compulsive personality disorder sounds as if it is a milder version of the obsessive-compulsive type of anxiety disorder (OCD, described in Chapter 15); family studies provide some empirical support for this. However, the diagnostic criteria for these two disorders are quite different. The compulsion of OCD is a single sequence of specific behaviors, whereas the personality disorder is more pervasive, involving a general preoccupation with trivial details that leads to difficulties in making decisions and getting anything accomplished. Only one small twin study of diagnosed obsessive-compulsive personality disorder has been reported, and it found substantial genetic influence (Torgersen et al., 2000). However, twin studies of obsessional symptoms in unselected samples of twins suggest modest heritability (Kendler, Aggen, et al., 2008; Torgersen, 1980; Young, Fenton, & Lader, 1971). Family studies indicate that obsessional traits are more common (about 15 percent) in relatives of probands with obsessive-compulsive disorder than in controls (5 percent) (Rasmussen & Tsuang, 1984). Furthermore, results obtained from a recent twin study examining symptoms of obsessive-compulsive disorder and obsessive-compulsive personality traits suggest common genetic influences (Taylor, Asmundson, & Jang, 2011). This finding implies that obsessive-compulsive personality disorder might be part of the spectrum of the obsessive-compulsive type of anxiety disorder.

Antisocial Personality Disorder and Criminal Behavior

Much more genetic research has focused on antisocial personality disorder (ASPD) than on other personality disorders. DSM-IV criteria for ASPD include such chronic behaviors as breaking the law, lying, and conning others for personal profit or pleasure but also include more cognitive and personality-based criteria such as impulsivity, aggressiveness, disregard for safety of self and others, and lack of remorse for having hurt, mistreated, or stolen from others. Although antisocial personality disorder shows early roots, the vast majority of juvenile delinquents and children with conduct disorders do not develop antisocial personality disorder (Robins, 1978). For this reason, there is a need to distinguish conduct disorder that is limited to adolescence from antisocial behavior that persists throughout the life span (Caspi & Moffitt, 1995; Kendler, Aggen, & Patrick, 2012; Moffitt, 1993). As diagnosed by DSM-IV criteria, antisocial personality disorder affects about 1 percent of females and 4 percent of males from 13 to 30 years of age (American Psychological Association, 2000; Kessler et al., 1994). The prevalence of the disorder is much higher in selected populations, such as prisons, where there is a preponderance of violent offenders, with 47% of male prisoners and 21% of female prisoners having antisocial personality disorder

(Fazel & Danesh, 2002). Similarly, the prevalence of ASPD is higher among patients in alcohol or other drug abuse treatment programs than in the general population, suggesting a link between ASPD and substance abuse and dependence (Moeller & Dougherty, 2001).

Family studies show that ASPD runs in families (Nigg & Goldsmith, 1994), and an adoption study found that familial resemblance is largely due to genetic rather than to shared environmental factors (Schulsinger, 1972). The risk for ASPD is increased fivefold for first-degree relatives of ASPD males, whether living together or adopted apart. For relatives of ASPD females, risk is increased tenfold, a result suggesting that, to be affected for this disproportionately male disorder, females need a greater genetic loading. Although no twin studies of diagnosed ASPD are available, there are over 100 twin and adoption studies on antisocial behavior. A meta-analysis of 52 independent twin and adoption studies of antisocial behavior found evidence for significant shared environmental influences (15 percent) as well as significant genetic effects, including additive and nonadditive influences (40 percent), and non-shared environmental influences (16 percent) (Rhee & Waldman, 2002). More recent meta-analyses, though, suggest slightly higher heritabilities of 50 to 60 percent and similar magnitudes of shared environmental influences of about 15 percent (Burt, 2009a; Ferguson, 2010). However, both shared environmental influences and heritability were lower in parent-offspring studies than in twin and sibling studies, which could signal developmental changes between childhood (offspring) and adulthood (parents), in contrast to twins, who are exactly the same age. These meta-analyses agree that, while genetic influences are important to antisocial behavior in childhood through adulthood, the magnitude of familial effects (genetic and shared environmental influences) decreases somewhat with age and nonfamilial influences increase with age (Ferguson, 2010; Rhee & Waldman, 2002). Moreover, as is typically found in longitudinal genetic analyses, genetics and shared environment largely contribute to stability and nonshared environment contributes to change during development (Burt, McGue, & Iacono, 2010).

There have been questions about whether the criteria for ASPD reflect one true disorder (or a single dimension) or whether ASPD is better represented by multiple dimensions that capture variation in this personality domain (Burt, 2009a). A recent multivariate twin study of DSM-IV ASPD symptoms suggests that two factors comprise ASPD: aggressive-disregard and disinhibition (Kendler et al., 2012). Scores on the genetic aggressive-disregard factor are more strongly associated with risk for conduct disorder and early and heavy alcohol use; in contrast, scores on the genetic disinhibition factor are more strongly associated with novelty seeking and major depression (Kendler et al., 2012). Interestingly, both genetic factors predicted cannabis, cocaine, and alcohol dependence, which suggests two potential pathways that might explain the association between ASPD and substance use disorders, a topic we turn to shortly.

A type of antisocial personality disorder called psychopathy has recently become the target of genetic research because of its prediction of violent crime and recidivism. Although there is no precise equivalent in DSM-IV, psychopathic personality disorder involves a lack of empathy, callousness, irresponsibility, and manipulativeness (Hare, 1993; Viding, 2004). As mentioned in Chapter 16, psychopathic tendencies appear to be highly heritable in childhood, with no influence of shared environment (Viding et al., 2005; Viding, Jones, Frick, Moffitt, & Plomin, 2008). Similar results have been found in late adolescence (Larsson, Andershed, & Lichtenstein, 2006). A follow-up report also indicated that the overlap between psychopathic personality and antisocial behavior is largely genetic in origin (Larsson et al., 2007). Furthermore, psychopathic personality during adolescence predicts antisocial behavior in adults, and genetic factors contribute to this association (Forsman, Lichtenstein, Andershed, & Larsson, 2010).

ASPD is genetically correlated with both criminal behavior and substance use. Two adoption studies of birth parents with criminal records found increased rates of ASPD in their adopted offspring (Cadoret & Stewart, 1991; Crowe, 1974), suggesting that genetics contributes to the relationship between criminal behavior and ASPD. Most genetic research in this area has focused on criminal behavior itself, rather than on ASPD, because crime can be assessed objectively by using criminal records. However, criminal behavior, although important in its own right, is only moderately associated with ASPD. About 40 percent of male criminals and 8 percent of female criminals qualify for a diagnosis of ASPD (Robins & Regier, 1991). Clearly, breaking the law cannot be equated with psychopathology (Rutter, 1996).

A classic twin study of criminal behavior included male twins born in Denmark from 1881 to 1910 (Christiansen, 1977). For more than a thousand twin pairs, genetic influence was found for criminal convictions, with an overall concordance of 51 percent for male identical twins and 30 percent for male-male fraternal twins. In multiple twin studies of adult criminality, identical twins are consistently more similar than fraternal twins (Raine, 1993). The average concordances for identical and fraternal twins are 52 and 21 percent, respectively. In a U.S. twin study of self-reported arrests and criminal behavior involving more than 3000 male twin pairs in which both members served in the Vietnam War, genetics contributed to self-reports of being arrested and engaging in criminal behavior (Lyons, 1996). However, self-reported criminal behavior before age 15 showed negligible genetic influence. Shared environment made a major contribution to arrests and criminal behavior before age 15, but not later. These results before age 15 are similar to results for conduct disorder in adolescence (Chapter 16).

Adoption studies are also consistent with the hypothesis of significant genetic influence on adult criminality, although adoption studies point to less genetic influence than do twin studies. It has been hypothesized that twin studies overestimate genetic effects because identical twins are more likely to be partners in crime (Carey, 1992). Adoption studies include both the adoptees' study method (Cloninger, Sigvardsson, Bohman, & von Knorring, 1982; Crowe, 1972) and the adoptees' family method (Cadoret, O'Gorman, Heywood, et al., 1985). One of the best studies used the adoptees'

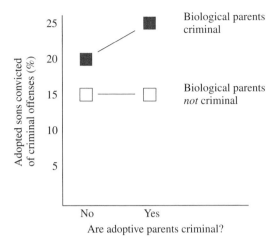

• **FIGURE 17.3** Evidence for genetic influence and genotype-environment interaction for criminal behavior in a Danish adoption study. (Adapted from Mednick et al., 1984.)

study method, beginning with more than 14,000 adoptions in Denmark between 1924 and 1947 (Mednick, Gabrielli, & Hutchings, 1984). Using court convictions as an index of criminal behavior, the researchers found evidence for genetic influence and for genotype-environment interaction, as shown in Figure 17.3. Adoptees were at greater risk for criminal behavior when their birth parents had criminal convictions, a finding implying genetic influence. Unlike the twin study just described, this adoption study (and others) found genetic influence for crimes against property but not for violent crimes (Bohman, Cloninger, Sigvardsson, & von Knorring, 1982; Brennan et al., 1996). Evidence for genotype-environment interaction was also suggested. Adoptive parents with criminal convictions had no effect on the criminal behavior of adoptees unless the adoptees' birth parents also had criminal convictions. A more recent study of adoptees included in the National Longitudinal Study of Adolescent Health found that those adoptees who had a birth father or birth mother who had ever been arrested were significantly more likely to be arrested, sentenced to probation, incarcerated, and arrested multiple times than adoptees whose birth parents had never been arrested (Beaver, 2011).

A Swedish adoption study of criminality using the adoptees' family method found evidence for genotype-environment interaction as well as interesting interactions with alcohol abuse, which greatly increases the likelihood of violent crimes (Bohman, 1996; Bohman et al., 1982). When adoptees' crimes did not involve alcohol abuse, their biological fathers were found to be at increased risk for nonviolent crimes. In contrast, when adoptees' crimes involved alcohol abuse, their biological fathers were not at increased risk for crime. These findings suggest that genetics contributes to criminal behavior but not to alcohol-related crimes, which are likely to be more violent.

Evidence from family, twin, and adoption studies consistently suggests a common underlying vulnerability to ASPD and substance use disorders. For example, relatives of alcohol-dependent individuals show significant familial aggregation of

ASPD (Nurnberger et al., 2004), and family history of alcohol use disorder is associated with ASPD. Large twin studies indicate that this familiality is due, in part, to genetic influences that contribute to the co-occurrence of ASPD and substance use disorders (Agrawal, Jacobson, Prescott, & Kendler, 2004; Fu et al., 2002; Hicks, Krueger, Iacono, McGue, & Patrick, 2004). Adoption studies provide additional support for a genetic link between ASPD and substance use. Male adoptees who were at increased biological risk for ASPD showed increased aggressiveness, conduct problems, ASPD, and eventual substance dependence (Cadoret, Yates, Troughton, Woodworth, & Stewart, 1995a), a finding that was replicated in female adoptees (Cadoret, Yates, Troughton, Woodworth, & Stewart, 1996).

Identifying Genes

In contrast to molecular genetic research on psychopathology, molecular genetic research on personality has received much less attention (Benjamin et al., 2002; Hamer & Copeland, 1998). The field began in 1996 with reports from two studies of an association between a DNA marker for a certain neuroreceptor gene (*DRD4*, dopamine D_4 receptor) and the personality trait of novelty seeking in unselected samples (Benjamin et al., 1996; Ebstein et al., 1996). *DRD4* is the gene mentioned in Chapter 16 that shows an association with attention-deficit hyperactivity disorder (ADHD). Novelty seeking is one of the four traits included in a theory of temperament developed by Cloninger (Cloninger, Svrakic, & Przybeck, 1993). Novelty seeking is very similar to the impulsive sensation-seeking dimension studied by Zuckerman (1994), and it is this impulsiveness that creates the genetic link with the impulsive component of ADHD. Individuals high in novelty seeking are characterized as impulsive, exploratory, fickle, excitable, quick-tempered, and extravagant. Cloninger's theory predicts that novelty seeking involves genetic differences in dopamine transmission.

The DNA marker consists of seven alleles involving 2, 3, 4, 5, 6, 7, or 8 repeats of a 48-base-pair sequence in a gene on chromosome 11 that codes for the D_4 receptor of dopamine and is expressed primarily in the brain limbic system. The number of repeats changes the receptor's structure, which has been shown to affect the receptor's efficiency in vitro. The shorter alleles (2, 3, 4, or 5 repeats) code for receptors that are more efficient in binding dopamine than are the receptors coded for by the larger alleles (6, 7, or 8 repeats). The theory is that individuals with the long-repeat *DRD4* allele are dopamine deficient and seek novelty to increase dopamine release. For this reason, the *DRD4* alleles are usually grouped as *short* (about 85 percent of alleles) or *long* (15 percent of alleles).

In both studies, individuals with longer *DRD4* alleles had significantly higher novelty-seeking scores than did individuals with the shorter alleles. Figure 17.4 shows the distributions of novelty-seeking scores for individuals with the short and the long *DRD4* alleles. However, many studies have failed to replicate the association with novelty seeking (Jang, 2005). A meta-analysis of 17 studies of extraversion

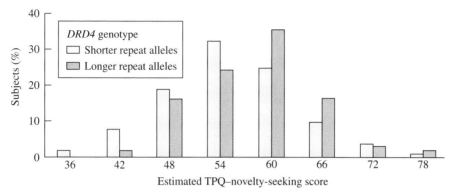

• **FIGURE 17.4** The longer alleles of the *DRD4* gene have been reported to be associated with increased novelty seeking. (From Benjamin et al., 1996, used with permission.)

rather than the narrow trait of novelty seeking found a weak association (Munafo et al., 2003). *DRD4* has also been reported to be associated with novelty seeking in the vervet monkey (Bailey, Breidenthal, Jorgensen, McCracken, & Fairbanks, 2007).

Neuroticism has been another focus of candidate gene studies of personality, in part because of the force of an early report of association with a polymorphism in a serotonin transporter gene (*5-HTTLPR;* Lesch et al., 1996), quickly followed by reports of its association with depression (Collier et al., 1996). A meta-analysis of 22 studies found some evidence for an association (Munafo et al., 2003; see also Munafo, Clark, & Flint, 2005), although a subsequent meta-analysis of more than 100,000 individuals found no association with either neuroticism or depression (Willis-Owen et al., 2005). Nonetheless, serotonin genes have been reported to show associations in studies of brain measures related to emotional responding (Chapter 15) and in interaction with an adverse childhood environment (Chapter 8) (Ebstein, 2006; Serretti, Calati, Mandelli, & De Ronchi, 2006).

As noted in Chapter 11, early reports of an association between XYY males and violence were overblown, although there seems to be some increase in hyperactivity and perhaps conduct problems (Ratcliffe, 1994). In a four-generation study of a Dutch family, a deficiency in a gene on the X chromosome that codes for an enzyme (monoamine oxidase A, MAOA), which is involved in the breakdown of several neurotransmitters, was associated with impulsive aggression and borderline mental retardation in males (Brunner, 1996; Brunner, Nelen, Breakefield, Ropers, & van Oost, 1993), although this genetic effect has not yet been found in any other families. However, as described in Chapter 8, the MAOA gene has been strongly associated with antisocial behavior in individuals who suffered severe childhood maltreatment, a genotype-environment interaction (Caspi et al., 2002). This finding has held up in a meta-analysis (Kim-Cohen et al., 2006). Two linkage studies of neuroticism suggested several, but different, linkages, and none included 17*q*, where the serotonin transporter gene is located (Fullerton et al., 2003; Nash et al., 2004).

More powerful methods for identifying such QTLs for personality are available in research on nonhuman animals, as described in Chapter 5 (see also Flint, 2004; Willis-Owen & Flint, 2007). For example, several QTLs for fearfulness have been localized in mice, as assessed in open-field activity (Flint et al., 1995; Henderson, Turri, DeFries, & Flint, 2004; Talbot et al., 1999). Also, transgenic knock-out gene studies in mice often find personality effects, such as increased aggression, when one or the other of two genes were knocked out, either the gene for a receptor for an important neurotransmitter (serotonin; Saudou et al., 1994) or the gene for an enzyme (neuronal nitric oxide synthase) that plays a basic role in neurotransmission (Nelson et al., 1995).

As noted in other chapters (see especially Chapter 9), candidate gene findings have a poor record of replicating, and this is also the case in the personality domain (Munafo & Flint, 2011). Several genomewide association studies have been reported for personality and have been summarized and reanalyzed in a recent meta-analysis that, in total, included over 20,000 individuals on the NEO Five-Factor Inventory (de Moor et al., 2012). Regions of significance were identified on chromosome 5 for openness to experience and on chromosome 18 for conscientiousness. An earlier genomewide association study of nearly 4000 individuals found associations for different genes, although none accounted for 1 percent or more of the variance and only one replicated in follow-up samples (Terracciano et al., 2010). A review of genetic research on personality noted that although there has not yet been clear identification of a gene or set of genes linked with personality, findings from animal studies may be used to guide human studies in ways more likely to yield replicable results (Flint & Willis-Owen, 2010). One study using the genome-wide complex trait analysis approach described in Box 7.2 examined about 12,000 individuals using genomewide SNP data and estimated heritabilities of approximately 6 percent for neuroticism and 12 percent for extraversion, thus recovering less than a quarter of the heritability estimated in most twin studies (Vinkhuyzen, Pedersen, et al., 2012).

Molecular genetic studies of antisocial personality disorder are similar to those for personality, with little evidence for specific genes or regions across studies. Genomewide linkage and genomewide association studies have used a variety of phenotypes related to antisocial behavior, primarily conduct disorder (see Chapter 16) (reviewed in Gunter, Vaughn, & Philibert, 2010). Two genomewide linkage studies focused specifically on antisocial behavior and ASPD in samples of families at high risk of alcohol dependence (Ehlers, Gilder, Slutske, Lind, & Wilhelmsen, 2008; Jacobson et al., 2008). These genomewide linkage studies, and others looking at variables in the antisocial spectrum, provide some clues about possible genomic locations (Gunter et al., 2010).

Candidate gene studies have also been conducted for ASPD, primarily for those genes related to serotonergic and dopaminergic activity. The monoamine oxidase (*MAOA*) gene codes for an enzyme that breaks down dopamine and serotonin. As noted earlier, *MAOA* has received considerable attention because a mutation in this gene was found to be associated with severe impulsive and aggressive behavior in males (Brunner et al., 1993) and because deletion of the gene coding for MAOA led to increased aggression in transgenic mice (Cases et al., 1995). Between 2004 and 2010, more than 25 studies on the *MAOA* gene and behaviors in the antisocial spectrum were reported (Gunter et al., 2010). Results are mixed, with some suggesting that the low-activity variant of *MAOA* is associated with antisocial behaviors, but only in combination with environmental factors such as maltreatment (e.g., Kim-Cohen et al., 2006), and others finding no such association (e.g., Young, Rhee, Stallings, Corley, & Hewitt, 2006). Other primary genes of interest are the serotonin transporter and the dopamine D_2 receptor, particularly within the context of the overlap between ASPD and substance dependence (Dick, 2007; Wu et al., 2008). However, although *MAOA* and various serotonergic and dopaminergic genes have surfaced as being potentially related to the risk for ASPD, they contribute only a small amount of variance.

Summary

More twin data are available from self-report personality questionnaires than from any other domain of psychology, and they consistently yield evidence for moderate genetic influence for dozens of personality dimensions. Most well studied are extraversion and neuroticism, with heritability estimates of about 50 percent for extraversion and about 40 percent for neuroticism across twin and adoption studies. Other personality traits assessed by personality questionnaire also show heritabilities ranging from 30 to 50 percent. There is no replicated example of zero heritability for any specific personality trait. Environmental influence is almost entirely due to nonshared environmental factors. These surprising findings are not limited to self-report questionnaires. For example, a twin study using peer ratings yielded similar results. Although the degree of genetic influence suggested by twin studies using parent ratings of their children's personalities appears to be inflated by contrast effects, more objective measures, such as behavioral ratings by observers, indicate genetic influence in twin and adoption studies.

New directions for genetic research include looking at personality continuity and change across situations and across time. Results indicate that genetic factors are largely responsible for continuity and that change is largely due to environmental factors. Other new findings include the central role that personality plays in producing genetic influence on measures of the environment. Another new direction for research lies at the border with social psychology. For example, genetic influence has been found for relationships, such as parent-offspring relationships and sexual orientation, but not romantic relationships. Other examples include evidence for genetic influence on self-esteem, attitudes, vocational interests, political affiliation and loyalty, and behavioral economics.

A major new direction for genetic research on personality is to consider its role in psychopathology. For example, depression and other internalizing forms of psychopathology are, to a large extent, the genetic extreme of normal variation in the major personality dimension of neuroticism. Personality disorders, which are at the border between personality and psychopathology, are another growth area for genetic research in personality. It is likely that some personality disorders are part of the genetic continuum of psychopathology: schizotypal personality disorder and schizo-phrenia, and obsessive-compulsive personality disorder and obsessive-compulsive anxiety disorder. Most genetic research on personality disorders has focused on an-tisocial personality disorder and its relationship to criminal behavior and substance abuse. From adolescence to adulthood, genetic influence increases and shared envi-ronmental influence decreases for symptoms of antisocial personality disorder, in-cluding juvenile delinquency and adult criminal behavior.

QTL associations have been reported for several candidate genes and personal-ity traits. However, similar to research on psychopathology, replication of associations has been difficult in part because effect sizes are much smaller than originally antici-pated. Genomewide association studies have also not yet yielded consistent results.

CHAPTER EIGHTEEN

Substance Use Disorders

Alcohol use disorders, nicotine use, and abuse of other drugs are major health-related behaviors. Externalizing behaviors, such as attention-deficit hyperactivity disorder (ADHD) and conduct disorder (see Chapter 16), have long been proposed as etiologic predictors of later alcohol and drug problems (Zucker, Heitzeg, & Nigg, 2011). More specifically, as discussed in Chapter 16, substance use is part of a general genetic factor of externalizing disorders, but alcohol and other drugs include significant disorder-specific genetic effects (Kendler, Prescott, et al., 2003; Lynskey, Agrawal, & Heath, 2010). Most behavioral genetic research in this area has focused on alcohol dependence or alcohol-related behavior and, to a somewhat lesser extent, nicotine dependence.

Alcohol Dependence

Twin and Adoption Research on Alcohol-Related Phenotypes

Clearly there are many steps on the path to alcohol dependence. For example, there is choice in whether or not to drink alcohol at all, in the amount one drinks, in the way one drinks, and in the development of tolerance and dependence. Each of these steps might involve different genetic mechanisms. For this reason, alcohol dependence is likely to be highly heterogeneous. Nonetheless, numerous family studies have shown that alcohol use disorders run in families, although the studies vary widely in the size of the effect and in diagnostic criteria. For males, alcohol dependence in a first-degree relative is by far the single best predictor of alcohol dependence. For example, a family study of 1212 alcohol-dependent probands and their 2755 siblings found an average risk for lifetime diagnosis of alcohol dependence of about 50 percent in male siblings and 25 percent in female siblings (Bierut et al., 1998). According to the U.S. Centers for Disease Control, the risk rates in the general population are about 17 percent for men and 8 percent for women. Assortative mating for alcohol use is substantial (correlation ranging from 0.38 to 0.45), which is thought to be caused by initial selection of the spouse rather than the effect of living with the spouse (Grant et al., 2007). Assortative mating of this magnitude could inflate estimates of shared environment and could also create a genotype-environment correlation in which children are more likely to experience both genetic and environmental risks. (See Chapter 12 for more discussion of assortative mating.)

Twin and adoption studies indicate that genetic factors play a major role in the familial aggregation of alcohol dependence. In a Danish adoption study, alcohol dependence in men was associated with alcohol dependence in birth parents but not adoptive parents (Goodwin, Schulsinger, Hermansen, Guze, & Winokur, 1973; Goodwin, Schulsinger, Knop, Mednick, & Guze, 1977). A similar association between alcohol dependence in adopted sons and their birth fathers was reported in Sweden (Cloninger, Bohman, & Sigvardsson, 1981; Sigvardsson, Bohman, & Cloninger, 1996). Likewise, the Iowa adoption studies (Cadoret, 1994; Cadoret, O'Gorman, Troughton, & Heywood, 1985; Cadoret, Troughton, & O'Gorman, 1987) showed a significantly elevated risk for alcohol dependence in adopted sons and daughters from an alcoholic birth family background, compared to control adoptees, consistent with a genetic influence on alcohol dependence. Numerous large twin studies on alcohol abuse, alcohol dependence, and other alcohol-related outcomes yield comparable results. The results of adult twin studies on various drinking-related behaviors are highly consistent, with genetic effects accounting for 40 to 60 percent of the variance across measures of quantity and frequency of use as well as problem use and dependence (Agrawal & Lynskey, 2008; Dick, Prescott, & McGue, 2009). Early twin studies suggested higher heritability for alcohol dependence in males (Legrand, McGue, & Iacono, 1999); however, this sex difference is not seen in more recent twin studies (Knopik et al., 2004; Prescott, 2002). Twin studies of adolescent alcohol-related variables, however, yield much more variable results. Studies of adolescent alcohol use disorders are uncommon because diagnostic criteria are typically not met until early adulthood (Lynskey et al., 2010). As such, the few genetic studies of adolescent alcohol dependence symptoms suggest small and nonsignificant genetic effects (Knopik, Heath, et al., 2009; Rose, Dick, Viken, Pulkkinen, & Kaprio, 2004), with shared environment playing a larger role. Regarding alcohol initiation in adolescence, results again suggest a large role for shared environment and a small yet significant role for genetic effects (Fowler et al., 2007). An interesting developmental finding is that shared environment appears to be related to the initial use of alcohol in adolescence and young adulthood but not to later alcohol abuse (Pagan et al., 2006).

Consistent with shared environment being important for adolescent alcohol-related outcomes, adoption studies yield some evidence for the influence of shared environment that is specific to siblings and not shared between parents and offspring. For example, in an adoption study of alcohol use and misuse among adolescents, the correlation between problem drinking in parents and adolescent alcohol use was

0.30 for biological offspring but only 0.04 for adoptive offspring (McGue, Sharma, & Benson, 1996). Despite the lack of resemblance between adoptive parents and their adopted offspring, adoptive sibling pairs who were not genetically related correlated 0.24. Moreover, the adoptive sibling correlation was significantly greater for same-sex siblings ($r = 0.45$) than for opposite-sex siblings ($r = 0.01$). These results suggest the reasonable hypothesis that sibling effects (or perhaps peer effects) may be more important than parent effects in the use of alcohol in adolescence. However, as mentioned earlier, assortative mating might be responsible, at least in part, for apparent shared environmental influences (Grant et al., 2007).

It is clear that both genes and environment play an important role in alcohol-related phenotypes; perhaps unsurprisingly, these genetic and environmental factors are likely to co-act in a complex fashion. Quantitative genetic research on alcohol use behaviors has provided several examples of genotype-environment interaction (see Chapter 8; see also Young-Wolff, Enoch, & Prescott, 2011, for a review). Heritability has been reported to be lower for those with later age of onset (Agrawal et al., 2009), for married individuals (Heath, Jardine, & Martin, 1989), for individuals with a religious upbringing (Koopmans, Slutske, van Baal, & Boomsma, 1999) and from stricter and closer families (Miles, Silberg, Pickens, & Eaves, 2005), in regions with lower alcohol sales (Dick, Rose, Viken, Kaprio, & Koskenvuo, 2001), and for individuals with peers who are less deviant (Dick et al., 2007; Kendler, Gardner, & Dick, 2011). These findings suggest that genetic risk for alcohol use is greater in more permissive environments (unmarried, nonreligious upbringing, greater alcohol availability, more peers reporting alcohol use). While adoption studies are fewer in number, they also suggest genotype-environment interaction. In studies of Swedish adoptees, adopted children who had both genetic risk (an alcohol-dependent birth parent) and environmental risk (an alcohol-dependent adoptive parent) were most likely to abuse alcohol (Sigvardsson et al., 1996). Additionally, having a birth father with a history of criminality (Chapter 17) interacted with unstable home environment to increase antisocial alcoholism in males (Cloninger et al., 1982). The Iowa adoption studies also suggest that birth family interacted with psychopathology in the adoptive parent and parental conflict in the adoptive home environment to increase risk for the development of alcoholism in females (Cutrona et al., 1994).

Animal Research on Alcohol-Related Phenotypes

Psychopharmacogenetics, which concerns the genetic effects on behavioral responses to drugs, is one of the most prolific areas of behavioral genetic research using animal models. The larger field of *pharmacogenetics* (Roses, 2000), often called ***pharmacogenomics*** in recognition of the ability to examine genetic effects on a genomewide basis, focuses on genetic differences in positive and negative effects of drugs in order to individualize and optimize drug therapy (Evans & Relling, 2004; Goldstein, Tate, & Sisodiya, 2003). Most research in psychopharmacogenetics involves alcohol (Bloom

& Kupfer, 1995; Broadhurst, 1978; Crabbe & Harris, 1991). For example, studies in *Drosophila* have examined susceptibility to the effects of alcohol by measuring the degree of sensitivity and tolerance to the sedative or motor-impairing effects of alcohol (e.g., Scholz, Ramond, Singh, & Heberlein, 2000). Recent work has also demonstrated that *Drosophila* can model many features of addiction, such as increased consumption over time, the overcoming of aversive stimuli in order to consume alcohol, and relapse after periods of alcohol deprivation (Devineni & Heberlein, 2009).

Using a mouse model, researchers discovered in 1959 that inbred strains of mice differ markedly in their preference for drinking alcohol, an observation that implies genetic influence (McClearn & Rodgers, 1959). Studies spanning more than 150 generations of mice find similar results, suggesting that this is a highly heritable trait that is very stable over time (Wahlsten et al., 2006). Moreover, research also suggests that preference drinking is a reasonable model for alcohol's reinforcing effects (Green & Grahame, 2008). Inbred strain differences have also been found for other behavioral responses to alcohol (see Crabbe et al., 2010, for a review).

Selection studies provide especially powerful demonstrations of genetic influence. For example, one classic study successfully selected for sensitivity to the effects of alcohol (McClearn, 1976). When mice are injected with the mouse equivalent of several drinks, they will "sleep it off" for various lengths of time. "Sleep time" in response to alcohol injections was measured by the time it took mice to right themselves after being placed on their backs in a cradle (Figure 18.1). Selection for this measure of alcohol sensitivity was successful, an outcome providing a powerful demonstration of the

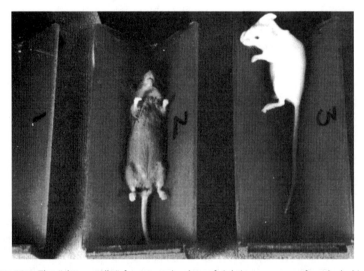

• **FIGURE 18.1** The "sleep cradle" for measuring loss of righting response after alcohol injections in mice. In cradle 2, a long-sleep mouse is still on its back, sleeping off the alcohol injection. In cradle 3, a short-sleep mouse has just begun to right itself. (Courtesy of E. A. Thomas.)

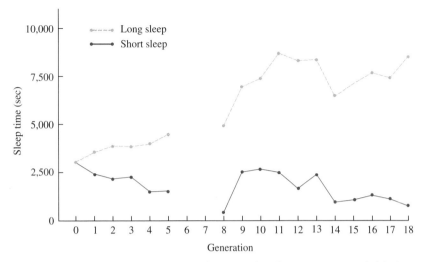

• **FIGURE 18.2** Results of alcohol sleep-time selection study. Selection was suspended during generations 6 through 8. (From G. E. McClearn, unpublished.)

importance of genetic factors (Figure 18.2). After 18 generations of selective breeding, the long-sleep (LS) animals "slept" for an average of two hours. Many of the short-sleep (SS) mice were not even knocked out, and their average "sleep time" was only about ten minutes. By generation 15, there was no overlap between the LS and SS lines (Figure 18.3). That is, every mouse in the LS line slept longer than any mouse in the SS line.

Alcohol has a combination of effects during consumption. Specifically, there are stimulatory effects during the first part of a drinking session that are rewarding, but

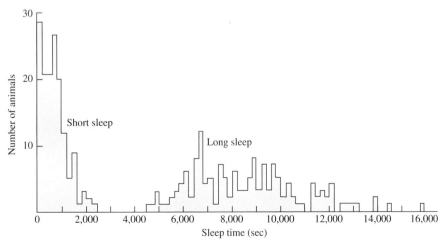

• **FIGURE 18.3** Distributions of alcohol sleep time after 15 generations of selection. (From G. E. McClearn, unpublished.)

after a peak alcohol level is reached, alcohol has sedating properties that are aversive. The extent to which genetic variation may disproportionately alter the balance between these effects may have profound implications on drinking behavior. If, due to genetic differences, individuals experience the rewarding effects of alcohol but not the aversive sedating effects, then they may be more likely to drink excessively in a fashion that may lead to dependence. By contrast, if an individual finds the sedating properties of alcohol particularly severe (consistent with the LS mice), this may lead to less drinking and a subsequent decrease in the risk for developing alcohol dependence. Since SS and LS mice are selectively bred and provide strong evidence for the genetic basis of these alcohol-related effects, they can serve as a critical translational bridge between animal research methods and understanding how genetic differences may influence the risk for alcohol dependence.

The steady divergence of the lines over 18 generations indicates that many genes affect this measure. If just one or two genes were involved, the lines would completely diverge in a few generations. Selected lines provide important animal models for additional research on pathways between genes and behavior. For example, the LS and SS lines have been extensively used as mouse models of alcohol sensitivity (Collins, 1981). Other selection studies include successful selection in mice for susceptibility to seizures during withdrawal from alcohol dependence and for voluntary alcohol consumption in rats (Crabbe, Kosobud, Young, Tam, & McSwigan, 1985; Green & Grahame, 2008). These are powerful genetic effects. For example, mice in the line selected for susceptibility to seizures are so sensitive to withdrawal that they show symptoms after a single injection of alcohol. Animal genetic models, including mice, rats, and *Drosophila,* continue to be widely used for behavioral genetic research on alcohol-related traits as well as for molecular genetic research (Awofala, 2011; Hitzemann et al., 2009; Phillips et al., 2010).

Molecular Genetic Research on Alcohol-Related Phenotypes

Alcohol dependence in humans has long been a target for molecular genetic studies in order to identify genes that contribute to the risk for developing the disorder. Whole-genome linkage studies using various populations, including Irish (Prescott et al., 2006), African-American (Gelernter et al., 2009), American Indian (Long et al., 1998), and Mission Indian families (Ehlers et al., 2004), as well as the Collaborative Genetics of Alcoholism (COGA) study (Foroud et al., 2000; Reich et al., 1998), have consistently reported linkage to a region on the long arm of chromosome 4 that contains the alcohol dehydrogenase (*ADH*) gene cluster family. A linkage region on the short arm of chromosome 4, close to the cluster of gamma-aminobutyric acid (GABA) receptors, has also been consistently reported (Long et al., 1998; Reich et al., 1998).

As the genes that code for alcohol metabolizing enzymes are well known (Lovinger & Crabbe, 2005), the aldehyde dehydrogenase gene (*ALDH2*) and the alcohol dehydrogenase (*ADH*) genes are the best-established genes in which polymorphisms may be associated with risk for alcohol dependence (see Kimura & Higuchi, 2011, for a review). Figure 18.4 presents a simplified model of these genetic influences on the role of alcohol dependence as well as several illustrative candidate genes, which are discussed below. One particularly interesting and consistent story based on the results of candidate gene studies surrounds the *ALDH2* polymorphism. There is evidence that the *ALDH2* polymorphism is associated with both the drinking behavior of healthy people and the risk for alcohol dependence. An *ALDH2* allele (*ALDH2*2*) that leads to inactivity of a key enzyme in the metabolism of alcohol occurs in 25 percent of Chinese and 40 percent of Japanese but is hardly ever found in Caucasians. The resulting buildup of acetaldehyde leads to unpleasant symptoms, such as flushing and nausea, when alcohol is consumed. This is an example of a mutant allele that protects against the development of alcoholism. This genetic variant results in reduced alcohol consumption and has been implicated as the reason why rates of alcoholism are much lower in Asian than in Caucasian populations. In fact, being homozygous for the *ALDH2*2* allele almost completely prevents individuals from becoming alcoholics (Higuchi et al., 2004). The same unpleasant symptoms described here are produced by the drug disulfiram (Antabuse), which is the basis for an alcoholism therapy used to deter drinking.

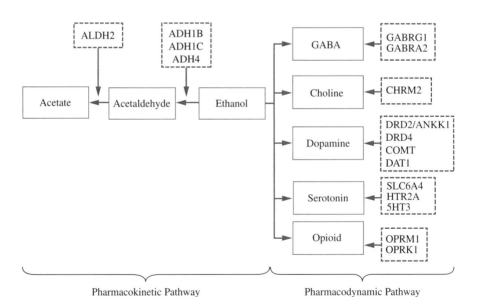

Pharmacokinetic Pathway   Pharmacodynamic Pathway

• **FIGURE 18.4** A network model of genes involved in alcohol dependence via alterations to ethanol's pharmacokinetics and pharmacodynamics. The left-hand side of the figure indicates the pharmacokinetic pathway that metabolizes ethanol into acetate. The right-hand side indicates the pharmacodynamic pathways that reflect ethanol's molecular pharmacological effects on multiple neurotransmitter systems. Dashed boxes contain a list of candidate genes that most likely affect the respective systems. See Palmer and colleagues (in press) for a review of these biological pathways.

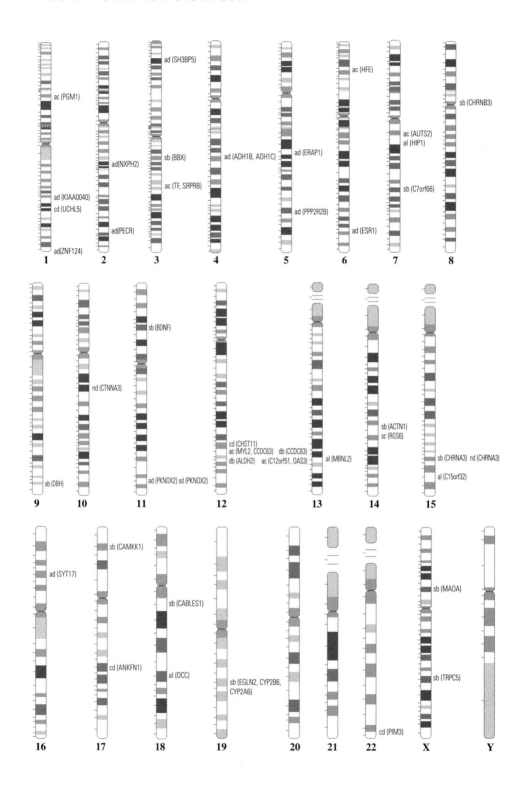

Associations for other candidate genes have been reported (see Kimura & Higuchi, 2011, for a review), especially genes that code for receptors for: GABA (Enoch et al., 2009), cholinergic muscarinic receptor-2 (*CHRM2;* Luo et al., 2005), dopamine (McGeary, 2009; van der Zwaluw et al., 2009), serotonin (Enoch, Gorodetsky, Hodgkinson, Roy, & Goldman, 2011), and opioid receptors (Anton et al., 2008). Efforts are now under way to test for moderation of specific gene effects by environmental risk factors. For example, three studies looking at three different genes have suggested that parental monitoring moderates the association between externalizing behavior, including alcohol use, and *GABRA2* (Dick, Latendresse, et al., 2009), *CHRM2* (Dick et al., 2011), and a dopaminergic pathway gene, catechol-O-methyl transferase (*COMT;* Laucht et al., 2012). More specifically, and supportive of quantitative genetic findings of G × E in alcohol use, these three studies suggest that the association between the genotype and externalizing behavior is stronger in environments with lower parental monitoring. Despite these interesting and encouraging results from candidate gene efforts, several genomewide association studies (GWAS) of alcohol dependence to date do not find a particular gene that consistently shows a significant association across studies. Rather, GWAS studies suggest that hundreds of genetic variants make modest contributions (0.25 percent of variance or less) to alcohol dependence risk (e.g., Heath et al., 2011). Figure 18.5 summarizes reported genetic associations with several drug-related phenotypes (www.genome.gov/26525384).

Phenotype	Gene
Alcohol Dependence [ad]	ADH1B, ADH1C, SH3BP5, NXPH2, PKNOX2, SYT17, KIAA0040, ZNF124, PECR, PPP2R2B, ERAP1, ESR1
Alcohol Consumption [ac]	TF, HFE, SRPRB,PGM1, AUTS2, C12orf51, MYL2, CCDC63, OAS3,
Alcoholism [al]*	HIP1, MBNL2, DCC, C15orf32
Drinking Behavior [db]	ALDH2, CCDC63
Smoking Behavior [sb]	C7orf66, CHRNA3, EGLN2, BDNF, DBH, CYP2A6, CHRNB3, CYP2B6, BBX, CABLES1, TRPC5, ACTN1, CAMKK1, MAOA
Substance Dependence [sd]	PNOX2
Cannabis Dependence [cd]	ANKFN1, UCHL5, PIM3, CHST11
Smoking Cessation [sc]	RGS6
Nicotine Dependence [nd]	CHRNA3, CTNNA3

*Alcoholism [al] was defined in four ways in the search criteria:
a. weekly consumption for 12 months
b. alcohol dependence factor score
c. alcohol use disorder factor score
d. heaviness of drinking

• **FIGURE 18.5** Significant genomewide association studies and candidate genes for drug-related outcomes. Information extracted from http://www.genome.gov/26525384. Chromosomes have been adjusted to be the same length; see Figure 4.4 for the relative lengths of the different chromosomes.

Pharmacogenomic studies of rodents have been successfully used to identify QTLs associated with alcohol-related behavior (Ehlers et al., 2010). For example, QTLs for alcohol preference drinking have been linked to mouse chromosome 9 (Phillips, Belknap, Buck, & Cunningham, 1998; Tabakoff et al., 2008) and to rat chromosome 4 (Spence et al., 2009). QTLs for acute alcohol withdrawal in mice have been mapped to mouse chromosome 1 (Kozell, Belknap, Hofstetter, Mayeda, & Buck, 2008). Other QTLs have been mapped for other alcohol-related responses in mice, such as alcohol-induced loss of righting reflex (Crabbe, Phillips, et al., 1999; Lovinger & Crabbe, 2005).

QTL research in animal models is especially exciting because it can nominate candidate QTLs that can then be tested in human QTL research (Lovinger & Crabbe, 2005). For example, over 90 percent of the mouse and human genomes include regions of conserved synteny. In other words, there are regions in the mouse and human genomes in which the gene order in the most common ancestor has been conserved in both species (Ehlers et al., 2010). Knock-out studies in mice also demonstrate the effects of specific genes on behavioral responses to alcohol. For example, knocking out a serotonin receptor gene in mice leads to increased alcohol consumption (Crabbe et al., 1996). Knocking out certain dopaminergic receptors results in supersensitivity to alcohol (Rubinstein et al., 1997) and reduced alcohol preference drinking (Savelieva, Caudle, Findlay, Caron, & Miller, 2002). Such differences in brain sensitivity to ethanol in human populations could be responsible for addiction in general (Martinez & Narendran, 2010) as well as the lethal consequences of binge drinking in some individuals (Heath et al., 2003).

Recent studies have used genomic approaches (Chapter 10) to shed light on the molecular pathways underlying alcohol response and addiction (Awofala, 2011; Tabakoff et al., 2009). Such approaches using mice, rats, and *Drosophila* combine genetic marker information, gene expression, and complex phenotypes to ascertain the candidate genes and gene product interaction pathways that significantly influence the variation in expression of a particular phenotype in animal models. Findings from animal models are then compared to what is known in humans. Examples of this approach in *Drosophila* (Awofala, 2011) and rats (Tabakoff et al., 2009) suggest that candidate pathways and networks of genes, rather than specific candidate genes, play an important role in determining the behavioral response to alcohol and that changes in gene expression in alcoholics are associated with widespread cellular functions (Awofala, 2011).

Nicotine Dependence

One of the most common and potentially hazardous environmental exposures that negatively influences health and development is exposure to cigarette smoke. The Centers for Disease Control and Prevention (CDC) has reported that almost 21 percent of adults in the United States—approximately 46 million people—smoked cigarettes in 2009 (CDC, 2010), and most of them are dependent on nicotine. Previous work has found over 4000 chemicals in cigarette smoke, including nicotine, benzo[a]pyrene, and carbon monoxide, and more than 40 of these chemicals have been established as known carcinogens (Thielen, Klus, & Mueller, 2008). Cigarette smoking has been linked to several diseases and disabling conditions, including heart disease and lung diseases (CDC, 2008). Further, for every individual who dies from a disease associated with smoking, 20 more people battle at least one major illness attributable to smoking (CDC, 2003). Several studies have singled out tobacco use as the world's leading preventable cause of death (CDC, 2008). By some estimates, up to 5 million deaths worldwide can be attributed to smoking, and current trend data predict that tobacco use will cause more than 8 million deaths a year by 2030 (World Health Organisation, 2009). In the United States, tobacco use has been implicated in 20 percent of deaths per year, or 443,000 deaths annually, and approximately 49,000 of these have been attributed to secondhand smoke exposure (CDC, 2008). On average, smokers die more than a decade earlier than nonsmokers (CDC, 2002). Although nicotine is an environmental agent, smoking behaviors aggregate in families and peer networks due to genetic predispositions and familial and extrafamilial influences (Rose, Broms, Korhonen, Dick, & Kaprio, 2009). Moreover, individual differences in susceptibility to nicotine's addictive properties and harmful effects are influenced by genetic factors.

Twin Research on Smoking-Related Phenotypes

Multiple phenotypes are associated with smoking and nicotine dependence, including smoking initiation, smoking persistence, tolerance to nicotine, smoking cessation, regular smoking, number of cigarettes smoked per day, and nicotine dependence (see Rose et al., 2009, for a detailed review). Considerable genetic research has investigated smoking initiation, which appears to be different from the reasons people persist or continue to smoke. While the heritability of nicotine dependence, smoking persistence, and regular smoking, for example, can be assessed only in those who have already started to smoke, the genetic effects on smoking initiation can be examined among all persons in the population (Rose et al., 2009). A meta-analysis of 17 twin cohorts from six studies of smoking initiation across three countries concluded that genetic factors play a more significant role in smoking initiation for adult women (heritability = 0.55) than for men (heritability = 0.37), although the range of estimates across studies suggests that this may not be a significant gender difference and that shared environment plays a more important role for smoking initiation in adult men (Li, Cheng, Ma, & Swan, 2003). This meta-analysis included studies from 1993 to 1999. Since that time, at least ten additional twin studies of over 60,000 twin pairs from four countries (Finland, Australia, the United States, and the Netherlands) have examined genetic effects on smoking initiation (e.g., Broms, Silventoinen, Madden, Heath, & Kaprio, 2006; Hamilton et al., 2006; Morley et al., 2007; Vink, Willemsen, & Boomsma, 2005). Among adult twins, genetic influences are substantial and explain, on average, about 50 percent or more of the variance. However, the estimates vary widely across studies. Studies on smoking initiation suggest heritabilities of about 0.20 to 0.75 for women and about 0.30 to 0.65 for men (reviewed in Rose et al., 2009). This range could be explained by various definitions of smoking initiation (e.g., age of first cigarette, age of initiation of regular smoking) as well as the likelihood that the magnitude of genetic effects varies with time and place (Chapter 7 on heritability; Kendler et al., 1999). More recent studies are also investigating genetic effects on reactions to first cigarette use, such as dizziness or headache. Evidence suggests that how people experience their initial few cigarettes is due to both heritable contributions and environmental experiences unique to the person (Haberstick, Ehringer, Lessem, Hopfer, & Hewitt, 2011). Multivariate genetic modeling identified a moderately heritable underlying factor that influenced the covariation of multiple subjective experiences and loaded most heavily on dizziness, suggesting a heritable sensitivity to the chemicals contained in an average cigarette that is best indexed by dizziness.

Smoking persistence also shows substantial genetic variance and very little influence of shared environment (Rose et al., 2009). Most studies that focus on smoking persistence test for a genetic correlation between smoking initiation and persistence using a special case of the liability-threshold model (Chapter 3) called a two-stage model. This model estimates the amount of genetic and environmental overlap between the first stage of initiation and the second stage of persistence (or dependence) and has been applied to other domains of substance use as well (Heath, Martin, Lynskey, Todorov, & Madden, 2002). In a sample of twins from Virginia, it was determined that the genetic influences that contribute to smoking initiation and persistence are not fully overlapping (Maes et al., 2004). Similar results were reported in a Finnish twin sample, with genetic effects influencing smoking initiation accounting for only about 3 percent of the variance in smoking persistence (Broms et al., 2006). Another interesting multivariate result is that the genetics of persistent smoking appears to be mediated by genetic vulnerability to nicotine withdrawal (Pergadia, Heath, Martin, & Madden, 2006).

When considering nicotine dependence, as defined by various diagnostic criteria, multiple large twin studies all point to genetic influence on adult nicotine dependence. A classic early study including 12,000 twin pairs from Sweden, of whom half smoked, suggested that if one twin currently smoked, the probability that the co-twin smoked was 75 percent for identical twins and 63 percent for fraternal twins (Medlund, Cederlof, Floderus-Myrhed, Friberg, & Sorensen, 1977). Subsequently, heritability estimates are even higher across several cultures, suggesting that about 60 percent of the risk for nicotine dependence is due to genetic influence. Studies also suggest that the time to first cigarette after waking, with a heritability of 55 percent, appears to tap a pattern of heavy, uninterrupted, and automatic smoking and may be a good single-item measure of nicotine dependence (Baker, Piper, et al., 2007) and genetic risk for nicotine dependence (Haberstick et al., 2007). It should be noted that these results refer to smoking cigarettes. An interesting study found that smoking tobacco in pipes and cigars showed no genetic influence and substantial shared environmental influence (Schmitt, Prescott, Gardner, Neale, & Kendler, 2005).

While there are many adult twin studies of smoking behavior, the literature on adolescent twin studies is less extensive. Unlike adolescent alcohol-related behaviors, in which shared environment appears significant, there is less evidence for the role of shared environmental influences on adolescent smoking-related behaviors (Lynskey et al., 2010). Rather, adolescent twin studies demonstrate the importance of genetic factors in smoking behaviors at this earlier developmental stage; however, the range of heritability estimates is large (25 to 80 percent) and, similar to adult studies, dependent on the smoking variable of interest. Nicotine withdrawal, however, shows remarkable similarity across adolescent and adult smokers, with genetic effects accounting for 50 percent of the variance in nicotine withdrawal (Pergadia et al., 2010).

Quantitative genetic research on genotype-environment interaction for smoking-related behaviors has not been as extensive as that for alcohol use. What little has been done has focused on adolescents. Perhaps unsurprisingly, similar to alcohol use, genetic influences on adolescent smoking decreased at higher levels of parental monitoring (Dick et al., 2007) and increased with self-reported religiosity (Timberlake et al., 2006).

Molecular Genetic Research on Smoking-Related Phenotypes

Despite the ranges in heritability estimates, there is consistent support for an important role of genetics for most smoking behaviors. However, estimates of heritability provide no information about what specific genes are involved. Figure 18.5 shows reported genetic associations with several drug-related phenotypes, including smoking outcomes. Early molecular genetic studies of smoking-related outcomes yielded inconsistent results (reviewed in Ho & Tyndale, 2007), perhaps because none were specifically designed to study nicotine dependence. However, recent efforts suggest a more consistent and compelling story. For example, the Nicotine Addiction Genetics (NAG) project has reported significant linkage to chromosome 22 in samples from Finland and Australia for the maximum number of cigarettes ever smoked in a 24-hour period (Saccone et al., 2007).

The strongest and most consistent genetic contributions to nicotine dependence come from genes that are associated with differences in nicotine's pharmacokinetics (i.e., the absorption, distribution, and metabolism of nicotine in the body) and with differences in pharmacodynamics (i.e., genetic variation that impacts nicotine's effects on an individual) (Bierut, 2011; MacKillop, Obasi, Amlung, McGeary & Knopick, 2010). A simplified model of these influences can be seen in Figure 18.6, which includes the primary metabolic pathways, neurotransmitter systems, and illustrative candidate genes.

Variation in nicotine metabolism plays an important role in cigarette consumption. Twin studies of nicotine metabolism suggest a heritability of 60 percent, and the major contributor to genetic variation in this metabolic pathway is the *CYP2A6* gene (Swan et al., 2005), whose enzyme is primarily responsible for the metabolism of nicotine to cotinine, a chemical that is made by the body as it processes nicotine. Recent

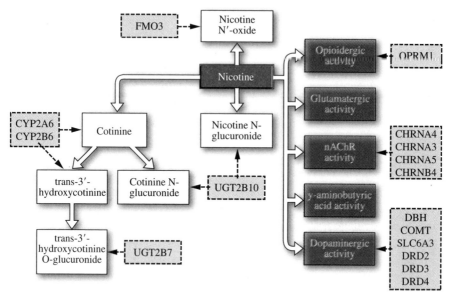

• **FIGURE 18.6** A model of genes involved in nictoine dependence via alterations to nicotine's pharmacokinetics and pharmacodynamics. Solid arrows indicate pharmacokinetic pathways that metabolize nicotine and pharmacodynamic pathways that reflect nicotine's molecular pharmacological effects on multiple neurotransmitter systems. Dashed arrows indicate candidate genes and their points of putative influence. (With kind permission from Springer Science+Business Media: *Current Cardiovascular Risk Reports,* "The role of genetics in nicotine dependence: Mapping the pathways from genome to syndrome," *4*, 2010, 446–453, J. MacKillop, E. Obasi, M. T. Amlung, J. E. McGeary, & V. S. Knopik, Figure 1.)

GWAS meta-analyses confirm the importance of the *CYP2A6* region on chromosome 19 as variants in this region were associated with number of cigarettes smoked per day (The Tobacco and Genetics Consortium, 2010; Thorgeirsson et al., 2010). Many of the multiple genes involved in the nicotine metabolism pathway are also promising (see MacKillop et al., 2010, for a review).

Nicotine's psychoactive effects on cognitive variables, such as attention, learning, and memory, are associated with nicotinic receptor stimulation (Benowitz, 2008). A robust finding suggests that genetic variation in the nicotinic receptor subunit cluster (for example, *CHRNA3*), located on chromosome 15, alters risk for becoming a heavy smoker (Bierut, 2011). There appear to be at least two distinct variants that contribute to heavy smoking in this region on chromosome 15 (Saccone et al., 2010; The Tobacco and Genetics Consortium, 2010). Other neurotransmitter systems, such as the endogenous opioid system, are also being investigated, although the effects are small and inconclusive.

Recent transgenetic studies in mice involving the deletion and replacement of nicotinic acetylcholine receptor subunits have begun to identify the molecular mechanisms underlying nicotine addiction (Changeux, 2010). Just as nicotine stimulates nicotine receptors and enhances cognitive functioning (e.g., attention), loss of receptor function actually impairs cognitive performance (Poorthuis, Goriounova, Couey, & Mansvelder, 2009). For example, mice lacking one of the subunits of the receptor show abnormalities in certain types of memory (Granon, Faure, & Changeux, 2003), social interaction (Granon et al., 2003), and decision making (Maubourguet, Lesne, Changeux, Maskos, & Faure, 2008). Molecular methods designed to turn specific genes "on" and "off" have revealed distinct contributions of certain subunits of the nicotinic acetylcholine receptor to the short-term effects of nicotine, including the acute behavioral effects (Changeux, 2010).

Other Drugs

Inbred strain and selection studies in mice have documented genetic influence on sensitivity to almost all drugs subject to abuse (Crabbe & Harris, 1991). Human studies are difficult to conduct because drugs such as amphetamines, heroin, and cocaine are illegal and exposure to these drugs changes over time (Seale, 1991). Although addictions such as cocaine or opiate dependence are less common, they can be more devastating socially, cause more physical illness, and be thought of as an extreme of addiction (Bierut, 2011). Family studies have shown about an eightfold increased risk of drug abuse in relatives of probands with drug abuse for a wide range of drugs such as cannabis, sedatives, opioids, and cocaine (Merikangas et al., 1998). Two major twin studies of a broad range of drug abuse have been conducted in the United States, one involving veterans of the U.S. war in Vietnam (Tsuang, Bar, Harley, & Lyons, 2001) and the other involving twins in Virginia (Kendler, Myers, & Prescott, 2007). Both studies yielded evidence of substantial heritabilities of liability (about 30 percent to 70 percent) and little evidence of shared environmental influence across various drugs of abuse. Similar results have been found in a more recent population twin study in Norway (Kendler, Aggen, Tambs, & Reichborn-Kjennerud, 2006). A focus of recent research has been on developmental issues (Zucker, 2006). For example, as found for alcohol and smoking, shared family environmental factors are more important for initiation, but genetic factors are largely responsible for subsequent use and abuse (Kendler & Prescott, 1998; Rhee et al., 2003). A slightly different picture is seen for cannabis initiation and problematic use in a meta-analysis of 28 studies of cannabis initiation and 24 studies of cannabis use. Genetic factors contribute to about 50 percent of the vulnerability for both initiation and problem use (Verweij et al., 2010).

Multivariate genetic analyses indicate that the same genes largely mediate vulnerability across different drugs, with additive genetic factors explaining more than 60 percent of the common liability to drug dependence (Palmer et al., 2012) but shared environmental influence in adolescence being more drug specific (Young et al., 2006). A systematic review of the literature also supports a common liability to multiple facets of substance dependence, particularly etiological factors, such as genetics (Vanyukov et al., 2012). This common liability model of addiction has gained more consistent support than the gateway hypothesis—the theory that the use of less deleterious drugs may lead to a future risk of using more dangerous hard drugs (Gelernter & Kranzler, 2010; Vanyukov et al., 2012). The gateway hypothesis has been tested using various approaches. A novel method, called Mendelian randomization (Davey Smith & Ebrahim, 2003), uses Mendel's second law of independent assortment as a means of examining the causal effect of environmental exposure, such as exposure to drugs of abuse. For example, a recent study used this method and the *ALDH2* gene, which, as described earlier, has a strong effect on alcohol use (Irons, McGue, Iacono, & Oetting, 2007). The essence of the approach is that an *ALDH2*-normal group and an *ALDH2*-deficient group should be similar genetically because of Mendelian randomization, except for their *ALDH2* genotypic difference. The gateway hypothesis would predict that the *ALDH2*-deficient genotypic group, which was much less exposed to alcohol, would be less likely to use other drugs if alcohol exposure is a gateway to the use of other drugs. The results of the study strongly disconfirmed this gateway hypothesis because the *ALDH2*-deficient genotypic group was just as likely to use other drugs despite using alcohol much less than the *ALDH2*-normal group.

The molecular genetics of drug-related behaviors has been examined in mice, especially for transgenic models of responses to opiates, cocaine, and amphetamine. More than three dozen transgenic mouse models have been established for responses to these drugs (Sora, Li, Igari, Hall, & Ikeda, 2010). Much QTL research in mice has also been conducted (Crabbe et al., 2010), including genes involved in reward mechanisms as well as drug preference and response (Goldman, Oroszi, & Ducci, 2005).

Genomewide linkage or association scans of use of drugs other than alcohol and nicotine have only recently been reported (see Figure 18.5 for significant findings). A QTL linkage study in adolescents suggested two linkage regions for vulnerability to substance abuse (Stallings et al., 2003); these two regions also show linkage to general antisocial behavior (Stallings et al., 2005). As is the case for alcoholism, recent reviews of the handful of genome scans for addiction to other drugs, such as heroin and methamphetamine, report many associations with small effect sizes but no large effects (Gelernter & Kranzler, 2010; Yuferov, Levran, Proudnikov, Nielsen, & Kreek, 2010). The first GWAS for cannabis use yielded similar results (Agrawal et al., 2011).

Complexities of Studying the Genetics of Substance Use

It is often implicitly assumed that there is substantial specificity between genetic factors and specific types of substance dependence, but there is a strong empirical basis for believing that most of the genetic variance is shared (MacKillop et al., 2010). For example, nicotine dependence and alcoholism are both comorbid with depression, smoking co-occurs with schizophrenia, alcohol use co-occurs with antisocial behavior, and, as outlined above, various types of substance use tend to occur together, such as alcohol use and smoking or cigarette smoking and cannabis use (Grant, Hasin, Chou, Stinson, & Dawson, 2004; Rose et al., 2009). All of these behaviors are genetically influenced and would be expected to affect motivation to use specific substances. It is both plausible and probable that the pathways from genes to behavior are not a result of independent and additive effects but rather reflect a much more complex system than typically considered, including interactions between genes and pleiotropic effects (MacKillop et al., 2010).

Summary

When taken together, results of twin and adoption studies of alcohol-related behaviors suggest moderate heritability and little evidence for shared environmental influences. Several examples of genotype-environment interaction have been reported in which genetic risk for alcohol-related outcomes is greater in more permissive environments. As is the case for alcohol-related behaviors, moderate genetic influence and little shared environmental influence have been found for smoking and other drug use, although shared environmental influence plays a larger role for initiation of smoking. Multivariate studies also suggest that common genes mediate vulnerability across various drugs. Pharmacogenetics has been a very active area of research, using animal models of drug use and abuse, especially for alcohol. For example, selection studies have documented genetic influence on many behavioral responses to drugs. Many QTLs for alcohol-related behavior in mice have been identified. In human populations, linkage and association studies are beginning to yield some consistent findings for alcohol, smoking, and, to a lesser extent, other drugs, such as cannabis, methamphetamine, and heroin.

CHAPTER NINETEEN

Health Psychology and Aging

G enetic research in psychology has focused on cognitive disabilities and abilities (Chapters 11–13), psychopathology (Chapters 14–16), personality (Chapter 17), and substance use (Chapter 18). The reason for this focus is that these are the areas of psychology that have had the longest history of research on individual differences. Much less is known about the genetics of other major domains of psychology that have not traditionally emphasized individual differences, such as perception, learning, and language.

The purpose of this chapter is to provide an overview of genetic research in other areas of the behavioral sciences. One such area is health psychology, sometimes called psychological or behavioral medicine because it lies at the intersection between psychology and medicine. Specifically, health psychology is concerned with understanding how biological, psychological, environmental, and cultural factors are involved in physical health and illness. Research in this area focuses on the role of behavior in promoting health and in preventing and treating disease. Although genetic research in this area is relatively new, some conclusions can be drawn about relevant topics such as body weight and subjective well-being.

The second area is aging. Behavioral genetic research has produced interesting results, especially about issues unique to aging, such as cognitive aging and quality of life in the later years. The explosion of molecular genetic research on cognitive decline, dementia, and longevity in the elderly has added momentum to genetic research on behavioral aging as well as to the relevance of **genetic counseling** for both health psychology and aging.

Health Psychology

Most of the central issues about the role of behavior in promoting health and in preventing and treating disease have only just begun to be addressed in genetic research. For example, the first book on genetics and health psychology was not published until

1995 (Turner, Cardon, & Hewitt, 1995). However, in the past 10 years, thousands of papers have been published related to health psychology, suggesting that this is an area of exponential growth. We will focus on two areas relevant to health psychology: body weight and subjective well-being.

Body Weight and Obesity

Obesity and overweight are becoming more widespread and are worldwide clinical and public health burdens (Kelly, Yang, Chen, Reynolds, & He, 2008). Obesity is a major health risk for several medical disorders, including diabetes, heart disease, and cancer, as well as for mortality (Flegal, Graubard, Williamson, & Gail, 2007; Pischon et al., 2008). Although it is often assumed that individual differences in weight are largely due to environmental factors, twin and adoption studies consistently lead to the conclusion that genetics accounts for the majority of the variance for weight (Grilo & Pogue-Geile, 1991), body mass index, and other measures of obesity and regional fat distribution (such as skinfold thickness and waist circumference) (Herrera, Keildson, & Lindgren, 2011). For example, as illustrated in Figure 19.1, twin correlations for weight based on thousands of pairs of twins are 0.80 for identical twins and 0.43 for fraternal twins. Identical twins reared apart correlate 0.72. Biological parents and their adopted-away offspring are almost as similar in weight (0.23) as are nonadoptive parents and their offspring (0.26), who share both nature and nurture. Adoptive parents and their offspring, and adoptive siblings, who share nurture but not nature, do not resemble each other at all for weight.

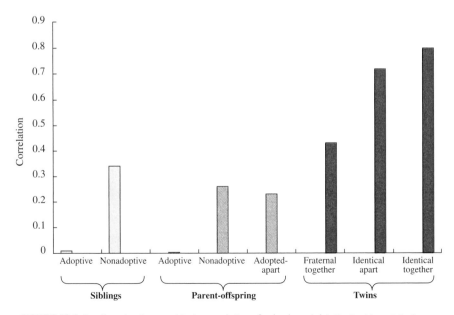

• **FIGURE 19.1** Family, adoption, and twin correlations for body weight. (Derived from Grilo & Pogue-Geile, 1991.)

Together, the results in Figure 19.1 imply a heritability of about 70 percent for body weight. Similar results have been found across eight European countries despite average differences in weight, with some suggestion of greater shared environmental influence for women (Schousboe et al., 2003). Similar results are also found for body mass index (BMI), which corrects weight on the basis of height (i.e., weight [kg]/height [m²]), and for skinfold thickness, which is an index of fatness (Grilo & Pogue-Geile, 1991; Maes, Neale, & Eaves, 1997). There are few genetic studies of overweight or obesity, in part because weight shows a continuous distribution, a situation rendering diagnostic criteria somewhat arbitrary (Bray, 1986). For both children and adults, overweight and obesity classifications are typically based on BMI. In general, BMI between the 5th and 85th percentiles is considered normal, and BMI greater than the 95th percentile is considered overweight or, more recently, obese (Krebs et al., 2007).

Using an obesity cutoff based on BMI, twin studies have indicated similarly high heritabilities for obesity in childhood (Dubois et al., 2012; Silventoinen, Rokholm, Kaprio, & Sorensen, 2010) and adulthood (Silventoinen & Kaprio, 2009). A parent-offspring family study indicates that, although assortative mating (see Chapter 12) for body weight is modest (a correlation of about 0.20), the risk of obesity in adult offspring is 20 percent if both parents are obese, 8 percent if only one parent is obese, and only 1 percent if neither parent is obese (Jacobson, Torgerson, Sjostrom, & Bouchard, 2007).

The dramatic increase in obesity throughout the world is sometimes thought to deny a role for genetics, but, as discussed in Chapter 7, the causes of population means and variances are not necessarily related. That is, the mean population increase in weight is probably due to the increased availability and reduced costs of energy-dense food, increased portion sizes, increased consumption of added sugars, and a reduction in physical activity (Skelton, Irby, Grzywacz, & Miller, 2011). However, despite our increasingly "obesogenic" environments, a wide range of variation in weight remains—many people are still thin. Obesogenic environments could shift the entire distribution upward while the causes of individual differences, including genetic causes, could remain unchanged (Wardle, Carnell, Haworth, & Plomin, 2008).

As also emphasized in Chapter 7, finding genetic influence does not mean that the environment is unimportant. Anyone can lose weight if they stop eating. The issue is not what *can* happen but rather what *does* happen. That is, to what extent are the obvious differences in weight among people due to genetic and environmental differences that exist in a particular population at a particular time? The answer provided by the research summarized in Figure 19.1 (which is consistent with more recent studies) is that genetic differences largely account for individual differences in weight. If everyone ate the same amount and exercised the same amount, people would still differ in weight for genetic reasons.

This conclusion was illustrated dramatically in an interesting study of dietary intervention in 12 pairs of identical twins (Bouchard et al., 1990). For three months, the twins were given excess calories and kept in a controlled sedentary environment. Individuals differed greatly in how much weight they gained, but members of identical twin pairs correlated 0.50 in weight gain. Similar twin studies show that the effects on weight of physical activity and exercise are also influenced by genetic factors (Fagard, Bielen, & Amery, 1991; Heitmann et al., 1997).

Such studies do not indicate the mechanisms by which genetic effects occur. For example, even though genetic differences occur when calories and exercise are controlled, in the world outside the laboratory, genetic contributions to individual differences might be mediated by individual differences in proximal processes such as food intake and metabolism (Silventoinen et al., 2010). In other words, individual differences in eating habits and in the tendency to exercise, although typically assumed to be environmental factors responsible for body weight, are influenced by genetic factors. Twin studies suggest that genetic factors do affect many aspects of eating, such as appetite (Carnell, Haworth, Plomin, & Wardle, 2008); the number, timing, and composition of meals; degree of hunger and sense of fullness after eating (de Castro, 1999; Llewellyn, van Jaarsveld, Johnson, Carnell, & Wardle, 2010); eating styles, such as emotional eating and uncontrolled eating (Tholin, Rasmussen, Tynelius, & Karlsson, 2005); speed of eating and enjoyment of food (Llewellyn et al., 2010); and food preferences in general (Breen, Plomin, & Wardle, 2006).

Previous chapters have indicated that environmental variance is of the nonshared variety for most areas of behavioral research. This is also the case for body weight. As noted in relation to Figure 19.1, adoptive parents and their adopted children and adoptive siblings do not resemble each other at all for weight. This finding is surprising because theories of weight and obesity have largely focused on weight control by means of dieting, yet individuals growing up in the same families do not resemble each other for environmental reasons (Grilo & Pogue-Geile, 1991). Attitudes toward eating and weight also show substantial heritability and no influence of shared family environment (Rutherford, McGuffin, Katz, & Murray, 1993). In other words, environmental factors that affect individual differences in weight are factors that do not make children growing up in the same family similar. The next step in this research is to identify environmental factors that differ for children growing up in the same family. For example, although it is reasonable to assume that children in the same family share similar diets, this may not be the case. The difficulty lies in the fact that the biological and environmental determinants of weight and obesity are intertwined. Many of these determinants or predictors of obesity can be seen in Figure 19.2, which includes diverse child, family and community characteristics.

The prevalence of overweight and obesity has increased over time and also increases with age (Flegal, Carroll, Kit, & Ogden, 2012; Ogden, Carroll, Kit, & Flegal, 2012). Thus, it is important to examine the relative contributions of genetics and environment to BMI over time, as this could potentially provide valuable insight into the causes of the obesity epidemic (Duncan et al., 2009). Genetic factors that affect body weight begin to have their effects in early childhood (Meyer, 1995). In fact, in a recent study of 23 twin birth cohorts from four countries, BMI was found to be strongly influenced by genetic factors in both males and females as early as five

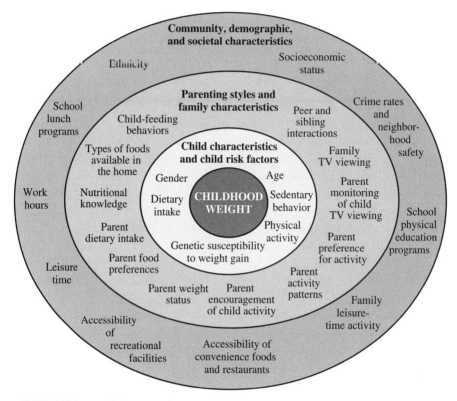

• **FIGURE 19.2** Simplified model of predictors of childhood obesity. (Adapted from *Pediatric Clinics of North America, 58*, J. A. Skelton, M. B. Irby, J. G. Grzywacz, & G. Miller, Etiologies of obesity in children: Nature and nurture, 1333–1354, Copyright 2011, with permission from Elsevier.)

months of age (Dubois et al., 2012). Longitudinal genetic studies are especially informative. The first longitudinal twin study from birth through adolescence found no heritability for birth weight, increasing heritability during the first year of life, and stable heritabilities of 60 to 70 percent thereafter (Figure 19.3; see Matheny, 1990). These results have consistently been replicated in other twin studies in childhood (e.g., Dellava, Lichtenstein, & Kendler, 2012; Estourgie-van Burk, Bartels, van Beijsterveldt, Delemarre-van de Waal, & Boomsma, 2006; Pietilainen et al., 1999) and more recently in early adulthood (Dubois et al., 2012; Duncan et al., 2009; Haberstick et al., 2010; Ortega-Alonso, Pietilainen, Silventoinen, Saarni, & Kaprio, 2012). Parent-offspring adoption research suggests that there is substantial genetic continuity from childhood to adulthood (Cardon, 1994), and twin studies in adulthood report heritabilities of 60 to 80 percent (Romeis, Grant, Knopik, Pedersen, & Heath, 2004). Longitudinal twin studies that examined the change in BMI from adolescence to young adulthood indicate that, while the magnitude of genetic influences is largely stable, different sets of genes may underlie the rate of change during this developmental period (Ortega-Alonso et al., 2012).

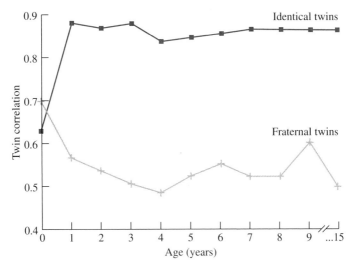

• **FIGURE 19.3** Identical and fraternal twin correlations for weight from birth to 15 years of age. (Derived from Matheny, 1990.)

Similar to most other behaviors and phenotypes discussed earlier in this book, there is keen interest in the role of gene-environment interplay in the risk for obesity. For example, heritability estimates may vary depending on certain environmental factors. Heritability of BMI has been reported to be lower among adults with higher income levels (Johnson & Krueger, 2005) and among young adults who exercise frequently (Mustelin, Silventoinen, Pietilainen, Rissanen, & Kaprio, 2009; Silventoinen et al., 2009). It has also been suggested that genetic and common environmental effects on BMI may be moderated by parental education level, with lower heritability if parental education was limited (i.e., not having completed high school) or mixed (one parent with limited education and one parent with a higher educational level). Common environment did not affect variation of adolescent BMI in highly educated families but did influence BMI in families with limited parental education (Lajunen, Kaprio, Rose, Pulkkinen, & Silventoinen, 2012). As mentioned previously, many of these ostensible environmental measures are heritable. For example, individual differences in physical activity during adulthood are due in part to genetic influences (Mustelin et al., 2012). Thus, despite the increased information that is now available about the predictors of BMI, the picture is becoming increasingly complex.

Molecular genetic studies Obesity has become the target of intense molecular genetic research in part because of the so-called obese gene in mice. Mouse models have historically been very important in uncovering the genetic architecture of obesity and related traits, and advances in these models continue to provide insight into the etiology of weight-related diseases (see Mathes, Kelly, & Pomp, 2011, for a review). In the 1950s, a recessive mutation that caused obesity was discovered in mice. When these obese mice were given blood from a normal mouse, they lost weight, a result suggesting that the obese mice were missing some factor important in control of weight. The gene was cloned and was found to be similar to a human gene (Zhang et al., 1994). The gene's product, a hormone called leptin, was shown to reduce weight in mice by decreasing appetite and increasing energy use (Halaas et al., 1995). However, with rare exceptions (Montague et al., 1997), obese humans do not appear to have defects in the leptin gene. The gene that codes for the leptin receptor in the brain has also been cloned from another mouse mutant (Chua et al., 1996). Mutations in this gene might contribute to genetic risk for obesity. Up to 3 percent of patients with severe obesity have been found to have a loss-of-function mutation in the leptin receptor (Farooqi et al., 2007). Interestingly, the obesity phenotype in individuals with defects in the leptin gene or its receptor is very similar, illustrating that leptin is a key piece of the body weight and obesity puzzle (Ramachandrappa & Farooqi, 2011).

Another biological system that has received interest is the melanocortin system. Many of the effects of leptin on the body are mediated by the central nervous system, particularly the hypothalamus. When leptin binds to leptin receptors in this area of the brain, it stimulates the melanocortin system. It is this stimulation that actually suppresses food intake (Ramachandrappa & Farooqi, 2011). A particular gene in this system, *MC4R,* has been associated with obesity in humans (Vaisse, Clement, Guy-Grand, & Froguel, 1998; Yeo et al., 1998), and targeted disruption of *MC4R* in mice leads to increased food intake and increased lean mass and growth (Huszar et al., 1997). It is believed that these hypothalamic pathways interact with other brain centers to coordinate appetite, regulate metabolism, and influence energy expenditure (Ramachandrappa & Farooqi, 2011). In other words, obesity-related traits are highly complex and are likely to be regulated by multiple genes that impact many systems, and these genes are likely to interact not only with one another but also with environmental stimuli (Mathes et al., 2011).

As with most complex traits, major single-gene effects on human obesity are rare and often involve severe disorders. In addition, hundreds of genes in mice have been shown to affect body weight when mutated or otherwise altered (Mathes et al., 2011; Rankinen et al., 2006). However, multiple genes of various effect sizes are likely to be responsible for the substantial genetic contribution to common overweight and obesity. More than 60 genomewide linkage scans have reported at least 250 linkages, and more than 50 regions have been supported by two or more studies (Rankinen et al., 2006). Candidate gene studies have also produced a welter of results, with reports of positive associations involving more than 120 candidate genes; encouragingly, 22 of these genes have been supported in at least five studies (Rankinen et al., 2006). Genomewide association approaches have identified genes that increase risk for common forms (i.e., not due to a single gene) of obesity, as defined by BMI, waist circumference, waist:hip ratio, and body fat percentage. More than 15 genomewide association studies (GWAS) have been published that, when combined, have yielded over 50 loci associated with obesity (see Herrera et al., 2011, for a review). The gene that has been consistently associated with common obesity, *FTO,* explains about 1 percent of the heritability of BMI (Frayling et al., 2007). As predicted by quantitative genetic research, the SNP in the *FTO* gene is associated with body weight throughout the distribution, not just with the obese end of the distribution. Also as predicted by quantitative genetic research, the SNP is not associated with birth weight but shows correlations with body weight beginning at 7 years of age.

MC4R, which was suggested initially through candidate gene studies, as discussed above, has also been identified via multiple GWAS to be associated with BMI (Zeggini et al., 2007), waist circumference (Chambers et al., 2008), higher energy and fat intake (Qi, Kraft, Hunter, & Hu, 2008), and early-onset obesity (Farooqi et al., 2003). A meta-analysis of 250,000 individuals confirmed 14 of the previously identified obesity genes, including *FTO* and *MC4R,* and also identified 18 new loci related to obesity (Speliotes et al., 2010).

Epigenetics and obesity-related outcomes Epigenetic modifications, such as DNA methylation and imprinting (see Chapter 10), have also been suggested to affect obesity. Recall that genomic imprinting influences the genetic expression of alleles as a function of whether the allele came from the father or the mother. One example is Prader-Willi syndrome (Chapter 11), which results from a paternal deletion at 15*q*11-13 and is characterized by severe early-onset obesity due to satiety dysfunction (Shapira et al., 2005). Epigenetic variation can also be induced by early environmental influences, and DNA methylation has been suggested to affect fetal growth, later metabolism, and risk for other chronic diseases (Herrera et al., 2011; Maccani & Marsit, 2009).

Although obese mothers tend to have obese children (Dabelea et al., 2008), maternal weight loss prior to pregnancy via clinical intervention can reduce the risk of obesity in children by providing a less obesogenic prenatal environment (Smith et al., 2009). However, in obese women, it is difficult to distinguish genetic and environmental contributions to offspring obesity. Animal models of maternal obesity have begun to shed some light on the possible interaction between the environment and the epigenetic mechanisms that might affect expression of genes associated with increased BMI and other obesity-related traits (see Li, Sloboda, & Vickers, 2011, for a review). For example, the *MC4R* gene shows reduced methylation following long-term exposure to a high-fat diet in mice (Widiker, Kaerst, Wagener, & Brockmann, 2010). A high fat diet also modifies methylation of the leptin promoter in rats (Milagro et al., 2009). Importantly, genetic and epigenetic factors are intimately intertwined. As more becomes known about the role of genetics and epigenetics in obesity, that information can be combined with known environmental risks in order to gain a more comprehensive picture of the etiology of obesity-related outcomes (Herrera et al., 2011).

New directions in genetics of obesity and weight gain The vast majority of studies, only a small fraction of which are discussed above, have focused on the observable outcome, or phenotype, of weight. Research is now emerging that attempts to uncover the effect of our genetic makeup on our gut microbiome (Mathes et al., 2011). The gut microbiome is a population of microbial species that interact with gastrointestinal tissues and may ultimately affect body weight, obesity, and other nutritionally relevant traits. The hypothesis is that lean and obese individuals have different gut microbial populations that affect energy extraction and later deposit of fat stores from consumed food, which could influence the host's weight gain environmentally (Turnbaugh & Gordon, 2009). However, the host's genome could also affect the function of the gut microbiome. Studies using animal models have begun to investigate these questions and suggest that genetic variations found in the host affect the function of the gut microbiome, which then influences the development of obesity (see Mathes et al., 2011, for a review). Recent studies of obese and lean twins have begun to dissect the relative contributions of host genotype and environmental exposures, such as diet, to shaping the microbial and viral landscape of our gut microbiota (Reyes et al., 2010; Turnbaugh et al., 2009). Results suggest that, although the human gut microbiome is shared to some extent among family members, gut microbiomes also contain a variety of specific (i.e., not shared) bacteria that affect individuals' ability to extract energy from their diet and deposit it into fat, in part as a function of the individuals' genotypes (Hansen et al., 2011; Turnbaugh et al., 2009).

Subjective Well-Being and Health

Subjective well-being, life satisfaction, and their relation to health constitute a growing area of research in behavioral genetics. Research suggests, perhaps unsurprisingly, that a lower subjective well-being is associated with chronic health problems (Strine, Chapman, Balluz, Moriarty, & Mokdad, 2008), depression (Koivumaa-Honkanen, Kaprio, Honkanen, Viinamaki, & Koskenvuo, 2004), poorer quality of life, increased health care costs, early retirement, and mortality (Gill et al., 2006; Katon et al., 2004). Positive well-being, on the other hand, is related to longevity and may add several years to the life span (Diener & Chan, 2011).

Twin studies suggest that about 30 to 60 percent of the variance in subjective well-being is due to genetic influences (e.g., Caprara et al., 2009). Moreover, continuity of subjective well-being over time also appears to be influenced by genetic factors (Roysamb, Tambs, Reichborn-Kjennerud, Neale, & Harris, 2003). The phenotypic relationships between subjective well-being and self-reported health, sleep, and physical activity are due, at least in part, to genetic overlap (Mosing, Zietsch, Shekar, Wright, & Martin, 2009; Paunio et al., 2009; Waller, Kujala, Kaprio, Koskenvuo, & Rantanen, 2010). The positive effects of exercise on subjective well-being are also thought to be attributable to common genetic factors (Bartels, de Moor, van der Aa, Boomsma, & de Geus, 2012).

Less is known about the molecular genetic underpinnings of subjective well-being or self-rated health. A genomewide linkage scan for subjective happiness suggested QTLs of interest on chromosomes 1 and 19 (Bartels et al., 2010); however, replication and additional studies are needed. Recent genomewide association efforts have yielded no significant findings for self-rated health (Mosing et al., 2010). Consistent with other phenotypes discussed in this book, it appears that self-rated health and subjective well-being are likely to be due to the contribution of multiple genes of small effect rather than a few genes of major effect.

Increasing interest is being paid to the relationships between subjective well-being, happiness, and healthy aging (Steptoe & Wardle, 2012). Mental health is increasingly defined not only by the absence of illness but also by the presence of subjective well-being (Sadler, Miller, Christensen, & McGue, 2011). It is clear that subjective well-being predicts favorable life outcomes, including better mental and somatic health, as well as longevity. Further, this body of research has prompted interventions and public health initiatives that are focused on increasing happiness and well-being, particularly among older adults. We now turn our attention to research related to healthy aging.

Psychology and Aging

Like health psychology, aging is an area of great social significance. The average age in most societies is increasing, primarily as a result of improvements in health care. For example, in the United States, the number of people age 65 and older will double from 10 to 20 percent by 2030 (Kinsella & He, 2009). Those 85 and older are projected to increase in number from 5.5 million in 2010 to 6.6 million in 2020. Worldwide, this group is growing nearly twice as fast as the population as a whole (Kinsella & He, 2009). Although obvious changes occur later in life, it is not possible to lump older individuals into a category of "the elderly" because older adults differ greatly biologically and psychologically. The question for genetics is the extent to which genetic factors contribute to individual differences in functioning later in life (Figure 19.4).

Genetic research in the behavioral sciences is increasingly being directed toward the last half of the life span. There are over two dozen twin studies investigating physical, psychological, and social aspects of aging (Bergeman, 2007). Chapter 11 described genetic research on dementia, for which moderate genetic influence has been found. Dementia is a focal area for molecular genetic research. Several genes have been identified that account for most cases of a rare form of dementia that occurs in middle adulthood. The best example of a QTL in behavioral genetics is the association between the apolipoprotein E gene (*APOE*) and typical late-onset dementia. *APOE* has also been associated with change in working memory in a sample of Swedish twins without dementia (Reynolds et al., 2006).

Another interesting finding about genetics and cognitive aging was described in Chapter 12: The heritability of general cognitive ability increases throughout the life span. In later life, heritability estimates reach 80 percent, one of the highest heritabilities reported for behavioral traits (Finkel & Reynolds, 2010; Pedersen, 1996). Although there is some evidence in the very oldest individuals that heritability may decline again, a recent meta-analysis suggests that this is inconclusive due to

• **FIGURE 19.4** Ninety-three-year-old MZ twins participating in a twin study of cognitive functioning late in life and photos of them going back to childhood (McClearn et al., 1997). Not only do MZ twins continue to look physically similar late in life, they also continue to perform similarly on measures of cognitive ability. (Reproduced with permission from *Science,* June 6, 1997. Copyright 1997 American Association for the Advancement of Science.)

heterogeneity across study samples (i.e., inclusion and exclusion criteria), confounding between cognitive measures examined, and the use of different measures to represent a particular cognitive domain (e.g., memory) in different samples (Lee, Henry, et al., 2010). A review of behavioral genetic research on cognitive aging suggests that it is possible that the genetic and environmental factors influencing *level* of cognitive functioning are not the same as those influencing *change* with age (Finkel & Reynolds, 2010). In contrast to general cognitive ability, less is known about specific cognitive abilities throughout the life span. However, evidence suggests that the pattern of genetic and environmental influences over time is similar to that seen for general cognitive ability. For example, the heritability for level of verbal performance is high (50 to 90 percent, depending on the specific verbal measure), but heritability is much less for rate of decline in verbal ability (10 to 30 percent) (Finkel & Reynolds, 2010). It is also important to note that specific cognitive abilities are not independent of one another. In fact, genetic influences on specific cognitive abilities largely overlap with genetic influence on general cognitive ability (Chapter 13).

Not mentioned in the discussion of multivariate genetic analysis in Chapter 13 is a distinction made in the field of cognition and aging between "fluid" abilities, such as spatial ability, which decline with age, and "crystallized" abilities, such as vocabulary, which increase with age (Baltes, 1993). Decades of gerontological research suggest that spatial, fluid, and memory abilities decrease over time, and it has been hypothesized that fluid abilities are more biologically based and crystallized abilities more culturally based (e.g., Lindenberger, 2001). However, genetic research so far has found that fluid and crystallized abilities are equally heritable (Finkel & Reynolds, 2010; Pedersen, 1996).

For psychopathology and personality, the few genetic studies in later life yield results similar to those described in Chapters 14–17 for research earlier in life (Bergeman, 1997). For example, for depression in later life, twin studies indicate modest heritabilities similar to those found earlier in life (Gatz, Pedersen, Plomin, Nesselroade, & McClearn, 1992; Johnson, McGue, Gaist, Vaupel, & Christensen, 2002). For personality, Type A behavior—hard-driving and competitive behavior that is of special interest because of its reputed link with heart attacks—shows moderate heritability typical of other personality measures in older twins (Pedersen, Lichtenstein, et al., 1989). Another interesting personality domain is locus of control, which refers to the extent that outcomes are believed to be due to one's own behavior or chance. For some older individuals, this sense of control declines, and the decline is linked to declines in psychological functioning and poor health. A twin study later in life found moderate genetic influence for two aspects of locus of control: sense of responsibility and life direction (Pedersen, Gatz, Plomin, Nesselroade, & McClearn, 1989). However, the key variable of the perceived role of luck in determining life's outcomes showed no genetic influence and substantial shared environmental influence. This finding, although in need of replication, stands out from the usual finding in personality research of moderate genetic influence and no shared environmental influence. The high stability of personality in later life is largely mediated by genetic factors (Johnson et al., 2005; Read et al., 2006).

The famous U.S. Supreme Court Justice Oliver Wendell Holmes quipped that "those wishing long lives should advertise for a couple of parents, both belonging to long-lived families" (cited by Cohen, 1964, p. 133). Research, however, indicates only modest genetic influence on longevity, with heritabilities of about 25 percent (Bergeman, 1997), although genetic influence on longevity may increase at the most advanced ages (Hjelmborg et al., 2006). The most consistent evidence from molecular genetic studies suggests that polymorphisms in the *APOE* (e.g., Novelli et al., 2008) and *FOXOA3* genes (Flachsbart et al., 2009; Li et al., 2009; Wilicox et al., 2008) are associated with longer life (see Wheeler & Kim, 2011, for a review). *APOE* is hypothesized to be associated with individual differences in human longevity, probably because of its links with cardiovascular disease rather than dementia (Christensen, Johnson, & Vaupel, 2006). The *FOXOA3* gene is part of the insulin signaling pathway.

Much genetic research in nonhuman species—especially mice, fruit flies, and nematode worms—has shown that mutations in the insulin signaling pathway affect the life span (reviewed in Martin, 2011; Wheeler & Kim, 2011). Animal models are continuing to aid in efforts to identify genes associated with longevity (Kenyon, 2010). For example, according to the Human Aging Genomic Resources (De Magalhães et al., 2009), 68 genes in the mouse have been identified as being related to aging. In the fruit fly *Drosophila melanogaster*, selective breeding, QTL analysis, and mutational analysis have identified 75 genes related to the aging process. In the nematode worm (*C. elegans*), more than 500 genes have been found to influence life span.

Longevity research also presents an excellent example of gene-by-environment interaction. A diet restricted in calories has been shown, across multiple organisms, to extend the life span. This finding was first reported in the 1930s, when it was observed that rats that were underfed, or had restricted caloric intake, lived significantly longer than their normally fed counterparts (McCay, Crowell, & Maynard, 1935). Research in this area has expanded so much that dietary restriction is currently considered a robust life-extending intervention. In fact, members of the Calorie Restriction Society practice self-imposed caloric restriction in an effort to extend their lives. However, research suggesting life extension due to reduced caloric intake has not been without contradictory findings. Some researchers have reported that restricted diets actually decrease the life span in certain strains of rodents (Harper, Leathers, & Austad, 2006). Recently, researchers have attempted to address this inconsistency by examining the efficacy of caloric restriction on life span across a range of genotypes (Liao, Rikke, Johnson, Diaz, & Nelson, 2010). Across 41 recombinant inbred strains of mice (Chapter 5), it was reported that dietary restriction shortened the life span in more strains than it increased the life span. Moreover, strain-specific "lengths of life span" under restricted or normal diets were not correlated, meaning that genetic determinants of longevity differ under the two dietary conditions (Liao et al., 2010). Thus, it appears that dietary restriction might not be a universal intervention for increasing the life span because it is dependent on the genetic background of the individual or organism.

As discussed above, psychologists are interested in subjective well-being, and this is especially the case in later life. That is, what is important is not just how long we live but how well we live—not just adding years to our life but adding life to our years. Health and functioning in daily life show moderate genetic influence later in life, as do the relationships among health and psychological well-being (Harris, Pedersen, Stacey,

McClearn, & Nesselroade, 1992), life satisfaction (Plomin & McClearn, 1990), and longevity (Sadler et al., 2011). Another aspect of quality of life is self-perceived competence. One study of older twins found that six dimensions of self-perceived competence—including interpersonal skills, intellectual abilities, and domestic skills—show heritabilities of about 50 percent (McGue, Hirsch, & Lykken, 1993).

Health Psychology and Genetic Counseling

It is clear that we are at the dawn of a new era in which behavioral genetic research is moving beyond the demonstration of the importance of heredity to the identification of specific genes. In clinics and research laboratories, behavioral scientists of the future will routinely collect saliva or blood and send the samples to a laboratory for DNA extraction (if we do not yet each have a memory key with our complete DNA sequence). Trait-specific sets of hundreds of genes will be available on microarrays that can genotype even large samples at a modest cost; these "gene set" data will be incorporated into behavioral research as genetic risk indicators. In the past, this type of information was available for single-gene disorders, such as fragile X mental retardation, as well as for the QTL association between apolipoprotein E and late-onset dementia. However, there are now companies that offer the ease of obtaining genetic risk prediction at a low cost for anyone willing to send a saliva sample.

As is the case with most important advances, identifying genes for behavior will raise new ethical issues. These issues are already beginning to affect genetic counseling (Box 19.1). Genetic counseling is expanding from the diagnosis and prediction of rare, untreatable single-gene conditions to the prediction of common, often treatable or preventable conditions (Karanjawala & Collins, 1998). Although there are many unknowns in this uncharted terrain, the benefits of identifying genes for understanding the etiology of behavioral disorders and dimensions seem likely to outweigh the potential abuses. The judicious use of genetic and genomic information has significant, but as yet untested, potential to enhance the clinical care and prevention of chronic diseases. That is, it can help us to understand the etiology of disease and also aid in providing treatment recommendations for patients' health behaviors (Cho et al., 2012; Green & Guyer, 2011). Health psychologists are at the forefront of research investigating the effects of genetic testing on patient attitudes, beliefs, and health-related behaviors (McBride, Koehly, Sanderson, & Kaphingst, 2010). For example, there is some evidence that when patients are provided with genetic testing results, their preventative behavior increases (Taylor & Wu, 2009). A recent systematic review of the impact of genetic risk information on chronic adult diseases found some psychological benefits of including genetic information in treatment of chronic diseases, but it concluded that many gaps in knowledge must be addressed before genetic science can be effectively translated into clinical practice (McBride et al., 2010). New studies are being designed that try to address these gaps in order to increase the clinical and personal utility of genetic testing (e.g., Cho et al., 2012).

BOX 19.1 • Genetic Counseling

Genetic counseling is an important interface between the behavioral sciences and genetics and goes well beyond simply conveying information about genetic risks and burdens. It helps individuals come to terms with the information by dispelling mistaken beliefs and allaying anxiety in a nondirective manner that aims to inform rather than to advise. In the United States, over 3000 health professionals have been certified as genetic counselors, and about half of these were trained in two-year master's programs (Mahowald, Verp, & Anderson, 1998). For more information about genetic counseling as a profession, including practice guidelines and perspectives, see the National Society of Genetic Counselors (www. nsgc.org/), which sponsors the *Journal of Genetic Counseling* and has a useful link called *How to Become a Genetic Counselor.* For more general information about professional education in genetic counseling, see the National Coalition for Health Professional Education in Genetics (www.nchpeg.org/).

Until recently, most genetic counseling was requested by parents who had an affected child and were concerned about risk for other children. Now genetic risk is often assessed directly by means of DNA testing. As more genes are identified for disorders, genetic counseling is increasingly involved in issues related to prenatal diagnoses, prediction, and intervention. This new information will create new ethical dilemmas. Huntington disease provides a good example. If you had a parent with the disease, you would have a 50 percent chance of developing the disease. However, with the discovery of the gene responsible for Huntington

disease, in almost all cases it is now possible to diagnose whether a fetus or an adult will have the disease. Would you want to take the test? It turns out that the majority of people at risk choose not to take the test, largely because there is as yet no cure (Maat-Kievit et al., 2000). If you did take the test, the results would likely affect knowledge of risk for your relatives. Do your relatives have the right to know, or is their right not to know more important? One generally accepted rule is that informed consent is required for testing; moreover, children should not be tested before they become adults unless a treatment becomes available.

Another increasingly important problem concerns the availability of genetic information to employers and insurance companies. These issues are most pressing for single-gene disorders like Huntington disease, in which a single gene is necessary and sufficient to develop the disorder. For most behavioral disorders, however, genetic risks will involve QTLs that are probabilistic risk factors rather than certain causes of the disorder. A major new dilemma concerns the burgeoning industry of marketing genetic tests directly to consumers (Biesecker & Marteau, 1999; Wade & Wilfond, 2006). Although genetic counseling has traditionally focused on single-gene and chromosomal disorders, increasingly the field is encompassing complex disorders including behavioral disorders (Finn & Smoller, 2006). Despite the ethical dilemmas that arise with the new genetic information, it should also be emphasized that these findings have the potential for profound improvements in the prediction, prevention, and treatment of diseases.

Summary

Two areas of psychology from which interesting genetic results are emerging are health psychology and aging. One example of genetic research on health psychology concerns body weight and obesity. Although most theories of weight gain are environmental, genetic research consistently shows substantial genetic influence on individual differences in body weight, with heritabilities of about 70 percent. Also interesting in light of environmental theories is the consistent finding that shared family environment does not affect weight. Longitudinal studies indicate that genetic influences on weight are surprisingly stable after infancy, although there is some evidence for genetic change even during adulthood. Body weight and obesity are the target of much molecular genetic research in mice and humans, with increasing success. Subjective well-being is another example of an area where genetic research, both quantitative and molecular, is beginning to expand.

Much genetic research now addresses the last half of the life span. Dementia and cognitive decline in later life are intense areas of molecular genetic research. For general cognitive ability, twin and adoption studies indicate that heritability increases during adulthood. Psychopathology and personality generally show results similar to those for younger ages: moderate heritability and no shared family environment. The molecular genetics of longevity and aging is an expanding field, with promising results from animal models and limited success with human data.

CHAPTER TWENTY

Evolution and Behavior

E volution is the environment writ large, but it is written in the genes. Although its roots lie firmly with Darwin's ideas of more than a century ago, evolutionary thinking has only recently established itself in the behavioral sciences. This chapter offers an overview of evolutionary theory and two related fields. Population genetics provides a quantitative basis for investigating forces, especially evolutionary forces, that change gene and genotype frequencies. The second related field is evolutionary psychology, which considers behavioral adaptations on an evolutionary time scale.

Charles Darwin

One of the most influential books ever written is Charles Darwin's 1859 *On the Origin of Species* (Figure 20.1). Darwin's famous 1831–1836 voyage around the world on the *Beagle* led him to observe the remarkable adaptation of species to their environments. For example, he made particularly compelling observations about 14 species of finches found in a small area on the Galápagos Islands. The principal differences among these finches were in their beaks, and each beak was exactly appropriate for the particular eating habits of the species (Figure 20.2).

Theology of the time proposed an "argument from design," which viewed the adaptation of animals and plants to the circumstances of their lives as evidence of the Creator's wisdom. Such exquisite design, so the argument went, implied a "Designer." Darwin was asked to serve as naturalist on the surveying voyage of the *Beagle* in order to provide more examples for the "argument from design." However, during his voyage, Darwin began to realize that species, such as the Galápagos finches, were not designed once and for all. This realization led to his heretical theory that species evolve one from another: "Seeing this gradation and diversity of structure in one small, intimately related group of birds, one might really fancy that from an original paucity of birds in this archipelago, one species had been taken

POODLESROCK/CORBIS

⊙ **FIGURE 20.1** Charles Darwin as a young man.

and modified for different ends" (Darwin, 1896, p. 380). For over 20 years after his voyage, Darwin gradually and systematically marshaled evidence for his theory of evolution.

Darwin's theory of evolution begins with variation within a population. Variation exists among individuals in a population due, at least in part, to heredity. If the likelihood of surviving to maturity and reproducing is influenced even to a slight degree by a particular trait, offspring of the survivors will show more of the trait than their parents' generation. In this way, generation after generation, the characteristics of a population can gradually change. Over a sufficiently long period, the cumulative changes can be so great that populations become different species, no longer capable of interbreeding successfully.

For example, the different species of finches that Darwin saw on the Galápagos Islands may have evolved because individuals in a progenitor species differed slightly in the size and shape of their beaks. Certain individuals with slightly more powerful beaks may have been more able to break open hard seeds. Such individuals could survive and reproduce when seeds were the main source of food. The beaks of other individuals may have been better at catching insects, and this shape gave those individuals a selective advantage at certain times. Generation after generation, these slight differences led to other differences, such as different habitats. For instance, seed eaters made their living on the ground and insect eaters lived in the trees. Eventually, the differences became so great that offspring of the seed eaters and insect eaters rarely interbred. Different species were born. A Pulitzer Prize–winning account of

• **FIGURE 20.2** The 14 species of finches in the Galápagos Islands and Cocos Island. (a) A woodpecker-like finch that uses a twig or cactus spine instead of its tongue to dislodge insects from tree-bark crevices. (b–e) Insect eaters. (f, g) Vegetarians. (h) The Cocos Island finch. (i–n) The birds on the ground eat seeds. Note the powerful beak of (i), which lives on hard seeds. (From "Darwin's finches" by D. Lack. ©1953 by Scientific American, Inc. All rights reserved.)

25 years of repeated observations of Darwin's finches, *The Beak of the Finch* (Weiner, 1994), shows natural selection in action.

Although this is the way the story is usually told, another possibility is that behavioral differences in habitat preference led the way to the evolution of beaks rather than the other way around. That is, heritable individual differences in habitat preference may have existed that led some finches to prefer life on the ground and others to prefer life in the trees. The other differences, such as beak size and shape, may have been secondary to these habitat differences. Although this proposal may seem to involve splitting hairs, this alternative story makes two points. First, it is difficult to know the mechanisms driving evolutionary change. Second, although behavior is not as well preserved as physical characteristics, it is likely that behavior was often at the cutting edge of natural selection. Artificial selection studies (Chapter 5) show that behavior can be changed through selection, as seen in the dramatic behavioral differences between breeds of dogs (see Figure 5.1), and that form often follows function.

Darwin's most notable contribution to the theory of evolution was his principle of *natural selection:*

> Owing to this struggle [for life], variations, however slight and from whatever cause proceeding, if they be in any degree profitable to the individuals of a species, in their infinitely complex relations to other organic beings and to their physical conditions of life, will tend to the preservation of such individuals, and will generally be inherited by the offspring. The offspring, also, will thus have a better chance of surviving, for, of the many individuals of any species which are periodically born, but a small number can survive. (Darwin, 1859, pp. 51–52)

Although Darwin used the phrase "survival of the fittest" to characterize this principle of natural selection, it could more appropriately be called reproduction of the fittest. Mere survival is necessary, but it is not sufficient. The key to the spread of alleles in a population is the relative number of surviving and reproducing offspring.

Darwin convinced the scientific world that species evolved by means of natural selection. *Origin of Species* is at the top of most scientists' lists of books of the millennium—his theory has changed how we think about all the life sciences. Nonetheless, outside science, controversy continues (Bolhuis, Brown, Richardson, & Laland, 2011; Pinker, 2010). For instance, in the United States, boards of education in several states have attempted to curtail the teaching of evolution in response to pressure from creationists who believe in a literal biblical interpretation of creation. Advocates of creationism have lost every major U.S. federal court case for the past 40 years (Berkman & Plutzer, 2010). Nevertheless, recent research investigating the evolution-creationism battle in state governments and classrooms has revealed the reluctance of teachers to teach evolutionary biology. In fact, 60 percent of teachers are strong advocates neither for evolution nor for nonscientific alternatives. Interestingly, much of this hesitancy appears to be due, at least in part, to a lack of confidence in their ability to defend evolution, perhaps because of their own lack of exposure to courses in evolution (Berkman & Plutzer, 2011). However, most people, but not everyone—see, for example, Dawkins (2006) versus Collins (2006)—accept the notion that science and religion occupy distinctly different realms, with science operating in the realm of verifiable facts and religion focused on purpose, meaning, and values. "Respectful noninterference" between science and religion is needed (Gould, 2011).

Scientifically, Darwin's theory of evolution had serious gaps, mainly because the mechanism for heredity, the gene, was not yet understood. Gregor Mendel's work was not published until seven years after the publication of the *Origin of Species,* and even then it was ignored until the turn of the century. Mendel provided the answer to the riddle of inheritance, which led to an understanding of how variability arises through mutations and how genetic variability is maintained generation after generation (Chapter 2). A rewrite of the *Origin of Species* is interesting in pointing out how evolutionary theory and research have changed since Darwin, as well as showing how prescient Darwin was (Jones, 1999).

Darwin considered behavioral traits to be just as subject to natural selection as physical ones. In the *Origin of Species,* an entire chapter is devoted to instinctive behavior patterns. In a later book, *The Descent of Man and Selection in Relation to Sex,* Darwin (1871) discussed intellectual and moral traits in animals and humans, concluding that the difference between the mind of a human being and the mind of an animal "is certainly one of degree and not of kind" (p. 101). Over 150 years after the publication of the *Origin of Species,* Darwin's influential theory is still highly relevant for the study of human behavior. As one example, Pinker (2010) recently proposed that intelligence, sociality, and language coevolved via natural selection:

> According to this theory, hominids evolved to specialize in the cognitive niche, which is defined by: reasoning about the causal structure of the world, cooperating with other individuals, and sharing that knowledge and negotiating those agreements via language. This triad of adaptations coevolved with one another and with life-history and sexual traits such as enhanced parental investment from both sexes and multiple generations, longer childhoods and lifespans, complex sexuality, and the accumulation of local knowledge and social conventions in distinct cultures. (p. 899)

Several principles that underlie the evolution and coevolution of various social behaviors are described below.

Inclusive Fitness

Darwin's theory of individual fitness has been extended to consider a measure called *inclusive fitness,* which is defined as the fitness of an individual plus part of the fitness of kin that is genetically shared by the individual (Hamilton, 1964). Inclusive fitness and kin selection explain altruistic acts that do not directly benefit the individual. If the net result of an altruistic act helps more of that individual's genes to survive and to be transmitted to future generations, the act is adaptive even if it results in the death of the individual.

The founder of quantitative genetics, R. A. Fisher, long ago suggested an example of kin selection and inclusive fitness that involves the distastefulness of some butterfly larvae (Fisher, 1930). A bird will learn that certain larvae taste bad, but the lesson costs the larva its life. However, sibling eggs are laid in a cluster, and inclusive fitness is served by the sacrifice of one larva if two siblings (the genetic equivalent of the sacrificed larva) are saved. Inclusive fitness switches the focus from the individual to the gene, which explains the title of a classic book in this area, *The Selfish Gene* (Dawkins, 1976). Acts that appear to be altruistic can be interpreted in terms of "selfish" genes that are maximizing their reproduction through inclusive fitness.

Inclusive fitness was popularized by a book in 1975 called *Sociobiology: The New Synthesis,* which promoted evolutionary thinking as a unifying theme for all of the life sciences, including the behavioral sciences (Wilson, 1975). **Sociobiology** has offered novel and interesting hypotheses that stem from the simple principle of inclusive fitness and kin selection.

One general theory is that of parental investment (Trivers, 1972, 1985). For example, why do mothers provide most of the care of offspring in the vast majority of mammalian species, including humans? Unless a species is completely monogamous (as eagles are, for instance), males have less invested in their offspring. Males can have many offspring by many females, but each female must devote large amounts of energy to each pregnancy and, in mammals, provide sustenance after birth. In terms of inclusive fitness, the fitness of females is better served by increased care of each offspring because females must make a substantial investment in each one of them. In many cases, however, the male's investment is little more than copulation, and he can maximize his inclusive fitness by having more offspring by different females.

A related reason for the relative investments of mothers and fathers in the care of their offspring is that females can always be sure that they share half of their genes with their young. Males, however, cannot be sure that offspring are theirs. The theory of parental investment led to two predictions that have received considerable support: (1) The sex that invests more in offspring (typically, but not always, the female) will be more discriminating about mating, and (2) the sex that invests less (typically the male) will compete more for sexual access (Platek & Shackelford, 2006; Trivers, 1985). It may, however, not be so simple. Recent theories suggest that the mating system of a species is characterized, at least in part, by social interactions that influence mating, fertilization, and parental investment and that these three key aspects of the mating system are intrinsically connected (Alonzo, 2010). For example, multiple traits in males and females coevolve simultaneously such that the parental effort and mating behavior of one sex affects selection of parental effort and mating behavior of the opposite sex. Thus, there may be social, behavioral, and coevolutionary feedback loops that are often ignored (Alonzo, 2010).

Despite the complexity of these evolutionary traits, in most species, males court and females choose, and dads are often cads (Miller, 2000; Symons, 1979). The greater altruism of mothers toward their offspring is no less selfish from the point of view of genes than that of fathers (Hrdy, 1999). Although there are many hypotheses of this sort in which the "selfish altruism" of genes evolved through kin selection, it is also likely that some positive social behaviors evolved through the less convoluted mechanism of individual selection (de Waal, 1996).

Evolutionary thinking is making major inroads in the behavioral sciences, a field called *evolutionary psychology* (Buss, 2011; Gangestad & Simpson, 2007). Before we turn to evolutionary psychology, however, an overview of the quantitative foundation of evolution, the field of population genetics, is in order.

Population Genetics

Darwin's evidence for the evolution of species, such as the beaks of the Galápagos finches, relied on qualitative descriptions. Population genetics provides evolution with a quantitative basis. Its unique contribution is to describe allelic and genotypic frequencies in populations and to study the forces that change these frequencies, such as natural selection. Increasingly, population genetics involves analyzing DNA rather than inferring genotypes from phenotypes.

In the absence of opposing forces, the frequencies of alleles and genotypes remain the same, generation after generation. As explained in Box 2.2, this stability is called Hardy-Weinberg equilibrium. Population geneticists investigate the forces that change this equilibrium (Hartl, 2004; Hartl & Clark, 2006; Lachance, 2009). For example, selection against a rare recessive allele is very slow, and it is for this reason that most deleterious alleles are recessive. Suppose that a recessive allele is lethal when homozygous and that the frequency of the allele is 2 percent in a population. If no homozygous recessive individuals were to reproduce for 50 generations, the frequency of this undesirable allele would only change from 2 to 1 percent. In contrast, complete selection against a dominant allele would wipe out the allele in a single generation. As mentioned in Chapter 2, the dominant allele responsible for Huntington disease persists because its lethal effect is not expressed until after the reproductive years.

Natural selection is often discussed in terms of *directional selection* of this sort, a process in which a deleterious allele is selected against. For simplicity, a form of selection acting on advantageous alleles is called *positive selection* (Akey, 2009). When an advantageous allele fixes in a population, it does so on a specific haplotype background, described in Chapter 10. This linked variation is then swept through a population along with the advantageous mutation, a process called a *selective sweep*. Selective sweep events tend to reduce genetic diversity in a particular chromosomal region (Vitti, Cho, Tishkoff, & Sabeti, 2012).

Another type of selection maintains different alleles rather than favoring one allele over another, a process that is especially interesting because genetic variability within a species is the focus of behavioral genetics. In contrast to directional selection, this type of selection is called *stabilizing selection* because it leads to *balanced polymorphisms*. Suppose that selection operated against both dominant and recessive homozygotes for a particular gene. In this process, heterozygotes would reproduce relatively more than the two homozygous genotypes. However, heterozygotes always produce homozygotes as well as heterozygotes (see Box 2.2). Genetic variability is thus maintained.

Sickle-cell anemia in humans is a specific example of this kind of balanced polymorphism. This single-gene autosomal recessive disease is caused by a single nucleotide mutation that damages red blood cell membranes, rendering the cells unable to fulfill their function of transporting oxygen. As a result, the reproductive fitness of individuals with sickle-cell anemia (recessive homozygotes) is lowered. However, there is a puzzle: The allele is maintained in relatively high frequency in some African populations and among African Americans. The high frequency of this debilitating recessive allele is due to the higher relative fitness of heterozygotes (carriers). Heterozygotes are more resistant than normal homozygotes to a form of malaria, prevalent in certain parts of Africa, that infects between 300 and 500 million people each year. It causes between 1 million and 3 million deaths annually, mostly among young children in sub-Saharan Africa. In other words, the decreased relative fitness of homozygotes with sickle-cell anemia is balanced by the increased relative fitness of heterozygote carriers.

Another sort of stabilizing selection involves environmental diversity. As noted in relation to Darwin's finches, if environments encountered by a species are diverse, selection pressures can differ and foster genetic variability. A balanced polymorphism can also occur if selection depends on the frequency of a genotype. For example, selection that favors rare alleles produces genetic variability. Individuals with a rare genotype might use resources that are not used by other members of the species and thus gain a selective edge. Predator-prey relationships can also be frequency dependent: Predatory birds and mammals tend to attack more common types of prey.

Another type of frequency-dependent selection involves mate selection in which rare genotypes have an edge. For instance, in fruit flies, females are more likely to mate with a rare male (Ehrman, 1972; Knoppien, 1985). Like the other types of stabilizing selection, frequency-dependent sexual selection maintains genetic variability in a species.

What evolutionary forces maintain harmful genetic variants, such as those for schizophrenia, that lower reproductive fitness? As just mentioned (see also Chapter 14, using schizophrenia as an example), balanced selection is a possibility. However, a new theory called *polygenic mutation-selection balance* may have general relevance to highly polygenic behavioral traits (Keller & Miller, 2006). If behavioral disorders like schizophrenia are influenced by hundreds or thousands of genes, natural selection would have a very difficult time screening out new deleterious mutations. More generally, selection may stabilize genetic variation because complex traits entail trade-offs of different fitness benefits and costs (Nettle, 2006).

In addition to considering forces that change allelic frequency, population genetics also investigates systems of mating—inbreeding and assortative mating—that change genotypic frequencies without changing allelic frequencies. Inbreeding involves matings between genetically related individuals. If inbreeding occurs, offspring are more likely than average to have the same alleles at any locus; therefore, recessive traits are more likely to be expressed. Inbreeding reduces heterozygosity and increases **homozygosity**. In relation to the derivation of inbred strains mentioned in Chapter 5, population genetics shows that, after 20 generations of brother-sister matings, at least 98 percent of all loci are homozygous. Inbreeding often leads to a higher rate of congenital disorders and reduction in viability and fertility, called *inbreeding depression*.

Inbreeding depression is caused by the increase in homozygosity for deleterious recessive alleles. Although inbreeding reduces genetic variability, its overall effect on genetic variability in natural populations is negligible because it is relatively rare. However, rates of inbreeding in human populations across the world are not rare. Offspring from second-cousin or closer marriages are estimated to account for about 10 percent of the global human population (Keller, Visscher, & Goddard, 2011) and can have important public health consequences (Bittles & Black, 2010). Historically, inbreeding has been studied using large pedigrees; however, with the advance of molecular technologies, dense SNP data can now be used to estimate inbreeding from distant common ancestors (Keller et al., 2011).

The other side of the coin is *hybrid vigor*, or *heterosis*. These terms refer to an increase in viability and performance when different inbred strains are crossed. Outbreeding reintroduces heterozygosity and masks the effects of deleterious recessive alleles. *Assortative mating*, phenotypic similarity between mates, is another system of mating that changes genotypic, not allelic, frequency. As discussed in Chapter 12, assortative mating for a particular trait increases genotypic variance for that trait in a population. Although assortative mating is modest for most behavioral traits, increases in genetic variance due to assortative mating accumulate over generations. In other words, even a small amount of assortative mating can greatly increase genetic variability after many generations.

Evolutionary Genomics

Both our access to the human genome and recent advances in computational methods have revolutionized evolutionary research (Vitti et al., 2012). For example, whole-genome microarray analysis has been used to demonstrate parallels between scouting behavior in honey bees and novelty seeking in humans, as well as to suggest that the genes associated with these behaviors are part of a basic "tool kit" used repeatedly in the evolution of behavior (Liang et al., 2012). Furthermore, advances in technology have allowed the development of the first human brain atlas based solely on genetically informative data; these data suggest that the human cerebral cortex is built on the foundation of primary functional areas of the brain that are shared among mammals (Chen et al., 2012).

New computational methods that focus on the whole genome can isolate genetic signatures left by selection events. For example, exome sequencing led to the discovery of a family of genes selected in Tibetan populations that may help them adjust to living at high altitudes (Yi et al., 2010). Results such as these can then be used to suggest the most promising loci for further investigation (Vitti et al., 2012). Another example is the use of a combination of approaches, including analyses of whole-genome data, to identify genes involved in skin pigmentation. Independent selection for different pigmentation gene sets has been found among Asian, European, and African populations, and functional testing of variant alleles has begun to account for these population differences (Sturm, 2009).

It is important to keep in mind the complexity of these systems. That is, the molecular mechanism of action, the interplay of multiple genes, epigenetic factors (Vitti et al., 2012), and the role of the environment (Roesti, Hendry, Salzburger, & Berner, 2012) all need to be considered in order to understand how evolution affects allelic variation and results in a diversity of phenotypes in human populations (Sturm, 2009).

Evolutionary Psychology

Thinking about behavior from an evolutionary perspective has brought new insights to the behavioral sciences, as Darwin predicted in *On the Origin of Species* (Buss, 2011). Evolutionary thinking is essential to the behavioral sciences because it paints the portrait of our species in broad strokes that show the similarities to and differences from other species. For example, the fact that we are mammals, defined in terms of the mammary gland, means that we have evolved a system in which mothers care for their young after birth. The fact that we are primates has many evolutionary implications, such as extremely slow postnatal development, which requires long-term care by parents. Also fundamental for understanding our species are facts such as these: Our species uses language naturally, walks upright on two feet, and has eyes in the front of the head that permit depth perception.

Evolutionary psychology seeks to understand the adaptive value of species-wide aspects of human behavior, such as our natural use of language, our similar facial expressions for basic emotions, and similarities in mating strategies across cultures. It also addresses differences between groups within a species, most notably differences between the sexes. This is a different level of analysis from that found in most behavioral genetics research, which typically focuses on differences among individuals within a species rather than species-wide aspects of behavior. However, the definition of behavioral genetics as the genetic analysis of behavior includes all levels of analysis, from individuals to species.

It is important to remember that the causes of the typical behavior of a species are not necessarily related to the causes of individual differences within a species. For example, young mammals typically bond to their caregivers. This behavior is an adaptation that has presumably evolved to protect them while they continue to develop after birth. But the evolution of bonding does not mean that individual differences in attachment are due to genetic factors. In fact, as noted in Chapter 16, attachment is one of the few traits that show little evidence of genetic influence. The role of genetic factors in the origins of individual differences in behavior is an empirical issue that requires quantitative genetic analysis. It is much more difficult to investigate genetic mechanisms in an evolutionary time frame and to pin down genes responsible for particular adaptations if the genes vary little within a species. Knock-out technology in mice (see Chapter 5) is one approach to studying the role of such genes. Nonetheless, it is difficult to glean information about evolution from knock-outs. There is a need to build bridges between evolutionary psychology and behavioral genetic theory based on individual differences (Bolhuis et al., 2011; Buss & Greiling, 1999; Nettle, 2006; Segal & MacDonald, 1998).

Instincts

Evolutionary psychologists have brought back the word *instinct*, which had been effectively banned from psychology. One example is an influential book on the evolution of language called *The Language Instinct* (Pinker, 1994). Instincts, which refer to evolved behavioral adaptations, were accepted by psychologists early in this century. William James (1890), the founder of American psychology, presented a long list of instincts that begin at birth. However, instincts were largely rejected as psychology moved more toward environmental explanations of behavior. For half a century, the only instinct discussed was a general ability to learn. During this period, cultural anthropologists focused on differences between cultures that are presumably learned, rather than on their similarities that might be due to evolution.

The ease with which all members of our species learn a language suggests that language is **innate**, an instinct. Although the dictionary defines *innate* as "inborn," in evolutionary psychology the word refers to the ease with which certain things but not others are learned. That is, the word *innate* refers to evolved capacities and constraints rather than rigid hard-wiring that is impervious to experience. *Instinct* means an innate behavioral tendency, not an inflexible pattern of behavior. Although language is now generally accepted as innate, debate continues about what exactly is innate—whether it is a general predisposition to learn language or to learn specific "modules" such as grammatical structures (Pinker, 1994).

Another example involves instinctive fears, such as fear of spiders and snakes, which protects against receiving poisonous bites, and fear of heights, which makes us wary of situations in which we might fall. Such instinctive fears are defined as a normal emotional response to realistic danger. Phobias, on the other hand, are fears that are out of proportion to realistic danger and overgeneralized. What is interesting from an evolutionary perspective is that phobias are not random—they typically involve overblown adaptive responses, such as fears of snakes and spiders, heights, and crowded places. Fear of snakes and spiders was adaptive in our evolutionary past, even though automobiles and guns are far more likely to harm us nowadays (Marks & Nesse, 1994; Ohman & Mineka, 2001). Darwin predicted this when he suggested: "May we not suspect that the fears of children, which are quite independent of experience, are the inherited effects of real dangers during ancient savage time?" (Darwin, 1877, p. 290). Moreover, such fears and phobias emerge in development when they are needed. For instance, fear of heights and fear of strangers emerge at about six months of age, when infants begin to crawl. A field called Darwinian or **evolutionary psychiatry** considers psychopathology to be the adaptive responses of a Stone Age brain to modern times (McGuire & Troisi, 1998; Stevens & Price, 2004).

Empirical Evidence

It is easy to make up stories about how anything might be adaptive, called "just-so stories" after Kipling's book for children that includes whimsical parables like how the elephant got its trunk. The danger is starting with a known phenomenon and working backward to propose an explanation rather than making a prediction whose accuracy is unknown until it is tested (de Waal, 2002). For example, we know that most mammalian fathers are less involved in rearing than mothers. As mentioned earlier, this behavioral difference can be explained in terms of differential parental investment—offspring cost mothers more. But if fathers had happened to be equally invested in the care of their offspring, it could have been argued that this evolved because supportive fathers perpetuate their genes. Evolutionary psychologists try to make testable predictions that tease apart evolutionary and cultural explanations (Buss, 2005; Gangestad, Haselton, & Buss, 2006), although it is difficult to rule out cultural explanations completely (Buller, 2005a, 2005b; Fehr & Fischbacher, 2003; Neher, 2006; Wood & Eagly, 2002). Some of the most interesting predictions are those in which behaviors that we think are pathological reflect adaptations, such as fears, men's aggressive violence, and overeating in a world of fast food.

Quantitative genetic methods, such as twin and adoption designs for addressing the origins of individual differences, are not available to evolutionary analyses. (Remember that showing genetic influence on individual differences in behavior does not imply that species-typical behavior is due to genetic adaptation.) The revolution in DNA analysis has transformed population genetics and other areas of behavioral genetics. DNA can also be used to peer into our species' distant past, for example, showing that between 1 and 4 percent of the genes in modern humans were derived from Neandertals (Green et al., 2010). DNA can also be used to shed light on migrations of ancient peoples and patterns of human genetic diversity, such as tracing the Saxon, Viking, and Celt origins of the British (McKie, 2007; Oppenheimer, 2006; Sykes, 2007). Modern humans are thought to have evolved in Africa over 200,000 years ago. In fact, genomewide data from both contemporary populations and extinct hominids strongly support a single dispersal of modern humans from Africa, followed by two admixture events: one with Neandertals somewhere outside Africa and a second with Denisovans, another possible human species that coexisted with

Neandertals (Stoneking & Krause, 2011). DNA evidence paired with information about glacial cycles and climate oscillations also promise to reveal new stories about human population history (Stewart & Stringer, 2012). However, the DNA revolution has not been fully incorporated into the field of evolutionary psychology. One interesting area in which DNA has been used involves the role of olfaction in the perceived sexual attractiveness of mates (Box 20.1).

In addition to comparisons across species, evolutionary psychologists also consider comparisons between groups within the human species. The most influential data are from different cultures. For example, females were found to be choosier than

---

BOX 20.1 • Mate Preference and the Major Histocompatibility Complex

The major histocompatibility complex (MHC), which is the most gene-dense region of the mammalian genome, plays an important role in the immune system. In humans, the MHC is on chromosome 6 and includes 3.6 million base pairs and 140 highly polymorphic genes—some genes have hundreds of alleles. Why is this gene region so highly polymorphic? The answer is balanced selection: Greater diversity of the MHC makes the immune system more adaptable in its response to infections.

Balanced selection for the MHC is driven by frequency-dependent selection. We prefer mates who are most different from us in MHC alleles (Roberts & Little, 2008; see Havlicek & Roberts, 2009, for a review). We do this on the basis of smell: Specific olfaction neurons in the nose function to detect body odors caused by the MHC locus (Boehm & Zufall, 2006). In a study known as the T-shirt experiment, female and male college students were genotyped for several MHC genes. Male students wore a T-shirt for two consecutive nights. The next day, each female student rated the pleasantness of the odors of the men's T-shirts. T-shirts were rated as more pleasant smelling by a woman when the man's MHC genotype was most different from hers (Wedekind, Seebeck, Bettens, & Paepke, 1995).

Choosing mates who differ from us in MHC genotypes is a type of frequency-dependent selection that leads to greater diversity of MHC genes between parents and thus produces stronger immune systems in offspring (Havlicek & Roberts, 2009). The MHC incompatibility also has been reported to predict the sexual compatibility of a couple. Greater MHC differences within a couple predict greater sexual responsivity of the woman and less attraction to other men (Garver-Apgar, Gangestad, Thornhill, Miller, & Olp, 2006). This association is strongest when the women are fertile during the middle of the menstrual cycle. The *ovulatory-shift hypothesis* proposes that mate preferences of women are generally accentuated when they are fertile (Gangestad, Thornhill, & Garver-Apgar, 2010a).

males in selecting mates in a study of more than 10,000 people in 37 cultures (Buss, 1994a). An early example that began with Darwin involved basic facial expressions of emotions that can be recognized in all cultures, such as expressions of happiness, anger, grief, disgust, and surprise (Eckman, 1973).

Differences between the sexes are often examined. For instance, the hypothesis of differential parental investment predicts that men have evolved adaptations that increase their chances of paternity. Support for this hypothesis comes from data showing that across many cultures, men are much more likely than women to be jealous about signs of sexual infidelity (Buss, 2003). A plausible adaptive explanation is that natural selection shaped sexual jealousy in men as a mechanism to prevent cuckoldry because women always know their child is theirs but men cannot be sure. Describing such sex differences and their plausible adaptive value supports but does not prove the hypothesis that these sex differences are caused by evolved genetic adaptations that differ between the sexes (Harris, 2003).

Consider mate selection. There is a growing body of literature examining mate choice. In fact, more than 75 percent of the research on human sexual selection concerns mate choice (Puts, 2010). Historically, the focus has been on what traits men prefer in women (e.g., high hips-to-waist ratio, facial attractiveness). More recently, evolutionary thinking has provided some new insights into questions about what women prefer in men, either as short-term or long-term partners. Some of these insights are discussed in Box 20.2.

In addition to considering differences between males and females within species, evolutionary psychology often makes comparisons across species, since differences between species can be assumed to have evolved. As mentioned earlier, the theory of differential parental investment (that mothers invest more in their offspring than do fathers) led to the hypothesis that the parent who invests more should be choosier about selecting a mate. If this is an evolved adaptation, we would expect that in most species females will be more discriminating than males in choice of mate because most mothers invest more in offspring than do fathers. Confirming this prediction, females are choosier than males in selecting mates in most species (Buss, 1994b), as noted earlier. Moreover, in the few species in which males invest more than females, males are choosier. For instance, in the pipefish seahorse, the male receives eggs from the female and nurtures them in a kangaroo-like pouch. Male pipefish seahorses are choosier than females in selecting mates. Comparisons of this sort across species support the hypothesis that behavioral differences in maternal and paternal investment are evolved adaptations.

Another example supporting the hypothesis that parental behaviors are in part evolutionary adaptations involves murder. Most murders are committed by family members, a fact that seems to violate the principle of inclusive fitness. Evolutionary psychologists predicted on the basis of inclusive fitness that family murders would primarily involve genetically unrelated stepfathers who harm their stepchildren rather than biological fathers and their own offspring, a hypothesis that has been

BOX 20.2 • Mate Selection for Facial Characteristics

In the 1990s, it was suggested that beauty or physical attractiveness, particularly in women, is a cultural construction (Wolf, 1992). However, several years later, it was hypothesized that beauty is not just "skin deep." Rather, physical appearance may provide cues to female fertility, parental investment, male dominance, and other fitness-related characteristics (Etcoff, 1999).

In the United States and many other developed countries, beauty, fashion, and physical fitness are multibillion-dollar industries. Moreover, because men and women are now much freer to select their mates, preference for the physical attractiveness of potential mates may be one of the primary forces that shape our phenotypes (Puts, 2010). However, is this really what drives human sexual selection?

We discussed an important genomic region, the MHC, in Box 20.1. Because diversity in the MHC genomic region is associated with stronger immune function, as humans, we tend to prefer MHC-dissimilar mates. This dissimilarity within human couples may produce attractive, healthy-looking offspring (Lie, Rhodes, &

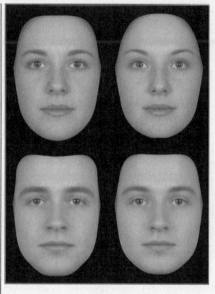

• Examples of masculinized (left) and feminized (right) male and female faces. (With kind permission from Springer Science+Business Media: *Genetica*. "Good genes, complimentary genes and human mate preferences," *134*, 2008, Roberts, SC & Little, AC, Figure 1; and from the authors.)

Simmons, 2008). Moreover, heterozygosity at MHC loci has been associated with facial attractiveness and healthy-looking skin (Lie et al., 2008; Puts, 2010). These

confirmed (Daly & Wilson, 1999; Tooley, Karakis, Stokes, & Ozanne-Smith, 2006), although cultural explanations cannot be completely excluded (Burgess & Drais, 1999).

As a final example, consider morning sickness, which during the first three months of pregnancy includes food aversions in addition to nausea. Rather than thinking about morning sickness as something bad, evolutionists have suggested that it may be an adaptation that prevents mothers from consuming toxins damaging to the developing fetus (Nesse & Williams, 1996). Food aversions during pregnancy typically involve foods that contain toxins, such as alcohol, coffee, and meat, but findings suggest that physical attractiveness, both in men and women, does indeed go much deeper, all the way down to our genes (Etcoff, 1999). In fact, there is a growing body of evidence suggesting that both men and women choose mates partly on "genetic quality" (Puts, 2010; Roberts & Little, 2008).

As another example, consider facial masculinity versus femininity (see the figure). During certain phases of the ovarian cycle, women are more attracted to men who possess a variety of masculinized traits including, but not limited to, facial masculinity (Gangestad, Thornhill, & Garver-Apgar, 2010b). Masculinity versus femininity in adult facial structures seems to be based primarily on the extent to which the jaw and chin are developed and secondarily on the extent to which the brow ridges are developed. It is thought that androgen-dependent masculine traits indicate increased dominance and competitive ability in contests and combat (Puts, Jones, & DeBruine, 2012). These androgen-dependent traits might also be indicators of "good genes." Multiple studies suggest that women prefer more masculine male faces for short-term relationships and relatively feminine faces for long-term relationships (e.g., Little, Jones, Penton-Voak, Burt, & Perrett, 2002). Additionally, when researchers have asked women to attribute personality to certain physical characteristics, they have associated high facial masculinity with high mating and low parenting effort (e.g., Kruger, 2006).

Facial attractiveness is another example of a characteristic of mate selection in both men and women. A meta-analysis of research on facial attractiveness across cultures suggests that averageness, symmetry, and sexual dimorphism set standards of what we consider attractive (Rhodes, 2006). Facial symmetry, a component of facial attractiveness, has also been suggested to be an indicator of "good genes" (Puts, 2010).

Physical appearance is not the only trait that provides cues to mating success. Recent evidence suggests that a good sense of humor is sexually attractive. Perhaps this is because it reveals cues about intelligence, creativity, and other "good genes" or "good parent" traits (Greengross & Miller, 2011).

hardly ever do they include foods that do not contain toxins, such as bread or cereals (Flaxman & Sherman, 2000). Moreover, the food aversions usually disappear after the first three months of pregnancy, which is the most sensitive period of fetal organ development. One piece of evidence in support of this hypothesis is that women who do not have morning sickness during the first trimester are three times more likely to experience a spontaneous abortion than women who do have morning sickness (Profet, 1992).

Evolutionary psychologists have studied many other behavioral adaptations, such as parent-offspring conflict, preference for particular habitats, and cooperation and conflict in groups. As indicated in recent books, evolutionary psychologists tackle big issues such as why we kill (Buss, 2011), why we lie (Smith, 2004), why we think (Gardenfors, 2006; Geary, 2005), why we suffer posttraumatic stress disorder (Cantor, 2005), why we believe in gods (Atran, 2005; Dawkins, 2006; Dennett, 2006; Wolpert, 2007), and public policy implications of evolutionary psychology (Crawford & Salmon, 2004). Books on evolutionary psychology also address entertaining issues, such as why we shop (Saad, 2007), why we like art (Pinker, 2002), why we like literature (Barash & Barash, 2005; Gottschall & Wilson, 2005), and why we should consider having more children (Caplan, 2011). Books by critics of evolutionary psychology have also appeared (Buller, 2005a; Fisher, 2004; Wallace, 2010), and, in response, evolutionary psychologists have attempted to clarify their stance on common criticisms of and concerns about the field (Confer et al., 2010).

Evolutionary psychology originated in the early 1980s, when our knowledge of the human genome was limited (Bolhuis et al., 2011). Recent developments in human genetics, evolutionary biology, cognitive neuroscience, developmental psychology, and paleontology may challenge some of the early concepts proposed by evolutionary psychology (Buss, 2011; Gangestad & Simpson, 2007; Goetz & Shackelford, 2006). Rather than rendering evolutionary psychology outmoded, these developments have led to a call for "a modern evolutionary psychology that will embrace a broader, more open, and multidisciplinary theoretical framework, drawing on, rather than being isolated from, the full repertoire of knowledge and tools available in adjacent disciplines" (Bolhuis et al., 2011, p. 6).

Summary

Charles Darwin's 1859 book on the origin of species convinced the scientific world that species evolved one from the other rather than being created once and for all. Reproductive fitness is the key to natural selection. Gaps in Darwin's theory of evolution occurred because the mechanism for heredity, the gene, was not understood at that time. Darwin's theory has been extended to consider inclusive fitness and kin selection, studies that go beyond his focus on individual reproductive fitness and lead to hypotheses such as differences in parental investment for mothers and fathers as well as kinship theories that address cooperation or competition for resources. Although Darwin noted that natural selection affected behavior as much as bones, evolutionary thinking has only entered the mainstream of the behavioral sciences in recent years.

Population genetics investigates forces that change allelic and genotypic frequencies. Because behavioral genetics focuses on genetic variability within a species, types of natural selection that increase genetic variation in a population are especially interesting, such as balanced polymorphisms due to heterozygote advantage or frequency-dependent selection. However, most advances in computational methods for selection tend to be focused on positive selection, which, if successful, will decrease genetic variation within a certain chromosomal region. Inbreeding and assortative mating change genotypic but not allelic frequencies. Inbreeding reduces genetic variability; assortative mating increases genotypic variability for many behavioral traits.

Evolutionary thinking is becoming increasingly influential in psychology. Most work in evolutionary psychology considers average differences between species on an evolutionary time scale. This level of analysis is different from that of most behavioral genetics, which focuses on contemporary individual differences. Although it is more difficult to test evolutionary genetic hypotheses, support has been found for hypotheses about behavioral adaptations (instincts), such as fears and phobias, and different mating strategies for males and females.

CHAPTER TWENTY-ONE

The Future of Behavioral Genetics

P redicting the future of behavioral genetics is not a matter of crystal ball gazing. The momentum of recent developments makes the field certain to thrive, especially as behavioral genetics continues to flow beyond psychology and psychiatry into the mainstream of research in diverse fields from neuroscience to economics. This momentum is propelled by new findings, methods, and projects, both in quantitative genetics and in molecular genetics.

Another reason for optimism about the continued growth of genetics in the behavioral sciences is that so many more researchers have incorporated genetic strategies into their studies. This trend has grown much stronger now that the price of admission to genetic research is just some saliva from which DNA is extracted, not difficult-to-obtain samples of twins or adoptees. Although caution is also warranted (Chapter 9), this easy access to genetics is important because the best behavioral genetic research is likely to be done by behavioral scientists who are not primarily geneticists. Experts from behavioral domains will focus on traits and theories that are pivotal to those domains and interpret their research findings in ways that will achieve the most impact. As described in the Preface, the goal of this book is to share with you our excitement about behavioral genetics and to whet your appetite for learning about genetics in the behavioral sciences. We hope that this introduction will inspire some readers to contribute to the field. Although we believe that the field of behavioral genetics has made some of the most important discoveries in the behavioral sciences, there is much left to do.

Quantitative Genetics

Following up on the exciting breakthroughs in molecular genetics and epigenetics, quantitative genetics will also continue to make important advances for at least three reasons. First, quantitative genetic methods estimate the cumulative effect of genetic influence regardless of the number of genes involved or the magnitude or complexity of their effects. If we could find all the genes responsible for heritability, there would no longer be any need for quantitative genetic research because genetic influence could be assessed directly from each individual's DNA rather than being assessed indirectly by genetic relatedness, as in twin and adoption studies. However, it seems highly unlikely that most—let alone all—of the genes responsible for the heritability for any complex trait will be identified in the foreseeable future (Chapter 9).

The second reason is that quantitative genetics is as much about the environment as it is about genetics, whereas molecular genetics is fundamentally about genetics. Just as quantitative genetic methods can be used to estimate the cumulative effect of genetic influences without identifying the individual genes involved, these methods can also estimate the cumulative effect of environmental influences without identifying the specific factors that are responsible for the environmental influence. Quantitative genetics can investigate environmental influences while controlling for genetics as well as study genetic influences while controlling for environmental influences (Chapter 8). For this reason, quantitative genetics provides the best available evidence for the importance of the environment in the behavioral sciences (Chapters 11–19). It has also made some of the most important discoveries about how the environment affects behavior. One example is the finding that environmental influences typically operate on an individual-by-individual basis, not generally on a family-by-family basis (Chapter 7). Another example is the finding that many putative environmental measures show substantial genetic influence (Chapter 8).

The third reason is that a completely new quantitative genetic technique has been developed recently that estimates genetic influence from chance genetic similarity among unrelated individuals. Genome-wide complex trait analysis (GCTA), described in Chapter 7, will be increasingly used because it does not require special relatives such as twins or adoptees. Although GCTA requires thousands of individuals genotyped on hundreds of thousands of DNA markers, these are also the requirements of genomewide association analysis (Chapter 9), which means that many studies are available that meet these requirements for GCTA.

The future will no doubt witness the application of quantitative genetic research to other behavioral traits. Behavioral genetics has only scratched the surface of possible applications, even within the domains of cognitive disabilities (Chapter 11), cognitive abilities (Chapters 12 and 13), psychopathology (Chapters 14–16), personality (Chapter 17), and substance abuse (Chapter 18). For example, for cognitive abilities, most research has focused on general cognitive ability and major group factors of specific cognitive abilities. The future of quantitative genetic research in this area lies in more fine-grained analyses of cognitive abilities and in the use of information-processing, cognitive psychology, and neuroimaging approaches to cognition. For psychopathology, genetic research has just begun to consider disorders other than schizophrenia, the major mood disorders, and substance use disorders. Much remains to be learned about disorders in childhood, for example. Approaching psychopathology as quantitative traits rather than qualitative disorders is a major new direction for quantitative genetic research. Personality and substance abuse are such complex domains that they can keep researchers busy for decades, especially as they go beyond self-report questionnaires and interviews to other measures such as neuro-imaging. A rich territory for future exploration is the link between psychopathology and personality.

Cognitive disabilities and abilities, psychopathology, personality, and substance abuse have been the targets for the vast majority of genetic research in the behavioral sciences because these areas have traditionally considered individual differences. Two other areas that are beginning to be explored genetically were described in Chapters 19 and 20: health psychology and evolutionary psychology. Some of the oldest areas of psychology—perception, learning, and language, for example—as well as some of the newest areas of research, such as neuroscience, have not emphasized individual differences and as a result are only beginning to be explored systematically from a genetic perspective. Other disciplines in the social and behavioral sciences are beginning to catch on to genetics, most notably economics, with other fields—such as demography, education, political science, and sociology—sure to follow.

Genetic research in the behavioral sciences will continue to move beyond simply demonstrating that genetic factors are important. The questions *whether* and *how much* genetic factors affect behavioral dimensions and disorders represent important first steps in understanding the origins of individual differences. But these are only first steps. The next steps involve the question *how*—that is, determining the mechanisms by which genes have their effect. How do genetic effects unfold developmentally? What are the biological pathways between genes and behavior? How do nature and nurture interact and correlate? Examples of these three directions for genetic research in psychology—developmental genetics, multivariate genetics, and "environmental" genetics—have been presented throughout the preceding chapters, especially Chapters 8 and 10. The future will see more research of this type as behavioral genetics continues to move beyond merely documenting genetic influence.

Developmental genetic analysis considers change as well as continuity during development throughout the human life span. Two types of developmental questions can be asked. First, do genetic and environmental components of variance change during development? The most striking example to date involves general cognitive ability (Chapter 12). Genetic effects become increasingly important throughout the life span. Shared family environment is important in childhood, but its influence becomes negligible after adolescence. The second question concerns the role of genetic and environmental factors in age-to-age change and continuity during development. Using general cognitive ability again as an example, we find a surprising degree of genetic continuity from childhood to adulthood. However, some evidence has been found for genetic change as well, for example, during the transition from early to middle childhood, when formal schooling begins. Interesting developmental discoveries are not likely to be limited to cognitive development or childhood—it just so happens that most developmental genetic research so far has focused on children's cognitive development, although aging will increasingly be the target for developmental research (Chapter 19).

Multivariate genetic research addresses the covariance between traits rather than the variance of each trait considered by itself. A surprising finding in relation to specific cognitive abilities is that the same genetic factors affect most cognitive abilities and disabilities (Chapter 13). For psychopathology, a key question is why so many disorders co-occur. Multivariate genetic research suggests that genetic overlap between disorders may be responsible for this comorbidity (Chapter 15). Another basic question in psychopathology involves heterogeneity. Are there subtypes of disorders that are genetically distinct? Multivariate genetic research is critical for investigating the causes of comorbidity and heterogeneity as well as for identifying the most heritable constellations (comorbidity) and components (heterogeneity) of psychopathology (Chapters 14–16), an area of inquiry that could impact treatment efforts such as drug design and discovery as well as diagnosis. Another fundamental question is the extent to which genetic and environmental effects on disorders are merely the quantitative extremes of the same genetic and environmental factors that affect the rest of the distribution. Or are disorders qualitatively different from the normal range of behavior? The goal is to test the validity of current symptom-based diagnostic schemes and ultimately to create an etiology-based scheme that recognizes quantitative dimensions as well as qualitative diagnoses.

Another general direction for multivariate genetic research is to investigate the mechanisms by which genetic factors influence behavior by identifying genetic correlations between behavior and biological processes such as those assessed by neuroimaging. It cannot be assumed that the nexus of associations between biology and behavior is necessarily genetic in origin. Multivariate genetic analysis is needed to investigate the extent to which genetic factors mediate these associations.

"Environmental" genetics will continue to explore the interface between nature and nurture. As mentioned earlier, genetic research has made some of the most important discoveries about the environment in recent decades, especially nonshared environment and the role of genetics in experience (Chapter 6). One of the major challenges for behavioral genetics is to identify the specific environmental factors responsible for the widespread influence of nonshared environment. MZ twins provide an especially sharp scalpel to dissect nonshared environment because MZ co-twins differ only for reasons of nonshared environment. An even broader topic is understanding how genes influence experience, which is part of the biggest question of all: How do genetic and environmental influences covary and interact to influence behavior? More discoveries about environmental mechanisms can be predicted, as the environment continues to be investigated in the context of genetically sensitive designs. New multivariate quantitative genetic methods have recently been developed that aim to distinguish environmental causation from correlation (Chapter 8). Much remains to be learned about interactions and correlations between nature and nurture.

In summary, no crystal ball is needed to predict that quantitative genetic research will continue to flourish as it turns to other areas of behavior and, especially, as it goes beyond the rudimentary questions of *whether* and *how much* to ask the question *how*. Such research will become increasingly important as it guides molecular genetic research to the most heritable components and constellations throughout the human life span as they interact and correlate with the environment. In return, developmental, multivariate, and "environmental" behavioral genetics will be transformed by molecular genetics.

Molecular Genetics

Molecular genetics has begun to revolutionize behavioral genetic research by identifying some of the specific genes that contribute to genetic variance for complex dimensions and disorders. The quest is to find not *the* gene for a trait, but rather the multiple genes (quantitative trait loci, QTLs) that are associated with the trait in a probabilistic rather than a predetermined manner. The breathtaking pace of molecular genetics (Chapter 9) leads us to predict that behavioral scientists will increasingly use DNA markers as a tool in their research to identify the relevant genetic differences among individuals. Even if the DNA markers individually predict only a small amount of variance of a trait, they could be incorporated as a set into any research that considers individual differences without the need for special family-based samples such as twins or adoptees. This is already happening in research on dementia and cognitive decline in the elderly and, increasingly, in prevention research. It is now standard practice for research in this area to take advantage of the genetic risk information provided by the DNA marker for apolipoprotein E (Chapter 11), even when researchers are interested primarily in psychosocial risk mechanisms. Aiding this prediction that behavioral scientists will routinely use DNA in their research is the fact that DNA is inexpensive to obtain and DNA microarrays make genotyping increasingly inexpensive, even for complex traits for which hundreds or thousands of DNA markers are genotyped.

To answer questions about how genes influence behavior, nothing can be more important than identifying specific genes responsible for the widespread genetic influence on behavior. Although one of the most far-reaching challenges will be to solve the missing heritability problem (Chapter 9), it seems unlikely that all of the missing heritability will be found to be due to common DNA variants with very small individual effects. Nevertheless, we predict that polygenic predictors using DNA markers will account for more of the genetic variance of behavioral traits than current family-based predictions that use information from first-degree relatives. Moreover, polygenic predictors using DNA markers would have the distinct advantage of making predictions for specific individuals rather than a general prediction for all members of a family. Such polygenic predictions will eventually transform quantitative genetic research, especially when whole-genome sequencing becomes more available, and will take the developmental, multivariate, and gene-environment interplay issues discussed throughout this book to the next level.

Even if polygenic predictors need hundreds or thousands of genes to reach the target, they can be used to begin to understand links among the genome, epigenome, transcriptome, proteome, neurome, and eventually behavior (Chapter 10). In contrast to bottom-up functional genomics research, top-down behavioral genomics research is likely to pay off more quickly in terms of prediction, diagnosis, intervention, and prevention of behavioral disorders. Behavioral genomics represents the long-term future of behavioral genetics, when we are likely to have polygenic predictors that account for some of the ubiquitous genetic influence on behavioral dimensions and disorders. Bottom-up functional genomics will eventually meet top-down behavioral genomics in the brain. The grandest implication for science is that DNA will serve as a common denominator integrating diverse disciplines. Clinically, polygenic predictors will be key to personalized genomics, which hopes to predict risk, identify treatment interactions, and propose interventions to prevent problems before they appear. A particularly promising area is the prediction of responses to drug treatments.

As indicated in Chapter 9, it has been predicted that in the next few years, rather than screening newborns for just a few known genetic mutations like phenylketonuria, we will sequence all of the 3 billion nucleotide base pairs of their genomes. Sequencing whole genomes will yield all DNA variants. We predict that some of these variants will be altogether different from traditional ones, deletions and duplications of long stretches of DNA being a recent example (Chapter 9); such rare DNA variants may have relatively large effects that will account for at least some of the "missing heritability." When whole-genome sequences become available, it will cost little to use this information. The promise and problems of these developments were discussed in the section on genetic counseling in Chapter 19. The impact on behavioral genetics is that this same whole-genome sequence information would also be available for use in behavioral research.

One of the great strengths of DNA analysis is that it can be used to predict risk long before a disorder appears. This predictive ability will allow research on interventions that can prevent the disorder rather than trying to reverse a disorder once it appears and has already caused collateral damage. Molecular genetics may also eventually lead to personalized genomics—individualized gene-based diagnoses and treatment programs.

For these reasons, it is crucial that behavioral scientists be prepared to take advantage of the exciting developments in molecular genetics. In the same way that we now assume that computer literacy is an essential goal to be achieved during elementary and secondary education, students in the behavioral sciences must be taught about genetics in order to prepare them for this future. Otherwise, this opportunity for behavioral scientists will slip away by default to geneticists, and genetics is much too important a topic to be left to geneticists! Clinicians use the acronym "DNA" to note that a client "did not attend"—it is critical to the future of the behavioral sciences that DNA mean deoxyribonucleic acid rather than "did not attend."

Implications of Nature and Nurture

The controversy that swirled around behavioral genetic research during the 1970s, especially in relation to cognitive abilities (Chapter 12), has largely faded, as indicated, for example, by the increase in publications in mainstream journals (Chapter 1). Indeed, the acceptance of genetic influence in the behavioral sciences is growing into a tidal wave that threatens to engulf the second message coming from behavioral genetic research. The first message is that genes play a surprisingly important role across all behavioral traits. The second message is just as important: Individual differences in complex behavioral traits are due at least as much to environmental influences as they are to genetic influences.

The first message will become more prominent during the next decade as more genes are identified that contribute to the widespread influence of genetics in the behavioral sciences. As explained in Chapter 7, it should be emphasized that genetic effects on complex traits describe *what is*. Such findings do not predict *what could be* or prescribe *what should be*. Genes are not destiny. Genetic effects on complex traits represent probabilistic propensities, not predetermined programming. A related point is that, for complex traits such as behavioral traits, QTL effects refer to average effects in a population, not to a particular individual. For example, one of the strongest DNA associations with a complex behavioral disorder is the association between allele 4 of the gene encoding apolipoprotein E and late-onset dementia (Chapter 11). Unlike simple single-gene disorders, this QTL association does not mean that allele 4 is necessary or sufficient for the development of dementia. Many people with dementia do not have the allele, and many people with the allele do not have dementia. A particular gene may be associated with a large average increase in risk for a disorder, but it is likely to be a weak predictor at an individual level. The importance of this point concerns the dangers of labeling individuals on the basis of population averages.

The relationship between genetics and equality is an issue that lurks in the shadows, causing a sense of unease about genetics. The main point is that finding genetic differences among individuals does not compromise the value of social equality. The essence of a democracy is that all people should have legal equality *despite* their genetic differences. Knowledge alone by no means accounts for societal and political decisions. Values are just as important as knowledge in the decision-making process. Decisions, both good and bad, can be made with or without knowledge. Nonetheless, scientific findings are often misused, and scientists, like the rest of the population, need to be concerned with reducing such misuse. We firmly believe, however, that better decisions can be made with knowledge than without. There is nothing to be gained by sticking our heads in the sand and pretending that genetic differences do not exist.

Finding widespread genetic influence creates new problems to consider. For example, could evidence for genetic influence be used to justify the status quo? Will people at genetic risk be labeled and discriminated against? As genetic variants are found that predict behavioral traits, will parents use them prenatally to select "designer" children? (See Chapter 19.) New knowledge also provides new opportunities. For example, identifying genes associated with a particular disorder could make it more likely that environmental preventions and interventions that are especially effective for the disorder can be found. Knowing that certain children have increased genetic risk for a disorder could make it possible to prevent or ameliorate the disorder before it appears, rather than trying to treat the disorder after it appears and causes other problems. Moreover, it should not be assumed that once a gene associated with some disorder is found, the logical next step is to get rid of it. For example, genes that persist in the population may be the result of stabilizing selection (Chapter 20), which might mean that the genes have good as well as bad effects.

Two other points should be made in this regard. First, most powerful scientific advances create new problems. For example, consider prenatal screening for genetic defects. This advance has obvious benefits in terms of detecting chromosomal and genetic disorders before birth. Combined with abortion, prenatal screening can relieve parents and society of the tremendous burden of severe birth defects. However, it also raises ethical problems concerning abortion and creates the possibility of abuses, such as compulsory screening and mandatory abortion. Despite the problems created by advances in science, we would not want to cut off the flow of knowledge and its benefits in order to avoid having to confront such problems.

The second point is that it is wrong to assume that environmental explanations are good and that genetic explanations are dangerous. Tremendous harm was done by the environmentalism that prevailed until the 1960s, when the pendulum swung back to a more balanced view that recognized genetic as well as environmental influences. For example, environmentalism led to blaming children's problems on what their parents did to them in the first few years of life. Imagine that, in the 1950s, you were among the 1 percent of parents who had a child who became schizophrenic in late adolescence. You faced a lifetime of concern. And then you were told that the schizophrenia was caused by what you did to the child in the first few years. The sense of guilt would be overwhelming. Worst of all, such parent blaming was not correct. There is no evidence that early parental treatment causes schizophrenia. Although the environment is important, whatever the salient environmental factors might be, they are not shared family environmental factors. Most important, we now know that schizophrenia is substantially influenced by genetic factors and individual-specific environmental factors.

Our hope for the future is that the next generation of behavioral scientists will wonder what the nature-nurture fuss was all about. We hope they will say, "Of course, we need to consider nature and nurture to understand behavior." The conjunction between nature and nurture is truly *and*, not *versus*.

The basic message of behavioral genetics is that each of us is an individual. Recognition of, and respect for, individual differences is essential to the ethic of individual worth. Proper attention to individual needs, including provision of the environmental circumstances that will optimize the development of each person, is a utopian ideal and no more attainable than other utopias. Nevertheless, we can approach this ideal more closely if we recognize, rather than ignore, individuality. Acquiring the requisite knowledge regarding the genetic and environmental etiologies of individual differences in behavior warrants a high priority because human individuality is the fundamental natural resource of our species.

Statistical Methods in Behavioral Genetics

Shaun Purcell

1 Introduction

Quantitative genetics offers a powerful theory and various methods for investigating the genetic and environmental etiology of any characteristic that can be measured, including both continuous and discrete traits. As discussed in Chapter 9, quantitative genetics and molecular genetics are coming together in the study of complex quantitative traits. In both fields, powerful statistical and epidemiological methods have been developed to address a series of related questions:

- Do genes influence this outcome?
- What types of genetic effects are at work?
- Can genetic effects explain the relationships between this and other outcomes?
- Where are the genes located?
- What specific form(s) of the genes cause certain outcomes?
- Do genetic effects operate similarly across different populations and environments?

This Appendix introduces some of the methods behind these research questions, in a manner designed to provide the rationale behind the methods as well as an appreciation of the directions in which the field is developing, including molecular genetics. Both quantitative genetics (with an emphasis on the components of variance model-fitting approaches to complex traits) and molecular genetics (with an emphasis on linkage and association approaches to gene mapping) are covered.

We begin with a brief overview of some of the statistical tools that are commonly used in behavioral genetic research: variance, covariance, correlation, regression, and matrices. Although one need not be a fully trained statistician to use most behavioral

BOX A.1 • Behavioral Genetic Interactive Models

The *Behavioral Genetic Interactive Modules* are a series of freely available interactive computer programs with accompanying textual guides designed to convey a sense of the methods of modern behavioral genetic analysis to students and researchers new to the field. Currently, 11 modules covering the material in this Appendix can be accessed from the website at http://pngu.mgh.harvard.edu/purcell/bgim/. Taken together, the modules listed below lead from the basic statistical foundations of quantitative genetic analysis to an introduction to some of the more advanced analytical techniques.

Variance is designed to introduce the concept of variance: what it represents, how it is calculated, and how it can be used to assess individual differences in any quantitative trait. Standardized scores are also introduced.

Covariance demonstrates how the covariance statistic can be used to represent association between two measures.

Correlation & Regression is an exploration of the relationship among variance, covariance, correlation, and regression coefficients.

Matrices provides a simple matrix calculator.

Single Gene Model introduces the basic biometrical model used to describe the effects of individual genes, in terms of additive genetic values and dominance deviations.

Variance Components: ACE illustrates the partitioning of variance into additive genetic, shared environmen-

tal, and nonshared environmental components in the context of MZ and DZ twins.

Families demonstrates the relationship between additive and dominance genetic variance, shared and nonshared environmental variance, and expected familial correlations for different types of relatives.

Model Fitting 1 defines a simple path diagram to model the covariance between observed variables and allows the user to manually adjust path coefficients to find the best-fitting model; it includes a twin ACE model and nested models that can be compared with the full ACE model.

Model Fitting 2 performs a maximum-likelihood analysis of univariate twin data and presents the parameter estimates for nested submodels.

Multivariate Analysis models the genetic and environmental etiology of two traits.

Extremes Analysis illustrates DF extremes analysis as well as individual differences analysis, in order to explore how these two methods can inform us about links between normal variation and extreme scores.

For individuals wishing to take their study of statistical analysis further, a guide is provided to help you get started on analyzing your own data as well as simulated data sets that can be used to explore these methods further. Behavioral genetic analyses using widely available statistics packages such as *Stata* are described, as well as an introduction to *Mx*, a powerful, freely available model-fitting package by Mike Neale.

About the Author

Shaun Purcell develops statistical and computational tools for the design of genetic studies, the detection of gene variants influencing complex human traits, and the dissection of these effects within the larger context of other genetic and environmental factors. He is currently an associate professor at Mount Sinai School of Medicine, in New York, and is on the faculty at Harvard Medical School, based at the Analytic and Translational Genetics Unit, Massachusetts General Hospital. He is also an associate member of the Broad Institute of Harvard and MIT, and the Stanley Center for Psychiatric Research. As an undergraduate from 1992 to 1995, he studied experimental psychology at Oxford University; in 1996, he had the opportunity to develop an interest in statistical methods while working toward a master's of science degree at University College London. In 1997, he joined the Social, Genetic and Developmental Psychiatry (SGDP) Research Centre at the Institute of Psychiatry in London, to embark on a Ph.D. with Pak Sham and Robert Plomin, working on a project designed to map quantitative trait loci for anxiety and depression. His current work involves whole genome association and whole exome sequencing studies of bipolar disorder and schizophrenia, and the development of statistical and computational tools for such studies.

genetic methods, understanding the main statistical concepts that underlie quantitative genetic research enables one to appreciate the ideas, assumptions, and limitations behind the methods.

Next, the classical quantitative genetic model is introduced, which relates the properties of a single gene to variation in a quantitative phenotype. This relatively simple model forms the basis for the majority of quantitative genetic methods. We then examine how the analysis of familial correlations can be used to infer the underlying etiological nature of a trait, given our knowledge of the way genes work. The basic model partitions the variance of a single trait into portions attributable to additive genetic effects, shared environmental effects, and nonshared environmental effects. The tools of model fitting and path analysis are introduced in this context. Extensions to the basic model are also considered: multivariate analysis, analysis of extremes, and interactions between genes and environments, for example.

Finally, we see how molecular genetic information on specific loci can be incorporated. In this way, the chromosomal positions of genes can be mapped. This work leads the way to the study of gene function at a molecular level—the vital next step if we really want to know *how* our genes make us what we are.

1.1 Variation and Covariation: Statistical Descriptions of Individual Differences

Behavioral genetics is concerned with the study of individual differences: detecting the factors that make individuals in a population different from one another. As a first step, it is concerned with gauging the relative importance of genetic and environmental factors that cause individual differences. To assess the importance of these factors, we need to be able to *measure* individual differences. This task requires some elementary statistical theory.

A population is defined as the complete set of all individuals in a group under study. Examples of populations would include sets such as all humans, all female Americans aged 20 to 25 in the year 2000, or all the stars in a galaxy. We might measure a characteristic, such as talkativeness, intelligence, weight, or temperature, for each of the individuals in a population. We are concerned with assessing how these characteristics vary both *within* populations (e.g., among 2-year-old males) and *between* them (e.g., male versus female infants).

If all the individuals in a set are studied, population statistics such as the average or the variance can be calculated exactly. However, it is usually not practical to measure every individual in the population, so we resort to *sampling* individuals from the population. A key concept in sampling is that, ideally, it should be conducted at *random*. A nonrandom sample, such as only the tallest 20 percent of 11-year-old girls, would give an inflated (biased) estimate of the average height of 11-year-old girls. An estimate of the average height in the population gathered from a random sample would not, *on average*, be biased. However, it is important to recognize that an estimate of the population mean made on the basis of a random sample will vary somewhat from the population mean. The amount of this variation will depend on the sample size and on chance. We need to know how much we expect this variation to be so that we know how accurate estimates of the population parameters are. This assessment of accuracy is critical when we want to compare populations.

Once we have defined a population, various *parameters* such as the mean, range, and variance can be described for the trait that we wish to study. Similarly, when we have a sample of the population, we can calculate *statistics* from the sample that correspond to the parameters of the population. It is not always the case that the measure of the sample statistic is the best estimate of the corresponding population parameter. This discrepancy distinguishes descriptive from inferential statistics. Descriptive statistics simply describe the sample; inferential statistics are used to get estimates of the parameters of the entire population.

1.1.1 The mean The arithmetic mean is one of the simplest and most useful statistics. It is a measure of the center of a distribution and is the familiar average statistic used in everyday speech. It is very simple to compute, being the sum of all observations' values divided by the number of observations in the sample:

$$\mu = \Sigma x/N$$

where Σx is the sum of all observations in the set of size N. Strictly speaking, the mean is only labeled μ (pronounced "mu") if it is computed from the entire population. Usually, the mean will be calculated from a sample, and the mean of a variable, say x, is written as $\bar{x}$ (read as "x bar").

The mean is especially useful when comparing groups. Given an estimate of how accurate the means are, it becomes possible to compare means for two or more groups. Examples might include whether women are more verbally skilled than men, whether albino mice are less active than other mice, or whether light moves faster than sound.

Some physical measures, such as the number of inches of rainfall per year, are obviously well ordered, such that the difference between 15 and 16 inches is the same as the difference between 21 and 22 inches—namely, 1 inch. Many physical measurements have the same scale throughout the distribution, which is called an *interval* scale. In behavioral research, however, it is often difficult to get measures that are on an interval scale. Some measures are *binary*, consisting simply of the presence or absence of a disease or symptom. The mean of a binary variable scored 0 for absent and 1 for present indicates the proportion of the sample that has the symptom or disease present, so again the mean is a useful summary. However, not all measures can be effectively summarized as means. The trouble starts when there are several ordered categories, such as "Not at all/Sometimes/Quite often/Always." Even if these items are scored 0, 1, 2, and 3, the mean tells us little about the frequencies in each category. This problem is even greater when the categories cannot be ordered, such as religious affiliation. Here the mean would be of no use at all.

1.1.2 Variance Variance is a statistic that tells us how spread out scores are. This is a measure of individual differences in the population, the focus of most behavioral genetic analyses. Variances are also important when assessing differences between group means. Behavioral genetic analyses are typically less focused on group differences, although such analyses are central to most quantitative sciences. For example, a researcher may wish to ask whether a control group significantly differs from an experimental one on a measure, or whether boys and girls differ in the amount that they eat. Testing differences between means is often carried out with a statistical method called the analysis of variance (ANOVA). In fact, individual differences are treated as the "error" term in ANOVA.

The usual approach to calculating the variance, established by R. A. Fisher (1922), one of the founders of quantitative genetics, is to take the average of the squared deviations from the mean. Fisher showed that the squared deviations from the mean had more desirable statistical properties than other measures of variance that might be considered, such as the average absolute difference. In particular, the average squared deviation is the most accurate statistic.

Calculating the variance (often written as s^2) is straightforward:

1. Calculate the mean.
2. Express the scores as deviations from the mean.
3. Square the deviations and sum them.
4. Divide the sum by the number of observations minus 1.

Or, written as a formula:

$$s^2 = \frac{\sum(x - \bar{x})^2}{N - 1}$$

A second commonly used approach involves computing the contribution of each observation to the variance and correcting for the mean at the end. This alternative method produces the same answer, but it can be more efficient for computers to use. Note that $N - 1$ instead of N is used to calculate the average squared deviation in order to produce unbiased estimates of variance—for technical, statistical reasons.

Variances range from zero upward: There is no such thing as a negative variance. A variance of zero would indicate no variation in the sample (i.e., all individuals would have to have exactly the same score). The greater the spread in scores, the greater the variance.

With binary "yes/no" or "affected/unaffected" traits, measuring the variance is difficult. We may imagine that a binary trait is observed because there is an underlying normal distribution of *liability* to the trait, caused by the additive effects of a large number of factors, each of small effect. The binary trait that we observe arises because there is a threshold, and only those with liability above threshold express the trait. We cannot directly observe the underlying liability, so typically we assume that it has variance of unity (1). If the variance of the underlying distribution were increased, it would simply change the proportion of subjects that are above threshold. That is, changing the variance is equivalent to changing the threshold. Distinguishing between mean changes and variance changes is not generally possible with binary data, but it is possible if the data are ordinal, with at least three ordered categories.

Having measured the variance, quantitative genetic analysis aims to partition it—that is, to divide the total variance into parts attributable to genetic and environmental components. This task requires the introduction of another statistical concept, covariance. Before turning to covariance, we will take a brief digression to consider another way of expressing scores that facilit*ates comparisons of means and variances.

1.1.3 Standardized scores Different types of measures have different scales, which can cause problems when making comparisons between them. For example, differences in height could be expressed in either metric or common (imperial) terms. In a population, the absolute value of variance in height will depend on the scale used to measure it—the unit of variance will be either centimeters squared or inches squared. If we take the square root of variance, we obtain a measure of spread that has the same unit of the observed trait, called the *standard deviation (s)*. The standard deviation also has several convenient statistical properties. If a trait is normally distributed (bell-shaped curve), then 95 percent of all observations will lie within two standard deviations on either side of the mean.

The example of measuring height demonstrates the difficulties that may be encountered when we wish to compare differently scaled measures. In the case of metric and common measurements of height, which both measure the same thing, the problem of scale can be easily overcome by using standard conversion formulas. In psychology, however, measurements often will have no fixed scale. A questionnaire that measures extraversion might have a scale from 0 to 12, from 1 to 100, or from −4 to +4. If scale is arbitrary, it makes sense to make all measures have the same, standardized scale.

Suppose we have data on two reliable questionnaire measures of extraversion, *A* and *B*, each from a different population. Say measure *A* has a range of 0 to 12 and a mean score of 6.4, whereas measure *B* has a range of 0 to 50 with a mean of 24. If we were to assess two individuals, one scoring 8 on measure *A* and the other scoring 30 on measure *B*, how could we tell which person is the more extraverted? The most commonly used technique is to *standardize* our measures. The formula for calculating a standardized score *z* from a raw score *x* is

$$z = \frac{x - \bar{x}}{\sqrt{s_x^2}}$$

where s_x^2 is the variance of *x*. That is, we reexpress the scores in standard deviation units. For example, if we calculate that measure *A* has a variance of 4, then the standard deviation is $\sqrt{4} = 2$. If we express scores as the number of standard deviations away from the mean, then a score of 2 raw-score units above the mean on measure *A* is +1 standard deviation units. Raw scores equaling the raw-score mean will become 0 in standard deviation units. A raw score of 2 will become $(2 - 6.4)/2 = -2.2$. Therefore, a score of 8 on measure *A* corresponds to a standardized score of $(8 - 6.4)/2 = 0.8$ standard deviation units above the mean.

We can also do the same for measure B, to be able to make scale-independent comparisons between our two measures of extraversion. If measure B is found to have a variance of 8 (and therefore a standard deviation of $\sqrt{8}$), then a raw score of 30 corresponds to a standardized score of $(30 - 24)/\sqrt{8} = 2.1$. We can therefore conclude that individual B is more extraverted than individual A (i.e., 2.1 > 0.8) (Figure A.1, next page). Converting the measures into standardized scores also allows statistical tests of the significance of such differences (the *z*-test).

Standardized scores are said to have *zero sum* property (they will always have a mean of 0) and unit standard deviation (i.e., a standard deviation of 1). As we have seen, standardizing is useful when comparing different measures of the same thing. Indeed, standardizing can be used to compare different measures of different things (e.g., whether a particular individual is more extreme in height or in extraversion).

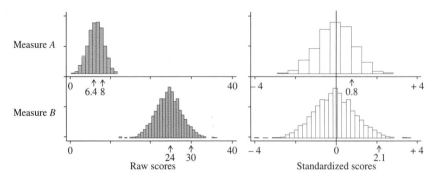

• **FIGURE A.1** Standardized scores. Raw scores on the two measures cannot be directly equated. Standardizing both measures to have a mean of 0 and a standard deviation of 1 facilitates the comparison of measures A and B.

However, there are some situations in which standardized scores can be misleading. Standardizing within groups (i.e., using the estimates of the mean and standard deviation from that group) will destroy between-group differences. All groups will end up with means of zero, which will hide any true between-group variation. Note that it was implicit in the example above that measures A and B are both reliable, and that the two populations are equivalent with respect to the distribution of "true" extraversion.

1.1.4 Covariance Another fundamental statistic that underlies behavioral genetic theory is *covariance*. Covariance is a statistic that informs us about the relationship between two characteristics (e.g., height and weight). Such a statistic is called a *bivariate* statistic, in contrast to the mean and variance, which are both *univariate* statistics. If two variables are associated (i.e., they *covary* together), we may have reason to believe that this covariation occurs because one characteristic influences the other. Alternatively, we might suspect that both characteristics have a common cause. Covariance, by itself, however, cannot tell us *why* two variables are associated: It is only a measure of the magnitude of association. Figure A.2 shows four possible relationships between two variables, X and Y, each of which could result in a similar covariance between the two variables. For example, it is clearly wrong to think of an individual's weight as *causing* his or her height, whereas it is fair to say that an individual's height does, in part, determine that person's weight—it should be noted that care is needed in the interpretation of *all* statistics. The methods of path analysis (as reviewed later) do offer an opportunity to begin to "tease apart" causation from "mere" correlation, especially when applied to data sets that differ in genetic or environmental factors.

A sensible first step when investigating the relationship between two continuous variables is to begin with a *scatterplot*. The scatterplot shown in Figure A.3 represents 200 observations. In this example, it is apparent that the two measures are not independent. As X increases (the scale for X increases toward the right), we see that the

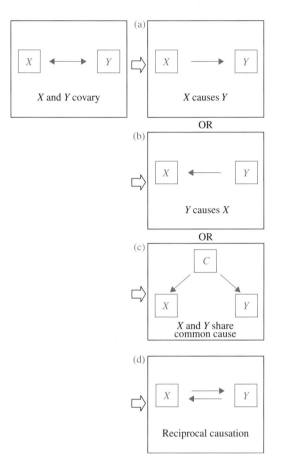

• **FIGURE A.2** Causes of covariation. Two variables can covary for a number of reasons: (a, b) One variable might cause the other, or (c) both variables might be influenced by a third variable (C), or (d) both variables might influence the other. The covariance statistic cannot by itself discriminate among these alternatives.

scores on Y also tend to increase. Covariance is a measure that attempts to quantify this kind of relationship (as do *correlation* and *regression coefficients*, introduced later).

Calculating the covariance proceeds in much the same way as calculating the variance. However, instead of squaring the deviations from the mean, we calculate

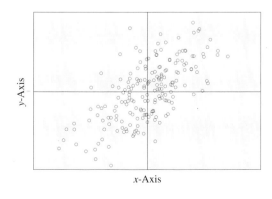

• **FIGURE A.3** Scatterplot representing 200 observations measured on two variables, X and Y. As can be seen, X and Y are not independent, because observations with higher values for X also tend to have higher values for Y.

the cross-product of the deviations of the first variable with those of the second. To compute the covariance, we would

1. Calculate the mean of X.
2. Calculate the mean of Y.
3. Express the scores as deviations from the means.
4. Calculate the product of the deviations for each data pair and sum them.
5. Divide by $N - 1$ to obtain an estimate of the covariance.

Written as a formula, the covariance is

$$\text{Cov}_{XY} = \frac{\sum (X - \bar{X})(Y - \bar{Y})}{N - 1}$$

Covariance values can range between plus and minus infinity. Negative values imply that high scores on one measure tend to be associated with low scores on the other measure. A covariance of 0 implies that there is no *linear* relationship between the two measures.

That covariance measures only linear association is an important issue: Consider the two scatterplots in Figure A.4. Neither of these two bivariate data sets displays any *linear* association between the two variables, so both have a covariance of zero. However, there is a clear difference between the two data sets: in one, the observations are truly *independent*, whereas it is clear that the variables in the other are related but not in a linear way.

A key to understanding covariance is to understand what the formula for its calculation is really doing. Figure A.5 represents the four quadrants of a scatterplot. The lines intersecting in the middle represent the mean value for each variable. When the scores are expressed as deviations from the mean, all those to the left of the vertical line (or below the horizontal line) will become negative; all values to the right of the vertical line (or above the horizontal line) will become positive. As we have seen,

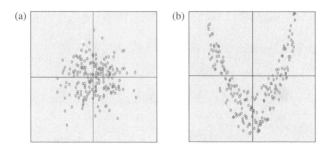

(a)   (b)

• **FIGURE A.4** Covariance and independence. The covariance statistic represents linear association. Both scatterplots represent data sets with a covariance of zero. (a) The two variables in this data set are truly independent; that is, the average value of one variable is independent of the value of the other. (b) The variables in this data set are not linearly related, but they are clearly not independent.

$(-,+) \rightarrow -$	$(+,+) \rightarrow +$
$(-,-) \rightarrow +$	$(+,-) \rightarrow -$

• **FIGURE A.5** Calculating covariance. The contribution each observation makes to the covariance will depend on the quadrant in which it falls.

covariance is calculated by summing the products of these deviations. Therefore, because both the product of two positive numbers and the product of two negative numbers are positive whereas the product of one positive and one negative number is always negative, the contribution each observation makes to the covariance will depend on which quadrant it falls in. Observations in the top-right and bottom-left quadrants (both numbers above the mean and both numbers below the mean, respectively) will make a positive contribution to the covariance. The farther away from the origin (the bivariate point where the two means intersect), the larger this contribution will be. Observations in the other two quadrants will tend to decrease the covariance. If all bivariate data points were evenly distributed across this space, the positive contributions to the covariance would tend to be canceled out by an equal number of negative contributions, resulting in a near zero covariance statistic. A large positive covariance would imply that the bulk of data points fall in the bottom-left and top-right quadrants; a large negative covariance would imply that the bulk of data points fall in the top-left and bottom-right quadrants.

1.1.5 Variance of a sum Covariance is also important for calculating the variance of a sum of two variables. This statistic is relevant to our later discussion of the basic quantitative genetic model. Say you have variables X and Y and you know their variances and the covariance between them. What would the variance of $(X + Y)$ be? If all the data are available, you may decide to calculate a new variable that is the sum of the two variables and then calculate its variance in the ordinary manner. Alternatively, if you know only the summary statistics, you can use the formula:

$$Var(X + Y) = Var(X) + Var(Y) + 2Cov(X, Y)$$

In other words, the variance of a sum is the sum of the two variances plus twice the covariance between the two measures. If two variables are uncorrelated, then the covariance term will be zero and the variance of the sum is simply the sum of the variances. As will be seen later, the mathematics of variance is critical in the formulation of the genetic model for describing complex traits.

1.1.6 Correlation and regression We have seen how using standardized scores can help when working with measures that have different scales. When creating a standardized score, we use information about the variance of a measure to rescale the raw data. As mentioned earlier, the covariance between two measures is dependent on the scales of the raw data and can range from plus to minus infinity. We can use information about the variance of two measures to standardize their covariance statistic, in a manner analogous to creating standardized scores. A covariance statistic standardized in this way is called a *correlation.*

The correlation is calculated by dividing the covariance by the square root of the product of the two variances for each measure. Therefore, the correlation between X and Y (r_{XY}) is

$$r_{XY} = \frac{\text{Cov}_{XY}}{\sqrt{s_X^2 s_Y^2}}$$

where Cov_{XY} is the covariance and s_X^2 and s_Y^2 are the variances. If both X and Y are standardized variables (i.e., s_X and s_Y, and therefore also s_X^2 and s_Y^2, both equal 1), then the correlation will be the same as the covariance (as can be seen in the formula above).

Correlations (typically labeled r) always range from $+1$ to -1. A correlation of $+1$ indicates a perfect positive *linear* relationship between two variables. A correlation of -1 represents a perfect negative linear relationship. A correlation of 0 implies no linear relationship between the two variables (in the same way that a covariance of 0 implies no linear relationship). The kind of correlations we might expect to observe in the real world are likely to fall somewhere between 0 and $+1$. How exactly do we interpret correlations of intermediate values? Does, for example, a correlation of 0.4 mean that the two measures are the same 40 percent of the time? In short, no. What it reflects, as seen in the equation above, is the proportion of variance that is shared by the two measures. (The square of a correlation, r^2, is a commonly used statistic that indicates the proportion of variance in one variable that can be predicted by the other. For correlations between relatives, the unsquared correlation, representing the proportion of variance common to both family members, is more useful.)

Regression is related to correlation in that it also examines the relationship between two variables. Regression is concerned with *prediction* in that it asks whether knowing the value of one variable for an individual helps us to guess what the value of another variable will be.

Regression coefficients (often called b) can be calculated by using a method similar to that used to calculate correlation coefficients. The regression coefficient of "y on x" (i.e., given X, what is our best guess for the value of Y) divides the covariance between X and Y by the variance of the variable (X) from which we are making the prediction (rather than standardizing the covariance by dividing by the product of the standard deviations of X and Y):

$$b = \frac{\text{Cov}_{XY}}{s_X^2}$$

Given this regression coefficient, an equation relating X and Y can be written:

$$\hat{Y} = bX + c$$

where c is called the regression constant. As plotted in Figure A.6, this equation describes a straight line (the *least squares regression line*) that can be drawn through the observed points and represents the best prediction of Y given information on X ($\hat{Y}$, pronounced "y hat"). The equation implies that for an increase of one unit in X, Y will increase an average of b units.

Regression equations can also be used to analyze more complicated, nonlinear relationships between two variables. For example, the variable Y might be a function of the square of X as well as of X itself. We would therefore include this higher-order term in the equation to describe the relationship between X and Y:

$$\hat{Y} = b_1 X^2 + b_2 X + c$$

This equation describes a nonlinear least squares regression line (i.e., a parabolic curve if b_1 doesn't equal zero).

It is possible to calculate the discrepancy, or error, between the predicted values of Y given X and the actual values of Y observed in the sample ($Y - \hat{Y}$). These discrepancies are called the *residuals*, and it is often useful to calculate the variance of

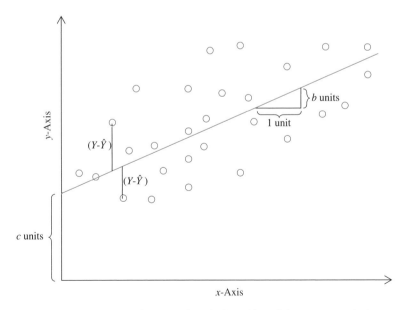

• **FIGURE A.6** Linear regression. The regression of a line of best fit between X and Y is represented by the equation $\hat{Y} = bX + c$. For each unit increase in X, we expect Y to increase b units. The two vertical lines represent the deviations between the expected and actual values of Y. The sum of the deviations squared is used to calculate the residual variance of Y, that is, the variance in Y not accounted for by X. The regression constant c represents the value of Y when X is zero (at the origin).

the residuals. From the first regression equation given above, if X and Y were totally unrelated, then b would be estimated near zero and c would be the mean of Y (because this value represents the *best guess* of Y if you don't have any other information). In this case, residual error variance would be the same as the variance of Y. To the extent that knowing X actually does help you guess Y, the regression coefficient will become significantly nonzero and the error term will decrease.

We can partition the variance in a variable, Y, into the part that is associated with another variable X and the part that is independent of X. In terms of a regression of Y on X, this partitioning is reflected in the variance of the predicted Y values (the variance of $\hat{Y}$) as opposed to the variance of the residuals (the variance of $(Y - \hat{Y})$). The correlation between the two variables can actually be used to estimate these values in a straightforward way:

$$s_{\hat{Y}}^2 = r^2 s_Y \text{ and } s_{Y-\hat{Y}}^2 = (1 - r^2)\, s_{\hat{Y}}^2.$$

A common regression-based technique can be used to "regress out" or "adjust for" the effects of one variable on another. For example, we may wish to study the relationship between verbal ability and gender in children. However, we also know that verbal ability is age related, and we do not want the effects of age to confound this analysis. We can calculate an age-adjusted measure of verbal ability by performing a regression of verbal ability on age. For every individual, we subtract their predicted value (given their age) from their observed value to create a new variable that reflects verbal ability without the effects of age-related variation: The new variable will not correlate with age. If there were any mean differences in age between boys and girls in the sample, then the effects of these on verbal ability have been effectively removed.

1.1.7 Matrices Reading behavioral genetic journal articles and books, one is likely to come across *matrices* sooner or later: "In QTL linkage the variance-covariance *matrix* for the sibship is modeled in terms of alleles shared identical-by-descent" or "The *matrix* of genotypic means can be observed. . . ." What are matrices and why do we use them? This section presents a brief introduction to matrices that will place such sentences in context.

Matrices are commonly used in behavioral genetics to represent information in a concise and easy-to-manipulate manner. A matrix is simply a block of *elements* organized in rows and columns. For example,

$$\begin{bmatrix} 34 & 23 \\ 56 & 17 \\ 65 & 38 \end{bmatrix}$$

is a matrix with three rows and two columns. Typically, a matrix will be organized such that each row and column has an associated meaning. In this example, the matrix might reflect scores for three students (each row representing one student) on English and French exams (the first column representing the score for English, the second for French). Elements are often indexed by their row and column: s_{ij} refers to the ith student's score on the jth test.

The matrix above represents raw data. In a similar way, the spreadsheet of values in statistical programs such as SPSS can be thought of as one large matrix. Perhaps the most commonly encountered form of matrix is the *correlation matrix,* which is used to represent descriptive statistics of raw data (correlations) in an orderly fashion. In a correlation matrix, the element in the ith row and jth column represents the pair-wise correlation between the ith and jth variables.

Here is a correlation matrix between three different variables:

$$\begin{bmatrix} 1.00 & 0.73 & 0.14 \\ 0.73 & 1.00 & 0.37 \\ 0.14 & 0.37 & 1.00 \end{bmatrix}$$

Correlation matrices have several easily recognizable properties. First, a correlation matrix will always be *square*—having the same number of rows as columns. For n variables, the correlation matrix will be an $n \times n$ matrix. The *diagonal* of a square matrix is the set of elements for which the row number equals the column number, so in terms of correlations, these elements represent the correlation of a variable with itself, which will always be 1. Additionally, correlation matrices will always be *symmetric* about the diagonal—that is, element r_{ij} equals r_{ji}. This symmetry represents the simple fact that the correlation between A and B is the same as the correlation between B and A. It is common practice not to write the redundant upper off-diagonal elements if a matrix is known to be symmetric. Our correlation matrix would be written

$$\begin{bmatrix} 1.00 & & \\ 0.73 & 1.00 & \\ 0.14 & 0.37 & 1.00 \end{bmatrix}$$

Correlation matrices are often presented in journal articles in tabular form to summarize correlational analyses.

A closely related type of matrix that occurs more often in behavioral genetic analysis is the *variance-covariance* matrix. In place of correlations, the elements of an $n \times n$ variance-covariance matrix are n variances along the diagonal and $(n - 1)n/2$ covariances in the lower off-diagonal. A correlation matrix is a *standardized* variance-covariance matrix, just as a correlation is a standardized covariance. The variance-covariance matrix for the three variables in the correlation matrix above might be

$$\begin{bmatrix} 2.32 & & \\ 1.43 & 1.64 & \\ 0.43 & 0.98 & 4.21 \end{bmatrix}$$

A variance-covariance matrix can be transformed into a correlation matrix: $r_{ij} = v_{ij}/\sqrt{v_{ii}v_{jj}}$, where r_{ij} are the new elements of the correlation matrix and v_{ij} are the elements of the variance-covariance matrix. (This is essentially a reformulation of the equation for calculating correlations given above in matrix notation.) Note that information is lost about the relative magnitude of variances among the different variables in a correlation matrix (because they are all standardized to 1). As mentioned earlier, because correlations are not scale dependent, however, they are easier to interpret than covariances and therefore better for descriptive purposes.

Matrices can be added to or subtracted from each other as long as both matrices have the same number of rows and the same number of columns:

$$\begin{bmatrix} 4 & -5 \\ 1 & 2 \end{bmatrix} + \begin{bmatrix} -2 & x \\ 0 & y \end{bmatrix} = \begin{bmatrix} 2 & x-5 \\ 1 & y+2 \end{bmatrix}$$

Note that here the elements of the sum matrix are not simple numerical terms—elements of matrices can be as complicated as you want. The beauty of matrix notation is that we can label matrices so that we can refer to many elements with a simple letter, say, **A**. (Matrices are generally written in bold type.)

$$\mathbf{A} = \begin{bmatrix} 4 & -5 \\ 1 & 2 \end{bmatrix}$$

$$\mathbf{B} = \begin{bmatrix} -2 & x \\ 0 & y \end{bmatrix}$$

$$\mathbf{A} - \mathbf{B} = \begin{bmatrix} 6 & -5-x \\ 1 & 2-y \end{bmatrix}$$

The other common matrix algebra operations are multiplication, inversion, and transposition. Matrix multiplication does not work in the same way as matrix addition (that kind of element-by-element multiplication is actually called a *Kronecker product*). Unlike normal multiplication, where $ab = ba$, in matrix multiplication $\mathbf{AB} \neq \mathbf{BA}$. For **A** to be multiplied by **B**, matrix **A** must have the same number of columns as **B** has rows. The resulting matrix has as many rows as **A** and as many columns as **B**. Each element is the sum of products across each row of **A** and each column of **B**. Following are two examples:

$$\begin{bmatrix} a & c & e \\ b & d & f \end{bmatrix} \begin{bmatrix} g & h & i \\ j & k & l \\ m & n & o \end{bmatrix} = \begin{bmatrix} ag+cj+em & ah+ck+en & ai+cl+eo \\ bg+dj+fm & bh+dk+fn & bi+dl+fo \end{bmatrix}$$

$$\begin{bmatrix} 3 & 3 & 0 \\ 1 & -2 & 5 \end{bmatrix} \begin{bmatrix} 7 & 2 \\ 3 & 2 \\ 8 & 4 \end{bmatrix} = \begin{bmatrix} 30 & 12 \\ 41 & 18 \end{bmatrix}$$

The equivalent to division is called matrix inversion and is complex to calculate, especially for large matrices. Only square matrices have an inverse, written $\mathbf{A}^{-1}$. Matrix inversion plays a central role in solving model-fitting problems.

Finally, the transpose of a matrix, $\mathbf{A}'$, is matrix $\mathbf{A}$ but with rows and columns swapped. Therefore, if $\mathbf{A}$ were a 3×2 matrix, then $\mathbf{A}'$ will be a 2×3 matrix (note that rows are given first):

$$\begin{bmatrix} 2 & 3 \\ 0 & -1 \\ -2 & 1 \end{bmatrix}' = \begin{bmatrix} 2 & 0 & -2 \\ 3 & -1 & 1 \end{bmatrix}$$

There is a great deal more to matrix algebra than the simple examples presented here. Basic familiarization with the types of matrices and matrix operations is useful, however, if only to realize that when behavioral genetic articles and books refer to matrices they are not necessarily talking about anything particularly complicated. The main utility of matrices is their convenience of presentation—it is the actual meaning of the elements that is important.

2 Quantitative Genetics

2.1 The Biometric Model

When we say that a trait is *heritable* or *genetic,* we are implying that at least one gene has a measurable effect on that trait. Although most behavioral traits appear to depend on many genes, it is still important to review the properties of a single gene because the more complex models are built upon these foundations. We will begin by examining the basic quantitative genetic model that mathematically describes the genetic and environmental underpinnings of a trait.

2.1.1 Allele and genotype The pair of alleles that an individual carries at a particular locus constitutes what we call the *genotype* at that locus. Imagine that, at a particular locus, two forms of a gene, labeled A_1 and A_2 (this would be called a *biallelic* locus), exist in the population. Because individuals have two copies of every gene (one from their father, one from their mother), individuals will possess one of three genotypes: They may have either two A_1 alleles or two A_2, in which case they are said to be *homozygous* for that particular allele. Alternatively, they may carry one copy of each allele, in which case they are said to be *heterozygous* at that locus. We would write the three genotypes as A_1A_1, A_1A_2, and A_2A_2 (or, using different notation, *AA, Aa,* and *aa*).

For biallelic loci, the two alleles will occur in the population at particular frequencies. If we counted all the alleles in a population and three-fourths were A_1, then we say that A_1 has an allelic frequency of 0.75. Because these frequencies must sum to 1, we know that the A_2 allele has a frequency of 0.25. It is common practice to denote the allelic frequencies of a biallelic locus as p and q (so here, p is the allelic frequency of the A_1 allele, 0.75, and q is the frequency of A_2, 0.25). Given these, we can predict the genotypic frequencies. Formally, if the two alleles A_1 and A_2 have allelic frequencies p and q, then, with random mating, we would expect to observe the three genotypes A_1A_1, A_1A_2, and A_2A_2 at frequencies p^2, $2pq$, and q^2, respectively. (See Box 2.2 and Chapter 20.)

2.1.2 Genotypic values Next, we need a way to describe any effects of the alleles at a locus on whatever trait we are interested in. A locus is said to be *associated with* a trait if some of its alleles are associated with different mean levels of that trait in the population. For qualitative diseases (i.e., diseases that are either present or not present), a single allele may be necessary and sufficient to develop the disease. In this case, the disease-predisposing allele acts in either a dominant or a recessive manner. Carrying a dominant allele will result in the disease irrespective of the other allele at that locus; conversely, if the disease-predisposing allele is recessive, then the disease will only develop in individuals homozygous for that allele.

For a quantitative trait, however, we need some way of specifying *how much* an allele affects the trait. Considering only a locus with two alleles, A_1 and A_2, we define the average value of one of the homozygotes (say, A_1A_1) as a and the average value of the other homozygote (A_2A_2) as $-a$. The value of the heterozygote (A_1A_2) is labeled d and is dependent on the mode of gene action. If there is no dominance, d will be zero (i.e., the midpoint of the two homozygotes' scores). If the A_1 allele is dominant to A_2, then d will be greater than zero. If dominance is complete (i.e., if the observed value for A_1A_2 equals that of A_1A_1), then $d = +a$.

2.1.3 Additive effects Observed genotypic values for a single locus can be defined in terms of an *additive genetic value* and a *dominance deviation*. The additive genetic value of a locus relates to the average effect of an allele. As illustrated in Figure A.7, the additive genetic value is the genotypic value expected from the number of a particular allele (say, A_1) at that locus, either 0, 1, or 2 (each A_1 allele increases an individual's score by A units).

Additive genetic values are important in behavioral genetics because they represent the extent to which genotypes "breed true" from parents to offspring. If a parent has one copy of a certain allele, say, A_1, then each offspring has a 50 percent chance of receiving an A_1 allele. If an offspring receives an A_1 allele, then its additive effect will contribute to the phenotype to exactly the same extent as it did to the parent's phenotype. That is, it will lead to increased parent-offspring resemblance on the phenotype, irrespective of other alleles at that locus or at other loci.

2.1.4 Dominance deviation Dominance is the extent to which the effects of alleles at a locus do not simply "add up" to produce genotypic values. The *dominance deviation* is the difference between actual genotypic values and what would be expected under a strictly additive model. Figure A.8 represents the deviations (labeled D) of the expected (or additive) genotypic values from the actual genotypic values that occur if there is an effect of dominance at the locus.

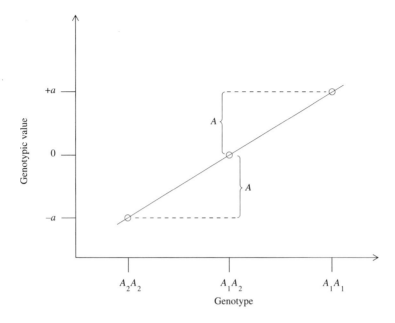

• **FIGURE A.7** Additive genetic values. The number of A_1 alleles predicts additive genetic values. Because there is no dominance (and assuming equal allelic frequency), the additive genetic values equal the genotypic values. (*A*, value added by each A_1 allele.)

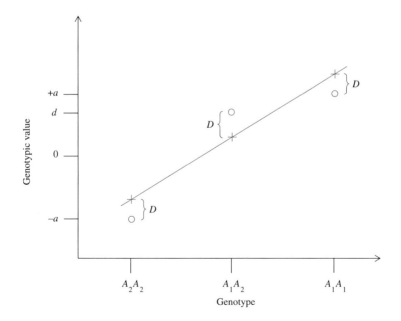

• **FIGURE A.8** Dominance deviations. The genotypic values (circles) deviate from the expected values under an additive model (crosses) when there is dominance (i.e., d ≠ 0). (D, deviation from expected attributed to dominance.)

Dominance genetic variance represents genetic influence that does not "breed true." Saying that the effect of a locus involves dominance is equivalent to saying that an individual's genotypic value results from the *combination* of alleles at that particular locus. However, offspring receive only one allele from each parent, not a combination of two alleles. Genetic influence due to dominance will not be transmitted from parent to offspring, therefore. In this way, additive and dominance genetic values are defined so as to be independent of each other.

2.1.5 Polygenic model Not only can we consider the additive and nonadditive effects at a single locus, we can also sum these effects across loci. This concept is the essence of the polygenic extension of the single-gene model. Just as additive genetic values are the summation of the average effects of two alleles at a single locus, they can also be summed across the many loci that may influence a particular phenotypic character. Similarly, dominance deviations from additive genetic values can also be summed for all the loci influencing a character. Thus, it is relatively easy to generalize the single-gene model to a polygenic one with many loci, each with its own additive and nonadditive effects. Under an *additive* polygenic model, the genetic effect G on the phenotype represents the sum of effects from different loci.

$$G = G_1 + G_2 + \ldots + G_N$$

This expression implies that the effects of different alleles simply add up—that is, there is no interaction between alleles where the effect of one allele, say, G_1 is modified by the presence of the allele with effect G_2. The polygenic model needs to consider the possibility that the effects of different loci do not add up independently but interact with each other—an interaction called *epistasis*. For example, imagine two loci, each with an allele that increases an individual's score by one point on a particular trait. If there were no epistasis, having a risk allele at both loci would increase the score by two points. If there were epistasis, however, having risk alleles at both loci might possibly lead to a ten-point increase. Epistasis therefore complicates analysis, but there is evidence that such phenomena might be quite prevalent for certain complex traits. In other words, dominance is *intralocus* interaction between alleles, whereas epistasis is *interlocus* interaction, that is, between loci.

The total genetic contribution to a phenotype is G, which is the sum of all additive genetic effects A, all dominance deviations D, and all epistatic interaction effects I:

$$G = A + D + I$$

2.1.6 Phenotypic values and variance components model Quantitative genetic theory states that every individual's phenotype is made up of genetic and environmental contributions. No behavioral phenotype will be entirely determined by genetic effects, so we should always expect an environmental effect, E, which also includes measurement error, on the phenotype P. In algebraic terms,

$$P = G + E$$

where, for convenience, we assume that P represents an individual's deviation from the population mean rather than an absolute score. In any case, behavioral genetics is not primarily interested in the score of any one individual. Rather, the focus is on explaining the causes of phenotypic differences in a population—why some individuals are more extraverted than others, or why some individuals are alcoholic, for example.

In fact, there is often no direct way of determining the relative magnitude of genetic and environmental deviations for any one individual, certainly if one has not obtained DNA from individuals. However, in a sample of individuals, especially of genetically related individuals, it is possible to estimate the variances of the terms P, G, and E. This approach is called the *variance components approach*, and it relies on the equation that showed us how to calculate the variance of a sum.

Recall that

$$\text{Var}(X + Y) = \text{Var}(X) + \text{Var}(Y) + 2\text{Cov}(X,Y)$$

Turning this expression around, it gives us a method for partitioning the variance of a variable that is a composite of constituent parts. That is, our goal is to "decompose the variance" of a trait into the constituent parts of genetic and environmental sources of variation.

For simplicity, we will assume no epistasis, so $P = G + E = A + D + E$. The variance of P is equal to the sum of the variance of the separate components, A, D, and E, plus twice the covariance between them:

$$\begin{aligned} \text{Var}(P) &= \text{Var}(A + D + E) \\ &= \text{Var}(A) + \text{Var}(D) + \text{Var}(E) + 2\text{Cov}(A, D) + 2\text{Cov}(A, E) + 2\text{Cov}(D, E) \end{aligned}$$

which begins to look unmanageable until we realize that we can use some theoretical assumptions of our model to constrain this equation. By definition, additive genetic influences are independent of dominance deviations. That is, $\text{Cov}(A, D)$ will necessarily equal zero, so this term can be dropped from the model. Another assumption that we may wish to make (but one that is not necessarily true) is that genetic and environmental influences are uncorrelated. This is equivalent to saying that $\text{Cov}(A, E)$ and $\text{Cov}(D, E)$ equal zero and can be dropped from the model. We will see later that there are detailed reasons why this assumption might not hold (what is called a *gene-environment correlation*) (see also Chapter 8). For the time being, however, our simplified model reads:

$$\text{Var}(P) = \text{Var}(A) + \text{Var}(D) + \text{Var}(E)$$

A note on notation: Variances are often written in other ways. For example, above we have denoted additive genetic variance as $\text{Var}(A)$. As we will see, this term is often written differently, depending on the context (mainly for historical reasons). In formal model fitting, the lowercase Greek letter sigma squared with a subscript (σ_A^2) might be used. A similar value calculated in the context of comparing familial correlations (narrow-sense heritability, introduced later) is typically labeled h^2, whereas it is written as a^2 in the context of path analysis (also introduced later). Under most circumstances, however, these all refer to roughly the same thing.

In conclusion, it might not seem that we have achieved very much in simply considering variances instead of values. However, as will be discussed, quantitative genetic methods can use these models to estimate the relative contribution of genetic and environmental influences to phenotypic variance.

2.1.7 Environmental variation Because the nature of environmental effects is more varied and changeable than the underlying nature of genetic influence, it is not possible to decompose this term into constituent parts in a straightforward way. That is, if we detect genetic influence, then we know that this effect must result from at least one gene—and we know something about the properties of genes.

However, if we detect environmental influence on a trait, we cannot assume any one mechanism. But behavioral genetics is able to investigate environmental influences in two main ways. As we will see later, family-based studies using twins or adoptive relatives allow environmental influences to be partitioned into those shared between relatives (i.e., those that make relatives resemble each other) and those that are nonshared (i.e., those that do not make relatives resemble each other). This type of analysis is not at the level of specific, measured environmental variables.

A second approach is to actually measure a specific aspect of the environment (e.g., parental socioeconomic status, or nutritional content of diet) and incorporate it into genetic analysis. For example, we may wish to partition out the variation in the trait due to a measured environmental source if we consider it to represent a cause of *nuisance* or *noise* variance in the trait (i.e., to treat it as a covariate). Alternatively, we may believe that an environment is important in the expression of genetic influence. For example, we might suspect that stress might bring out genetic vulnerabilities toward depression. Therefore, depression might be expected to show greater genetic influence for individuals experiencing stress. In this case, we would not want to adjust for the effects of the environmental variable. Such a circumstance is named a gene-environment interaction ($G \times E$ interaction). In terms of the quantitative genetic model,

$$P = G + E + (G \times E)$$

where ($G \times E$) does not necessarily represent a multiplication effect but rather any interactive effect of genes and environment that is independent of their main effects.

2.2 Estimating Variance Components

In the previous section we outlined a simple biometric model, describing the variation in observed phenotypes in terms of various genetic and environmental sources of variation. In this section, we consider how we can use family data to estimate some of the key parameters of such models, with a focus on heritability as estimated from the classical twin study, introducing maximum likelihood estimation and model fitting.

2.2.1 Genes and families Until now, we have built a general genetic model of the etiology of variation in a trait among individuals. A major step in quantitative genetics is to incorporate knowledge of basic laws of heredity to allow us to extend our model to include the covariance between relatives. Conceptually, most behavioral genetic analysis contrasts phenotypic similarity between related individuals (which is measured) with their genetic similarity (which is known from genetics). If individuals who are more closely related genetically also tend to be more similar on a measured trait, then this tendency is evidence for that trait being heritable—that is, the trait is at least partially influenced by genes.

When we study families, we are not only interested in the variance of a trait—the main focus is on the covariance between relatives. Earlier we saw how we can study two variables, such as height and weight, and ask whether they are associated with each other. In a similar way, covariances and correlations can also be used to ask whether a single variable is associated between family members. For example, do brothers and sisters tend to be similar in height or not? If we measured height in sibling pairs, we could calculate the covariance between an individual's height and the sibling's height. If the covariance equaled zero, this would imply that brothers and sisters are no more likely to have similar heights than any two unrelated individuals picked at random from the population. If the covariance is greater than zero, this would imply that taller individuals tend to have taller brothers and sisters. Quantitative genetic analysis attempts to determine the factors that can make relatives similar—their shared nature or their shared nurture.

2.2.2 Genetic relatedness in families An individual has two copies of every gene, one paternally inherited and one maternally inherited. When an individual passes one copy of each gene to its offspring, there is an equal chance that either the paternally inherited gene or the maternally inherited gene will be transmitted. From these two simple facts, we can calculate the expected proportion of gene sharing between individuals of different genetic relatedness. Siblings who share both biological parents will share either zero, one, or two alleles at each locus. For autosomal loci, there is a 50 percent chance that siblings will share the same paternal allele (two ways of sharing, two ways of not sharing, all with equal probability) and, correspondingly, a 50 percent chance of sharing the same maternal allele. Therefore, siblings stand a $0.50 \times 0.50 = 0.25$ (25 percent) chance of sharing both paternal and maternal alleles; a $(1.00 - 0.50) \times (1.00 - 0.50) = 0.25$ (25 percent) chance of sharing no alleles; a $1.00 - 0.25 - 0.25 = 0.50$ (50 percent) chance of sharing one allele. The average, or expected, alleles shared is therefore $(0 \times 0.25) + (1 \times 0.5) + (2 \times 0.25) = 1$. Therefore, in the average case, siblings will share half of the additive genetic variation that could potentially contribute to phenotypic variation because they share one out of two alleles. Because siblings stand only a 25 percent chance of sharing *both* alleles, in the average case, siblings will share a quarter of the dominance genetic variation that could potentially contribute to phenotypic variation.

For other types of relatives, we can work out their expected genetic relatedness in terms of genetic components of variance. Parent-offspring pairs always share precisely one allele: They will share half of the additive genetic effects that contribute to variation in the population but none of the dominance genetic effects. Half siblings, who have only one parent in common, share only a quarter of additive genetic variance but no dominance variance (because they can never inherit two alleles at the same locus from the same parent).

The majority of behavioral genetic studies focus on twins. Genetically, full sibling pairs and DZ twin pairs are equivalent. So, whereas DZ twins will only share half the additive genetic variance and one-fourth of the dominance variance, MZ twins share all their genetic makeup, so additive and dominance genetic variance components will be completely shared.

These coefficients of genetic relatedness are summarized in Table A.1. Sharing additive and dominance genetic variance contributes to the phenotypic correlation between relatives. As mentioned earlier, correlations between relatives directly estimate the proportion of variance shared between them. So we can think of the familial correlation as the sum of all the shared components of variance between two relatives.

Not only genes are shared between most relatives, however. Individuals that are genetically related are more likely to experience similar environments than unrelated individuals are. If an environmental factor influences a variable, then sharing this environment will also contribute to the phenotypic correlation between relatives. As explained in Chapter 8, behavioral genetics conceptually divides environmental influences into two distinct types with regard to their impact on families. Environments that are shared by family members *and* that tend to make members more similar on a particular trait are called *shared environmental* influences. In contrast, *nonshared environmental* influences do not result in family members becoming more alike for a given trait.

TABLE A.1
Coefficients of Genetic Relatedness

Related Pair	Proportion of Additive Genetic Variation Shared	Proportion of Dominance Genetic Variation Shared
Parent and offspring (PO)	$\frac{1}{2}$	0
Half siblings (HS)	$\frac{1}{4}$	0
Full siblings (FS)	$\frac{1}{2}$	$\frac{1}{4}$
Nonidentical twins (DZ)	$\frac{1}{2}$	$\frac{1}{4}$
Identical twins (MZ)	1	1

Most behavioral genetic analysis focuses on three components of variance: additive genetic, shared environmental, and nonshared environmental. As we will see, this tripartite approach underlies the estimation of heritability by comparing twin correlations and is the basic model used in more sophisticated model-fitting analysis. This model is often referred to as the ACE model. (A stands for additive genetic effects, C for common (shared) environment, and E for nonshared environment.)

2.2.3 Heritability As explained in Chapter 6, heritability is the proportion of phenotypic variance that is attributable to genotypic variance. There are two types of heritability: *broad-sense heritability* refers to all sources of genetic variance, whether the genes operate in an additive manner or not. *Narrow-sense heritability* refers only to the proportion of phenotypic variance explained by additive genetic effects. Narrow-sense heritability therefore gives an indication of the extent to which a trait will "breed true"—that is, the degree of parent-offspring similarity that is expected. Broad-sense heritability, on the other hand, gives an indication of the extent to which genetic factors of any kind are responsible for trait variation in the population.

We are able to estimate the heritability of a trait by comparing correlations between certain types of family members. For simplicity, we will assume that the only influences on a trait are additive genetic effects and environmental effects that are either shared or nonshared between family members. We can describe the correlation we observe between different types of relatives in terms of the components of variance they share. For example, we expect the correlation between full siblings to represent half the additive genetic variance and, by definition, all the shared environmental variance but none of the nonshared environmental variance. As mentioned earlier, additive genetic variance is typically labeled h^2 in this context (representing narrow-sense heritability). The shared environmental variance is labeled c^2 (nonshared environment is e^2). Therefore,

$$r_{FS} = \frac{h^2}{2} + c^2$$

Suppose we observed for full siblings a correlation of 0.45 for a trait. We would not be able to work out what h^2 and c^2 are from this information alone because, as reflected in the equation above, nature and nurture are shared by siblings. However, by comparing sets of correlations between certain different types of relatives, we are able to estimate the relative balance of genetic and environmental effects. The most common study design uses MZ and DZ twin pairs. The correlations expressed in terms of shared variance components are therefore

$$r_{MZ} = h^2 + c^2$$

$$r_{DZ} = \frac{h^2}{2} + c^2$$

Subtracting the second equation from the first gives

$$r_{MZ} - r_{DZ} = h^2 - \frac{h^2}{2} + c^2 - c^2$$

$$= \frac{h^2}{2}$$

$$h^2 = 2(r_{MZ} - r_{DZ})$$

That is, narrow-sense heritability is calculated as twice the difference between the correlations observed for MZ and DZ twin pairs. The proportion of variance attributable to shared environmental effects can easily be estimated as the difference between the MZ correlation and the heritability ($c^2 = r_{MZ} - h^2$). Because we have estimated these two variance components from correlations, which are standardized, h^2 and c^2 represent *proportions* of variance. The final component of variance we are interested in is nonshared environmental variance, e^2. This statistic does not appear in the equations describing the correlations between relatives, of course. However, we know that h^2, c^2, and e^2 must sum to 1 if they represent proportions, so

$$h^2 + c^2 + e^2 = 1$$

$$[2(r_{MZ} - r_{DZ})] + [r_{MZ} - 2(r_{MZ} - r_{DZ})] + e^2 = 1$$

$$\therefore r_{MZ} + e^2 = 1$$

$$\therefore e^2 = 1 - r_{MZ}$$

This conclusion is intuitive: Because MZ twins are genetically identical, any variance that is not shared between them (i.e., the extent to which the MZ twin correlation is not 1) must be due to nonshared environmental sources of variance.

Let's consider an example: Suppose we observe a correlation of 0.64 in MZ twins and 0.44 in DZ twins. Taking twice the difference between the correlations, we can conclude that the trait has a heritability of 0.4 [= 2 × (0.64 − 0.44)]. That is, 40 percent of variation in the population from which we sampled is attributable to the additive effects of genes. The shared family environment therefore accounts for 24 percent ($c^2 = 0.64 - 0.4 = 0.24$) of the variance; the nonshared environment accounts for 36 percent ($e^2 = 1 - 0.64 = 0.36$).

A pattern of results such as those just described would suggest that genes play a significant role in individual differences for this trait, differences between people being roughly half due to nature, half due to nurture. We have made several assumptions, however, in order to arrive at this conclusion. These assumptions will be considered more fully in the context of model fitting, but we will mention two immediate assumptions. First, we have assumed that dominance is not important for this trait (not to mention other more complex interactions such as epistasis). We have assumed that all genetic effects are additive (which is why h^2 represents narrow-sense heritability). If this assumption were not true, the heritability estimate would be biased.

Second, we have assumed that MZ and DZ twins only differ in terms of the genetic relatedness. That is, the same shared environment term, c^2, appears in both MZ and DZ equations. If parents treat identical twins more similarly than they treat nonidentical twins, this assumption could result in higher MZ correlations relative to DZ correlations. This assumption, which is in theory testable, is called the *equal environments assumption* (see Chapter 6). Violations of this assumption would overestimate the importance of genetic effects.

Other types of relatives can be studied to calculate heritability; for example, we could compare correlations for full siblings and half siblings. Not all comparisons will be informative, however. Comparing the correlation for full siblings and the correlation for parent and offspring will not help to estimate heritability (because these relatives do not differ in terms of shared additive genetic variance). It is preferable to study twins for several reasons. It can be shown that for statistical reasons, twins afford greater accuracy in determining heritability because larger proportions of variance are shared by MZ twins. Additionally, twins are more closely matched for age, familial, and social influences than are half siblings or parents and offspring. The interpretation of the shared environment is much less clear for parents and offspring.

Quantitative genetic studies can also contrast family members who are genetically similar but have not shared any environmental influences. This comparison is the basis of the adoption study. The simplest form of adoption study is that of MZ twins reared apart. Because MZ twins reared apart are genetically identical but do not share any environmental influences, the correlation directly estimates heritability. That is, if there has been no selective placement, any tendency for MZ twins reared apart to be similar must be attributable to the influences of shared genes.

2.2.4 Model fitting and the classical twin design Simple comparisons between twin correlations can indicate whether genetic influences are important for a trait. This is the important first question that any quantitative genetic analysis must ask. Here we will examine some of the more formal statistical techniques that can be used to analyze genetically informative data and to ask other, more involved questions.

Model-fitting approaches involve constructing a model that describes some observed data. In the quantitative genetic studies, the observed data that we model are typically the variance-covariance matrices for family members. The model will then consist of a variance-covariance matrix formulated in terms of various *parameters*. These will typically be the variance components (additive genetic and so on) we encountered earlier. Various combinations of different values for the model parameters will generate different expected variance-covariance matrices. The goal of model fitting is twofold: (1) to select the model with the smallest number of parameters that (2) generates expectations that match the observed data as closely as possible. As we will see, there is a payoff between the number of parameters in a model and the accuracy with which it can model the observed data.

If we were to fit the ACE model to observed MZ and DZ twin data, the three parameter estimates selected to match the expected variance-covariance matrices with the observed ones would correspond directly to the estimates of heritability, and of shared and nonshared environmental influences that we calculated earlier in a relatively straightforward manner. Why would we ever want to perform more complicated model fitting? There are several good reasons: First, these calculations are only valid *if* the ACE model is a true reflection of reality. Model fitting allows different types of models to be explicitly tested and compared. Model fitting also facilitates the calculation of confidence intervals around the parameter estimates. It is common to read something such as "$h^2 = 0.35 \ (0.28 - 0.42)$," which means that the heritability was estimated at 35 percent, but there is a 95 percent chance that, even if it is not exactly 35 percent, it at least lies within the range of 28 to 42 percent. Model fitting can also incorporate many different types of family structures, model multivariate data, and include any *measured* genetic or environmental information we may have, in order to improve our estimates and explore potential interactions of genetic and environmental effects, or to test whether specific loci are associated with the trait or not.

Let's start from basics. Imagine that we have measured a trait in a population of twins. We have not measured any DNA, nor have we measured any other environmental factors that might influence the trait. We summarize our data as two variance-covariance matrices, one for MZ twin pairs and one for DZ twin pairs; so our "observed data" are six unique statistics:

$$\begin{bmatrix} Var_1^{MZ} & \\ Cov_{12}^{MZ} & Var_2^{MZ} \end{bmatrix}$$

$$\begin{bmatrix} Var_1^{DZ} & \\ Cov_{12}^{DZ} & Var_2^{DZ} \end{bmatrix}$$

Using our knowledge of the quantitative genetic model as outlined earlier, we can begin to construct a model that describes the two variance-covariance matrices for the twins. That is, we assume that observed trait variation is due to a certain mixture of additive genetic, dominance genetic, shared environmental, and nonshared environmental effects (we will ignore epistasis and other interactions).

Model fitting begins by creating an explicit model for the variance-covariance matrix for families, in terms of genetic and environmental variance components. Returning to the basic genetic model, phenotype, P, is a function of additive, A, and dominance, D, genetic effects. Additionally, we include environmental effects, which are either shared, C, or nonshared, E. (*Note:* The basic model did not make this distinction because it is primarily formulated to describe variation in a population of *unrelated* individuals, i.e., E referred to *all* environmental effects.)

$$P = A + D + C + E$$

In terms of variances, therefore, remembering all the assumptions outlined under the single-gene model that apply at this step (no gene-environment correlation, for example), we obtain

$$\sigma_P^2 = \sigma_A^2 + \sigma_D^2 + \sigma_C^2 + \sigma_E^2$$

where, using the model-fitting notation, $\sigma_{A/D/C/E}^2$ (pronounced "sigma") stands for the components of variance associated with the four types of effect and σ_P^2 is the phenotypic variance.

To construct our twin model, we need to explicitly write out every element of the variance-covariance matrices in terms of the parameters of the model. We have already defined the trait variance in terms of the variance components:

$$\sigma_A^2 + \sigma_D^2 + \sigma_C^2 + \sigma_E^2$$

We will write this term for all four variance elements in the model. Note that we are modeling variances and covariances instead of correlations; this is often done in model fitting because it captures more information (the variance and covariance) than a correlation does. The σ_A^2 parameter will not directly estimate narrow-sense heritability—we need to divide the additive genetic variance component by the total variance:

$$\sigma_A^2/(\sigma_A^2 + \sigma_D^2 + \sigma_C^2 + \sigma_E^2)$$

We make the assumption that components of variance are identical for all individuals. That is, we write the same expression for all four variance elements. This assumption implies that the effects of genes and environments on an individual are not altered by that individual being a member of an MZ or DZ twin pair. Additionally, it assumes that individuals were not assigned a Twin 1 or Twin 2 label in a way that might make Twin 1's variance differ from Twin 2's variance. For example, if the first-born twin was always coded as Twin 1, then, depending on the nature of the trait, this assumption might not be warranted. (This problem is sometimes avoided by "double-entering" twin pairs so that each individual is entered twice, once as Twin 1 and once as Twin 2, when calculating the observed variance-covariance matrices. This method will, of course, ensure that Twin 1 and Twin 2 have equal variances.)

The covariance term between twins is also a function of the components of variance, in terms of the extent to which they are shared between twins, as stated earlier. All additive and dominance genetic variance, as well as shared environmental variance, is shared by MZ twins. These components contribute to the covariance between MZ twins fully. DZ twins share one-half the additive genetic variance, one-fourth the dominance genetic variance, all the shared environmental variance, and none of the nonshared environmental variance. The contributions of these components to the DZ covariance are in proportion to these coefficients of sharing.

Therefore, for MZ twin pairs, the variance-covariance matrix is modeled as

$$\begin{bmatrix} \sigma_A^2 + \sigma_D^2 + \sigma_C^2 + \sigma_E^2 & \\ \sigma_A^2 + \sigma_D^2 + \sigma_C^2 & \sigma_A^2 + \sigma_D^2 + \sigma_C^2 + \sigma_E^2 \end{bmatrix}$$

whereas, for DZ twins, it is

$$\begin{bmatrix} \sigma_A^2 + \sigma_D^2 + \sigma_C^2 + \sigma_E^2 & \\ \dfrac{\sigma_A^2}{2} + \dfrac{\sigma_D^2}{4} + \sigma_C^2 & \sigma_A^2 + \sigma_D^2 + \sigma_C^2 + \sigma_E^2 \end{bmatrix}$$

These two matrices represent our model. Different values of σ_A^2, σ_D^2, σ_C^2, and σ_E^2 will result in different *expected* variance-covariance matrices. These matrices are "expected," in the sense that, *if* the values of the model parameters were true, then these are the averaged matrices we would expect to observe if we repeated the experiment a very large number of times.

As an example, consider a trait with a variance of 5. Imagine that variation in this trait was entirely due to an equal balance of additive genetic effects and nonshared environmental effects. In terms of the model, this assumption is equivalent to saying that σ_A^2 and σ_E^2 both equal 2.5, whereas σ_D^2 and σ_C^2 both equal 0. If this were true, then what variance-covariance matrices would we *expect* to observe for MZ and DZ twins? Simply substituting these values, we would expect to observe for MZ twins,

$$\begin{bmatrix} 2.5+0+0+2.5 & \\ 2.5+0+0 & 2.5+0+0+2.5 \end{bmatrix} = \begin{bmatrix} 5 & \\ 2.5 & 5 \end{bmatrix}$$

and for DZ twins,

$$\begin{bmatrix} 2.5+0+0+2.5 & \\ \dfrac{2.5}{2} + \dfrac{0.0}{4} + 0 & 2.5+0+0+2.5 \end{bmatrix} = \begin{bmatrix} 5 & \\ 1.25 & 5 \end{bmatrix}$$

To recap, we have seen how a specific set of parameter values will result in a certain expected set of variance-covariance matrices for twins. This result is, in itself, not very useful. We do not know the true values of these parameters—these are the very values we are trying to discover! Model fitting helps us to estimate the parameter values most likely to be true by evaluating the expected values produced by very many sets of parameter values. The set of parameter values that produces expected matrices that most closely match the observed matrices are selected as the *best-fit parameter estimates*. These represent the best estimates of the true parameter values. Because of the iterative nature of model fitting (evaluating very many different sets of parameter values), it is a computationally intensive technique that can only be performed by using computers.

2.2.5 An example of the model-fitting principle Suppose that, for a certain trait, we observe the following variance-covariance matrices for MZ and DZ pairs, respectively (note that the observed variances are similar although not identical):

$$\begin{bmatrix} 2.81 & \\ 2.13 & 3.02 \end{bmatrix}$$

$$\begin{bmatrix} 3.17 & \\ 1.54 & 3.06 \end{bmatrix}$$

The model fitting would start by substituting *any* set of parameters to generate the expected matrices. Suppose we substituted $\sigma_A^2 = 0.7$, $\sigma_D^2 = 0.2$, $\sigma_C^2 = 1.2$, and $\sigma_E^2 = 0.8$. These values only represent a "first guess" that will be evaluated and improved on by the model-fitting process. These values imply that 24 percent $[0.7/(0.7 + 0.2 + 1.2 + 0.8)]$ of phenotypic variation is attributable to additive genetic effects. If these were the true values, the variance-covariance matrix we would expect to observe for MZ twins is

$$\begin{bmatrix} 0.7+0.2+1.2+0.8 & \\ 0.7+0.2+1.2 & 0.7+0.2+1.2+0.8 \end{bmatrix} = \begin{bmatrix} 2.9 & \\ 2.1 & 2.9 \end{bmatrix}$$

whereas, for DZ twins, it is

$$\begin{bmatrix} 0.7+0.2+1.2+0.8 & \\ \dfrac{0.7}{2}+\dfrac{0.2}{4}+1.2 & 0.7+0.2+1.2+0.8 \end{bmatrix} = \begin{bmatrix} 2.9 & \\ 1.6 & 2.9 \end{bmatrix}$$

Comparing these expectations with the observed statistics, we can see that they are numerically similar but not exactly the same. We need an exact method for determining *how good* the fit between the expected and observed matrices is. Model fitting can therefore proceed, changing the parameter values to increase the *goodness of fit* between the model-dependent expected values and the sample-based observed values. When a set of values has been found that cannot be beaten for goodness of fit, these will be presented as the "output" from the model-fitting programs, the best-fit estimates. This process is called *optimization.* It would be very inefficient to evaluate *every* possible set of parameter values. For most models, evaluating every set would in fact be virtually impossible, given current computing technology. Rather, optimization will try to change the parameters in an intelligent way. One way of thinking about this process is as a form of a "hotter-colder" game: The aim is to increasingly refine your guess as to where the hidden object is, rather than exhaustively searching every inch of the room.

There are many indices of fit—one simple one is the chi-squared (χ^2, pronounced "ki," as in *kite*) goodness-of-fit statistic. This statistic essentially evaluates the magnitude of the discrepancies between expected and observed values by comparing how likely the observed data are under the model. The χ^2 goodness-of-fit statistic can be formally tested for significance in order to indicate whether or not the model provides a good approximation of the data. If the χ^2 goodness-of-fit statistic is low (i.e., nonsignificant), it indicates that the observed values *do not signifi cantly deviate* from the expected values. However, a low χ^2 value does not necessarily mean that the parameter values being tested are the best-fit estimates. As we have mentioned, different values for the four parameters might provide a better fit (i.e., an even lower χ^2 goodness-of-fit-statistic).

Just because we can write down a model that we believe to be an accurate description of the real-world processes affecting a trait, it does not necessarily mean that we can derive values for its parameters. In the preceding example, we would not be able to estimate the four parameters (additive and dominance genetic variances, shared and nonshared environmental variances) from our twin data. In simple terms, we are asking too many questions of too little information.

Consider what happens when we change the parameter values to see whether we can improve the fit of the model. Try substituting $\sigma_A^2 = 0.1$, $\sigma_D^2 = 0.6$, $\sigma_C^2 = 1.4$, and $\sigma_E^2 = 0.8$ instead, and you will notice that we obtain the same two expected variance-covariance matrices for both MZ and DZ twins as we did under the previous set of parameters. Both sets of parameters would therefore have an identical fit, so we would not be able to distinguish these two alternative explanations of the observations. This phenomenon can make model fitting very difficult or even impossible. This is an instance of a model not being *identified*.

2.2.6 The ACE model Although we will not follow the proof here, researchers have demonstrated that we cannot ask about additive genetic effects, dominance genetic effects, *and* shared environmental effects simultaneously if the only information we have is from MZ and DZ twins reared together.

In virtually every circumstance, we will wish to retain the nonshared environmental variance component in the model. We wish to retain it partly because random measurement error is modeled as a nonshared environmental effect and we do not wish to have a model that assumes no measurement error (it is unlikely to fit very well). Most commonly, we would then model additive genetic variance and shared environmental variance. As mentioned earlier, such a model is called the ACE model.

If we had reason to suspect that dominance genetic variance might be affecting a trait, then we might fit an ADE model instead. If the MZ twin correlation is more than twice the DZ twin correlation, one explanation is that dominance genetic effects play a large role for that trait (an explanation that might suggest fitting an ADE model).

The ACE model (and the ADE model) is an identified model. That is, the best fit between the expected and observed matrices is produced by one and only one set of parameter values. As long as the twin covariances are both positive and the MZ covariance is not smaller than the DZ covariance (both of which are easily justified biologically as reasonable demands), the ACE model will always be able to select a unique set of parameters that best account for the observed statistics.

If we were to model standardized scores (so that differences in the observed variance elements could not reduce fit), then under the ACE model the best-fitting parameters will always have a χ^2 goodness of fit of precisely zero. Such a model is called a *saturated* model. Imagine that, for a standardized trait (i.e., one with a variance of 1), we found an MZ covariance of 0.6 (this can be considered as the MZ twin correlation, of course) and a DZ covariance of 0.4. There is, in fact, one and only one set of values for the three parameters of the ACE model that will produce expected values that exactly match these observed values. In this case, these are $\sigma_A^2 = 0.4$, $\sigma_C^2 = 0.2$, and $\sigma_E^2 = 0.4$. Substituting these into the model, we obtain for MZ twins,

$$\begin{bmatrix} 0.4+0.2+0.4 & \\ 0.4+0.2 & 0.4+0.2+0.4 \end{bmatrix} = \begin{bmatrix} 1.0 & \\ 0.6 & 1.0 \end{bmatrix}$$

and for DZ twins,

$$\begin{bmatrix} 0.4+0.2+0.4 & \\ \dfrac{0.4}{2}+0.2 & 0.4+0.2+0.4 \end{bmatrix} = \begin{bmatrix} 1.0 & \\ 0.4 & 1.0 \end{bmatrix}$$

There are no other values that σ_A^2, σ_C^2, and σ_E^2 can take to produce the same expected variance-covariance matrices. This property does not mean that these values will necessarily reflect the true balance of genetic and environmental effects—they will only reflect the true values if the model (ACE or ADE or whatever) is a good one. All parameter estimates are model dependent: We can only conclude that, *if* the ACE model is a good model, then this result is the balance of genetic and environmental effects. We are able to test different models relative to one another, however, in order to get a sense of whether or not the model is a fair approximation of the underlying reality. We can only compare models if they are *nested,* however. A model is nested in another model if and only if that model results from constraining to zero one or more of the variance components in the larger model. For example, we may suspect that the shared environment plays no significant role for a given trait. We can test this supposition by fixing the shared environment variance component to zero and comparing the fit of the full model with the fit of this reduced model. Nesting is important because it forms the basis for testing and selecting between different models of our data.

A general principle of science is parsimony: to always prefer a simpler theory if it accounts equally well for the observations. This concept, often referred to as *Occam's razor,* is explicit in model fitting. Having derived estimates for genetic and environmental variance components under an ACE model, we might ask whether we could drop the shared environment term from the model. Might our simpler AE model provide a comparable fit to the data? Instead of estimating the shared environment variance component, we assume that it is zero (which is equivalent to ignoring it or removing it from the model). The AE model is therefore nested in the ACE model. We are able to calculate the goodness of fit of the ACE model, which estimates three parameters to explain the data, and the goodness of fit for the AE model, which only estimates two parameters to explain the same data. Any model with fewer parameters will not fit as well as a sensible model with more parameters. The question is whether or not the reduction in fit is *significantly* worse relative to the "advantage" of having fewer parameters in a more parsimonious model.

In our example, the ACE model will estimate $\sigma_A^2 = 0.4$, $\sigma_C^2 = 0.2$, and $\sigma_E^2 = 0.4$. As we saw earlier, substituting these values and *only* these values will produce expected variance-covariance matrices that match the observed perfectly (because we are modeling standardized scores, or correlations). In contrast, consider what happens under the AE model with the same data. Table A.2 shows that the AE model is unable to account for this particular set of observed values. Such a model is said to be *underidentified*. This condition is not necessarily problematic: In general, underidentified models are to be favored. Because a saturated model will *always* be able to fit the observed data perfectly, the goodness of fit does not really mean anything. However, if an underidentified model *does* fit the data, then we should take notice—it is not fitting out of mere statistical necessity. Perhaps it is a better, more parsimonious model of the data. Table A.2 represents three different sets of the values for the two parameters that attempt to explain the observed data. As the table shows, the AE model does not seem able to model our observed statistics quite as well as the ACE model.

If we run a model-fitting program such as *Mx*, we can formally determine which values for σ_A^2 and σ_E^2 give the best fit for the AE model and whether or not this fit is significantly worse than that of the saturated ACE model. Additionally, we can fit a CE model (which implies that any covariation between twins is not due to genetic factors) and an E model (which implies that there is no significant covariation between twins in any case). The results are presented in Table A.3, showing the optimized parameter values for the different models.

Because these models are not saturated, they cannot necessarily guarantee a perfect fit to the data. Adjusting one parameter to perfectly fit the MZ twin covariance pulls the DZ twin covariance or the variance estimate out of line, and vice versa. We see here that the AE model has estimated the variance and MZ covariance quite ac-

TABLE A.2
Fit of AE Model to Three Parameter Value Sets

Parameters		Variance	MZ Covariance	DZ Covariance
σ_A^2	σ_E^2			
OBSERVED				
—	—	1.0	0.6	0.4
EXPECTED				
0.6	0.4	1.0	0.6	0.3
0.7	0.3	1.0	0.7	0.35
0.8	0.2	1.0	0.8	0.4

TABLE A.3

Best-Fit Univariate Parameter Estimates

Parameters		Variance	MZ Covariance	DZ Covariance	χ^2	df^a
AE Model						
σ^2_A	σ^2_E					
0.609	0.382	0.991	0.609	0.304	1.91	4
CE Model						
σ^2_C	σ^2_E					
0.5	0.5	1.000	0.500	0.500	6.75	4
E Model	σ^2_E					
	1.000	1.000	0.000	0.000	92.47	5

^{a}df, degrees of freedom.

curately in selecting the optimized parameters $\sigma^2_A = 0.609$ and $\sigma^2_E = 0.382$ but the expected DZ covariance departs substantially from the observed value of 0.4. But is this departure significant? The last two columns give the χ^2 and associated *degrees of freedom (df)* of the test. Because we have six observed statistics, from which we are estimating two parameters under the AE model, we say that we have $6 - 2 = 4$ degrees of freedom. The degrees of freedom therefore represent a measure of how simple or complex a model is—we need to know this when deciding which is the most parsimonious model. The E model, for example, estimates only one parameter and so has $6 - 1 = 5$ degrees of freedom.

The test of whether a nested, simpler model is more parsimonious is quite simple: We look at the difference in χ^2 goodness of fit between the two models. The difference in degrees of freedom between the two models is used to determine whether or not the difference in fit is significant. If the difference is significant, then we say that the nested submodel does *not* provide a good account of the data when compared with the goodness of fit of the fuller model. The χ^2 statistics calculated in our example in Table A.3 are dependent on sample size—these figures are based on 150 MZ twins and 150 DZ twins.

The ACE model estimates three parameters from the six observed statistics, so it has three degrees of freedom; the χ^2 is always 0.0 because the model is saturated. Therefore, the difference in fit between the ACE and AE models is $1.91 - 0 = 1.91$ with $4 - 3 = 1$ degree of freedom. Looking up this χ^2 value in significance tables tells us that it is not significant at the $p = 0.05$ level (in fact, $p = 0.17$). A p value lower than 0.05 indicates that the observed results would be expected to arise less than 5 percent of the time by chance alone, if there were in reality no effect. This is commonly accepted to be sufficient evidence to reject a null hypothesis, which states that no effect is present. Therefore, because the AE model does not show a significant reduction in fit relative to the ACE model, this result provides evidence that the shared environment is not important (i.e., that σ_C^2 is not substantially greater than 0.0) for this trait.

What about the CE and E models, though? The CE model fit is reduced by a χ^2 value of 6.75, also for a gain of one degree of freedom. This reduction in fit is significant at the $p = 0.05$ level ($p = 0.0093$). This significant reduction in goodness of fit suggests that additive genetic effects are important for this trait (i.e., that $\sigma_A^2 > 0.0$). Unsurprisingly, the E model shows an even greater reduction in fit ($\Delta\chi^2 = 92.47$ for two degrees of freedom: $p < 0.00001$), thus confirming the obvious fact that the members of both types of twins do in fact show a reasonable degree of resemblance to each other.

2.2.7 *Path analysis* The kind of model fitting we have described so far is intimately related to a field of statistics called *path analysis.* Path analysis provides a visual and intuitive way to describe and explore any kind of model that describes some observed data. The *paths,* drawn as arrows, reflect the statistical effect of one variable on another, independent of all the other variables—what are called *partial regression coefficients.* The *variables* can be either measured traits (squares) or the *latent* (unmeasured; circles) variance components of our model. The twin ACE model can be represented as the path diagram in Figure A.9.

The curved, double-headed arrows between latent variables represent the covariance between them. The 1.0/0.5 on the covariance link between the two *A* latent variables indicates that for MZ twins, this covariance link is 1.0; for DZ twins, 0.5. The covariance links between the *C* and *E* terms therefore represent the previously defined sharing of these variance components between twins (i.e., no link implies a 0 covariance). The double-headed arrow loops on each latent variable represent the

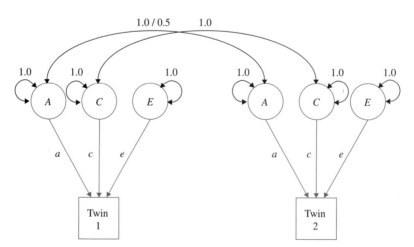

• **FIGURE A.9** ACE path diagram. This path diagram is equivalent to the matrix formulation of the ACE model. Path coefficients (*a*, *c*, and *e*) rather than variance components (which are assumed to be 1) are estimated.

variance of that variable. In our previous model fitting, we estimated the variances of these latent variables, calling them σ_A^2, σ_C^2, and σ_E^2. In our path diagram, we have fixed all the variances to 1.0. Instead, we estimate the path coefficients, which we have labeled as a, c, and e. The differences here are largely superficial: The diagram and the previous models are mathematically identical.

To understand a path diagram and how it relates to the kind of models we have discussed, we need to acquaint ourselves with a few basic rules of path analysis. The covariance between two variables is represented by tracing along all the paths that connect the two variables. There are certain rules about the directions in which paths can or cannot be traced, how loops in paths are dealt with, and so on, but the principle is simple. For each path, we multiply all the path coefficients together with the variances of any latent variables traced through. We sum these paths to calculate the expected covariance. The variance for the first twin is therefore a (up the first path) times 1.0 (the variance of latent variable A) times a (back down the path) plus the same for the paths to latent variables C and E. This equals $(a \times 1.0 \times a) + (c \times 1.0 \times c) + (e \times 1.0 \times e) = a^2 + c^2 + e^2$. So instead of estimating the variance components, we have written the model to estimate the path coefficients. This approach is used for practical reasons (e.g., it means that estimates of variance always remain positive, being the square of the path coefficient). The covariance between twins is derived in a similar way. When we trace the two paths between the twins, we get $(a \times 1.0 \times a) + (c \times 1.0 \times c)$ for MZ twins and $(a \times 0.5 \times a) + (c \times 1.0 \times c)$ for DZ twins. That is, $a^2 + c^2$ for MZ twins and $0.5a^2 + c^2$ for DZ twins, as before.

So we have seen how a properly constructed path diagram implies an expected variance-covariance (or correlation) matrix for the observed variables in the model. As noted, it is standard for the parameters in path diagrams to be path coefficients instead of variance components, although, for most basic purposes, this substitution makes very little difference. Any path diagram can be converted into a model that can be written down as algebraic terms in the elements of variance-covariance matrices, and vice versa.

2.2.8 Multivariate analysis So far we have focused on the analysis of only one phenotype at a time. This method is often called a *univariate* approach—studying the genetic-environmental nature of the *variance of one trait*. If multiple measures have been assessed for each individual, however, a model-fitting approach easily extends to analyze the genetic-environmental basis of the *covariance between multiple traits*. Is, for example, the correlation between depression and anxiety due to genes that influence both traits, or is it largely due to environments that act as risk factors for both depression and anxiety? If we think of a correlation as essentially reflecting shared causes somewhere in the etiological pathways of the two traits, multivariate genetic analysis can tell us something about the nature of these shared causes. The development of multivariate quantitative genetics is one of the most important advances in behavioral genetics during the past two decades.

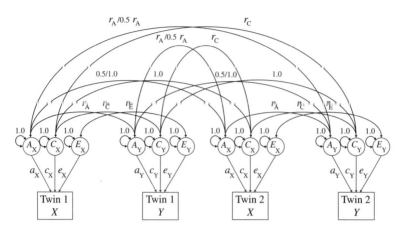

• **FIGURE A.10** Multivariate ACE path diagram. This path diagram represents a multivariate ACE model. The expected variance-covariance matrix (given in Table A.4) can be derived from this diagram by tracing the paths.

The essence of multivariate genetic analysis is the analysis of *cross-covariance* in relatives. That is, we can ask whether trait X is associated with another family member's trait Y. Path analysis provides an easy way to visualize multivariate analysis. The path diagram for a multivariate genetic analysis of two measures is shown in Figure A.10. The new parameters in this model are r_A, r_C, and r_E. These symbols represent the *genetic correlation*, the *shared environmental correlation*, and the *nonshared environmental correlation*, respectively. A genetic correlation of 1.0 would imply that all additive genetic influences on trait X also impact on trait Y. A shared environmental correlation of 0 would imply that the environmental influences that make twins more similar on measure X are independent of the environmental influences that make twins more similar on measure Y. The phenotypic correlation between X and Y can therefore be dissected into genetic and environmental constituents. A high genetic correlation implies that if a gene were found for one trait, there is a reasonable chance that this gene would also influence the second trait.

Multivariate analysis can model more than two variables—as many measures as we wish can be included. In matrix terms, instead of modeling a 2×2 matrix, we model a $2n \times 2n$ matrix, where n is the number of variables in the model. In a bivariate case, if we call the measures X and Y in Twins 1 and 2 (such that X_1 represents measure X for Twin 1), then the variance-covariance matrix would be

$$\begin{bmatrix} \mathrm{Var}(X_1) & & & \\ \mathrm{Cov}(X_1 X_2) & \mathrm{Var}(X_2) & & \\ \mathrm{Cov}(X_1 Y_1) & \mathrm{Cov}(X_2 Y_1) & \mathrm{Var}(Y_1) & \\ \mathrm{Cov}(X_1 Y_2) & \mathrm{Cov}(X_2 Y_2) & \mathrm{Cov}(Y_1 Y_2) & \mathrm{Var}(Y_2) \end{bmatrix}$$

giving us ten unique pieces of information. Along the diagonal, we have four variances—each measure in each twin. The terms $\mathrm{Cov}(X_1 Y_1)$ and $\mathrm{Cov}(X_2 Y_2)$ are the phenotypic covariances between X and Y for the first and second twin, respectively. The terms $\mathrm{Cov}(X_1 X_2)$ and $\mathrm{Cov}(Y_1 Y_2)$ are the univariate cross-twin covariances; the final two terms $\mathrm{Cov}(X_1 Y_2)$ and $\mathrm{Cov}(X_2 Y_1)$ are the cross-twin cross-trait covariances.

The corresponding multivariate ACE model for the expected variance-covariance matrix would be written in terms of univariate parameters as before (three parameters for measure X and three for measure Y) as well as three parameters for the genetic, shared environmental, and nonshared environmental correlations between the two measures (where G is the coefficient of relatedness; i.e., either 1.0 or 0.5 for MZ or DZ twins). Table A.4 presents the elements of this matrix in tabular form.

The shaded area in Table A.4 represents the cross-trait part of the model, which looks more complex than it really is. In path diagram terms, the phenotypic (within-individual) cross-trait covariance results from three paths. The first path includes the additive genetic path for measure X (a_X) multiplied by the genetic correlation between the two traits (r_A) and the additive genetic path for measure Y (a_Y). The shared environmental and nonshared environmental paths are constructed in a similar way. The cross-twin cross-trait correlations are identical, except that there are no nonshared environmental components (by definition) and there is a coefficient of relatedness, G, to determine the magnitude of shared additive genetic variance for MZ and DZ twins. Be careful in the interpretation of a *nonshared environmental correlation;* remember that this term means *nonshared* between family members, not *trait-specific.* Any environmental effect that family members do not have in common and that influences more than one trait will induce a nonshared environmental correlation between these traits.

Genetic, shared environmental, and nonshared environmental correlations are independent of univariate heritabilities. That is, two traits might both have low heritabilities but a high genetic correlation. This would mean that, although there are

TABLE A.4
Variance-Covariance Matrix for a Multivariate Genetic Model

	Twin 1 Measure X	Twin 2 Measure X	Twin 1 Measure Y	Twin 2 Measure Y
Twin 1 Measure X	$a_X^2 + c_X^2 + e_X^2$			
Twin 2 Measure X	$Ga_X^2 + c_X^2$	$a_X^2 + c_X^2 + e_X^2$		
Twin 1 Measure Y	$r_A a_X a_Y +$ $r_C c_X c_Y +$ $r_E e_X e_Y$	$Gr_A a_X a_Y +$ $r_C c_X c_Y$	$a_Y^2 + c_Y^2 + e_Y^2$	
Twin 2 Measure Y	$Gr_A a_X a_Y +$ $r_C c_X c_Y$	$r_A a_X a_Y +$ $r_C c_X c_Y +$ $r_E e_X e_Y$	$Ga_Y^2 + c_Y^2$	$a_Y^2 + c_Y^2 + e_Y^2$

probably only a few genes of modest effect that influence both these traits, whichever gene influences one trait is very likely to influence the other trait also. In this way, the analysis of these three etiological correlations can begin to tell us not just *whether* two traits are correlated but also *why* they are correlated.

Imagine that we have measured three traits, X, Y, and Z, in a sample of MZ and DZ twins (400 MZ pairs, 400 DZ pairs). What might a multivariate genetic analysis be able to tell us about the relationships between these traits? Looking at the phenotypic correlations, we observe that each trait is moderately correlated with the other two:

$$\begin{bmatrix} 1.00 & & \\ 0.42 & 1.00 & \\ 0.30 & 0.45 & 1.00 \end{bmatrix}$$

Naturally, we would be interested in the twin correlations for these measures— both the univariate and cross-trait twin correlations. For MZ twins, we might observe

$$\begin{bmatrix} 0.78 & & \\ 0.44 & 0.91 & \\ 0.08 & 0.39 & 0.70 \end{bmatrix}$$

whereas for DZ twins, we might see

$$\begin{bmatrix} 0.40 & & \\ 0.23 & 0.61 & \\ 0.04 & 0.23 & 0.58 \end{bmatrix}$$

The twin correlations along the diagonal therefore represent univariate twin correlations. For example, we can see that the correlation between MZ twins for trait Y is 0.91. The off-diagonal elements represent the cross-twin cross-trait correlations. For example, the correlation between an individual's trait X with their co-twin's trait Y is 0.23 for DZ twins. Submitting our data to formal model-fitting analysis gives optimized estimates for the univariate parameters (heritability, proportion of variance attributable to shared environment, proportion of variance attributable to nonshared environment) shown in Table A.5.

TABLE A.5
Best-Fit Univariate Parameter Estimates

Trait	Optimized Estimate (%)[a]		
	h^2	c^2	e^2
X	74	4	22
Y	60	31	9
Z	23	47	30

[a]h^2, heritability or additive genetic variance; c^2, shared environmental variance; e^2, nonshared environmental variance.

That is, traits X and Y both appear to be strongly heritable. Trait Z appears less heritable, although one-fourth of the variation in the population of twins is still due to genetic factors. The more interesting results emerge when the multivariate structure of the data is examined. The best-fitting parameter estimates for the genetic correlation matrix, the shared environment correlation matrix, and the nonshared environment correlation matrix, respectively, are presented in the following matrices:

$$\begin{bmatrix} 1.00 & & \\ 0.44 & 1.00 & \\ 0.11 & 0.75 & 1.00 \end{bmatrix} \quad \begin{bmatrix} 1.00 & & \\ 0.98 & 1.00 & \\ 0.17 & 0.26 & 1.00 \end{bmatrix} \quad \begin{bmatrix} 1.00 & & \\ 0.10 & 1.00 & \\ 0.89 & 0.46 & 1.00 \end{bmatrix}$$

Genetic correlation   Shared environmental   Nonshared environmental
matrix   correlation matrix   correlation matrix

These correlations tell an interesting story about the underlying nature of the association between the three traits. Although on the surface, traits X, Y, and Z appear to be all moderately intercorrelated, behavioral genetic analysis has revealed a nonuniform pattern of underlying genetic and environmental sources of association.

The genetic correlation between traits Y and Z is high ($r_A = 0.75$), so any genes impacting on Y are likely to also affect Z, and vice versa. The contribution of shared genetic factors to the phenotypic correlation between two traits is called the *bivariate heritability*. This statistic is calculated by tracing the genetic paths that contribute to the phenotypic correlation: in this case, a_Y and r_A (Y-Z correlation) and a_Z. In other words, the bivariate heritability is the product of the square root of both univariate heritabilities multiplied by the genetic correlation. In the case of traits Y and Z, this statistic equals $\sqrt{0.60} \times 0.75 \times \sqrt{0.23} = 0.28$. As shown in an earlier matrix, the phenotypic correlation between traits Y and Z is 0.45. Therefore, over half (62 percent $= 0.28/0.45$) of the correlation between traits Y and Z can be explained by shared genes. Note that we take the square root of the univariate heritabilities because, in path analysis terms, we only trace up the path once—in calculating the univariate heritability, we would come back down that path, therefore squaring the estimate.

The same logic can be applied to the environmental influences. Focusing on traits Y and Z, tracing the paths for shared and nonshared environmental influences yields values of 0.10 ($\sqrt{0.31} \times 0.26 \times \sqrt{0.47}$) and 0.07 ($\sqrt{0.09} \times 0.46 \times \sqrt{0.30}$) for the bivariate estimates. Note that these add up to the phenotypic correlation, as expected ($0.28 + 0.10 + 0.07 = 0.45$).

In contrast, the correlation between traits X and Z ($r = 0.30$) is not predominantly mediated by shared genetic influence: $\sqrt{0.74} \times 0.11 \times \sqrt{0.23} = 0.04$; only 13 percent of this phenotypic correlation is due to genes.

An interesting aspect of this kind of analysis is that it could potentially reveal a strong genetic overlap between two heritable traits even when the phenotypic correlation is near 0. This scenario could arise if there were, for example, a negative nonshared environmental correlation (i.e., certain environments [nonshared between family members] tend to make individuals dissimilar for two traits). Consider the following example: Two traits both have univariate heritabilities of 0.50 and no shared environmental influences, so the nonshared environment will account for the remaining 50 percent of the variance. If the traits had a genetic correlation of 0.75 but a nonshared environmental correlation of -0.75, then the phenotypic correlation would be 0. The phenotypic correlation is the sum of the chains of paths ($\sqrt{0.5} \times 0.75 \times \sqrt{0.5}$) + ($\sqrt{0.5} \times -0.75 \times \sqrt{0.5}$) = 0.0. This example shows that the phenotypic correlation by itself does not necessarily tell you very much about the shared etiologies of traits.

The preceding model is just one form of multivariate model. Different models that make different assumptions about the underlying nature of the traits can be fitted to test whether a more parsimonious explanation fits the data. For example, the *common-factor independent-pathway* model assumes that each measure has specific (subscript "S") genetic and environmental effects as well as general (subscript "C") genetic and environmental effects that create the correlations between all the measures. Figure A.11 shows a schematic path diagram for a three-trait version of this model. (*Note:* The diagram represents only one twin for convenience—the full model would have the three traits for both twins and the *A* and *C* latent variables would have the appropriate covariance links between twins.) In this path diagram, the general factors are at the bottom.

A similar but more restricted model, the *common-factor common-pathway* model, assumes that the common genetic *and* environmental effects load onto a latent variable, *L,* that in turn loads onto all the measures in the model. This model is said to be more restricted in that, because fewer parameters are estimated, the expected variance-covariance is not as free to model any pattern of phenotypic, cross-twin same-trait

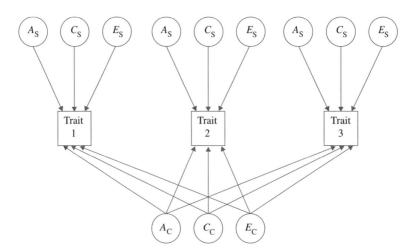

• **FIGURE A.11** Common-factor independent-pathway multivariate path diagram. This is a partial diagram, for one twin. *A,* additive genetic effects; *C,* shared environmental effects; *E,* nonshared environmental effects; *S* (subscript), specific effects; *C* (subscript), general effects.

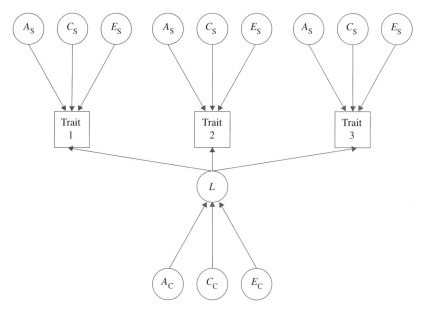

• **FIGURE A.12** Common-factor common-pathway multivariate path diagram. This is a partial diagram, for one twin. *A*, additive genetic effects; *C*, shared environmental effects; *E*, non-shared environmental effects; S (subscript), specific effects; C (subscript), general effects; *L*, latent variable.

and cross-twin cross-trait, correlations. Figure A.12 represents this model (again, for only one twin).

The common-factor independent-pathway model is nested in the more general multivariate model presented earlier; the common-factor common-pathway model is nested in both. These models can therefore be tested against each other to see which provides the most parsimonious explanation of the observations. Note that these multivariate models can also vary in terms of whether they are ACE, ADE, CE, AE, or E models.

A more specific form of multivariate model that has received a lot of interest is the *longitudinal* model. This model is appropriate for designs that take repeated measures of a trait over a period of time (say, IQ at 5, 10, 15, and 20 years of age). Such models can be used to unravel the etiology of continuity and change in a trait over time and are especially powerful for studying the interaction of genetic makeup and environment.

2.2.9 Complex effects including gene-environment interaction For the sake of simplicity (and parsimony), all the ACE-type models we have looked at so far have made various assumptions about the nature of the genetic and environmental influences that operate on the trait. Nature does not always conform to our expectations, however. In this section, we will briefly review some of the "complexities" that can be incorporated into models of genetic and environmental influence.

As mentioned earlier, an important feature of the model-fitting approach is that, as well as being flexible, it tends to make the assumptions of the model quite apparent. One such assumption is the *equal environments assumption* that MZ and DZ twins receive equally similar environments (see Chapter 6). The assumption is implicit in the model—we estimate the same parameter for shared environmental effects for MZ and DZ twins. This assumption might not always be true in practice. Can we account for potential inequalities of environment in our model? Unfortunately, not without collecting more information. The model-fitting approach is flexible, but it cannot do everything—this problem is an example of how experimental design and analysis should work hand-in-hand to tackle such questions. For example, research has compared MZ twins who have been mistakenly brought up as DZ twins, and vice versa, to study whether MZ twins are in fact treated more similarly, as indicated in Chapter 6.

Another assumption of the models used so far is random mating in the population. When nonrandom (or assortative) mating occurs (Chapter 12), then loci for a trait will be correlated between spouses. This unexpected correlation will lead to siblings and DZ twins sharing more than half their genetic variation, a situation that will bias the estimates derived from our models. In model fitting, the effects of assortative mating can be modeled (and therefore accounted for) if appropriate parental information is gathered.

Covariance between relatives on any trait can arise from a number of different sources that are not considered in our basic models. As mentioned earlier, shared causation is not the only process by which covariation can arise. The phenotype of one twin might *directly* influence the phenotype of the other, for example, because the co-twin is very much part of a twin's environment. Having an aggressive co-twin may influence levels of aggression as a result of the direct exposure to the co-twin's aggressive behavior. Such an effect is called *sibling interaction*. In the context of multivariate analysis, it is possible that trait *X* actually causes trait *Y* in the same individual, rather than a gene or environment impacting on both. These situations can be modeled by using fairly standard approaches. If such factors are important but are ignored in model fitting, they will bias estimates of genetic and environmental influence.

Another way in which the basic model might be extended is to account for possible *heterogeneity* in the sample. Genetic and environmental influences may be different for boys and girls on the same trait, or for young versus old people. Heritability is only a sample-based statistic: A heritability of 70 percent means that 70 percent of the variation *in the sample* can be accounted for by genetic effects. This outcome could be because the trait is completely heritable in 70 percent of the sample and not at all heritable in 30 percent. Such a sample would be called *heterogeneous*—there is something different and potentially interesting about the 30 percent that we may wish to study. The standard model-fitting approaches we have studied so far would leave the researcher oblivious to such effects.

To uncover heterogeneity, various approaches can be taken. Potential indices of heterogeneity (e.g., sex or age) can be incorporated into a model, for example. We could ask, Does heritability increase with age? Or we could test a model having separate parameter estimates for boys and girls for genetic effects against the nested model with only one parameter for both sexes. Same-sex and opposite-sex DZ twins can be modeled separately to test for quantitative and qualitative etiological differences between males and females. This design is called a *sex-limitation model,* and it can ask whether the magnitude of genetic and environmental effects are similar in males and females. Additionally, such designs are potentially able to test whether the *same* genes are important for both sexes, irrespective of magnitude of effect.

Other complications include *nonadditivity,* such as epistasis, gene-environment interaction, and gene-environment correlation. These three types of effects were defined under the preceding biometric model section. Epistasis is any gene-gene interaction; $G \times E$ interaction is the interaction between genetic effects and environments; G-E correlation occurs when certain genes are associated with certain environments. As an example of epistasis, imagine that an allele at locus A only predisposes toward depression if that individual also has a certain allele at locus B. As an example of gene-environment interaction, the allele at locus A may have an effect only for individuals living in deprived environments. These types of effects complicate model fitting because there are many forms in which they could occur. Normal twin study designs do not offer much hope for identifying them. An MZ correlation that is much higher than twice the DZ correlation would be suggestive of epistasis, but the models cannot really go any further in quantifying such effects.

Although model fitting can often be extended to incorporate more complex effects, it is not generally possible to include *all* these "modifications" at the same time. Successful approaches will typically select specific types of models that should be fitted a priori, on the basis of existing etiological knowledge of the traits under study.

One exciting development in model fitting involves incorporating measured variables for individuals into the analysis. Measuring alleles at specific loci, or specific environmental variables, makes the detection of specific, complex, interactive effects feasible, as well as forming the basis for modern techniques for mapping genes, as we will review in the final section.

2.2.10 Environmental mediation Behavioral genetic studies have convincingly demonstrated that genes play a significant role in many complex human traits and diseases. As a result, rather than just estimating heritability and other genetic quantities of interest, an increasingly important application of genetically informative designs, such as the twin study, is to shed light on the nature of *environmental effects.*

Although we might know that an environment and an outcome show a statistical correlation, we often do not understand the true nature of that association. For example, an association might be causal if the environment directly affects the outcome. Alternatively, the association might only arise as a reflection of some other underlying shared, possibly genetic, factor that influences both environment and the outcome. As illustrated in detail in Chapter 8, many "environmental" measures do indeed show genetic influence. By using a genetically informative design to control for genetic factors, researchers are able to make stronger inferences about environmental factors. A simple but powerful design is to focus on environmental measures that predict phenotypic differences between MZ twins.

2.2.11 Extremes analysis When we partition the variance of a trait into portions attributable to genetic or environmental effects, we are analyzing the sources of *individual differences* across the entire range of the trait. When looking at a quantitative trait, we may be more interested in one end, or extreme, of that trait. Instead of asking what makes individuals different for a trait, we might want to ask what makes individuals score high on that trait.

Consider a trait such as reading ability. Low levels of reading ability have clinical significance; individuals scoring very low will tend to be diagnosed as having reading disability. We may want to ask what makes people reading disabled, rather than what influences individuals' reading ability. We could perform a qualitative analysis where the dependent variable is simply a *Yes* or a *No* to indicate whether or not individuals are reading disabled (i.e., low scoring). If we have used a quantitative trait measure (such as a score on a reading ability task) that we believe to be related to reading disability, we may wish to retain this extra information. Indeed, we can ask whether reading disability is etiologically related to the continuum of reading ability or whether it represents a distinct syndrome. In the latter case, the factors that tend to make individuals score lower on a reading ability task in the entire population will not be the same as the factors that make people reading disabled. A regression-based method for analyzing twin data, DF (DeFries-Fulker) extremes analysis, addresses such questions, by analyzing means as opposed to variances. The methodology for DF extremes analysis is described in Chapter 11.

3 Molecular Genetics

Mapping genes for quantitative traits (quantitative trait loci, or QTLs) and diseases is a fast-developing area in behavioral genetics. The goal is to identify either the chromosomal region in which a QTL resides (via *linkage analysis*) or to pinpoint the specific variants or genes involved (via *association analysis*). The starting point for both of these molecular genetic approaches is to collect DNA, either from families or samples of unrelated individuals, and directly measure the genotype (one or more variants) to study their relationship with the phenotype. The process of measuring genotypes is called *genotyping*, where we obtain the genotype for one or more *markers* (DNA variants) in each individual. Genotyping technology has evolved rapidly over the past few decades: Whereas early studies might have considered only a handful of markers, modern molecular genetic studies can now genotype a million variants or more in *genomewide association studies*, the current state-of-the-art.

Here we will briefly review the two complementary techniques of linkage and association analysis. Linkage tests whether or not the pattern of inheritance within families at a specific locus correlates with the pattern of trait similarity. Association, on the other hand, directly tests whether specific alleles at specific markers are correlated with increased or decreased scores on a trait or with prevalence of disease.

Although there are other molecular techniques that can be applied to complex behavioral traits, we restrict our focus in this section to approaches that correlate genotype marker data to phenotype. Other approaches not covered here include *expression analysis* using microarrays (to see whether patterns of gene expression, the amount of RNA produced in particular cell types, is related to phenotype), *DNA sequencing* (to study a region's entire DNA code for each individual, for example, to see whether rare mutations, that are not represented by common, polymorphic markers, are related to phenotype), and *epigenetics* (looking at features of the genome other than the standard inherited variation of DNA bases, such as methylation patterns).

3.1 Linkage Analysis

As described in Chapter 2, Mendel coined two famous "laws," based on his studies with garden peas. His first law, the "law of segregation," basically states that each person gets a paternal and a maternal copy of each gene, and which copy they pass on to each of their offspring is random. His second law, the "law of independent assortment," further states that which copy (i.e., the paternal or maternal) of a particular gene an individual passes on to his or her child does not depend on which copy of any other gene is passed on. In other words, Mendel believed that the transmission of any two genes is statistically independent, in the same way two coin tosses are, implying four equally likely possible combinations.

Mendel did not get it 100 percent right, however. There is an important exception, which is when the two genes, let's call them *A* and *B*, are close to each other on the same chromosome. In this case, we would say that *A* and *B* are *linked* or *in linkage.* Importantly, we can exploit the property of linkage (that nearby genes tend to be cotransmitted from a parent to its offspring) to localize genes that affect phenotypes, in *linkage analysis,* as described below.

3.1.1 Patterns of gene flow in families If genes *A* and *B* were on different chromosomes, then Mendel's second law would hold. But consider what happens when they are not, as shown in Figure A.13 (next page). This figure shows a possible set of transmissions from a father and mother to their child for a stretch of this chromosome, which contains both genes *A* and *B*, very close to each other. For this whole region, the father transmits to his child the copy he received from his own father. In contrast, we see that during meiosis (the process of forming the sex cells) the mother's paternal and maternal chromosomes have experienced a *recombination event,* such that the mother transmits a mosaic of her own mother's and father's chromosomes.

Whether or not a recombination occurs at any one position is more or less a random process. Importantly, the farther away two points on a chromosome are, the more likely they are to be separated by a recombination event (technically, separated by *an odd number of recombination events,* as more than one can occur per chromosome).

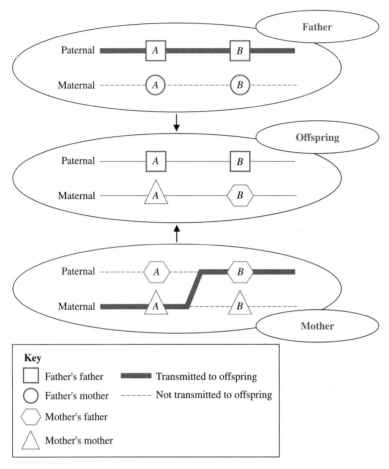

• **FIGURE A.13** Recombination of chromosomes during meiosis.

Two genes that are very close to each other on the same chromosome will tend not to be separated by recombination, however, and so they will tend to be cotransmitted from parent to offspring (i.e., either both are transmitted, or neither is). As mentioned above, this tendency is called linkage.

3.1.2 Genome scans using linkage But what is the relevance of linkage to gene mapping? How does it help us find genes that influence particular phenotypes? First, linkage analysis has been centrally important in creating maps of the genome: By studying whether or not particular DNA variants are cotransmitted in families, researchers were able to infer the relative order and positions of these markers along each chromosome. Second, linkage analysis can help to detect genotype-phenotype correlations. Instead of considering markers at two genes, *A* and *B*, it is possible to consider linkage between a marker and a phenotype. If the marker and the phenotype are similarly cotransmitted in families, we can infer the presence of a pheno-type-influencing gene that is linked to the marker.

A typical linkage analysis might involve genotyping a couple of hundred highly informative microsatellite markers (ones with many alleles) spaced across the genome, in a collection of families with multiple generations or multiple offspring. Often, the markers that are tested are not themselves assumed to be functional for the trait—they are merely selected because they are polymorphic in the population. The markers are used to statistically reconstruct the pattern of gene flow within these families for all positions along a chromosome. Such a study, often called a *genome scan,* provides an elegant way to search the entire genome for regions that might harbor phenotype-related genes. For disease traits, the simplest form of linkage analysis is to consider families with at least two affected siblings. If a region is linked with disease, we would expect the two siblings to have inherited the exact same stretch of chromosome from their parents more often than expected by chance, as a consequence of their sharing the same disease.

In practice, there are many complexities and many flavors of linkage analysis (e.g., for larger families, for continuous as well as disease traits, using different statistical models and assumptions, including variance components frameworks as described above that also incorporate marker data). Classical (parametric) linkage analysis relies on small numbers of large families (pedigrees) and explicitly models the distance between a test marker locus and a putative disease locus. The term *disease locus* (as opposed to QTL) reflects the fact that classical linkage is primarily concerned with mapping genes for dichotomous disease-like traits. Classical linkage requires that a model for the disease locus be specified a priori, in terms of allelic frequencies and mode of action (recessive or dominant). Figure A.14 shows an example of a pedigree in which a dominant gene is causing disease.

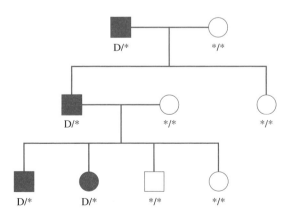

• **FIGURE A.14** A pedigree for a dominant disease (D) allele transmitted by the father. The asterisks refer to alleles other than D.

The approach of classical linkage is not so well suited to complex traits, however, for it is hard to specify any one model if we expect a large number of loci of small effect to impact on a trait. The alternative, nonparametric, or allele sharing, approach to linkage simply tests whether allele sharing at a locus correlates with trait similarity, as described above for affected sibling pairs. For quantitative traits, linkage analysis is often performed in nuclear families using a variance components framework similar to that described above for twin analysis. Using marker data, we can partition a sample of sibling pairs into those that share 0, 1, or 2 copies of the exact same parental DNA at any particular position along each chromosome. If the test locus is linked to the trait, then the sibling correlation should increase with the amount of sharing. Considering any one position, it is as if we are effectively splitting the siblings into unrelated pairs (those sharing 0 at that particular position), parent-offspring pairs (those sharing 1) and MZ twins (those sharing 2) and fitting the kind of quantitative genetic model described above, comparing these groups in the same way we compare MZ and DZ twins.

In general, linkage analysis has proven spectacularly successful in mapping many rare disease genes of large effect (for example, see Chapters 9 and 11). For many complex traits (which are often highly heritable but not influenced by only one or two major genes), linkage analysis has been less directly useful. Although linkage analysis can effectively search the entire genome with a relatively small number of markers, it lacks power to detect genes of small effect and has limited resolution. In many cases, collecting enough informative families might also be difficult.

3.2 Association Analysis

Over the past decade, association analysis has become the approach of choice for many researchers attempting to map genes of small effect for complex traits. In many ways, association analysis asks a simpler question compared to linkage analysis. Whereas linkage analysis dissects patterns of genotypic and phenotypic sharing between related individuals, association analysis directly tests whether there is a genotype-phenotype correlation. Association is typically more powerful than linkage analysis to detect small effects, but it is necessary to genotype a much greater number of markers to cover the same genomic area. Traditionally, researchers would tend to restrict association analysis to a few "candidate" genes, or regions of the genome implicated by previous linkage studies. Modern advances in genotyping technology, which allow a million or more markers to be genotyped per individual, have made very large scale studies feasible.

3.2.1 Population-based association analysis Imagine that a particular gene with two alleles, A_1 and A_2, is thought to be a QTL for a quantitative measure of cognitive ability. To test this hypothesis, a researcher might collect a sample of unrelated individuals, measured for this phenotype and genotyped for this particular locus (so we know whether an individual has A_1A_1, A_1A_2, or A_2A_2 genotype), and then ask whether the phenotype depends on genotype. The actual analysis might be a regression of phenotype (the dependent variable) on genotype (the independent variable, coded as the number of A_1 alleles an individual has, i.e., 0, 1, or 2). Similarly, if the phenotype was, instead, a disease, one might perform a *case-control study* in which a sample of cases (people with a particular disease, for example) are ascertained along with a control sample (people without the disease, but ideally who are otherwise well-matched to the case sample). If the frequency of a particular allele or genotype is significantly higher (or lower) in cases relative to controls, one would conclude that the gene shows an association with disease. For example, as discussed in Chapter 11, the frequency of the *ApoE4* allele of the gene that encodes apolipoprotein E is about 40 percent in individuals with Alzheimer disease and about 15 percent in controls.

Consider the following example of a disease-based association analysis. The basic data for a single biallelic marker can be presented in a 3×2 contingency table of disease status by genotype. In this case, the cell counts refer to the number of individuals in each of the six categories.

	Case	Control
A_1A_1	64	41
A_1A_2	86	88
A_2A_2	26	42

One could perform a test of association based on a chi-squared test of independence for a contingency table. Often, however, such data are instead collapsed into allele counts, as opposed to genotype counts. In this case, each individual contributes twice (if the marker is autosomal): A_1A_1 individuals contribute two A_1 alleles, A_2A_2 individuals contribute two A_2 alleles, and A_1A_2 individuals contribute one of each. The 2×2 contingency table now represents the number of "case alleles" and "control alleles." A test based on this table implicitly assumes a simple dosage model for the effect of each allele, which will be more powerful, if true, than a genotypic analysis.

	Case	Control
A_1	$64 \times 2 + 86 = 214$	$41 \times 2 + 88 = 170$
A_2	$26 \times 2 + 86 = 138$	$42 \times 2 + 88 = 172$

Pearson's chi-squared statistic for this table is 8.63 (which has an associated *p*-value of 0.003, as this is a 1 degree of freedom test). Standard statistical software packages can be used to calculate this kind of association statistic. Often the effect will be described as an *odds ratio*, where a value of 1 indicates no effect, a value significantly greater than 1 represents a risk effect (of A_1 in this case), and a value significantly less than 1 represents a protective effect. If the four cells of a 2×2 table are labeled a, b, c and d:

	Case	Control
A_1	a	c
A_2	b	d

then the odds ratio is calculated ad/bc. In this example, the odds ratio is therefore $(214 \times 172)/(138 \times 170) = 1.57$, indicating that A_1 increases risk for disease. For many complex traits, researchers expect very small odds ratios, such as 1.2 or 1.1, for individual markers; such small effects are statistically hard to detect. If the disease is rare, an odds ratio can be interpreted as a *relative risk*, meaning, in this example, that each extra copy of the A_1 allele an individual possesses increases his or her risk of disease by a factor of 1.57. So if A_2A_2 individuals have a baseline risk of disease of 1%, then A_1A_2 individuals would have an expected risk of 1.57% and A_1A_1 individuals would have a risk of $1.57\% \times 1.57\% = 2.46\%$.

3.2.2 Population stratification and family-based association In the previous section, we noted that samples should be well-matched. In any association study, it is particularly critical that samples be well-matched in terms of ethnicity. Failure to adequately match can result in *population stratification* (a type of confounding) which causes spurious results in which between-group differences confound the search for biologically relevant within-group effects. For example, imagine a case/control study where the sample actually comes from two distinct ethnic groups. Further, imagine that one group is overrepresented in cases versus controls (this might be because the disease is more prevalent in one group, or it might just reflect differences in how cases and controls were ascertained). Any gene that is more common in one of the ethnic groups than in the other will now show an obligatory statistical association with disease because of this third, confounding variable, ethnicity. Almost always, these associations will be completely spurious (i.e., the gene has no causal association with disease).

That correlation does not imply causation is, of course, a maxim relevant to any epidemiological study. But often in genetics we are less concerned with proving causality, per se, than we are with having *useful* correlational evidence (i.e., that could be used in locating a nearby causal disease gene, as described below in the section on indirect association). The problem with population stratification is that it will tend to throw up a very large number of red herrings that have absolutely no useful interpretation.

Luckily there are a number of ways to avoid the possible confounding due to population stratification in association studies. The most obvious is to apply sound experimental and epidemiological principles of randomization and appropriate sampling protocols. Another alternative is to use families to test for association, as most family members are necessarily well-matched for ethnicity. For example, for siblings discordant for Alzheimer disease, we would expect that the affected siblings would have a higher frequency of the *APoE4* allele of the gene encoding apolipoprotein E than would the unaffected members of the sibling pairs. Note that this is distinct from linkage analysis, which is based on sharing of chromosomal regions within families rather than testing the effects of specific alleles across families.

A common family-based association design is the *transmission/disequilibrium test* (TDT), which involves sampling affected individuals and their parents; in effect, the control individuals are created as "ghost-siblings" of cases, using the alleles that the *parents did not transmit* to their affected offspring. The test focuses only on parents who are heterozygous (e.g., have both an A_1 and an A_2 allele) and asks whether one allele was more often transmitted to affected offspring. If neither allele is associated with disease, we would expect 50:50 transmission of both alleles, as stated by Mendel's first law.

Although family-based association designs control against population stratification (and allow for some other specific hypotheses to be tested, for example, imprinting effects, in which the parental origin of an allele matters), they are in general less efficient, as more individuals must be sampled to achieve the same power as a population-based design. Recently, due to the increasing ability to genotype large numbers of markers, another approach to population stratification has emerged. By using markers randomly selected from across the genome, it is possible to empirically derive and control for ancestry in population-based studies using statistical methods.

3.2.3 Indirect association and haplotype analysis

In linkage analysis, the actual markers tested are not themselves assumed to be functional; they are merely proxies that provide information on the inheritance patterns of chromosomal regions. Similarly, in association analysis we do not necessarily assume the marker being tested is the functional, causal variant. This is because when we test any one marker, more often than not we are also implicitly testing the effects of surrounding markers, as alleles at nearby positions will be correlated at the population level. This phenomenon is closely related to linkage, described above, and is in fact called *linkage disequilibrium.*

A correlation between markers at the population level means that knowing a person's genotype at one marker tells you something about their genotype at a second marker. This correlation between markers, or linkage disequilibrium, actually reflects our shared ancestry. Over many generations, recombination has rearranged the genome, but like an imperfectly shuffled deck of cards, some traces of the previous order still exist. Because we inherit stretches of chromosomes that contain many alleles, certain strings of alleles will tend to be preserved by chance. Unless these strings are broken by recombination, the strings of alleles that sit on the same chromosomal stretch of DNA (called *haplotypes*) may become common at the population level. Considering three markers, A, B, and C (each with alleles coded 1 and 2), there may be only three common haplotypes in the population

$A_1 B_1 C_1$ 80%
$A_2 B_2 C_2$ 12%
$A_1 B_2 C_2$ 8%

In this example, possessing an A_2 allele makes you much more likely also to possess B_2 and C_2 alleles (100% of the time, in fact) than if you possess an A_1 allele (now only $8/(8 + 80) = 9\%$ of the time). We would, therefore, say that marker A is in linkage disequilibrium with B and C (and vice versa).

Linkage disequilibrium leads to indirect association; for example, if B were the true QTL, then performing an association analysis at A would still recover some of the true signal, due to the correlation in alleles, although it would be somewhat attenuated. In contrast, genotyping C instead of B would recover all the information, as it is a perfect proxy for B.

It is possible to use haplotype information in association analysis, by testing haplotypes instead of genotypes. In the above example, we might ask whether the number of copies of the $A_2B_2C_2$ haplotype that an individual possesses predicts the phenotype. By combining multiple markers in this way (called *haplotype-based association analysis*), one can extract extra information without extra genotyping. For example, imagine a fourth, ungenotyped locus, D. In this case, the $A_1B_2C_2$ haplotype is a perfect proxy for D (as it is completely correlated with the D_2 allele) whereas none of the three original individual markers are.

$A_1\ B_1\ C_1\ D_1$ 80%
$A_2\ B_2\ C_2\ D_1$ 12%
$A_1\ B_2\ C_2\ D_2$ 8%

Any one individual will possess two of these haplotypes (one paternally, one maternally inherited), for example, $A_1B_1C_1$ and $A_2B_2C_2$ if we consider just the three genotyped markers. We do not usually observe haplotypes directly, however. Instead, we observe genotypes, which in this case would be A_1A_2 for the first marker, B_1B_2 for the second, and C_1C_2 for the last. As illustrated in Figure A.15, in themselves, genotypes do not contain information about haplotypes, so it might not always be possible to determine unambiguously which haplotypes an individual has. (A particular combination of genotypes might be compatible with more than one pair of haplotypes.) However, statistical techniques can be used to estimate the frequencies of the different possible haplotypes, which in turn can be used to guess which pair of haplotypes is most likely given the genotypes for an individual (this process is called *haplotype phasing*).

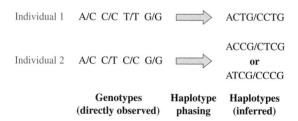

| | Genotypes (directly observed) | Haplotype phasing | Haplotypes (inferred) |

Individual 1   A/C C/C T/T G/G ⟹ ACTG/CCTG

Individual 2   A/C C/T C/C G/G ⟹ ACCG/CTCG or ATCG/CCCG

• **FIGURE A.15** Observed genotypes and inferred haplotypes.

3.2.4 The HapMap and genomewide association studies In the example above, wouldn't it have been great if we knew in advance that markers *B* and *C* were perfect proxies for each other, or that marker *D* could be predicted by a haplotype of *A, B* and *C*? Knowing that, we would probably not want to waste money genotyping all the markers, when genotyping a subset would give exactly the same information. In fact, now we usually do know in advance, thanks to the HapMap Project (http://www.hapmap.org/). This was a large, international survey of patterns of linkage disequilibrium across the genome, performed in a number of different populations, focusing on single nucleotide polymorphisms (SNPs), the most common form of variation in the human genome. SNPs are biallelic markers, with the alleles being two of A, C, G and T (i.e., the four nucleotide bases of DNA).

For many common variants, the HapMap shows that there are lots of perfect proxies in the genome; that is, there is a lot of redundancy. This means that it is possible to measure almost all common variation in the human genome using a much smaller set of SNPs. This concept is called *tagging* and effectively determines how to optimally choose which markers to genotype.

Based on large-scale genomic efforts such as the HapMap and new genotyping technologies, association analysis has recently been taken to its logical conclusion: the *genomewide association study* (GWAS). As the name suggests, this involves genotyping hundreds of thousands of markers, usually in large case/control samples. The hope is that such studies combine the power of association analysis with the genomewide, unbiased coverage of the previous generation of linkage genome scans.

Web Sites

Associations

The Behavior Genetics Association, with links to its journal, *Behavior Genetics:*
>   http://www.bga.org/

The International Society for Twin Studies is an international, multidisciplinary scientific organization whose purpose is to further research and public education in all fields related to twins and twin studies. Its Web site is linked to the society's journal, *Twin Research and Human Genetics:*
>   http://www.ists.qimr.edu.au/

The International Society of Psychiatric Genetics is a worldwide organization that aims to promote and facilitate research in the genetics of psychiatric disorders, substance use disorders, and allied traits. With links to associated journals, *Psychiatric Genetics* and *Neuropsychiatric Genetics:*
>   http://www.ispg.net/

The American Society of Human Genetics, with links to its journal, *American Journal of Human Genetics:*
>   http://www.ashg.org/

The European Society of Human Genetics, with links to its journal, *European Journal of Human Genetics:*
>   http://www.eshg.org/

The Human Genome Organization (HUGO), the international organization of scientists involved in human genetics:
>   http://www.hugo-international.org/

The International Behavioral and Neural Genetics Society (IBANGS) works to promote the field of neurobehavioral genetics. With links to its journal *Genes, Brain and Behavior.*
>   http://www.ibngs.org/

Databases and Genome Browsers

EMBL-EBI, the European Molecular Biology Laboratory's (EMBL) European Bioinformatics Institute (EBI), is the European node for globally coordinated efforts to collect and disseminate biological data:
>   http://www.ebi.ac.uk/

NCBI, the National Center for Biotechnology Information, is the U.S. node of the European Bioinformatics Institute:
>   http://www.ncbi.nlm.nih.gov/

Ensembl, the EBI and Wellcome Trust Sanger Institute's genome browser:
>   http://www.ensembl.org/

The genome browser maintained by the University of California Santa Cruz (UCSC) is an interactive open source Web site offering graphical access to genome sequence data from a variety of vertebrate and invertebrate species and major model organisms:
>   http://genome.ucsc.edu/

NCBI's Online Mendelian Inheritance in Man (OMIM) database is a catalog of human genes and

genetic disorders. The database contains textual information, pictures, and reference material:
    http://www.ncbi.nlm.nih.gov/omim/

Resources

Behavioral Genetic Interactive Modules based on the Appendix to this text by Shaun Purcell:
    http://pngu.mgh.harvard.edu/purcell/bgim/

Mx is freely available software widely used in quantitative genetic analysis:
    http://www.vcu.edu/mx/

Mx script library:
    http://www.psy.vu.nl/mxbib/

Open *Mx* forum contains lots of useful information on twin model fitting in *Mx:*
    http://openmx.psyc.virginia.edu/forums/

The Jackson Laboratory, Mouse Genome Informatics, is an excellent resource for mouse genetics:
    http://www.informatics.jax.org/

NIH Science provides the latest information on animal models used in genetic research:
    http://www.nih.gov/science/models/

The DNA Learning Center of Cold Spring Harbor Laboratory is a science center devoted entirely to genetics and provides much information online, including an animated primer on the basics of DNA, genes, and heredity:
    http://www.dnalc.org/

The Allen Brain Atlas is a collection of online public resources integrating extensive gene expression and neuroanatomical data, with a novel suite of search and viewing tools:
    http://www.brain-map.org/

R is a free software environment for statistical computing and graphics. *Bioconductor* is an open source and open development software project for the analysis and comprehension of genomic data using the *R* environment:
    http://www.r-project.org/

    http://www.bioconductor.org/

A catalog of published genome-wide association studies:
    http://www.genome.gov/gwastudies/

Microarray Technology

Affymetrix and Illumina are two of the leading suppliers of microarray technology:
    http://www.affymetrix.com/estore/

    http://www.illumina.com/

Public Understanding of Genetics

www.yourgenome.org is a Web site intended to help people understand genetics and genomic science and its implications curated by the Wellcome Trust's Sanger Institute:
    http://www.yourgenome.org/

The Genetic Science Learning Center is an outreach education program at the University of Utah. Its aim is to help people understand how genetics affects their lives and society. This is an introductory guide to molecular genetics:
    http://learn.genetics.utah.edu/index.html

The Genetics Home Reference provides consumer-friendly information about the effects of genetic variations on human health:
    http://ghr.nlm.nih.gov/

Information about genetic counseling is available from the Web site of the National Society of Genetic Counselors:
    http://www.nsgc.org/

Glossary

additive genetic variance Individual differences caused by the independent effects of alleles or loci that "add up." In contrast to *nonadditive genetic variance,* in which the effects of alleles or loci interact.

adoption studies A range of studies that use the separation of biological and social parentage brought about by adoption to assess the relative importance of genetic and environmental influences. Most commonly, the strategy involves a comparison of adoptees' resemblance to their biological parents, who did not rear them, and to their adoptive parents. May also involve the comparison of genetically related siblings and genetically unrelated (adoptive) siblings reared in the same family.

adoptive siblings Genetically unrelated children adopted by the same family and reared together.

affected sib-pair linkage design A QTL linkage design that involves pairs of siblings who meet criteria for a disorder. Linkage with DNA markers is assessed by allele sharing within the pairs of siblings—whether they share 0, 1, or 2 alleles for a DNA marker. (See Box 9.2.)

allele An alternative form of a gene at a locus, for example, A_1 versus A_2.

allele sharing Presence of zero, one, or two of the parents' alleles in two siblings (a sibling pair, or sib pair).

allelic association An association between allelic frequencies and a phenotype. For example, the frequency of allele *4* of the gene that encodes apolipoprotein E is about 40 percent for individuals with Alzheimer disease and 15 percent for control individuals who do not have the disorder.

allelic frequency Population frequency of an alternate form of a gene. For example, the frequency of the PKU allele is about 1 percent. (In contrast, see *genotypic frequency.*)

alternative splicing The process by which mRNA is reassembled to create different transcripts that are then translated into different proteins. More than half of human genes are alternatively spliced.

amino acid One of the 20 building blocks of proteins, specified by a triplet code of DNA.

amniocentesis A medical procedure used for prenatal diagnosis in which a small amount of amniotic fluid is extracted from the amnion surrounding a developing fetus. Because some of the fluid contains cells from the fetus, fetal chromosomes can be examined and fetal genes can be tested.

anticipation The severity of a disorder becomes greater or occurs at an earlier age in subsequent generations. In some disorders, this phenomenon is known to be due to the intergenerational expansion of DNA repeat sequences.

assortative mating Nonrandom mating that results in similarity between spouses. Assortative mating can be negative ("opposites attract") but is usually positive.

assortment Independent assortment is Mendel's second law of heredity. It states that the inheritance of one locus is not affected by the inheritance of another locus. Exceptions to the law occur when genes are inherited which are close together on the same chromosome. Such linkages make it possible to map genes to chromosomes.

autosome Any chromosome other than the X or Y sex chromosomes. Humans have 22 pairs of autosomal chromosomes and 1 pair of sex chromosomes.

balanced polymorphism Genetic variability that is maintained in a population, for example, by selecting against both dominant homozygotes and recessive homozygotes.

band (chromosomal) A chromosomal segment defined by staining characteristics.

base pair (bp) One step in the spiral staircase of the double helix of DNA, consisting of adenine bonded to thymine, or cytosine bonded to guanine.

behavioral genomics The study of how genes in the genome function at the behavioral level of analysis. In contrast to *functional genomics*, behavioral genomics is a top-down approach to understanding how genes work in terms of the behavior of the whole organism.

bioinformatics Techniques and resources to study the genome, transcriptome, and proteome, such as DNA sequences and functions, gene expression maps, and protein structures.

candidate gene A gene whose function suggests that it might be associated with a trait. For example, dopamine genes are considered as candidate genes for hyperactivity because the drug most commonly used to treat hyperactivity, methylphenidate, acts on the dopamine system.

carrier An individual who is heterozygous at a given locus, carrying both a normal allele and a mutant recessive allele, and who appears normal phenotypically.

centimorgan (cM) Measure of genetic distance on a chromosome. Two loci are 1 cM apart if there is a 1 percent chance of recombination due to crossover in a single generation. In humans, 1 cM corresponds to approximately 1 million base pairs.

centromere A chromosomal region without genes where the chromatids are held together during cell division.

chorion Sac within the placenta that surrounds the embryo. Two-thirds of the time, identical twins share the same chorion.

chromatid One of the two copies of DNA making up a duplicated chromosome, which are joined at their centromeres for the process of cell division (mitosis or meiosis). They are normally identical but may have slight differences in the case of mutations. They are called sister chromatids so long as they are joined by the centromeres, during which time they can recombine. When they separate, the strands are called daughter chromosomes.

chromosome A structure that is composed mainly of chromatin, which contains DNA, and resides in the nucleus of cells. Latin for "colored body" because chromosomes stain differently from the rest of the cell. (See also *autosome.*)

coding region The portion of a gene's DNA composed of exons that code for proteins.

codon A sequence of three base pairs that codes for a particular amino acid or the end of a chain.

comorbidity Presence of more than one disorder or disease in an individual.

concordance Presence of a particular condition in two family members, such as twins.

copy number variant (CNV) A structural variation that involves duplication or deletion of long stretches of DNA (one thousand to many thousands of base pairs in length), often encompassing protein-coding genes as well as noncoding genes. CNVs account for more than 10% of the human genome.

correlation An index of resemblance that ranges from -1.00 to $+1.00$, where 0.00 indicates no resemblance.

crossover See *recombination.*

developmental genetic analysis Analysis of change and continuity of genetic and environmental parameters during development. Applied to longitudinal data, assesses genetic and environmental influences on age-to-age change and continuity.

DF extremes analysis An analysis of familial resemblance that takes advantage of quantitative scores of the relatives of probands rather than just assigning a dichotomous diagnosis to the relatives and assessing concordance. (In contrast, see *liability-threshold model.*)

diallel design Complete intercrossing of three or more inbred strains and comparing all possible F_1 crosses between them.

diathesis-stress A type of genotype-environment interaction in which individuals at genetic risk for a disorder (diathesis) are especially sensitive to the effects of risky (stress) environments.

direct association An association between a trait and a DNA marker that is the functional polymorphism that causes the association. In contrast to *indirect association,* in which the DNA marker is not the functional polymorphism.

directional selection Natural selection operating against a particular allele, usually selection against a deleterious allele. (See also *balanced polymorphism* and *stabilizing selection.*)

dizygotic (DZ) Fraternal, or nonidentical, twins; literally, "two zygotes."

DNA (deoxyribonucleic acid) The double-stranded molecule that encodes genetic information. The two strands are held together by hydrogen bonds between two of the four bases, with adenine bonded to thymine, and cytosine bonded to guanine.

DNA marker A polymorphism in DNA itself, such as a single-nucleotide polymorphism (SNP) or copy number variant (CNV).

DNA sequence The order of base pairs on a single chain of the DNA double helix.

dominance The effect of one allele depends on that of another. A dominant allele produces the same phenotype in an individual regardless of whether one or two copies are present. (Compare with *epistasis,* which refers to nonadditive effects between genes at different loci.)

effect size The proportion of individual differences for the trait in the population accounted for by a particular factor. For example, heritability estimates the effect size of genetic differences among individuals.

electrophoresis A method used to separate DNA fragments by size. When an electrical charge is applied to DNA fragments in a gel, smaller fragments travel farther.

endophenotype An "inside" or intermediate phenotype that does not involve overt behavior.

environmentality Proportion of phenotypic differences among individuals that can be attributed to environmental differences in a particular population.

epigenetics DNA methylation or histone modifications that affect gene expression without changing DNA sequence. Can be involved in long-term developmental changes in gene expression.

epigenome Epigenetic events throughout the genome.

epistasis Nonadditive interaction between genes at different loci. The effect of one gene depends on that of another. (Compare with *dominance,* which refers to nonadditive effects between alleles at the same locus.)

equal environments assumption In twin studies, the assumption that environments are similar for identical and fraternal twins.

evolutionary psychology and psychiatry Fields that focus on the adaptive value of behavior as a function of natural selection.

exon DNA sequence transcribed into messenger RNA and translated into protein. (Compare with *intron.*)

expanded triplet repeat A repeating sequence of three base pairs, such as the CGG repeat responsible for fragile X, that increases in number of repeats over several generations.

expression QTL (eQTL) Treating gene expression as a phenotype, QTLs can be identified that account for genetic influence on gene expression.

F_1, F_2 The offspring in the first and second generations following mating between two inbred strains.

familial Resemblance among family members.

family study Assessing the resemblance between genetically related parents and offspring, and between siblings living together. Resemblance can be due to heredity or to shared family environment.

first-degree relative See *genetic relatedness.*

fragile X Fragile sites are breaks in chromosomes that occur when chromosomes are stained or cultured. Fragile X is a fragile site on the X chromosome that is the second most important cause after Down syndrome of mental retardation in males, and is due to an expanded triplet repeat.

full siblings Individuals who have both biological (birth) parents in common.

functional genomics The study of gene function that traces pathways between genes, brain, and behavior. Usually implies a bottom-up approach that begins with molecules in a cell, in contrast to *behavioral genomics.*

gamete Mature reproductive cell (sperm or ovum) that contains a haploid (half) set of chromosomes.

gene The basic unit of heredity. A sequence of DNA bases that codes for a particular product. Includes DNA sequences that regulate transcription. (See also *allele; locus.*)

gene expression Transcription of DNA into mRNA.

gene expression profiling Using microarrays to assess the expression of all genes in the genome simultaneously.

gene frequency Refers to frequency of alleles (e.g., A1 or A2) in a sample or population.

gene map Visual representation of the relative distances between genes or genetic markers on chromosomes.

gene silencing Suppression of gene expression.

gene targeting Mutations that are created in a specific gene and can then be transferred to an embryo.

genetical genomics Genes throughout the genome that affect gene expression. (See *transcriptome.*)

genetic anticipation See *anticipation.*

genetic counseling Conveys information about genetic risks and burdens, and helps individuals come to terms with the information and make their own decisions concerning actions.

genetic relatedness The extent to which relatives have genes in common. *First-degree relatives* of the proband (parents and siblings) are 50 percent similar genetically. *Second-degree relatives* of the proband (grandparents, aunts, and uncles) are 25 percent similar genetically. *Third-degree relatives* of the proband (first cousins) are 12.5 percent similar genetically.

genome All the DNA sequences of an organism. The human genome contains about 3 billion DNA base pairs.

genomewide association An association study that assesses DNA variation throughout the genome.

genome-wide complex trait analysis (GCTA) A technique to estimate the extent to which phenotypic variance for a trait can potentially be explained by all the single-nucleotide polymorphisms (SNPs) on a microarray. For a sample of thousands of individuals, overall genotypic similarity pair by pair is used to predict phenotypic similarity. Does not identify specific allelic associations.

genomic imprinting The process by which an allele at a given locus is expressed differently depending on whether it is inherited from the mother or the father.

genotype An individual's combination of alleles at a particular locus.

genotype-environment correlation Genetic influence on exposure to environment; experiences that are correlated with genetic propensities. In molecular genetic research, genotype-environment correlation refers to the actual correlation between genotype and an environmental measure.

genotype-environment interaction Genetic sensitivity or susceptibility to environments. Genotype-environment interaction is usually limited to statistical interactions, such as genetic effects that differ in different environments. For example, the association between a genotype for a particular gene and a phenotype might differ in different environments.

genotypic frequency The frequency of alleles considered two at a time as they are inherited in individuals. The genotypic frequency of PKU individuals (homozygous for the recessive PKU allele) is 0.0001. The genotypic frequency of PKU carriers (who are heterozygous for the PKU allele) is 0.02. In contrast, the *allelic frequency* of the recessive PKU allele is 0.01. (See Box 2.2.)

half siblings Individuals who have just one biological (birth) parent in common.

haploid genotype (haplotype) The DNA sequence on one chromosome. In contrast to *genotype,* which refers to a pair of chromosomes, the DNA sequence on one chromosome is called a *haploid genotype,* which has been shortened to *haplotype.*

haplotype block A series of single-nucleotide polymorphisms (SNPs) on a chromosome that are very highly correlated (i.e., seldom separated by recombination). The HapMap Project has systematized haplotype blocks for several ethnic groups (http://snp.cshl.org/index.html.en).

Hardy-Weinberg equilibrium Allelic and genotypic frequencies remain the same, generation after generation, in the absence of forces such as natural selection that change these frequencies. If a two-allele locus is in Hardy-Weinberg equilibrium, the frequency of genotypes is $p^2 + 2pq + q^2$, where p and q are the frequencies of the two alleles.

heritability The proportion of phenotypic differences among individuals that can be attributed to genetic differences in a particular population. *Broad-sense heritability* involves all additive and nonadditive sources of genetic variance, whereas *narrow-sense heritability* is limited to additive genetic variance.

heterosis See *hybrid vigor.*

heterozygosity The presence of different alleles at a given locus on both members of a chromosome pair.

homozygosity The presence of the same allele at a given locus on both members of a chromosome pair.

hybrid vigor The increase in viability and fertility that can occur during outbreeding, for example, when inbred strains are crossed. The increase in heterozygosity masks the effects of deleterious recessive alleles.

imprinting See *genomic imprinting.*

inbred strain A strain of animal that has been created by mating bothers and sisters for at least 20 generations, resulting in nearly genetically identical individuals.

inbred strain study Comparing inbred strains, for example on behavioral traits. Differences between strains can be attributed to their genetic differences when the strains are reared in the same laboratory environment. Differences within strains estimate environmental influences, because all individuals within an inbred strain are virtually identical genetically.

inbreeding Mating between genetically related individuals.

inbreeding depression A reduction in viability and fertility that can occur following inbreeding, which makes it more likely that offspring will have the same alleles at any locus and that deleterious recessive traits will be expressed. In contrast, see *hybrid vigor.*

inclusive fitness The reproductive fitness of an individual plus part of the fitness of kin that is genetically shared by the individual.

index case See *proband.*

indirect association An association between a trait and a DNA marker that is not itself the functional polymorphism that causes the association. In contrast to *direct association,* in which the DNA marker itself is the functional polymorphism.

innate Evolved capacities and constraints; not necessarily rigid hard-wiring that is impervious to experience.

instinct An innate behavioral tendency.

intron DNA sequence within a gene that is transcribed into messenger RNA but spliced out before the translation into protein. (Compare with *exon.*)

knock-out Inactivation of a gene by gene targeting.

latent class analysis A multivariate technique that clusters traits or symptoms into hypothesized underlying or latent classes.

liability-threshold model A model which assumes that dichotomous disorders are due to underlying genetic liabilities that are distributed normally. The disorder appears only when a threshold of liability is exceeded.

lifetime expectancy See *morbidity risk estimate.*

linkage Loci that are close together on a chromosome. Linkage is an exception to Mendel's second law of independent assortment, because closely linked loci are not inherited independently within families.

linkage analysis A technique that detects linkage between DNA markers and traits, used to map genes to chromosomes. (See also *DNA marker; linkage; mapping.*)

linkage disequilibrium A violation of Mendel's law of independent assortment in which genes are uncorrelated. It is most frequently used to describe how close DNA markers are together on a chromosome; linkage disequilibrium of 1.0 means that the alleles

of the DNA markers are perfectly correlated; 0.0 means that there is complete nonrandom association (linkage equilibrium).

locus (plural, loci) The site of a specific gene on a chromosome. Latin for "place."

major histocompatibility complex (MHC) A highly polymorphic region in a gene-dense chromosomal region (human chromosome 6) with genes particularly involved in immune functions.

mapping Linkage of DNA markers to a chromosome and to specific regions of chromosomes.

meiosis Cell division that occurs during gamete formation and results in halving the number of chromosomes, so that each gamete contains only one member of each chromosome pair.

messenger RNA (mRNA) Processed RNA that leaves the nucleus of the cell and serves as a template for protein synthesis in the cell body cytoplasm.

methylation An epigenetic process by which gene expression is inactivated by adding a methyl group to a chromosome region.

microarray Commonly known as gene chips, microarrays are slides the size of a postage stamp with hundreds of thousands of DNA sequences that serve as probes to detect gene expression (RNA microarrays) or single-nucleotide polymorphisms (DNA microarrays).

microRNA (miRNA) A class of non-coding RNA with just 21–25 nucleotides that can degrade or silence gene expression by binding with messenger RNA.

microsatellite marker Two, three, or four DNA base pairs that are repeated up to a hundred times. Unlike SNPs which generally have just two alleles, microsatellite markers often have many alleles that are inherited in a Mendelian manner.

missing heritability The difference between the total variance accounted for by known genomewide associations and heritability estimates from quantitative genetic studies.

mitosis Cell division that occurs in somatic cells in which a cell duplicates itself and its DNA.

model fitting A technique for testing the fit between a model of genetic and environmental relatedness against the observed data. Different models can be compared, and the best-fitting model is used to estimate genetic and environmental parameters.

molecular genetics The investigation of the effects of specific genes at the DNA level. In contrast to *quantitative genetics*, which partitions phenotypic variances and covariances into genetic and environmental components.

monozygotic (MZ) Identical twins; literally, "one zygote."

morbidity risk estimate The chance of being affected during one's lifetime.

multivariate genetic analysis Quantitative genetic analysis of the covariance between traits.

mutation A heritable change in DNA base-pair sequences.

natural selection The driving force in evolution in which the frequency of alleles change as a function of the differential reproduction of individuals and survival of their offspring.

neurome Effects of the genome throughout the brain.

nonadditive genetic variance Individual differences due to nonlinear interactions between alleles at the same (dominance) or different (epistasis) loci. (In contrast, see *additive genetic variance*.)

non-coding RNA (ncRNA) DNA that is transcribed into RNA but not translated into amino acid sequences. Examples include *introns* and *microRNA*.

nondisjunction Uneven division of members of a chromosome pair during meiosis.

nonshared environment Environmental influences that do not contribute to resemblance between family members.

nucleus The part of the cell that contains chromosomes.

odds ratio An effect size statistic for association calculated as the odds of an allele in cases divided by the odds of the allele in controls. An odds ratio of 1.0 means that there is no difference in allele frequency between cases and controls.

pedigree A family tree. Diagram depicting the genealogical history of a family, especially showing the inheritance of a particular condition in the family members.

pharmacogenetics and -genomics The genetics and genomics of responses to drugs.

phenotype An observed characteristic of an individual that results from the combined effects of genotype and environment.

pleiotropy Multiple effects of a gene.

polygenic trait A trait influenced by many genes.

polymerase chain reaction (PCR) A method to amplify a particular DNA sequence.

polymorphism A locus with two or more alleles. Greek for "multiple forms."

population genetics The study of allelic and genotypic frequencies in populations and the forces that change these frequencies, such as natural selection.

posttranslational modification Chemical change to polypeptides (amino acid sequences) after they have been translated from mRNA.

premutation Production of eggs or sperm with an unstable expanded number of repeats (up to 200 repeats for fragile X).

primer A short (usually 20-base) DNA sequence that marks the starting point for DNA replication. Primers on either side of a polymorphism mark the boundaries of a DNA sequence that is to be amplified by polymerase chain reaction (PCR).

proband The index case from whom other family members are identified.

proteome All the proteins translated from RNA (*transcriptome*).

psychopharmacogenetics The genetics of behavioral responses to drugs.

QTL linkage analysis Linkage analysis that searches for linkages of small effect size, quantitative trait loci (QTLs). Most widely used is the affected sib-pair QTL linkage design.

qualitative disorder An either-or trait, usually a diagnosis.

quantitative dimension Traits that are continuously distributed within a population, for example, general cognitive ability, height, and blood pressure.

quantitative genetics A theory of multiple-gene influences that, together with environmental variation, result in quantitative (continuous) distributions of phenotypes. Quantitative genetic methods (such as the twin and adoption methods for human analysis, and inbred strain and selection methods for nonhuman analysis) estimate genetic and environmental contributions to phenotypic variance and covariance in a population.

quantitative trait locus (QTL; plural: quantitative trait loci, QTLs) A gene in multiple-gene systems that contributes to quantitative (continuous) variation in a phenotype.

recessive An allele that produces its effect on a phenotype only when two copies are present.

recombinant inbred strains Inbred strains derived from brother-sister matings from an initial cross of two inbred progenitor strains. Called *recombinant* because, in the F_2 and subsequent generations, chromosomes from the progenitor strains recombine and exchange parts. Used to map genes.

recombination During meiosis, chromosomes exchange parts by a crossing over of chromatids.

recombinatorial hotspot Chromosomal location subject to much recombination. Often marks the boundaries of haplotype blocks.

repeat sequence Short sequences of DNA—two, three, or four nucleotide bases of DNA—that repeat a few times to a few dozen times. Used as DNA markers.

restriction enzyme Recognizes specific short DNA sequences and cuts DNA at that site.

ribosome A small dense structure in the cell body (cytoplasm) that assembles amino acid sequences in the order dictated by mRNA.

RNA interference (RNAi) The use of double-stranded RNA to change the expression of the gene that shares its sequence. Also called *small interfering RNA* (siRNA) because it degrades complementary RNA transcripts.

second-degree relative See *genetic relatedness.*

segregation The process by which two alleles at a locus, one from each parent, separate during heredity. Mendel's law of segregation is his first law of heredity.

selection study Breeding for a phenotype over several generations by selecting parents with high scores on the phenotype, mating them, and assessing their offspring to determine the response to selection. Bidirectional selection studies also select in the other direction, that is, for low scores.

selective placement Adoption of children into families in which the adoptive parents are similar to the children's biological parents.

sex chromosome See *autosome.*

shared environment Environmental factors that make family members similar.

single-nucleotide polymorphism (SNP) The most common type of DNA polymorphism which involves a mutation in a single nucleotide. SNPs (pronounced "snips") can produce a change in an amino acid sequence (called *nonsynonymous*, i.e., not synonymous).

small interfering RNA (siRNA) See *RNA interference (RNAi).*

sociobiology An extension of evolutionary theory that focuses on inclusive fitness and kin selection.

somatic cells All cells in the body except gametes.

stabilizing selection Selection that maintains genetic variation within a population, for example, selection for intermediate phenotypic values.

synapse A junction between two nerve cells, through which impulses pass by diffusion of a neurotransmitter, such as dopamine or serotonin.

synteny Loci on the same chromosome; related to *linkage*. *Synteny homology* refers to similar ordering of loci in chromosomal regions in different species.

targeted mutation A process by which a gene is changed in a specific way to alter its function, such as knock-outs. Called *transgenics* when the mutated gene is transferred from another species.

third-degree relative See *genetic relatedness*.

transcription The synthesis of an RNA molecule from DNA in the cell nucleus.

transcriptome RNA transcribed from all the DNA in the genome.

transgenic Containing foreign DNA. For example, gene targeting can be used to replace a gene with a nonfunctional substitute in order to knock out the gene's functioning.

translation Assembly of amino acids into peptide chains on the basis of information encoded in messenger RNA. Occurs on ribosomes in the cell cytoplasm.

triplet code See *codon*.

triplet repeat See *expanded triplet repeat*.

trisomy Having three copies of a particular chromosome due to nondisjunction.

twin correlation Correlation of twin 1 with twin 2. Typically computed separately for MZ and DZ twins. Used to estimate genetic and environmental influences.

twin study Comparing the resemblance of identical and fraternal twins to estimate genetic and environmental components of variance.

whole-genome amplification Using a few restriction enzymes in polymerase chain reactions (PCRs) to chop up and amplify the entire genome. This makes microarrays possible.

whole genome sequencing (also known as full genome sequencing) Determining the complete sequence of nucleotide base pairs for a genome.

X-linked trait A phenotype influenced by a gene on the X chromosome. X-linked recessive diseases occur more frequently in males because they only have one X chromosome.

zygote The cell, or fertilized egg, resulting from the union of a sperm and an egg (ovum).

References

Abelson, J. F., Kwan, K. Y., O'Roak, B. J., Baek, D. Y., Stillman, A. A., Morgan, T. M., . . . State, M. W. (2005). Sequence variants in SLITRK1 are associated with Tourette's syndrome. *Science, 310,* 317–320. DOI: 10.1126/science.1116502

Ackerman, P. L., Beier, M. E., & Boyle, M. O. (2005). Working memory and intelligence: The same or different constructs? *Psychological Bulletin, 131,* 30-60. DOI: 10.1037/0033–2909.131.1.30

Addington, A. M., Gornick, M., Duckworth, J., Sporn, A., Gogtay, N., Bobb, A., . . . Straub, R. E. (2005). GAD1 (2q31.1), which encodes glutamic acid decarboxylase (GAD67), is associated with childhood-onset schizophrenia and cortical gray matter volume loss. *Molecular Psychiatry, 10,* 581–588. DOI: 10.1038/sj.mp.4001599

Aebersold, R., & Mann, M. (2003). Mass spectrometry-based proteomics. *Nature, 422,* 198–207. DOI: 10.1038/nature01511

Agrawal, A., Jacobson, K. C., Prescott, C. A., & Kendler, K. S. (2002). A twin study of sex differences in social support. *Psychological Medicine, 32,* 1155–1164. DOI: 10.1017/S0033291702006281

Agrawal, A., Jacobson, K. C., Prescott, C. A., & Kendler, K. S. (2004). A twin study of personality and illicit drug use and abuse/dependence. *Twin Research, 7,* 72–81. DOI: 10.1375/13690520460741462

Agrawal, A., & Lynskey, M. T. (2008). Are there genetic influences on addiction: Evidence from family, adoption and twin studies. *Addiction, 103,* 1069–1081. DOI: 10.1111/j.1360-0443.2008.02213.x

Agrawal, A., Lynskey, M. T., Hinrichs, A., Grucza, R., Saccone, S. F., Krueger, R., . . . Bierut, L. (2011). A genome-wide association study of DSM-IV cannabis dependence. *Addiction Biology, 16,* 514–518. DOI: 10.1111/j.1369-1600.2010.00255.x

Agrawal, A., Sartor, C. E., Lynskey, M. T., Grant, J. D., Pergadia, M. L., Grucza, R., . . . Heath, A. C. (2009). Evidence for an interaction between age at first drink and genetic influences on DSM-IV alcohol dependence symptoms. *Alcoholism: Clinical and Experimental Research, 33,* 2047–2056. DOI: 10.1111/j.1530-0277.2009.01044.x

Agrawal, N., Sinha, S. N., & Jensen, A. R. (1984). Effects of inbreeding on Raven matrices. *Behavior Genetics, 14,* 579–585. DOI: 10.1007/BF01068128

Ainsworth, M. D. S., Blehar, M. C., Waters, E., & Wall, S. (1978). *Patterns of attachment: A psychological study of the strange situation.* Hillsdale, NJ: Erlbaum.

Akey, J. M. (2009). Constructing genomic maps of positive selection in humans: Where do we go from here? *Genome Research, 19,* 711–722. DOI: 10.1101/gr.086652.108

Alarcón, M., DeFries, J. C., Light, J. G., & Pennington, B. F. (1997). A twin study of mathematics disability. *Journal of Learning Disabilities, 30,* 617–623. DOI: 10.1177/002221949703000605

Alarcón, M., Plomin, R., Fulker, D. W., Corley, R., & DeFries, J. C. (1999). Molarity not modularity: Multivariate genetic analysis of specific cognitive abilities in parents and their 16-year-old children in the Colorado Adoption Project. *Cognitive Development, 14,* 175–193. DOI: 10.1016/S0885-2014(99)80023-9

Albright, F., Light, K., Light, A., Bateman, L., & Cannon-Albright, L. A. (2011). Evidence for a heritable predisposition to chronic fatigue syndrome. *BMC Neurology, 11,* 62. DOI: 10.1186/1471-2377-11-62

Allen, M. G. (1976). Twin studies of affective illness. *Archives of General Psychiatry, 33,* 1476–1478. DOI: 10.1001/archpsyc.1976.01770120080008

Allis, C. D., Jenuwein, T., & Reinberg, D. (2007). *Epigenetics*. Cold Spring Harbor, NY: Cold Spring Harbor Laboratory Press.

Alonzo, S. H. (2010). Social and coevolutionary feedbacks between mating and parental investment. *Trends in Ecology and Evolution, 25,* 99–108. DOI: 10.1016/j.tree.2009.07.012

Althoff, R. R., Faraone, S. V., Rettew, D. C., Morley, C. P., & Hudziak, J. J. (2005). Family, twin, adoption, and molecular genetic studies of juvenile bipolar disorder. *Bipolar Disorders, 7,* 598–609. DOI: 10.1111/j.1399-5618.2005.00268.x

Altmuller, J., Palmer, L. J., Fischer, G., Scherb, H., & Wjst, M. (2001). Genomewide scans of complex human diseases: True linkage is hard to find. *American Journal of Human Genetics, 69,* 936–950. DOI: 10.1086/324069

Altshuler, D. L., Durbin, R. M., Abecasis, G. R., Bentley, D. R., Chakravarti, A., Clark, A. G., ... 1000 Genomes Project Consortium. (2010). A map of human genome variation from population-scale sequencing. *Nature, 467,* 1061–1073. DOI: 10.1038/nature09534

Altshuler, D. M., Gibbs, R. A., Peltonen, L., Dermitzakis, E., Schaffner, S. F., Yu, F., ... International HapMap 3 Consortium. (2010). Integrating common and rare genetic variation in diverse human populations. *Nature, 467,* 52–58. DOI: 10.1038/nature09298

Alzheimer's Disease International. (2009). World Alzheimer Report 2009. Retrieved from http://www.alz.co.uk/research/world-report

American Psychological Association. (2000). *Diagnostic and statistical manual of mental disorders* (4th ed., text rev.) Washington, DC: American Psychiatric Association.

Amir, R. E., Van den Veyer, I. B., Wan, M., Tran, C. Q., Francke, U., & Zoghbi, H. Y. (1999). Rett syndrome is caused by mutations in X-linked MECP2, encoding methyl-CpG-binding protein 2. *Nature Genetics, 23,* 185–188. DOI: 10.1038/13810

Anderson, L. T., & Ernst, M. (1994). Self-injury in Lesch-Nyhan disease. *Journal of Autism and Developmental Disorders, 24,* 67–81. DOI: 10.1007/BF02172213

Andrews, G., Stewart, G., Allen, R., & Henderson, A. S. (1990). The genetics of six neurotic disorders: A twin study. *Journal of Affective Disorders, 19,* 23–29. DOI: 10.1016/0165-0327(90)90005-S

Anholt, R. R., & Mackay, T. F. (2004). Quantitative genetic analyses of complex behaviours in drosophila. *Nature Reviews Genetics, 5,* 838–849. DOI: 10.1038/nrg1472

Anokhin, A. P., Muller, V., Lindenberger, U., Heath, A. C., & Myers, E. (2006). Genetic influences on dynamic complexity of brain oscilla-

tions. *Neuroscience Letters, 397,* 93–98. DOI: 10.1016/j.neulet.2005.12.025

Anton, R. F., Oroszi, G., O'Malley, S., Couper, D., Swift, R., Pettinati, H., & Goldman, D. (2008). An evaluation of mu-opioid receptor (OPRM1) as a predictor of naltrexone response in the treatment of alcohol dependence: Results from the Combined Pharmacotherapies and Behavioral Interventions for Alcohol Dependence (COMBINE) Study. *Archives of General Psychiatry, 65,* 135–144. DOI: 10.1001/archpsyc.65.2.135

Archie, E. A., & Theis, K. R. (2011). Animal behaviour meets microbial ecology. *Animal Behaviour, 82,* 425–436. DOI: 10.1016/j.anbehav.2011.05.029

Ardiel, E. L., & Rankin, C. H. (2010). An elegant mind: Learning and memory in caenorhabditis elegans. *Learning and Memory, 17,* 191–201. DOI: 10.1101/lm.960510

Arseneault, L., Moffitt, T. E., Caspi, A., Taylor, A., Rijsdijk, F. V., Jaffee, S. R., ... Measelle, J. R. (2003). Strong genetic effects on cross-situational antisocial behaviour among 5-year-old children according to mothers, teachers, examiner-observers, and twins' self-reports. *Journal of Child Psychology and Psychiatry, 44,* 832–848. DOI: 10.1111/1469-7610.00168

Arvey, R. D., Bouchard, T. J., Jr., Segal, N. L., & Abraham, L. M. (1989). Job satisfaction: Environmental and genetic components. *Journal of Applied Psychology, 74,* 187–192. DOI: 10.1037/0021-9010.74.2.187

Asbury, K., Dunn, J., & Plomin, R. (2006a). The use of discordant MZ twins to generate hypotheses regarding non-shared environmental influence on anxiety in middle childhood. *Social Development, 15,* 564–570. DOI: 10.1111/j.1467-9507.2006.00356.x

Asbury, K., Dunn, J. F., & Plomin, R. (2006b). Birthweight-discordance and differences in early parenting relate to monozygotic twin differences in behaviour problems and academic achievement at age 7. *Developmental Science, 9,* F22–F31. DOI: 10.1111/j.1467-7687.2006.00469.x

Asbury, K., Wachs, T., & Plomin, R. (2005). Environmental moderators of genetic influence on verbal and nonverbal abilities in early childhood. *Intelligence, 33,* 643–661. DOI: 10.1016/j.intell.2005.03.008

Aslund, C., Nordquist, N., Comasco, E., Leppert, J., Oreland, L., & Nilsson, K. W. (2011). Maltreatment, MAOA, and delinquency: Sex differences in gene-environment interaction in a large population-based cohort of adolescents. *Behavior Genetics, 41,* 262–272. DOI: 10.1007/s10519-010-9356-y

Astrom, R. L., Wadsworth, S. J., Olson, R. K., Willcutt, E. G., & DeFries, J. C. (2011). Defries-Fulker analysis of longitudinal reading performance data from twin pairs ascertained for reading difficulties and from their nontwin siblings. *Behavior Genetics, 41,* 660–667. DOI: 10.1007/s10519-011-9445-6

Atran, S. (2005). *In gods we trust: The evolutionary landscape of religion.* Oxford: Oxford University Press. DOI: 10.1093/acprof:oso/9780195178036.001.0001

Awofala, A. A. (2011). Genetic approaches to alcohol addiction: Gene expression studies and recent candidates from drosophila. *Invertebrate Neuroscience, 11,* 1–7. DOI: 10.1007/s10158-010-0113-y

Aylor, D. L., Valdar, W., Foulds-Mathes, W., Buus, R. J., Verdugo, R. A., Baric, R. S., ... Churchill, G. A. (2011). Genetic analysis of complex traits in the emerging collaborative cross. *Genome Research, 21,* 1213–1222. DOI: 10.1101/gr.111310.110

Baddeley, A. D. (2007). *Working memory, thought, and action.* Oxford: Oxford University Press. DOI: 10.1093/acprof:oso/9780198528012.001.0001

Badner, J. A., & Gershon, E. S. (2002). Meta-analysis of whole-genome linkage scans of bipolar disorder and schizophrenia. *Molecular Psychiatry, 7,* 405–411. DOI: 10.1038/sj.mp.4001012

Bailey, A., Palferman, S., Heavey, L., & Le Couteur, A. (1998). Autism: The phenotype in relatives. *Journal of Autism and Developmental Disorders, 28,* 369–392. DOI: 10.1023/A:1026048320785

Bailey, A., Phillips, W., & Rutter, M. (1996). Autism: Towards an integration of clinical, genetic, neuropsychological, and neurobiological perspectives. *Journal of Child Psychology and Psychiatry, 37,* 89–126. DOI: 10.1111/j.1469-7610.1996.tb01381.x

Bailey, J. M., Dunne, M. P., & Martin, N. G. (2000). Genetic and environmental influences on sexual orientation and its correlates in an Australian twin sample. *Journal of Personality and Social Psychology, 78,* 524–536. DOI: 10.1037/0022-3514.78.3.524

Bailey, J. M., & Pillard, R. C. (1991). A genetic study of male sexual orientation. *Archives of General Psychiatry, 48,* 1089–1096. DOI: 10.1001/archpsyc.1991.01810360053008

Bailey, J. M., Pillard, R. C., Dawood, K., Miller, M. B., Farrer, L. A., Trivedi, S., & Murphy, R. L. (1999). A family history study of male sexual orientation using three independent samples. *Behavior Genetics, 29,* 79–86. DOI: 10.1023/A:1021652204405

Bailey, J. M., Pillard, R. C., Neale, M. C., & Agyei, Y. (1993). Heritable factors influence sexual orientation in women. *Archives of General Psychiatry, 50,* 217–223. DOI: 10.1001/archpsyc.1993.01820150067007

Bailey, J. N., Breidenthal, S. E., Jorgensen, M. J., McCracken, J. T., & Fairbanks, L. A. (2007). The association of DRD4 and novelty seeking is found in a nonhuman primate model. *Psychiatric Genetics, 17,* 23–27. DOI: 10.1097/YPG.0b013e32801140f2

Baker, J. H., Maes, H. H., Lissner, L., Aggen, S. H., Lichtenstein, P., & Kendler, K. S. (2009). Genetic risk factors for disordered eating in adolescent males and females. *Journal of Abnormal Psychology, 118,* 576–586. DOI: 10.1037/a0016314

Baker, L. A., Barton, M., Lozano, D. I., Raine, A., & Fowler, J. H. (2006). The Southern California Twin Register at the University of Southern California: II. *Twin Research and Human Genetics, 9,* 933–940. DOI: 10.1375/183242706779462912

Baker, L. A., Jacobson, K. C., Raine, A., Lozano, D. I., & Bezdjian, S. (2007). Genetic and environmental bases of childhood antisocial behavior: A multi-informant twin study. *Journal of Abnormal Psychology, 116,* 219–235. DOI: 10.1037/0021-843X.116.2.219

Baker, T. B., Piper, M. E., McCarthy, D. E., Bolt, D. M., Smith, S. S., ... Transdisciplinary Tobacco Use Research Center (TTURC) Tobacco Dependence Phenotype Workgroup (2007). Time to first cigarette in the morning as an index of ability to quit smoking: Implications for nicotine dependence. *Nicotine and Tobacco Research, 9,* S555–S570. DOI: 10.1080/14622200701673480

Bakermans-Kranenburg, M. J., van Uzendoorn, M. H., Bokhorst, C. L., & Schuengel, C. (2004). The importance of shared environment in infant-father attachment: A behavioral genetic study of the attachment Q-sort. *Journal of Family Psychology, 18,* 545–549. DOI: 10.1037/0893-3200.18.3.545

Bakwin, H. (1971). Enuresis in twins. *American Journal of Diseases in Children, 21,* 222–225. DOI: 10.1001/archpedi.1971.02100140088007

Ball, H. A., Arseneault, L., Taylor, A., Maughan, B., Caspi, A., & Moffitt, T. E. (2008). Genetic and environmental influences on victims, bullies and bully-victims in childhood. *Journal of Child Psychology and Psychiatry, 49,* 104–112. DOI: 10.1111/j.1469-7610.2007.01821.x

Baltes, P. B. (1993). The aging mind: Potential and limits. *Gerontologist, 33,* 580–594. DOI: 10.1093/geront/33.5.580

Bamshad, M. J., Ng, S. B., Bigham, A. W., Tabor, H. K., Emond, M. J., Nickerson, D. A., & Shendure, J. (2011). Exome sequencing as a tool for Mendelian disease gene discovery. *Nature Reviews Genetics, 12,* 745–755. DOI: 10.1038/nrg3031

Banaschewski, T., Becker, K., Scherag, S., Franke, B., & Coghill, D. (2010). Molecular genetics of attention-deficit/hyperactivity disorder: An overview. *European Child and Adolescent Psychiatry, 19,* 237–257. DOI: 10.1007/s00787-010-0090-z

Banerjee, K., Chabris, C. F., Johnson, V. E., Lee, J. J., Tsao, F., & Hauser, M. D. (2009). General intelligence in another primate: Individual differences across cognitive task performance in a New World monkey (Saguinus oedipus). *PLoS One, 4*: e5883. DOI: 10.1371/journal.pone.0005883

Barash, D. P., & Barash, N. R. (2005). *Madame Bovary's ovaries: A Darwinian look at literature.* New York: Delacorte Press.

Barash, Y., Calarco, J. A., Gao, W., Pan, Q., Wang, X., Shai, O., . . . Frey, B. J. (2010). Deciphering the splicing code. *Nature, 465,* 53–59. DOI: 10.1038/nature09000

Barclay, N. L., Eley, T. C., Buysse, D. J., Maughan, B., & Gregory, A. M. (2011). Nonshared environmental influences on sleep quality: A study of monozygotic twin differences. *Behavior Genetics, 42,* 234–244. DOI: 10.1007/s10519-011-9510-1

Barnea, A., Cronqvist, H., & Siegel, S. (2010). Nature or nurture: What determines investor behavior? *Journal of Financial Economics, 98,* 583–604. DOI: 10.1016/j.jfineco.2010.08.001

Baron, M., Freimer, N. F., Risch, N., Lerer, B., & Alexander, J. R. (1993). Diminished support for linkage between manic depressive illness and X-chromosome markers in three Israeli pedigrees. *Nature Genetics, 3,* 49–55. DOI: 10.1038/ng0193-49

Baron, M., Gruen, R., Asnis, L., & Lord, S. (1985). Familial transmission of schizotypal and borderline personality disorder. *American Journal of Psychiatry, 142,* 927–934.

Bartels, M. (2007). An update on longitudinal twin and family studies. *Twin Research and Human Genetics, 10,* 1–2. DOI: 10.1375/twin.10.1.1

Bartels, M., de Moor, M. H. M., van der Aa, N., Boomsma, D. I., & de Geus, E. J. C. (2012). Regular exercise, subjective wellbeing, and internalizing problems in adolescence: Causality or genetic pleiotropy? *Frontiers in Genetics, 3*: 4. DOI: 10.3389/fgene.2012.00004

Bartels, M., Rietveld, M. J., van Baal, G. C., & Boomsma, D. I. (2002a). Genetic and environmental influences on the development of intelligence. *Behavior Genetics, 32,* 237–249. DOI: 10.1023/A:1019772628912

Bartels, M., Rietveld, M. J., van Baal, G. C., & Boomsma, D. I. (2002b). Heritability of educational achievement in 12-year-olds and the overlap with cognitive ability. *Twin Research, 5,* 544–553. DOI: 10.1375/136905202762342017

Bartels, M., Saviouk, V., de Moor, M. H. M., Willemsen, G., van Beijsterveldt, T. C. E. M., Hottenga, J.-J., . . . Boomsma, D. I. (2010). Heritability and genome-wide linkage scan of subjective happiness. *Twin Research and Human Genetics, 13,* 135–142. DOI: 10.1375/twin.13.2.135

Bashi, J. (1977). Effects of inbreeding on cognitive performance. *Nature, 266,* 440–442. DOI: 10.1038/266440a0

Bassell, G. J., & Warren, S. T. (2008). Fragile X syndrome: Loss of local mRNA regulation alters synaptic development and function. *Neuron, 60,* 201–214. DOI: 10.1016/j.neuron.2008.10.004

Bates, G. P. (2005). History of genetic disease: The molecular genetics of Huntington disease—a history. *Nature Reviews Genetics, 6,* 766–773. DOI: 10.1038/nrg1686

Bates, T. C., Luciano, M., Castles, A., Coltheart, M., Wright, M. J., & Martin, N. G. (2007). Replication of reported linkages for dyslexia and spelling and suggestive evidence for novel regions on chromosomes 4 and 17. *European Journal of Human Genetics, 15,* 194–203. DOI: 10.1038/sj.ejhg.5201739

Bearden, C. E., & Freimer, N. B. (2006). Endophenotypes for psychiatric disorders: Ready for primetime? *Trends in Genetics, 22,* 306–313. DOI: 10.1016/j.tig.2006.04.004

Beauchamp, J. P., Cesarini, D., Johannesson, M., van der Loos, M. J. H. M., Koellinger, P. D., Groenen, P. J. F., . . . Christakis, N. A. (2011). Molecular genetics and economics. *Journal of Economic Perspectives, 25,* 57–82. DOI: 10.1257/jep.25.4.57

Beaujean, A. A. (2005). Heritability of cognitive abilities as measured by mental chronometric tasks: A meta-analysis. *Intelligence, 33,* 187–201. DOI: 10.1016/j.intell.2004.08.001

Beaver, K. M. (2011). Genetic influences on being processed through the criminal justice system: Results from a sample of adoptees. *Biological Psychiatry, 69,* 282–287. DOI: 10.1016/j.biopsych.2010.09.007

Beaver, K. M., Boutwell, B. B., Barnes, J. C., & Cooper, J. A. (2009). The biosocial underpinnings to adolescent victimization results from a longitudinal sample of twins. *Youth Violence and Juvenile Justice, 7,* 223–238. DOI: 10.1177/1541204009333830

Beaver, K. M., Shutt, J. E., Boutwell, B. B., Ratchford, M., Roberts, K., & Barnes, J. C. (2009). Genetic and environmental influences on levels of self-control and delinquent peer affiliation. *Criminal Justice and Behavior, 36,* 41–60. DOI: 10.1177/0093854808326992

Beck, J. A., Lloyd, S., Hafezparast, M., Lennon-Pierce, M., Eppig, J. T., Festing, M. F. W., & Fisher, E. M. C. (2000). Genealogies of mouse inbred strains. *Nature Genetics, 24,* 23–25. DOI: 10.1038/71641

Bekris, L. M., Yu, C.-E., Bird, T. D., & Tsuang, D. W. (2010). Genetics of Alzheimer disease. *Journal of Geriatric Psychiatry and Neurology, 23,* 213–227. DOI: 10.1177/0891988710383571

Bellack, A. S. (2006). Scientific and consumer models of recovery in schizophrenia: Concordance, contrasts, and implications. *Schizophrenia Bulletin, 32,* 432–442. DOI: 10.1093/schbul/sbj044

Benjamin, J., Ebstein, R., & Belmaker, R. H. (2002). *Molecular genetics and the human personality.* Washington, DC: American Psychiatric Press.

Benjamin, J., Li, L., Patterson, C., Greenburg, B. D., Murphy, D. L., & Hamer, D. H. (1996). Population and familial association between the D4 dopamine receptor gene and measures of novelty seeking. *Nature Genetics, 12,* 81–84. DOI: 10.1038/ng0196-81

Bennett, B., Carosone-Link, P., Zahniser, N. R., & Johnson, T. E. (2006). Confirmation and fine mapping of ethanol sensitivity quantitative trait loci, and candidate gene testing in the LXS recombinant inbred mice. *Journal of Pharmacology and Experimental Therapeutics, 319,* 299–307. DOI: 10.1124/jpet.106.103572

Benoit, C.-E., Rowe, W. B., Menard, C., Sarret, P., & Quirion, R. (2011). Genomic and proteomic strategies to identify novel targets potentially involved in learning and memory. *Trends in Pharmacological Sciences, 32,* 43–52. DOI: 10.1016/j.tips.2010.10.002

Benowitz, N. L. (2008). Neurobiology of nicotine addiction: Implications for smoking cessation treatment. *American Journal of Medicine, 121,* S3–S10. DOI: 10.1016/j.amjmed.2008.01.015

Bentwich, I., Avniel, A., Karov, Y., Aharonov, R., Gilad, S., Barad, O., . . . Bentwich, Z. (2005). Identification of hundreds of conserved and nonconserved human microRNAs. *Nature Genetics, 37,* 766–770. DOI: 10.1038/ng1590

Benyamin, B., St Pourcain, B., Davis, O. S. P., Davies, G., Hansell, N. K., Brion, M. A., . . . Visscher, P. M. (submitted). Childhood intelligence is heritable, highly polygenic and associated with FNBP1L. *Molecular Psychiatry.*

Benzer, S. (1973). Genetic dissection of behavior. *Scientific American, 229(6),* 24–37. DOI: 10.1038/scientificamerican1273-24

Bergeman, C. S. (1997). *Aging: Genetic and environmental influences.* Newbury Park, CA: Sage.

Bergeman, C. S. (2007). Behavioral genetics. In J. E. Birren (Ed.), *Encyclopedia of gerontology* (2nd ed.) New York: Academic Press.

Bergeman, C. S., Chipuer, H. M., Plomin, R., Pedersen, N. L., McClearn, G. E., Nesselroade, J. R., . . . McCrae, R. R. (1993). Genetic and environmental effects on openness to experience, agreeableness, and conscientiousness: An adoption/twin study. *Journal of Personality, 61,* 159–179. DOI: 10.1111/j.1467-6494.1993.tb01030.x

Bergeman, C. S., Plomin, R., McClearn, G. E., Pedersen, N. L., & Friberg, L. (1988). Genotype-environment interaction in personality development: Identical twins reared apart. *Psychology and Aging, 3,* 399–406. DOI: 10.1037/0882-7974.3.4.399

Bergeman, C. S., Plomin, R., Pedersen, N. L., & McClearn, G. E. (1991). Genetic mediation of the relationship between social support and psychological well-being. *Psychology and Aging, 6,* 640–646. DOI: 10.1037/0882-7974.6.4.640

Bergeman, C. S., Plomin, R., Pedersen, N. L., McClearn, G. E., & Nesselroade, J. R. (1990). Genetic and environmental influences on social support: The Swedish Adoption/Twin Study of Aging. *Journal of Gerontology, 45,* 101–106.

Bergen, S. E., & Petryshen, T. L. (2012). Genome-wide association studies of schizophrenia: Does bigger lead to better results? *Current Opinion in Psychiatry, 25,* 76–82. DOI: 10.1097/YCO.0b013e32835035dd

Berkman, M. B., & Plutzer, E. (2010). *Evolution, creationism, and the battle to control America's classrooms.* Cambridge: Cambridge University Press.

Berkman, M. B., & Plutzer, E. (2011). Defeating creationism in the courtroom, but not in the classroom. *Science, 331,* 404–405. DOI: 10.1126/science.1198902

Bernards, R. (2006). Exploring the uses of RNAi—gene knockdown and the Nobel Prize. *New England Journal of Medicine, 355,* 2391–2393. DOI: 10.1056/NEJMp068242

Bertelsen, A. (1985). Controversies and consistencies in psychiatric genetics. *Acta Psychiatrica Scandinavica, 71,* 61–75. DOI: 10.1111/j.1600-0447.1985.tb08523.x

Bertram, L., McQueen, M. B., Mullin, K., Blacker, D., & Tanzi, R. E. (2007). Systematic meta-analyses of Alzheimer disease genetic association studies: The AlzGene database. *Nature Genetics, 39,* 17–23. DOI: 10.1038/ng1934

Bessman, S. P., Williamson, M. L., & Koch, R. (1978). Diet, genetics, and mental retardation interaction between phenylketonuric heterozygous mother and fetus to produce non-specific dimunution of IQ: Evidence in support of the justification hypothesis. *Proceedings of the National Academy of Sciences (USA), 78,* 1562–1566. DOI: 10.1073/pnas.75.3.1562

Betjemann, R. S., Johnson, E. P., Barnard, H., Boada, R., Filley, C. M., Filipek, P. A., . . Pennington, B. F. (2010). Genetic covariation between brain volumes and IQ, reading performance, and processing speed. *Behavior Genetics, 40,* 135–145. DOI: 10.1007/s10519-009-9328-2

Betsworth, D. G., Bouchard, T. J., Jr., Cooper, C. R., Grotevant, H. D., Hansen, J. I. C., Scarr, S., & Weinberg, R. A. (1994). Genetic and environmental influences on vocational interests assessed using adoptive and biological families and twins reared apart and together. *Journal of Vocational Behavior, 44,* 263–278. DOI: 10.1006/jvbe.1994.1018

Bettens, K., Sleegers, K., & Van Broeckhoven, C. (2010). Current status on Alzheimer disease molecular genetics: From past, to present, to future. *Human Molecular Genetics, 19,* R4–R11. DOI: 10.1093/hmg/ddq142

Bezdjian, S., Baker, L. A., & Tuvblad, C. (2011). Genetic and environmental influences on impulsivity: A meta-analysis of twin, family and adoption studies. *Clinical Psychology Review, 31,* 1209–1223. DOI: 10.1016/j.cpr.2011.07.005

Bickle, J. (2003). *Philosophy and neuroscience: A ruthlessly reductive account.* Boston, MA: Kluwer Academic. DOI: 10.1007/978-94-010-0237-0

Biederman, J., Faraone, S. V., Keenan, K., Benjamin, J., Krifcher, B., Moore, C., . . . Steingard, R. (1992). Further evidence for family-genetic risk factors in attention deficit hyperactivity disorder. Patterns of comorbidity in probands and relatives psychiatrically and pediatrically referred samples. *Archives of General Psychiatry, 49,* 728–738. DOI: 10.1001/archpsyc.1992.01820090056010

Bienvenu, T., & Chelly, J. (2006). Molecular genetics of Rett syndrome: When DNA methylation goes unrecognized. *Nature Reviews Genetics, 7,* 415–426. DOI: 10.1038/nrg1878

Bierut, L. J. (2011). Genetic vulnerability and susceptibility to substance dependence. *Neuron, 69,* 618–627. DOI: 10.1016/j.neuron.2011.02.015

Bierut, L. J., Dinwiddie, S. H., Begleiter, H., Crowe, R. R., Hesselbrock, V., Nurnberger, J. I., . . . Reich, T. (1998). Familial transmission of substance dependence: Alcohol, marijuana, cocaine, and habitual smoking—a report from the Collaborative Study on the Genetics of Alcoholism. *Archives of General Psychiatry, 55,* 982–988. DOI: 10.1001/archpsyc.55.11.982

Biesecker, B. B., & Marteau, T. (1999). The future of genetic counselling: An international perspective. *Nature Genetics, 22,* 133–137. DOI: 10.1038/9641

Binder, E. B., & Holsboer, F. (2006). Pharmacogenomics and antidepressant drugs. *Annals of Medicine, 38,* 82–94. DOI: 10.1080/07853890600551045

Bird, A. (2007). Perceptions of epigenetics. *Nature, 447,* 396–398. DOI: 10.1038/nature05913

Bird, T. D. (2008). Genetic aspects of Alzheimer disease. *Genetics in Medicine, 10,* 231–239. DOI: 10.1097/GIM.0b013e31816b64dc

Bishop, D. V. M., Jacobs, P. A., Lachlan, K., Wellesley, D., Barnicoat, A., Boyd, P. A., . . . Scerif, G. (2011). Autism, language and communication in children with sex chromosome trisomies. *Archives of Disease in Childhood, 96,* 954–959. DOI: 10.1136/adc.2009.179747

Bittles, A. H., & Black, M. L. (2010). The impact of consanguinity on neonatal and infant health. *Early Human Development, 86,* 737–741. DOI: 10.1016/j.earlhumdev.2010.08.003

Blaser, R., & Gerlai, R. (2006). Behavioral phenotyping in zebrafish: Comparison of three behavioral quantification methods. *Behavior Research Methods, 38,* 456–469. DOI: 10.3758/BF03192800

Blokland, G. A. M., McMahon, K. L., Thompson, P. M., Martin, N. G., de Zubicaray, G. I., & Wright, M. J. (2011). Heritability of working memory brain activation. *Journal of Neuroscience, 31,* 10882–10890. DOI: 10.1523/JNEUROSCI.5334-10.2011

Bloom, F. E., & Kupfer, D. J. (1995). *Psychopharmacology: A fourth generation of progress.* New York: Raven Press.

Boardman, J. D. (2011). Gene-environment interplay for the study of political behaviors. In P. K. Hatemi & R. McDermot (Eds.), *Man is by nature a political animal: Evolution, biology, and politics.* Chicago: University of Chicago Press.

Boardman, J. D., Alexander, K. B., & Stallings, M. C. (2011). Stressful life events and depression among adolescent twin pairs. *Biodemography and Social Biology, 57,* 53–66. DOI: 10.1080/19485565.2011.574565

Bobb, A. J., Castellanos, F. X., Addington, A. M., & Rapoport, J. L. (2006). Molecular genetic studies of ADHD: 1991 to 2004. *American Journal of Medical Genetics. B: Neuropsychiatric Genetics, 141B,* 551–681. DOI: 10.1002/ajmg.b.30086

Boehm, T., & Zufall, F. (2006). MHC peptides and the sensory evaluation of genotype. *Trends in Neuroscience, 29,* 100–107. DOI: 10.1016/j.tins.2005.11.006

Bohman, M. (1996). Predisposition to criminality: Swedish adoption studies in retrospect. In G. R. Bock & J. A. Goode (Eds.), *Ciba Foundation Symposium 194—Genetics of criminal and antisocial behaviour* (pp. 99-114). Chichester, UK: John Wiley & Sons. DOI: 10.1002/9780470514825.ch6

Bohman, M., Cloninger, C. R., Sigvardsson, S., & von Knorring, A. L. (1982). Predisposition to petty criminality in Swedish adoptees. I. Genetic and environmental heterogeneity. *Archives of General Psychiatry, 39,* 1233–1241. DOI: 10.1001/archpsyc.1982.04290110001001

Bohman, M., Cloninger, C. R., von Knorring, A. L., & Sigvardsson, S. (1984). An adoption study of somatoform disorders. III. Cross-fostering analysis and genetic relationship to alcoholism and criminality. *Archives of General Psychiatry, 41,* 872–878. DOI: 10.1001/archpsyc.1984.01790200054007

Boker, S., Neale, M., Maes, H., Wilde, M., Spiegel, M., Brick, T., . . . Fox, J. (2011). OpenMx: An open source extended structural equation modeling framework. *Psychometrika, 76,* 306–317. DOI: 10.1007/s11336-010-9200-6

Bokhorst, C. L., Bakermans-Kranenburg, M. J., Fearon, R. M., van Ijzendoorn, M. H., Fonagy, P., & Schuengel, C. (2003). The importance of shared environment in mother-infant attachment security: A behavioral genetic study. *Child Development, 74,* 1769–1782. DOI: 10.1046/j.1467-8624.2003.00637.x

Bolhuis, J. J., Brown, G. R., Richardson, R. C., & Laland, K. N. (2011). Darwin in mind: New opportunities for evolutionary psychology. *PLoS Biology, 9:* e1001109. DOI: 10.1371/journal.pbio.1001109

Bolinskey, P. K., Neale, M. C., Jacobson, K. C., Prescott, C. A., & Kendler, K. S. (2004). Sources of individual differences in stressful life event exposure in male and female twins. *Twin Research, 7,* 33–38. DOI: 10.1375/13690520460741426

Bolton, D., Eley, T. C., O'Connor, T. G., Perrin, S., Rabe-Hesketh, S., Rijsdijk, F. V., & Smith, P. (2006). Prevalence and genetic and environmental influences on anxiety disorders in 6-year-old twins. *Psychological Medicine, 36,* 335–344. DOI: 10.1017/S0033291705006537

Bolton, D., & Hill, J. (2004). *Mind, meaning and mental disorder: The nature of causal explanation in psychology and psychiatry.* Oxford: Oxford University Press.

Böök, J. A. (1957). Genetical investigation in a North Swedish population: The offspring of first-cousin marriages. *Annals of Human Genetics, 21,* 191–221. DOI: 10.1111/j.1469-1809.1972.tb00282.x

Boomsma, D., Busjahn, A., & Peltonen, L. (2002). Classical twin studies and beyond. *Nature Reviews Genetics, 3,* 872–882. DOI: 10.1038/nrg932

Borkenau, P., Riemann, R., Spinath, F. M., & Angleitner, A. (2006). Genetic and environmental influences on person X situation profiles. *Journal of Personality, 74,* 1451–1480. DOI: 10.1111/j.1467-6494.2006.00416.x

Bornovalova, M. A., Hicks, B. M., Iacono, W. G., & McGue, M. (2010). Familial transmission and heritability of childhood disruptive disorders. *American Journal of Psychiatry, 167,* 1066–1074. DOI: 10.1176/appi.ajp.2010.09091272

Bouchard, T. J., Jr., & Loehlin, J. C. (2001). Genes, evolution, and personality. *Behavior Genetics, 31,* 243–273. DOI: 10.1023/A:1012294324713

Bouchard, T. J., Jr., Lykken, D. T., McGue, M., Segal, N. L., & Tellegen, A. (1990). Sources of human psychological differences: The Minnesota Study of Twins Reared Apart. *Science, 250,* 223–228. DOI: 10.1126/science.2218526

Bouchard, T. J., Jr., & McGue, M. (1981). Familial studies of intelligence: A review. *Science, 212,* 1055–1059. DOI: 10.1126/science.7195071

Bouchard, T. J., Jr., & Propping, P. (1993). *Twins as a tool of behavioral genetics.* Chichester, UK: John Wiley & Sons.

Bovet, D. (1977). Strain differences in learning in the mouse. In A. Oliverio (Ed.), *Genetics, environment and intelligence* (pp. 79–92). Amsterdam: North-Holland.

Bovet, D., Bovet-Nitti, F., & Oliverio, A. (1969). Genetic aspects of learning and memory in mice. *Science, 163,* 139–149. DOI: 10.1126/science.163.3863.139

Bowes, L., Maughan, B., Ball, H., Shakoor, S., Ouellet-Morin, I., Caspi, A., . . . Arseneault, L. (in press). Chronic bullying victimization across school transitions: The role of genetic and environmental influences. *Development and Psychopathology.*

Bratko, D., & Butkovic, A. (2007). Stability of genetic and environmental effects from adolescence to young adulthood: Results of Croatian longitudinal twin study of personality. *Twin Research and Human Genetics, 10,* 151–157. DOI: 10.1375/twin.10.1.151

Braungart, J. M., Fulker, D. W., & Plomin, R. (1992). Genetic mediation of the home environment during infancy: A sibling adoption study of the home. *Developmental Psychology, 28,* 1048–1055. DOI: 10.1037/0012-1649.28.6.1048

Braungart, J. M., Plomin, R., DeFries, J. C., & Fulker, D. W. (1992). Genetic influence on tester-rated infant temperament as assessed by Bayley's Infant Behavior Record: Nonadoptive and adoptive siblings and twins. *Developmental Psychology, 28,* 40–47. DOI: 10.1037/0012-1649.28.1.40

Bray, G. A. (1986). Effects of obesity on health and happiness. In K. E. Brownell & J. P. Foreyt (Eds.), *Handbook of eating disorders: Physiology, psychology and treatment of obesity anorexia and bulimia* (pp. 1–44). New York: Basic Books.

Breen, F., Plomin, R., & Wardle, J. (2006). Heritability of food preferences in young children. *Physiology and Behavior, 88,* 443–447. DOI: 10.1016/j.physbeh.2006.04.016

Breitner, J. C., Welsh, K. A., Gau, B. A., McDonald, W. M., Steffens, D. C., Saunders, A. M., . . . Folstein, M. F. (1995). Alzheimer's disease in the National Academy of Sciences—National Research Council registry of aging twin veterans. III. Detection of cases, longitudinal results, and observations on twin concordance. *Archives of Neurology, 52,* 763–771. DOI: 10.1001/archneur.1995.0054032 0035011

Brendgen, M., Boivin, M., Dionne, G., Barker, E. D., Vitaro, F., Girard, A., . . . Perusse, D. (2011). Gene-environment processes linking aggression, peer victimization, and the teacher-child relationship. *Child*

Development, 82, 2021–2036. DOI: 10.1111/j.1467-8624.2011.01644.x

Brendgen, M., Boivin, M., Vitaro, F., Girard, A., Dionne, G., & Perusse, D. (2008). Gene-environment interaction between peer victimization and child aggression. *Development and Psychopathology, 20,* 455–471. DOI: 10.1017/S0954579408000229

Brendgen, M., Vitaro, F., Boivin, M., Girard, A., Bukowski, W. M., Dionne, G., … Perusse, D. (2009). Gene-environment interplay between peer rejection and depressive behavior in children. *Journal of Child Psychology and Psychiatry, 50,* 1009–1017. DOI: 10.1111/j.1469-7610.2009.02052.x

Brennan, P. A., Mednick, S. A., & Jacobsen, B. (1996). Assessing the role of genetics in crime using adoption cohorts. In G. R. Bock & J. A. Goode (Eds.), *Ciba Foundation Symposium 194—Genetics of criminal and antisocial behaviour* (pp. 115–128). Chichester, UK: John Wiley & Sons. DOI: 10.1002/9780470514825.ch7

Brett, D., Pospisil, H., Valcárcel, J., Reich, J., & Bork, P. (2002). Alternative splicing and genome complexity. *Nature Genetics, 30,* 29–30. DOI: 10.1038/ng803

Broadhurst, P. L. (1978). *Drugs and the inheritance of behavior.* New York: Plenum. DOI: 10.1007/978-1-4613-3979-3

Brody, N. (1992). *Intelligence* (2nd ed.) New York: Academic Press.

Broms, U., Silventoinen, K., Madden, P. A., Heath, A. C., & Kaprio, J. (2006). Genetic architecture of smoking behavior: A study of Finnish adult twins. *Twin Research and Human Genetics, 9,* 64–72. DOI: 10.1375/twin.9.1.64

Brooker, R. J., Neiderhiser, J. M., Kiel, E. J., Leve, L. D., Shaw, D. S., & Reiss, D. (2011). The association between infants' attention control and social inhibition is moderated by genetic and environmental risk for anxiety. *Infancy, 16,* 490–507. DOI: 10.1111/j.1532-7078.2011.00068.x

Brouwer, S. I., van Beijsterveldt, T. C., Bartels, M., Hudziak, J. J., & Boomsma, D. I. (2006). Influences on achieving motor milestones: A twin-singleton study. *Twin Research and Human Genetics, 9,* 424–430. DOI: 10.1375/183242706777591191

Brown, S. D., Hancock, J. M., & Gates, H. (2006). Understanding mammalian genetic systems: The challenge of phenotyping in the mouse. *PLoS Genetics, 2:* e118. DOI: 10.1371/journal.pgen.0020118

Brumm, V. L., & Grant, M. L. (2010). The role of intelligence in phenylketonuria: A review of research and management. *Molecular Genetics and Metabolism, 99,* S18–S21. DOI: 10.1016/j.ymgme.2009.10.015

Brunner, H. G. (1996). MAOA deficiency and abnormal behaviour. In G. R. Bock & J. A. Goode (Eds.), *Ciba Foundation Symposium 194—Genetics of criminal and antisocial behaviour* (pp. 155–164). Chichester, UK: John Wiley & Sons.

Brunner, H. G., Nelen, M., Breakefield, X. O., Ropers, H. H., & van Oost, B. A. (1993). Abnormal behavior associated with a point mutation in the structural gene for monoamine oxidase A. *Science, 262,* 578–580. DOI: 10.1126/science.8211186

Bruun, K., Markkananen, T., & Partanen, J. (1966). *Inheritance of drinking behaviour: A study of adult twins.* Helsinki, Finland: Finnish Foundation for Alcohol Research.

Buchwald, D., Herrell, R., Ashton, S., Belcourt, M., Schmaling, K., Sullivan, P., . . . Goldberg, J. (2001). A twin study of chronic fatigue. *Psychosomatic Medicine, 63,* 936–943.

Buck, K. J., Rademacher, B. S., Metten, P., & Crabbe, J. C. (2002). Mapping murine loci for physical dependence on ethanol. *Psychopharmacology, 160,* 398–407. DOI: 10.1007/s00213-001-0988-8

Buizer-Voskamp, J. E., Muntjewerff, J.-W., Genetic Risk and Outcome in Psychosis (GROUP) Consortium, Strengman, E., Sabatti, C., Stefansson, H., … Ophoff, R. A. (2011). Genome-wide analysis shows increased frequency of copy number variation deletions in Dutch schizophrenia patients. *Biological Psychiatry, 70,* 655–662. DOI: 10.1016/j.biopsych.2011.02.015

Bulik, C. M. (2005). Exploring the gene-environment nexus in eating disorders. *Journal of Psychiatry and Neuroscience, 30,* 335–339.

Bulik, C. M., Sullivan, P. F., Tozzi, F., Furberg, H., Lichtenstein, P., & Pedersen, N. L. (2006). Prevalence, heritability, and prospective risk factors for anorexia nervosa. *Archives of General Psychiatry, 63,* 305–312. DOI: 10.1001/archpsyc.63.3.305

Bulik, C. M., Sullivan, P. F., Wade, T. D., & Kendler, K. S. (2000). Twin studies of eating disorders: A review. *International Journal of Eating Disorders, 27,* 1–20. DOI: 10.1002/(SICI)1098-108X(200001)27:1<1::AID-EAT1>3.0.CO;2-Q

Buller, D. J. (2005a). *Adapting minds: Evolutionary psychology and the persistent quest for human nature.* London: Bradford Books.

Buller, D. J. (2005b). Evolutionary psychology: The emperor's new paradigm. *Trends in Cognitive Sciences, 9,* 277–283. DOI: 10.1016/j.tics.2005.04.003

Bullock, B. M., Deater-Deckard, K., & Leve, L. D. (2006). Deviant peer affiliation and problem behavior: A test of genetic and environmental influences. *Journal of Abnormal Child Psychology, 34,* 29–41. DOI: 10.1007/s10802-005-9004-9

Burgess, R. L., & Drais, A. A. (1999). Beyond the "Cinderella effect": Life history theory and child maltreatment. *Human Nature, 10,* 373–398. DOI: 10.1007/s12110-999-1008-7

Burke, K. C., Burke, J. D., Roe, D. S., & Regier, D. A. (1991). Comparing age at onset of major depression and other psychiatric disorders by birth cohorts in five U.S. community populations. *Archives of General Psychiatry, 48,* 789–795. DOI: 10.100 1/archpsyc.1991.01810330013002

Burkhouse, K. L., Gibb, B. E., Coles, M. E., Knopik, V. S., & McGeary, J. E. (2011). Serotonin transporter genotype moderates the link between children's reports of overprotective parenting and their behavioral inhibition. *Journal of Abnormal Child Psychology, 39,* 783–790. DOI: 10.1007/s10802-011-9526-2

Burks, B. (1928). The relative influence of nature and nurture upon mental development: A comparative study on foster parent-foster child resemblance. *Yearbook of the National Society for the Study of Education (Part 1), 27,* 219–316.

Burt, C. (1966). Genetic determination of differences in intelligence—a study of monozygotic twins reared together and apart. *British Journal of Psychology, 57,* 137–153. DOI: 10.1111/j.2044-8295.1966.tb01014.x

Burt, S. A. (2008). Genes and popularity—evidence of an evocative gene-environment correlation. *Psychological Science, 19,* 112–113. DOI: 10.1111/j.1467-9280.2008.02055.x

Burt, S. A. (2009a). Are there meaningful etiological differences within antisocial behavior? Results of a meta-analysis. *Clinical Psychology Review, 29,* 163–178. DOI: 10.1016/j.cpr.2008.12.004

Burt, S. A. (2009b). Rethinking environmental contributions to child and adolescent psychopathology: A meta-analysis of shared environmental influences. *Psychological Bulletin, 135,* 608–637. DOI: 10.1037/a0015702

Burt, S. A., Krueger, R. F., McGue, M., & Iacono, W. (2003). Parent-child conflict and the comorbidity among childhood externalizing disorders. *Archives of General Psychiatry, 60,* 505–513. DOI: 10.1001/archpsyc.60.5.505

Burt, S. A., McGue, M., & Iacono, W. G. (2010). Environmental contributions to the stability of antisocial behavior over time: Are they shared or non-shared? *Journal of Abnormal Child Psychology, 38,* 327–337. DOI: 10.1007/s10802-009-9367-4

Burt, S. A., McGue, M., Krueger, R. F., & Iacono, W. G. (2005). How are parent-child conflict and childhood externalizing symptoms related over time? Results from a genetically informative cross-lagged study. *Development and Psychopathology, 17,* 145–165. DOI: 10.1017/S095457940505008X

Burt, S. A., & Neiderhiser, J. M. (2009). Aggressive versus nonaggressive antisocial behavior:

Distinctive etiological moderation by age. *Developmental Psychology, 45,* 1164–1176. DOI: 10.1037/a0016130

Busjahn, A. (2002). Twin registers across the globe: What's out there in 2002? *Twin Research, 5,* v–vi. DOI: 10.1375/twin.5.5.v

Buss, A. H., & Plomin, R. (1984). *Temperament: Early developing personality traits.* Hillsdale, NJ: Lawrence Erlbaum.

Buss, D. M. (1994a). *The evolution of desire: Strategies of human mating.* New York: Basic Books.

Buss, D. M. (1994b). The strategies of human mating. *American Scientist, 82,* 238–249.

Buss, D. M. (2003). *The evolution of desire: Strategies of human mating* (rev. ed.). New York: Basic Books.

Buss, D. M. (2005). *The handbook of evolutionary psychology.* New York: John Wiley & Sons.

Buss, D. M. (2011). *Evolutional psychology: The new science of the mind* (4th ed.) Boston, MA: Allyn & Bacon.

Buss, D. M., & Greiling, H. (1999). Adaptive individual differences. *Journal of Personality, 67,* 209–243. DOI: 10.1111/1467-6494.00053

Busto, G. U., Cervantes-Sandoval, I., & Davis, R. L. (2010). Olfactory learning in drosophila. *Physiology, 25,* 338–346. DOI: 10.1152/physiol.00026.2010

Butcher, L. M., Davis, O. S. P., Craig, I. W., & Plomin, R. (2008). Genome-wide quantitative trait locus association scan of general cognitive ability using pooled DNA and 500k single nucleotide polymorphism microarrays. *Genes, Brain and Behavior, 7,* 435–446. DOI: 10.1111/j.1601-183X.2007.00368.x

Butcher, L. M., Meaburn, E., Dale, P. S., Sham, P., Schalkwyk, L. C., Craig, I. W., & Plomin, R. (2005). Association analysis of mild mental impairment using DNA pooling to screen 432 brain-expressed single-nucleotide polymorphisms. *Molecular Psychiatry, 10,* 384–392. DOI: 10.1038/sj.mp.4001589

Butcher, L. M., Meaburn, E., Knight, J., Sham, P., Schalkwyk, L. C., Craig, I. W., & Plomin, R. (2005). SNPs, microarrays, and pooled DNA: Identification of four loci associated with mild mental impairment in a sample of 6000 children. *Human Molecular Genetics, 14,* 1315–1325. DOI: 10.1093/hmg/ddi142

Butler, R. J., Galsworthy, M. J., Rijsdijk, F., & Plomin, R. (2001). Genetic and gender influences on nocturnal bladder control: A study of 2900 3-year-old twin pairs. *Scandinavian Journal of Urology and Nephrology, 35,* 177–183. DOI: 10.1080/003655901750291917

Byrne, B., Coventry, W. L., Olson, R. K., Samuelsson, S., Corley, R., Willcutt, E. G., . . . DeFries, J. C. (2009). Genetic and environmental influences on aspects of literacy and language in early childhood: Continuity and change from preschool to Grade 2. *Journal of Neurolinguistics, 22,* 219–236. DOI: 10.1016/j.jneuroling.2008.09.003

Byrne, B., Samuelsson, S., Wadsworth, S., Hulslander, J., Corley, R., DeFries, J. C., . . . Olson, R. K. (2007). Longitudinal twin study of early literacy development: Preschool through Grade 1. *Reading and Writing, 20,* 77–102. DOI: 10.1007/s11145-006-9019-9

Byrne, B., Wadsworth, S., Corley, R., Samuelsson, S., Quain, P., DeFries, J. C., . . . Olson, R. K. (2005). Longitudinal twin study of early literacy development: Preschool and kindergarten phases. *Scientific Studies of Reading, 9,* 219–235. DOI: 10.1207/s1532799xssr0903_3

Cadoret, R. J. (1994). Genetic and environmental contributions to heterogeneity in alcoholism: Findings from the Iowa adoption studies. *Annals of the New York Academy of Sciences, 708,* 59–71.

Cadoret, R. J., Cain, C. A., & Crowe, R. R. (1983). Evidence from gene-environment interaction in the development of adolescent antisocial behavior. *Behavior Genetics, 13,* 301–310. DOI: 10.1007/BF01071875

Cadoret, R. J., O'Gorman, T. W., Heywood, E., & Troughton, E. (1985). Genetic and environmental factors in major depression. *Journal of Affective Disorders, 9,* 155–164. DOI: 10.1016/0165-0327(85)90095-3

Cadoret, R. J., O'Gorman, T. W., Troughton, E., & Heywood, E. (1985). Alcoholism and antisocial personality. Interrelationships, genetic and environmental factors. *Archives of General Psychiatry, 42,* 161–167. DOI: 10.1001/archpsyc.1985.01790250055007

Cadoret, R. J., & Stewart, M. A. (1991). An adoption study of attention deficit/hyperactivity/aggression and their relationship to adult antisocial personality. *Comprehensive Psychiatry, 32,* 73–82. DOI: 10.1016/0010-440X(91)90072-K

Cadoret, R. J., Troughton, E., & O'Gorman, T. W. (1987). Genetic and environmental factors in alcohol abuse and antisocial personality. *Journal of Studies on Alcohol, 48,* 1–8.

Cadoret, R. J., Yates, W. R., Troughton, E., Woodworth, G., & Stewart, M. A. (1995a). Adoption study demonstrating two genetic pathways to drug abuse. *Archives of General Psychiatry, 52,* 42–52. DOI: 10.1001/archpsyc.1995.03950130042005

Cadoret, R. J., Yates, W. R., Troughton, E., Woodworth, G., & Stewart, M. A. (1995b). Genetic-environmental interaction in the genesis of aggressivity and conduct disorders. *Archives of General Psychiatry, 52,* 916–924. DOI: 10.1001/archpsyc.1995.03950230030006

Cadoret, R. J., Yates, W. R., Troughton, E., Woodworth, G., & Stewart, M. A. (1996). An adoption study of drug abuse/dependency in females. *Comprehensive Psychiatry, 37,* 88–94. DOI: 10.1016/S0010-440X(96)90567-2

Caldwell, B. M., & Bradley, R. H. (1978). *Home observation for measurement of the environment.* Little Rock, AR: University of Arkansas.

Camp, N. J., Lowry, M. R., Richards, R. L., Plenk, A. M., Carter, C., Hensel, C. H., . . . Cannon-Albright, L. A. (2005). Genome-wide linkage analyses of extended Utah pedigrees identifies loci that influence recurrent, early-onset major depression and anxiety disorders. *American Journal of Medical Genetics. B: Neuropsychiatric Genetics, 135B,* 85–93. DOI: 10.1002/ajmg.b.30177

Cannon, T. D., Mednick, S. A., Parnas, J., Schulsinger, F., Praestholm, J., & Vestergaard, A. (1993). Developmental brain abnormalities in the offspring of schizophrenic mothers. I. Contributions of genetic and perinatal factors. *Archives of General Psychiatry, 50,* 551–564. DOI: 10.1001/archpsyc.1993.01820190053006

Cantor, C. (2005). *Evolution and posttraumatic stress.* London: Routledge.

Cantwell, D. P. (1975). Genetic studies of hyperactive children: Psychiatric illness in biological and adopting parents. In R. R. Fieve, D. Rosenthal, & H. Brill (Eds.), *Genetic research in psychiatry* (pp. 273–280). Baltimore, MD: Johns Hopkins University Press.

Capecchi, M. R. (1994). Targeted gene replacement. *Scientific American, 270(3),* 52–59. DOI: 10.1038/scientificamerican0394-52

Caplan, B. (2011). *Selfish reasons to have more kids: Why being a great parent is less work and more fun than you think.* New York: Basic Books.

Caprara, G. V., Fagnani, C., Alessandri, G., Steca, P., Gigantesco, A., Sforza, L. L. C., & Stazi, M. A. (2009). Human optimal functioning: The genetics of positive orientation towards self, life, and the future. *Behavior Genetics, 39,* 277–284. DOI: 10.1007/s10519-009-9267-y

Capron, C., & Duyme, M. (1989). Assessment of the effects of socioeconomic status on IQ in a full cross-fostering study. *Nature, 340,* 552–554. DOI: 10.1038/340552a0

Capron, C., & Duyme, M. (1996). Effect of socioeconomic status of biological and adoptive parents on WISC-R subtest scores of their French adopted children. *Intelligence, 22,* 259–275. DOI: 10.1016/S0160-2896(96)90022-7

Cardno, A. G., & Gottesman, I. I. (2000). Twin studies of schizophrenia: From bow-

and-arrow concordances to Star Wars MX and functional genomics. *American Journal of Medical Genetics, 97,* 12–17. DOI: 10.1002/(SICI)1096-8628(200021)97:1<12::AID-AJMG3>3.0.CO;2-U

Cardno, A. G., Jones, L. A., Murphy, K. C., Sanders, R. D., Asherson, P., Owen, M. J., & McGuffin, P. (1999). Dimensions of psychosis in affected sibling pairs. *Schizophrenia Bulletin, 25,* 841–850. DOI: 10.1093/oxfordjournals.schbul.a033423

Cardno, A. G., Rijsdijk, F. V., Sham, P. C., Murray, R. M., & McGuffin, P. (2002). A twin study of genetic relationships between psychotic symptoms. *American Journal of Psychiatry, 159,* 539–545. DOI: 10.1176/appi.ajp.159.4.539

Cardno, A. G., Rijsdijk, F. V., West, R. M., Gottesman, I. I., Craddock, N., Murray, R. M., & McGuffin, P. (2012). A twin study of schizoaffective-mania, schizoaffective-depression, and other psychotic syndromes. *American Journal of Medical Genetics. B: Neuropsychiatric Genetics, 159B,* 172–182. DOI: 10.1002/ajmg.b.32011

Cardon, L. R. (1994). Height weight and obesity. In J. C. DeFries, R. Plomin & D. W. Fulker (Eds.), *Nature and nurture during middle childhood* (pp. 165–172). Cambridge, MA: Blackwell.

Cardon, L. R., Fulker, D. W., DeFries, J. C., & Plomin, R. (1992). Multivariate genetic analysis of specific cognitive abilities in the Colorado Adoption Project at age 7. *Intelligence, 16,* 383–400. DOI: 10.1016/0160-2896(92)90016-K

Cardon, L. R., Smith, S. D., Fulker, D. W., Kimberling, W. J., Pennington, B. F., & DeFries, J. C. (1994). Quantitative trait locus for reading disability on chromosome 6. *Science, 266,* 276–279. DOI: 10.1126/science.7939663

Carey, G. (1986). Sibling imitation and contrast effects. *Behavior Genetics, 16,* 319–341. DOI: 10.1007/BF01071314

Carey, G. (1992). Twin imitation for anti-social behavior: Implications for genetic and family environmental research. *Journal of Abnormal Psychology, 101,* 18–25. DOI: 10.1037/0021-843X.101.1.18

Carmelli, D., Swan, G. E., & Cardon, L. R. (1995). Genetic mediation in the relationship of education to cognitive function in older people. *Psychology and Aging, 10,* 48–53. DOI: 10.1037/0882-7974.10.1.48

Carnell, S., Haworth, C. M. A., Plomin, R., & Wardle, J. (2008). Genetic influence on appetite in children. *International Journal of Obesity, 32,* 1468–1473. DOI: 10.1038/ijo.2008.127

Caron, M. G. (1996). Images in neuroscience. Molecular biology, II. A dopamine transporter mouse knockout. *American Journal of Psychiatry, 153,* 1515.

Carroll, J. B. (1993). *Human cognitive abilities.* New York: Cambridge University Press. DOI: 10.1017/CBO9780511571312

Carroll, J. B. (1997). Psychometrics, intelligence, and public policy. *Intelligence, 24,* 25–52. DOI: 10.1016/S0160-2896(97)90012-X

Cases, O., Seif, I., Grimsby, J., Gaspar, P., Chen, K., Pournin, S., … Demaeyer, E. (1995). Aggressive-behavior and altered amounts of brain-serotonin and norepinephrine in mice lacking MAOA. *Science, 268,* 1763–1766. DOI: 10.1126/science.7792602

Caspi, A., McClay, J., Moffitt, T. E., Mill, J., Martin, J., Craig, I. W., … Poulton, R. (2002). Role of genotype in the cycle of violence in maltreated children. *Science, 297,* 851–854. DOI: 10.1126/science.1072290

Caspi, A., & Moffitt, T. E. (1995). The continuity of maladaptive behavior: From description to understanding of antisocial behavior. In D. Cicchetti & D. J. Cohen (Eds.), *Developmental psychopathology* (pp. 472–511). New York: Wiley.

Caspi, A., Moffitt, T. E., Cannon, M., McClay, J., Murray, R., Harrington, H., … Craig, I. W. (2005). Moderation of the effect of adolescent-onset cannabis use on adult psychosis by a functional polymorphism in the catechol-O-methyltransferase gene: Longitudinal evidence of a gene × environment interaction. *Biological Psychiatry, 57,* 1117–1127. DOI: 10.1016/j.biopsych.2005.01.026

Caspi, A., Sugden, K., Moffitt, T. E., Taylor, A., Craig, I. W., Harrington, H., … Poulton, R. (2003). Influence of life stress on depression: Moderation by a polymorphism in the 5-HTT gene. *Science, 301,* 386–389. DOI: 10.1126/science.1083968

Cattell, R. B. (1982). *The inheritance of personality and ability.* New York: Academic Press.

Centers for Disease Control and Prevention. (2002). Annual smoking-attributable mortality, years of potential life lost, and economic costs—United States, 1995–1999. *Morbidity and Mortality Weekly Report, 51,* 300–303.

Centers for Disease Control and Prevention. (2003). Cigarette smoking-attributable morbidity—United States, 2000. *Morbidity and Mortality Weekly Report, 52,* 842–844.

Centers for Disease Control and Prevention. (2008). Smoking-attributable mortality, years of potential life lost, and productivity losses—United States, 2000–2004. *Morbidity and Mortality Weekly Report, 57,* 1226–1228.

Centers for Disease Control and Prevention. (2010). Vital signs: Current cigarette smoking among adults aged ≥ 18 years—United States, 2009. *Morbidity and Mortality Weekly Report, 59,* 1135–1140.

Cesarini, D., Dawes, C. T., Johannesson, M., Lichtenstein, P., & Wallace, B. (2009). Genetic variation in preferences for giving and risk taking. *Quarterly Journal of Economics, 124,* 809–842. DOI: 10.1162/qjec.2009.124.2.809

Cesarini, D., Johannesson, M., Lichtenstein, P., Sandewall, Ö., & Wallace, B. (2010). Genetic variation in financial decision-making. *Journal of Finance, 65,* 1725–1754. DOI: 10.1111/j.1540-6261.2010.01592.x

Chabris, C. F., Hebert, B. M., Benjamin, D. J., Beauchamp, J. P., Cesarini, D., van der Loos, M. J. H. M., ... Laibson, D. (2012). Most reported genetic associations with general intelligence are probably false positives. *Psychological Science.*

Chambers, J. C., Elliott, P., Zabaneh, D., Zhang, W., Li, Y., Froguel, P., ... Kooner, J. S. (2008). Common genetic variation near MC4R is associated with waist circumference and insulin resistance. *Nature Genetics, 40,* 716–718. DOI: 10.1038/ng.156

Champagne, F. A., & Curley, J. P. (2009). Epigenetic mechanisms mediating the long-term effects of maternal care on development. *Neuroscience and Biobehavioral Reviews, 33,* 593–600. DOI: 10.1016/j.neubiorev.2007.10.009

Changeux, J.-P. (2010). Nicotine addiction and nicotinic receptors: Lessons from genetically modified mice. *Nature Reviews Neuroscience, 11,* 389–401. DOI: 10.1038/nrn2849

Chen, C.-H., Gutierrez, E. D., Thompson, W., Panizzon, M. S., Jernigan, T. L., Eyler, L. T., ... Dale, A. M. (2012). Hierarchical genetic organization of human cortical surface area. *Science, 335,* 1634–1636. DOI: 10.1126/science.1215330

Chen, C.-H., Panizzon, M. S., Eyler, L. T., Jernigan, T. L., Thompson, W., Fennema-Notestine, C., ... Dale, A. M. (2011). Genetic influences on cortical regionalization in the human brain. *Neuron, 72,* 537–544. DOI: 10.1016/j.neuron.2011.08.021

Cherny, S. S., Fulker, D. W., Emde, R. N., Robinson, J., Corley, R. P., Reznick, J. S., ... DeFries, J. C. (1994). Continuity and change in infant shyness from 14 to 20 months. *Behavior Genetics, 24,* 365–379. DOI: 10.1007/BF01067538

Cherny, S. S., Fulker, D. W., & Hewitt, J. K. (1997). Cognitive development from infancy to middle childhood. In R. J. Sternberg & E. L. Grigorenko (Eds.), *Intelligence, heredity and environment* (pp. 463–482). Cambridge: Cambridge University Press.

Chesler, E. J., Lu, L., Shou, S. M., Qu, Y. H., Gu, J., Wang, J. T., ... Williams, R. W. (2005). Complex trait analysis of gene expression uncovers polygenic and pleiotropic networks that modulate nervous system function. *Nature Genetics, 37,* 233–242. DOI: 10.1038/ng1518

Chesler, E. J., Miller, D. R., Branstetter, L. R., Galloway, L. D., Jackson, B. L., Philip, V. M., ... Manly, K. F. (2008). The Collaborative Cross at Oak Ridge National Laboratory: Developing a powerful resource for systems genetics. *Mammalian Genome, 19,* 382–389. DOI: 10.1007/s00335-008-9135-8

Cheung, V. G., Conlin, L. K., Weber, T. M., Arcaro, M., Jen, K. Y., Morley, M., & Spielman, R. S. (2003). Natural variation in human gene expression assessed in lymphoblastoid cells. *Nature Genetics, 33,* 422–425. DOI: 10.1038/ng1094

Chiang, M.-C., Barysheva, M., Shattuck, D. W., Lee, A. D., Madsen, S. K., Avedissian, C., ... Thompson, P. M. (2009). Genetics of brain fiber architecture and intellectual performance. *Journal of Neuroscience, 29,* 2212–2224. DOI: 10.1523/JNEUROSCI.4184-08.2009

Chipuer, H. M., & Plomin, R. (1992). Using siblings to identify shared and non-shared home items. *British Journal of Developmental Psychology, 10,* 165–178. DOI: 10.1111/j.2044-835X.1992.tb00570.x

Chipuer, H. M., Plomin, R., Pedersen, N. L., McClearn, G. E., & Nesselroade, J. R. (1993). Genetic influence on family environment: The role of personality. *Developmental Psychology, 29,* 110–118. DOI: 10.1037/0012-1649.29.1.110

Chipuer, H. M., Rovine, M. J., & Plomin, R. (1990). LISREL modeling: Genetic and environmental influences on IQ revisited. *Intelligence, 14,* 11–29. DOI: 10.1016/0160-2896(90)90011-H

Cho, A. H., Killeya-Jones, L. A., O'Daniel, J. M., Kawamoto, K., Gallagher, P., Haga, S., ... Ginsburg, G. S. (2012). Effect of genetic testing for risk of type 2 diabetes mellitus on health behaviors and outcomes: Study rationale, development and design. *BMC Health Services Research, 12:* 16. DOI: 10.1186/1472-6963-12-16

Chow, B. W.-Y., Ho, C. S.-H., Wong, S. W.-L., Waye, M. M. Y., & Bishop, D. V. M. (2011). Genetic and environmental influences on Chinese language and reading abilities. *PLoS One, 6:* e16640. DOI: 10.1371/journal.pone.0016640

Christensen, K., Johnson, T. E., & Vaupel, J. W. (2006). The quest for genetic determinants of human longevity: Challenges and insights. *Nature Reviews Genetics, 7,* 436–448. DOI: 10.1038/nrg1871

Christensen, K., Petersen, I., Skytthe, A., Herskind, A. M., McGue, M., & Bingley, P. (2006). Comparison of academic performance of twins and singletons in adolescence: Follow-up study. *British Medical Journal, 333,* 1095–1097. DOI: 10.1136/bmj.38959.650903.7C

Christiansen, K. O. (1977). A preliminary study of criminality among twins. In S. Mednick & K. O. Christiansen (Eds.), *Biosocial bases of criminal behavior* (pp. 89-108). New York: Gardner Press, Inc.

Christiansen, L., Frederiksen, H., Schousboe, K., Skytthe, A., von Wurmb-Schwark, N., Christensen, K., & Kyvik, K. (2003). Age- and sex-differences in the validity of questionnaire-based zygosity in twins. *Twin Research, 6,* 275–278. DOI: 10.1375/136905203322296610

Chua, S. C., Jr., Chung, W. K., Wu-Peng, X. S., Zhang, Y., Liu, S. M., Tartaglia, L., & Leibel, R. L. (1996). Phenotypes of mouse diabetes and rat fatty due to mutations in the OB (leptin) receptor. *Science, 271,* 994–996. DOI: 10.1126/science.271.5251.994

Cicchetti, D., Rogosch, F. A., & Oshri, A. (2011). Interactive effects of corticotropin releasing hormone receptor 1, serotonin transporter linked polymorphic region, and child maltreatment on diurnal cortisol regulation and internalizing symptomatology. *Development and Psychopathology, 23,* 1125–1138. DOI: 10.1017/S0954579411000599

Cirulli, E. T., Kasperaviciute, D., Attix, D. K., Need, A. C., Ge, D., Gibson, G., & Goldstein, D. B. (2010). Common genetic variation and performance on standardized cognitive tests. *European Journal of Human Genetics, 18,* 815–819. DOI: 10.1038/ejhg.2010.2

Claridge, G., & Hewitt, J. K. (1987). A biometrical study of schizotypy in a normal population. *Personality and Individual Differences, 8,* 303–312. DOI: 10.1016/0191-8869(87)90030-4

Clasen, P. C., Wells, T. T., Knopik, V. S., McGeary, J. E., & Beevers, C. G. (2011). 5-HTTLPR and BDNF Val66Met polymorphisms moderate effects of stress on rumination. *Genes, Brain and Behavior, 10,* 740–746. DOI: 10.1111/j.1601-183X.2011.00715.x

Clementz, B. A., McDowell, J. E., & Zisook, S. (1994). Saccadic system functioning among schizophrenic patients and their first-degree biological relatives. *Journal of Abnormal Psychology, 103,* 277–287. DOI: 10.1037/0021-843X.103.2.277

Cloninger, C. R. (1987). A systematic method for clinical description and classification of personality variants—a proposal. *Archives of General Psychiatry, 44,* 573–588. DOI: 10.1001/archpsyc.1987.01800180093014

Cloninger, C. R. (2002). The relevance of normal personality for psychiatrists. In J. Benjamin, R. Ebstein, & R. H. Belmaker (Eds.), *Molecular genetics and human personality* (pp. 33-42). New York: American Psychiatric Press.

Cloninger, C. R., Bohman, M., & Sigvardsson, S. (1981). Inheritance of alcohol abuse: Cross-fostering analysis of adopted men. *Archives of General Psychiatry, 38,* 861–868. DOI: 10.1001/archpsyc.1981.01780330019001

Cloninger, C. R., Sigvardsson, S., Bohman, M., & von Knorring, A. L. (1982). Predisposition to petty criminality in Swedish adoptees. II. Cross fostering analysis of gene-environment interaction. *Archives of General Psychiatry, 39,* 1242–1247. DOI: 10.1001/archpsyc.1982.04290110010002

Cloninger, C. R., Svrakic, D. M., & Przybeck, T. R. (1993). A psychobiological model of temperament and character. *Archives of General Psychiatry, 50,* 975–990. DOI: 10.1001/archpsyc.1993.01820240059008

Cobb, J. P., Mindrinos, M. N., Miller-Graziano, C., Calvano, S. E., Baker, H. V., Xiao, W. Z., . . . Inflammation and Host Response to Injury Large-Scale Collaborative Research Program. (2005). Application of genome-wide expression analysis to human health and disease. *Proceedings of the National Academy of Sciences (USA), 102,* 4801–4806. DOI: 10.1073/pnas.0409768102

Cohen, B. H. (1964). Family patterns of mortality and life span. *Quarterly Review of Biology, 39,* 130–181. DOI: 10.1086/404164

Cohen, P., Cohen, J., Kasen, S., Velez, C. N., Hartmark, C., Johnson, J., . . . Streuning, E. L. (1993). An epidemiological study of disorders in late childhood and adolescence. I. Age- and gender-specific prevalence. *Journal of Child Psychology and Psychiatry, 34,* 851–867. DOI: 10.1111/j.1469-7610.1993.tb01094.x

Colantuoni, C., Lipska, B. K., Ye, T. Z., Hyde, T. M., Tao, R., Leek, J. T., . . . Kleinman, J. E. (2011). Temporal dynamics and genetic control of transcription in the human prefrontal cortex. *Nature, 478,* 519–523. DOI: 10.1038/nature10524

Collier, D. A., Stober, G., Li, T., Heils, A., Catalano, M., Di Bella, D., . . . Lesch, K. P. (1996). A novel functional polymorphism within the promoter of the serotonin transporter gene: Possible role in susceptibility to affective disorders. *Molecular Psychiatry, 1,* 453–460.

Collins, A. C. (1981). A review of research using short-sleep and long-sleep mice. In G. E. McClearn, R. A. Deitrich, & V. G. Erwin (Eds.), *Development of animal models as pharmocogenetic tools. USDHHS-NIAAA Research Monograph No. 6* (pp. 161-170). Washington, DC: U.S. Government Printing Office.

Collins, F. (2006). *The language of God: A scientist presents evidence for belief.* New York: Simon & Schuster.

Collins, F. S. (2010). *The language of life: DNA and the revolution in personalized medicine.* New York: Harper Collins.

Collins, J. S., Marvelle, A. F., & Stevenson, R. E. (2011). Sibling recurrence in intellectual disability of unknown cause. *Clinical Genetics, 79,* 498–500. DOI: 10.1111/j.1399-0004.2010.01601.x

Colom, R., Rebollo, I., Abad, F. J., & Shih, P. C. (2006). Complex span tasks, simple span tasks, and cognitive abilities: A reanalysis of key studies. *Memory and Cognition, 34,* 158–171. DOI: 10.3758/BF03193395

Compton, W. M., Conway, K. P., Stinson, F. S., & Grant, B. F. (2006). Changes in the prevalence of major depression and comorbid substance use disorders in the United States between 1991–1992 and 2001–2002. *American Journal of Psychiatry, 163,* 2141–2147. DOI: 10.1176/appi.ajp.163.12.2141

Confer, J. C., Easton, J. A., Fleischman, D. S., Goetz, C. D., Lewis, D. M. G., Perilloux, C., & Buss, D. M. (2010). Evolutionary psychology controversies, questions, prospects, and limitations. *American Psychologist, 65,* 110–126. DOI: 10.1037/a0018413

Conrad, D. F., Pinto, D., Redon, R., Feuk, L., Gokcumen, O., Zhang, Y. J., . . . Hurles, M. E. (2010). Origins and functional impact of copy number variation in the human genome. *Nature, 464,* 704–712. DOI: 10.1038/nature08516

Constantino, J. N., & Todd, R. D. (2003). Autistic traits in the general population: A twin study. *Archives of General Psychiatry, 60,* 524–530. DOI: 10.1001/archpsyc.60.5.524

Cook, E. H., Jr., Courchesne, R. Y., Cox, N. J., Lord, C., Gonen, D., Guter, S. J., . . . Courchesne, E. (1998). Linkage-disequilibrium mapping of autistic disorder, with 15q11-13 markers. *American Journal of Human Genetics, 62,* 1077–1083. DOI: 10.1086/301832

Coolidge, F. L., Thede, L. L., & Jang, K. L. (2001). Heritability of personality disorders in childhood: A preliminary investigation. *Journal of Personality Disorders, 15,* 33–40. DOI: 10.1521/pedi.15.1.33.18645

Cooper, A. J. L., & Blass, J. P. (2011). Trinucleotide-expansion diseases. *Neurochemical Mechanisms in Disease, 1,* 319–358. DOI: 10.1007/978-1-4419-7104-3_11

Cooper, R. M., & Zubek, J. P. (1958). Effects of enriched and restricted early environments on the learning ability of bright and dull rats. *Canadian Journal of Psychology, 12,* 159–164. DOI: 10.1037/h0083747

Corder, E. H., Saunders, A. M., Strittmatter, W. J., Schmechel, D. E., Gaskell, P. C., Small, G. W., . . . Pericak Vance, M. A. (1993). Gene dose of apolipoprotein E type 4 allele and the risk of Alzheimer's disease in late onset families. *Science, 261,* 921–923. DOI: 10.1126/science.8346443

Coren, S. (2005). *The intelligence of dogs: A guide to the thoughts, emotions, and inner lives of our canine companions.* New York: Simon & Schuster.

Correa, C. R., & Cheung, V. G. (2004). Genetic variation in radiation-induced expression phenotypes. *American Journal of Human Genetics, 75,* 885–890. DOI: 10.1086/425221

Costa, F. F. (2005). Non-coding RNAs: New players in eukaryotic biology. *Gene, 357,* 83–94. DOI: 10.1016/j.gene.2005.06.019

Costa, P. T., & McCrae, R. R. (1994). Stability and change in personality from adolescence through adulthood. In C. F. Haverson, Jr., G. A. Kohnstamm, & R. P. Martin (Eds.), *The developing structure of temperament and personality from infancy to adulthood* (pp. 139–150). Hillsdale, NJ: Erlbaum.

Crabbe, J. C., & Harris, R. A. (1991). *The genetic basis of alcohol and drug actions.* New York: Plenum.

Crabbe, J. C., Kosobud, A., Young, E. R., Tam, B. R., & McSwigan, J. D. (1985). Bidirectional selection for susceptibility to ethanol withdrawal seizures in mus musculus. *Behavior Genetics, 15,* 521–536. DOI: 10.1007/BF01065448

Crabbe, J. C., Phillips, T. J., & Belknap, J. K. (2010). The complexity of alcohol drinking: Studies in rodent genetic models. *Behavior Genetics, 40,* 737–750. DOI: 10.1007/s10519-010-9371-z

Crabbe, J. C., Phillips, T. J., Buck, K. J., Cunningham, C. L., & Belknap, J. K. (1999). Identifying genes for alcohol and drug sensitivity: Recent progress and future directions. *Trends in Neurosciences, 22,* 173–179. DOI: 10.1016/S0166-2236(99)01393-4

Crabbe, J. C., Phillips, T. J., Feller, D. J., Hen, R., Wenger, C. D., Lessov, C. N., & Schafer, G. L. (1996). Elevated alcohol consumption in null mutant mice lacking 5-HT1B serotonin receptors. *Nature Genetics, 14,* 98–101. DOI: 10.1038/ng0996-98

Crabbe, J. C., Phillips, T. J., Harris, R. A., Arends, M. A., & Koob, G. F. (2006). Alcohol-related genes: Contributions from studies with genetically engineered mice. *Addiction Biology, 11,* 195–269. DOI: 10.1111/j.1369-1600.2006.00038.x

Crabbe, J. C., Wahlsten, D., & Dudek, B. C. (1999). Genetics of mouse behavior: Interactions with laboratory environment. *Science, 284,* 1670–1672. DOI: 10.1126/science.284.5420.1670

Craddock, N., & Forty, L. (2006). Genetics of affective (mood) disorders. *European Journal of Human Genetics, 14,* 660–668. DOI: 10.1038/sj.ejhg.5201549

Craddock, N., O'Donovan, M. C., & Owen, M. J. (2005). The genetics of schizophrenia and bipolar disorder: Dissecting psychosis. *Journal of Medical Genetics, 42,* 193–204. DOI: 10.1136/jmg.2005.030718

Craddock, N., & Owen, M. J. (2005). The beginning of the end for the Kraepelinian dichotomy. *British Journal of Psychiatry, 186,* 364–366.

Craddock, N., Owen, M. J., & O'Donovan, M. C. (2006). The catechol-O-methyl transferase (COMT) gene as a candidate for psychiatric phenotypes: Evidence and lessons. *Molecular Psychiatry, 11,* 446–458. DOI: 10.1038/sj.mp.4001808

Crawford, C. B., & Salmon, C. A. (2004). *Evolutionary psychology, public policy and personal decisions.* Mahwah, NJ: Lawrence Erlbaum Associates.

Crawley, J. N. (2003). Behavioral phenotyping of rodents. *Comparative Medicine, 53,* 140–146.

Crawley, J. N. (2007). *What's wrong with my mouse: Behavioral phenotyping of transgenic and knockout mice* (2nd ed.) Wilmington, DE: Wiley-Liss.

Cronk, N. J., Slutske, W. S., Madden, P. A., Bucholz, K. K., Reich, W., & Heath, A. C. (2002). Emotional and behavioral problems among female twins: An evaluation of the equal environments assumption. *Journal of American Academic Child and Adolescent Psychiatry, 41,* 829–837. DOI: 10.1097/00004583-200207000-00016

Crow, T. J. (1985). The two syndrome concept: Origins and current states. *Schizophrenia Bulletin, 11,* 471–486.

Crowe, R. R. (1972). The adopted offspring of women criminal offenders: A study of their arrest records. *Archives of General Psychiatry, 27,* 600–603. DOI: 10.1001/archpsyc.1972.01750290026004

Crowe, R. R. (1974). An adoption study of antisocial personality. *Archives of General Psychiatry, 31,* 785–791. DOI: 10.1001/archpsyc.1974.01760180027003

Crusio, W. E. (2004). Flanking gene and genetic background problems in genetically manipulated mice. *Biological Psychiatry, 56,* 381–385. DOI: 10.1016/j.biopsych.2003.12.026

Culbert, K. M., Racine, S. E., & Klump, K. L. (2011). The influence of gender and puberty on the heritability of disordered eating symptoms. In R. A. H. Adan & W. H. Kaye (Eds.), *Current topics in behavioral neurosciences: Vol. 6. Behavioral neurobiology of eating disorders* (pp. 177–185). DOI: 10.1007/7854_2010_80

Cumings, J. L., & Benson, D. F. (1992). *Dementia: A clinical approach.* Boston, MA: Butterworth.

Cutrona, C. E., Cadoret, R. J., Suhr, J. A., Richards, C. C., Troughton, E., Schutte, K., & Woodworth, G. (1994). Interpersonal variables in the prediction of alcoholism among adoptees—evidence for gene-environment interactions. *Comprehensive Psychiatry, 35,* 171–179. DOI: 10.1016/0010-440X(94)90188-0

Czajkowski, N., Kendler, K. S., Tambs, K., Roysamb, E., & Reichborn-Kjennerud, T. (2011). The structure of genetic and environmental risk factors for phobias in women. *Psychological Medicine, 41,* 1987–1995. DOI: 10.1017/S0033291710002436

D'Angelo, M. G., Lorusso, M. L., Civati, F., Comi, G. P., Magri, F., Del Bo, R., . . . Bresolin, N. (2011). Neurocognitive profiles in Duchenne muscular dystrophy and gene mutation site. *Pediatric Neurology, 45,* 292–299. DOI: 10.1016/j.pediatrneurol.2011.08.003

D'Onofrio, B. M., Turkheimer, E., Emery, R. E., Slutske, W. S., Heath, A. C., Madden, P. A., & Martin, N. G. (2006). A genetically informed study of the processes underlying the association between parental marital instability and offspring adjustment. *Developmental Psychology, 42,* 486–499. DOI: 10.1037/0012-1649.42.3.486

D'Onofrio, B. M., Turkheimer, E. N., Eaves, L. J., Corey, L. A., Berg, K., Solaas, M. H., & Emery, R. E. (2003). The role of the children of twins design in elucidating causal relations between parent characteristics and child outcomes. *Journal of Child Psychology and Psychiatry, 44,* 1130–1144. DOI: 10.1111/1469-7610.00196

Dabelea, D., Mayer-Davis, E. J., Lamichhane, A. P., D'Agostino, R. B., Liese, A. D., Vehik, K. S., . . . Hamman, R. F. (2008). Association of intrauterine exposure to maternal diabetes and obesity with type 2 diabetes in youth: The SEARCH Case-Control Study. *Diabetes Care, 31,* 1422–1426. DOI: 10.2337/dc07-2417

Dale, P. S., Simonoff, E., Bishop, D. V. M., Eley, T. C., Oliver, B., Price, T. S., . . . Plomin, R. (1998). Genetic influence on language delay in two-year-old children. *Nature Neuroscience, 1,* 324–328. DOI: 10.1038/1142

Daly, M., & Wilson, M. (1999). *The truth about Cinderella.* New Haven, CT: Yale University Press.

Daniels, J. K., Owen, M. J., McGuffin, P., Thompson, L., Detterman, D. K., Chorney, K., . . . Plomin, R. (1994). IQ and variation in the number of fragile X CGG repeats: No association in a normal sample. *Intelligence, 19,* 45–50. DOI: 10.1016/0160-2896(94)90052-3

Darvasi, A. (1998). Experimental strategies for the genetic dissection of complex traits in animal models. *Nature Genetics, 18,* 19–24. DOI: 10.1038/ng0198-19

Darwin, C. (1859). *On the origin of species by means of natural selection, or the preservation of favoured races in the struggle for life.* London: John Murray.

Darwin, C. (1871). *The descent of Man and selection in relation to sex.* London: John Murray.

Darwin, C. (1877). A biographical sketch of an infant. *Mind, 2,* 285–294. DOI: 10.1093/mind/os-2.7.285

Darwin, C. (1896). *Journal of researches into the natural history and geology of the countries visited during the voyage of H. M. S. Beagle round the world under the command of Capt. Fitz Roy, R. N.* New York: Appleton.

Das, I., & Reeves, R. H. (2011). The use of mouse models to understand and improve cognitive deficits in Down syndrome. *Disease Models and Mechanisms, 4,* 596–606. DOI: 10.1242/dmm.007716

Davey Smith, G. (2011). Epidemiology, epigenetics and the 'gloomy prospect': Embracing randomness in population health research and practice. *International Journal of Epidemiology, 40,* 537–562. DOI: 10.1093/ije/dyr117

Davey Smith, G., & Ebrahim, S. (2003). 'Mendelian randomization': Can genetic epidemiology contribute to understanding environmental determinants of disease? *International Journal of Epidemiology, 32,* 1–22. DOI: 10.1093/ije/dyg070

Davidson, J. R. T., Hughes, D., Blazer, D. G., & George, L. (1991). Posttraumatic stress disorder in the community: An epidemiological study. *Psychological Medicine, 21,* 713–721. DOI: 10.1017/S0033291700022352

Davies, G., Tenesa, A., Payton, A., Yang, J., Harris, S. E., Liewald, D., . . . Deary, I. J. (2011). Genome-wide association studies establish that human intelligence is highly heritable and polygenic. *Molecular Psychiatry, 16,* 996–1005. DOI: 10.1038/mp.2011.85

Davis, M. H., Luce, C., & Kraus, S. J. (1994). The heritability of characteristics associated with dispositional empathy. *Journal of Personality, 62,* 369–391. DOI: 10.1111/j.1467-6494.1994.tb00302.x

Davis, O. S. P., Butcher, L. M., Docherty, S. J., Meaburn, E. M., Curtis, C. J. C., Simpson, A., . . . Plomin, R. (2010). A three-stage genome-wide association study of general cognitive ability: Hunting the small effects. *Behavior Genetics, 40,* 759–767. DOI: 10.1007/s10519-010-9350-4

Davis, O. S. P., Haworth, C. M. A., & Plomin, R. (2009a). Dramatic increase in heritability of cognitive development from early to middle childhood: An 8-year longitudinal study of 8700 pairs of twins. *Psychological Science, 20,* 1301–1308. DOI: 10.1111/j.1467-9280.2009.02433.x

Davis, O.S.P., Haworth, C.M.A., & Plomin, R. (2009b). Learning abilities and disabilities: Generalist genes in early adolescence. *Cognitive Neuropsychiatry, 14,* 312–331. DOI: 10.1080/13546800902797106

Davis, R. L. (2005). Olfactory memory formation in Drosophila: From molecular to systems neuroscience. *Annual Review of Neuroscience, 28,* 275–302. DOI: 10.1146/annurev.neuro.28.061604.135651

Davis, R. L. (2011). Traces of Drosophila memory. *Neuron, 70,* 8–19. DOI: 10.1016/j.neuron.2011.03.012

Dawkins, R. (1976). *The selfish gene.* New York: Oxford University Press.

Dawkins, R. (2006). *The God delusion.* London: Bantam Press.

de Castro, J. M. (1999). Behavioral genetics of food intake regulation in free-living humans. *Nutrition, 15,* 550–554. DOI: 10.1016/S0899-9007(99)00114-8

De Magalhães, J. P., Budovsky, A., Lehmann, G., Costa, J., Li, Y., Fraifeld, V., & Church, G. M. (2009). The Human Ageing Genomic Resources: Online databases and tools for biogerontologists. *Aging Cell, 8,* 65–72. DOI: 10.1111/j.1474-9726.2008.00442.x

de Moor, M. H. M., Costa, P. T., Terracciano, A., Krueger, R. F., de Geus, E. J. C., Toshiko, T., . . . Boomsma, D. I. (2012). Meta-analysis of genome-wide association studies for personality. *Molecular Psychiatry, 17,* 337–349. DOI: 10.1038/mp.2010.128

de Waal, F. (1996). *Good natured: The origins of right and wrong in humans and other animals.* Cambridge, MA: Harvard University Press.

de Waal, F. (2002). Evolutionary psychology: The wheat and the chaff. *Current Directions in Psychological Science, 11,* 187–191. DOI: 10.1111/1467-8721.00197

Deary, I. J. (2000). *Looking down on human intelligence: From psychometrics to the brain.* Oxford: Oxford University Press. DOI: 10.1093/acprof:oso/9780198524175.001.0001

Deary, I. J. (2012). Intelligence. *Annual Review of Psychology, 63,* 453–482. DOI: 10.1146/annurev-psych-120710-100353

Deary, I. J., Pattie, A., Wilson, V., & Whalley, L. J. (2005). The cognitive cost of being a twin: Two whole-population surveys. *Twin Research and Human Genetics, 8,* 376–383. DOI: 10.1375/twin.8.4.376

Deary, I. J., Penke, L., & Johnson, W. (2010). The neuroscience of human intelligence differences. *Nature Reviews Neuroscience, 11,* 201–211. DOI: 10.1038/nrn2793

Deary, I. J., Whiteman, M. C., Starr, J. M., Whalley, L. J., & Fox, H. C. (2004). The impact of childhood intelligence on later life: Following up the Scottish mental surveys of 1932 and 1947. *Journal of Personality and Social Psychology, 86,* 130–147. DOI: 10.1037/0022-3514.86.1.130

Deary, I. J., Yang, J., Davies, G., Harris, S. E., Tenesa, A., Liewald, D., . . . Visscher, P. M. (2012). Genetic contributions to stability and change in intelligence from childhood to old age. *Nature, 482,* 212–215. DOI: 10.1038/nature10781

Deater-Deckard, K., & O'Connor, T. G. (2000). Parent-child mutuality in early childhood: Two behavioral genetic studies. *Developmental Psychology, 36,* 561–570. DOI: 10.1037/0012-1649.36.5.561

Debener, S., Ullsperger, M., Siegel, M., & Engel, A. K. (2006). Single-trial EEG-fMRI reveals the dynamics of cognitive function. *Trends in Cognitive Sciences, 10,* 558–563. DOI: 10.1016/j.tics.2006.09.010

DeFries, J. C., & Alarcón, M. (1996). Genetics of specific reading disability. *Mental Retardation and Developmental Disabilities Research Reviews, 2,* 39–47. DOI: 10.1002/(SICI)1098-2779(1996)2:1<39::AID-MRDD7>3.0.CO;2-S

DeFries, J. C., & Fulker, D. W. (1985). Multiple regression analysis of twin data. *Behavior Genetics, 15,* 467–473. DOI: 10.1007/BF01066239

DeFries, J. C., & Fulker, D. W. (1988). Multiple regression analysis of twin data: Etiology of deviant scores versus individual differences. *Acta Geneticae Medicae et Gemellologicae, 37,* 205–216.

DeFries, J. C., Fulker, D. W., & LaBuda, M. C. (1987). Evidence for a genetic aetiology in reading disability of twins. *Nature, 329,* 537–539. DOI: 10.1038/329537a0

DeFries, J. C., Gervais, M. C., & Thomas, E. A. (1978). Response to 30 generations of selection for open-field activity in laboratory mice. *Behavior Genetics, 8,* 3–13. DOI: 10.1007/BF01067700

DeFries, J. C., & Gillis, J. J. (1993). Genetics and reading disability. In R. Plomin & G. E. McClearn (Eds.), *Nature, nurture and psychology* (pp. 121–145). Washington, DC: American Psychological Association. DOI: 10.1037/10131-006

DeFries, J. C., Johnson, R. C., Kuse, A. R., McClearn, G. E., Polovina, J., Vandenberg, S. G., & Wilson, J. R. (1979). Familial resemblance for specific cognitive abilities. *Behavior Genetics, 9,* 23–43. DOI: 10.1007/BF01067119

DeFries, J. C., Knopik, V. S., & Wadsworth, S. J. (1999). Colorado Twin Study of Reading Disability. In D. D. Duane (Ed.), *Reading and attention disorders: Neurobiological correlates* (pp. 17–41). Baltimore, MD: York Press.

DeFries, J. C., Vandenberg, S. G., & McClearn, G. E. (1976). Genetics of specific cognitive abilities. *Annual Review of Genetics, 10,* 179–207. DOI: 10.1146/annurev.ge.10.120176.001143

DeFries, J. C., Vogler, G. P., & LaBuda, M. C. (1986). Colorado Family Reading Study: An overview. In J. L. Fuller & E. C. Simmel (Eds.), *Perspectives in behavior genetics* (pp. 29–56). Hillsdale, NJ: Erlbaum.

DeLisi, L. E., Mirsky, A. F., Buchsbaum, M. S., van Kammen, D. P., Berman, K. F., Caton, C., . . . Karoum, F. (1984). The Genain quadruplets 25 years later: A diagnostic and biochemical followup. *Psychiatric Research, 13,* 59–76. DOI: 10.1016/0165-1781(84)90119-7

Dellava, J. E., Lichtenstein, P., & Kendler, K. S. (2012). Genetic variance of body mass index from

childhood to early adulthood. *Behavior Genetics, 42,* 86–95. DOI: 10.1007/s10519-011-9486-x

Demirkan, A., Penninx, B. W. J. H., Hek, K., Wray, N. R., Amin, N., Aulchenko, Y. S., . . . Middeldorp, C. M. (2011). Genetic risk profiles for depression and anxiety in adult and elderly cohorts. *Molecular Psychiatry, 16,* 773–783. DOI: 10.1038/mp.2010.65

Dempster, E. L., Pidsley, R., Schalkwyk, L. C., Owens, S., Georgiades, A., Kane, F., . . . Mill, J. (2011). Disease-associated epigenetic changes in monozygotic twins discordant for schizophrenia and bipolar disorder. *Human Molecular Genetics, 20,* 4786–4796. DOI: 10.1093/hmg/ddr416

Dennett, D. C. (2006). *Breaking the spell: Religion as a natural phenomenon.* St Albans: Allen Lane.

Derks, E. M., Dolan, C. V., & Boomsma, D. I. (2006). A test of the Equal Environment Assumption (EEA) in multivariate twin studies. *Twin Research and Human Genetics, 9,* 403–411. DOI: 10.1375/twin.9.3.403

Detterman, D. K. (1986). Human intelligence is a complex system of separate processes. In R. J. Sternberg & D. K. Detterman (Eds.), *What is intelligence? Contemporary viewpoints on its nature and measurement* (pp. 57–61). Norwood, NJ: Ablex.

DeVincenzo, J., Lambkin-Williams, R., Wilkinson, T., Cehelsky, J., Nochur, S., Walsh, E., . . . Vaishnaw, A. (2010). A randomized, double-blind, placebo-controlled study of an RNAi-based therapy directed against respiratory syncytial virus. *Proceedings of the National Academy of Sciences (USA), 107,* 8800–8805. DOI: 10.1073/pnas.0912186107

Devineni, A. V., & Heberlein, U. (2009). Preferential ethanol consumption in Drosophila models features of addiction. *Current Biology, 19,* 2126–2132. DOI: 10.1016/j.cub.2009.10.070

Dick, D., Prescott, C. A., & McGue, M. (2009). The genetics of substance use and substance use disorders. In Y.-K. Kim (Ed.), *Handbook of behavior genetics* (pp. 433–453). New York: Springer. DOI: 10.1007/978-0-387-76727-7_29

Dick, D. M. (2007). Identification of genes influencing a spectrum of externalizing psychopathology. *Current Directions in Psychological Science, 16,* 331–335. DOI: 10.1111/j.1467-8721.2007.00530.x

Dick, D. M., Agrawal, A., Schuckit, M. A., Bierut, L., Hinrichs, A., Fox, L., . . . Begleiter, H. (2006). Marital status, alcohol dependence, and GABRA2: Evidence for gene-environment correlation and interaction. *Journal of Studies on Alcohol, 67,* 185–194.

Dick, D. M., Aliev, F., Bierut, L., Goate, A., Rice, J., Hinrichs, A., . . . Hesselbrock, V. (2006). Linkage analyses of IQ in the Collaborative Study on the Genetics of Alcoholism (COGA) sample.

Behavior Genetics, 36, 77–86. DOI: 10.1007/s10519-005-9009-8

Dick, D. M., Latendresse, S. J., Lansford, J. E., Budde, J. P., Goate, A., Dodge, K. A., Bates, J. E. (2009). Role of GABRA2 in trajectories of externalizing behavior across development and evidence of moderation by parental monitoring. *Archives of General Psychiatry, 66,* 649–657. DOI: 10.1001/archgenpsychiatry.2009.48

Dick, D. M., Meyers, J. L., Latendresse, S. J., Creemers, H. E., Lansford, J. E., Pettit, G. S., . . . Huizink, A. C. (2011). CHRM2, parental monitoring, and adolescent externalizing behavior: Evidence for gene-environment interaction. *Psychological Science, 22,* 481–489. DOI: 10.1177/0956797611403318

Dick, D. M., Pagan, J. L., Viken, R., Purcell, S., Kaprio, J., Pulkkinen, L., & Rose, R. J. (2007). Changing environmental influences on substance use across development. *Twin Research and Human Genetics, 10,* 315–326. DOI: 10.1375/twin.10.2.315

Dick, D. M., Rose, R. J., Viken, R. J., Kaprio, J., & Koskenvuo, M. (2001). Exploring gene-environment interactions: Socioregional moderation of alcohol use. *Journal of Abnormal Psychology, 110,* 625–632. DOI: 10.1037/0021-843X.110.4.625

Dickson, S. P., Wang, K., Krantz, I., Hakonarson, H., & Goldstein, D. B. (2010). Rare variants create synthetic genome-wide associations. *PLoS Biology, 8:* e1000294. DOI: 10.1371/journal.pbio.1000294

Diener, E., & Chan, M. Y. (2011). Happy people live longer: Subjective well-being contributes to health and longevity. *Applied Psychology: Health and Well-Being, 3,* 1–43. DOI: 10.1111/j.1758-0854.2010.01045.x

DiLalla, L. F., & Gottesman, I. I. (1989). Heterogeneity of causes for delinquency and criminality: Lifespan perspectives. *Development and Psychopathology, 1,* 339–349. DOI: 10.1017/S0954579400000511

DiPetrillo, K., Wang, X., Stylianou, I. M., & Paigen, B. (2005). Bioinformatics toolbox for narrowing rodent quantitative trait loci. *Trends in Genetics, 21,* 683–692. DOI: 10.1016/j.tig.2005.09.008

Dobzhansky, T. (1964). *Heredity and the nature of man.* New York: Harcourt, Brace & World.

Docherty, S. J., Davis, O. S. P., Kovas, Y., Meaburn, E. L., Dale, P. S., Petrill, S. A., . . . Plomin, R. (2010). A genome-wide association study identifies multiple loci associated with mathematics ability and disability. *Genes, Brain and Behavior, 9,* 234–247. DOI: 10.1111/j.1601-183X.2009.00553.x

Docherty, S. J., Kovas, Y., Petrill, S. A., & Plomin, R. (2010). Generalist genes analysis of DNA markers associated with mathematical ability and disability reveals shared influence across ages and abilities. *BMC Genetics, 11:* 61. DOI: 10.1186/1471-2156-11-61

Dodge, K. A., & Rutter, M. (2011). *Gene-environment interactions in developmental psychopathology.* New York: Guilford Press.

Down, J. L. H. (1866). Observations on an ethnic classification of idiots. *Clinical Lecture Reports, London Hospital 3,* 259–262.

Doyle, A. E., Biederman, J., Ferreira, M. A. R., Wong, P., Smoller, J. W., & Faraone, S. V. (2010). Suggestive linkage of the Child Behavior Checklist Juvenile Bipolar Disorder phenotype to 1p21, 6p21, and 8q21. *Journal of the American Academy of Child and Adolescent Psychiatry, 49,* 378–387. DOI: 10.1016/j.jaac.2010.01.008

Dubnau, J., & Tully, T. (1998). Gene discovery in Drosophila: New insights for learning and memory. *Annual Review of Neuroscience, 21,* 407–444. DOI: 10.1146/annurev.neuro.21.1.407

Dubois, L., Ohm Kyvik, K., Girard, M., Tatone-Tokuda, F., Perusse, D., Hjelmborg, J., . . . Martin, N. G. (2012). Genetic and environmental contributions to weight, height, and BMI from birth to 19 years of age: An international study of over 12,000 twin pairs. *PLoS One, 7:* e30153. DOI: 10.1371/journal.pone.0030153

Duncan, A. E., Agrawal, A., Grant, J. D., Bucholz, K. K., Madden, P. A. F., & Heath, A. C. (2009). Genetic and environmental contributions to BMI in adolescent and young adult women. *Obesity, 17,* 1040–1043. DOI: 10.1038/oby.2008.643

Duncan, J. D. (2010). *How intelligence happens.* New Haven, CT: Yale University Press.

Duncan, L. E., & Keller, M. C. (2011). A critical review of the first 10 years of candidate gene-by-environment interaction research in psychiatry. *American Journal of Psychiatry, 168,* 1041–1049. DOI: 10.1176/appi.ajp.2011.11020191

Dunn, J. F., & Plomin, R. (1986). Determinants of maternal behavior toward three-year-old siblings. *British Journal of Developmental Psychology, 4,* 127–137. DOI: 10.1111/j.2044-835X.1986.tb01004.x

Dunn, J. F., & Plomin, R. (1990). *Separate lives: Why siblings are so different.* New York: Basic Books.

Duyme, M., Dumaret, A. C., & Tomkiewicz, S. (1999). How can we boost IQs of "dull children"?: A late adoption study. *Proceedings of the National Academy of Sciences (USA), 96,* 8790–8794. DOI: 10.1073/pnas.96.15.8790

Dworkin, R. H. (1979). Genetic and environmental influences on person-situation interactions. *Journal of Research in Personality, 13,* 279–293. DOI: 10.1016/0092-6566(79)90019-9

Dworkin, R. H., & Lenzenweger, M. F. (1984). Symptoms and the genetics of schizophrenia: Implications for diagnosis. *American Journal of Psychiatry, 14,* 1541–1546.

Dworzynski, K., Remington, A., Rijsdijk, F., Howell, P., & Plomin, R. (2007). Genetic etiology in cases of recovered and persistent stuttering in an unselected, longitudinal sample of young twins. *American Journal of Speech-Language Pathology, 16,* 169–178. DOI: 10.1044/1058-0360(2007/021)

Eaves, L., Foley, D., & Silberg, J. (2003). Has the "equal environments" assumption been tested in twin studies? *Twin Research, 6,* 486–489. DOI: 10.1375/136905203322686473

Eaves, L., Rutter, M., Silberg, J. L., Shillady, L., Maes, H., & Pickles, A. (2000). Genetic and environmental causes of covariation in interview assessments of disruptive behavior in child and adolescent twins. *Behavior Genetics, 30,* 321–334. DOI: 10.1023/A:1026553518272

Eaves, L. J. (1976). A model for sibling effects in man. *Heredity, 36,* 205–214. DOI: 10.1038/hdy.1976.25

Eaves, L. J., D'Onofrio, B., & Russell, R. (1999). Transmission of religion and attitudes. *Twin Research, 2,* 59–61. DOI: 10.1375/twin.2.2.59

Eaves, L. J., Eysenck, H., & Martin, N. G. (1989). *Genes, culture, and personality: An empirical approach.* London: Academic Press.

Eaves, L. J., & Eysenck, H. J. (1976). Genetical and environmental components of inconsistency and unrepeatability in twins' responses to a neuroticism questionaire. *Behavior Genetics, 6,* 145–160. DOI: 10.1007/BF01067145

Eaves, L. J., Heath, A. C., Martin, N. G., Maes, H., Neale, M., Kendler, K., . . . Corey, L. (1999). Comparing the biological and cultural inheritance of personality and social attitudes in the Virginia 30,000 study of twins and their relatives. *Twin Research, 2,* 62–80. DOI: 10.1375/136905299320565933

Eaves, L. J., Heath, A. C., Neale, M. C., Hewitt, J. K., & Martin, N. G. (1998). Sex differences and non-additivity in the effects of genes in personality. *Twin Research, 1,* 131–137.

Eaves, L. J., Kendler, K. S., & Schulz, S. C. (1986). The familial sporadic classification: Its power for the resolution of genetic and environmental etiological factors. *Journal of Psychiatric Research, 20,* 115–130. DOI: 10.1016/0022-3956(86)90011-7

Eaves, L. J., Prom, E. C., & Silberg, J. L. (2010). The mediating effect of parental neglect on adolescent and young adult anti-sociality: A longitudinal study of twins and their parents. *Behavior Genetics, 40,* 425–437. DOI: 10.1007/s10519-010-9336-2

Eaves, L. J., Silberg, J. L., Hewitt, J. K., Meyer, J., Rutter, M., Simonoff, E., . . . Pickles, A. (1993). Genes, personality and psychopathology: A latent class analysis of liability to symptoms of attention-deficit hyperactivity disorder in twins. In R. Plomin & G. E. McClearn (Eds.), *Nature, nurture, and psychology* (pp. 285-303). Washington, DC: American Psychological Association. DOI: 10.1037/10131-014

Eaves, L. J., Silberg, J. L., Meyer, J. M., Maes, H. H., Simonoff, E., Pickles, A., . . . Hewitt, J. K. (1997). Genetics and developmental psychopathology: 2. The main effects of genes and environment on behavioral problems in the Virginia Twin Study of Adolescent Behavioral Development. *Journal of Child Psychology and Psychiatry, 38,* 965–980. DOI: 10.1111/j.1469-7610.1997.tb01614.x

Ebstein, R. P. (2006). The molecular genetic architecture of human personality: Beyond self-report questionnaires. *Molecular Psychiatry, 11,* 427–445. DOI: 10.1038/sj.mp.4001814

Ebstein, R. P., Novick, O., Umansky, R., Priel, B., Osher, Y., Blaine, D., . . . Belmaker, R. H. (1996). Dopamine D4 receptor (D4DR) exon III polymorphism associated with the human personality trait of novelty-seeking. *Nature Genetics, 12,* 78–80. DOI: 10.1038/ng0196-78

Eckman, P. (1973). *Darwin and facial expression: A century of research in review.* New York: Academic Press.

Edwards, A. C., Dodge, K. A., Latendresse, S. J., Lansford, J. E., Bates, J. E., Pettit, G. S., . . . Dick, D. M. (2010). MAOA-uVNTR and early physical discipline interact to influence delinquent behavior. *Journal of Child Psychology and Psychiatry, 51,* 679–687. DOI: 10.1111/j.1469-7610.2009.02196.x

Edwards, S. L., Rapee, R. M., & Kennedy, S. (2010). Prediction of anxiety symptoms in preschool-aged children: Examination of maternal and paternal perspectives. *Journal of Child Psychology and Psychiatry, 51,* 313–321. DOI: 10.1111/j.1469-7610.2009.02160.x

Egeland, J. A., Gerhard, D. S., Pauls, D. L., Sussex, J. N., Kidd, K. K., Allen, C. R., . . . Housman, D. E. (1987). Bipolar affective disorders linked to DNA markers on chromosome 11. *Nature, 325,* 783–787. DOI: 10.1038/325783a0

Ehlers, C. L., Gilder, D. A., Slutske, W. S., Lind, P. A., & Wilhelmsen, K. C. (2008). Externalizing disorders in American Indians: Comorbidity and a genome wide linkage analysis. *American Journal of Medical Genetics. B: Neuropsychiatric Genetics, 147B,* 690–698. DOI: 10.1002/ajmg.b.30666

Ehlers, C. L., Gilder, D. A., Wall, T. L., Phillips, E., Feiler, H., & Wilhelmsen, K. C. (2004). Genomic screen for loci associated with alcohol dependence in mission Indians. *American Journal of Medical Genetics. B: Neuropsychiatric Genetics, 129B,* 110–115. DOI: 10.1002/ajmg.b.30057

Ehlers, C. L., Walter, N. A. R., Dick, D. M., Buck, K. J., & Crabbe, J. C. (2010). A comparison of selected quantitative trait loci associated with alcohol use phenotypes in humans and mouse models. *Addiction Biology, 15,* 185–199. DOI: 10.1111/j.1369-1600.2009.00195.x

Ehringer, M. A., Rhee, S. H., Young, S., Corley, R., & Hewitt, J. K. (2006). Genetic and environmental contributions to common psychopathologies of childhood and adolescence: A study of twins and their siblings. *Journal of Abnormal Child Psychology, 34,* 1–17. DOI: 10.1007/s10802-005-9000-0

Ehrman, L. (1972). A factor influencing the rare male mating advantage in Drosophila. *Behavior Genetics, 2,* 69–78. DOI: 10.1007/BF01066735

Eley, T. C., Bolton, D., O'Connor, T. G., Perrin, S., Smith, P., & Plomin, R. (2003). A twin study of anxiety-related behaviours in pre-school children. *Journal of Child Psychology and Psychiatry, 44,* 945–960. DOI: 10.1111/1469-7610.00179

Eley, T. C., Collier, D., & McGuffin, P. (2002). Anxiety and eating disorders. In P. McGuffin, M. J. Owen, & I. I. Gottesman (Eds.), *Psychiatric genetics and genomics* (pp. 303–340). Oxford: Oxford University Press.

Eley, T. C., Lichtenstein, P., & Stevenson, J. (1999). Sex differences in the aetiology of aggressive and non-aggressive antisocial behavior: Results from two twin studies. *Child Development, 70,* 155–168. DOI: 10.1111/1467-8624.00012

Eley, T. C., Napolitano, M., Lau, J. Y. F., & Gregory, A. M. (2010). Does childhood anxiety evoke maternal control? A genetically informed study. *Journal of Child Psychology and Psychiatry, 51,* 772–779. DOI: 10.1111/j.1469-7610.2010.02227.x

Eley, T. C., Rijsdijk, F. V., Perrin, S., O'Connor, T. G., & Bolton, D. (2008). A multivariate genetic analysis of specific phobia, separation anxiety and social phobia in early childhood. *Journal of Abnormal Child Psychology, 36,* 839–848. DOI: 10.1007/s10802-008-9216-x

Elia, J., Glessner, J. T., Wang, K., Takahashi, N., Shtir, C. J., Hadley, D., . . . Hakonarson, H. (2012). Genome-wide copy number variation study associates metabotropic glutamate receptor gene networks with attention deficit hyperactivity disorder. *Nature Genetics, 44,* 78–84. DOI: 10.1038/ng.1013

Elkins, I. J., McGue, M., & Iacono, W. G. (1997). Genetic and environmental influences on parent-son relationships: Evidence for increasing genetic influence during adolescence. *Developmental Psychology, 33,* 351–363. DOI: 10.1037/0012-1649.33.2.351

Enoch, M.-A., Hodgkinson, C. A., Yuan, Q., Albaugh, B., Virkkunen, M., & Goldman, D. (2009). GABRG1 and GABRA2 as independent predictors for alcoholism in two populations. *Neuropsychopharmacology, 34,* 1245–1254. DOI: 10.1038/npp.2008.171

Enoch, M. A., Gorodetsky, E., Hodgkinson, C., Roy, A., & Goldman, D. (2011). Functional genetic variants that increase synaptic serotonin and 5-HT3 receptor sensitivity predict alcohol and drug dependence. *Molecular Psychiatry, 16,* 1139–1146. DOI: 10.1038/mp.2010.94

Enoch, M. A., Steer, C. D., Newman, T. K., Gibson, N., & Goldman, D. (2010). Early life stress, MAOA, and gene-environment interactions predict behavioral disinhibition in children. *Genes Brain and Behavior, 9,* 65–74. DOI: 10.1111/j.1601-183X.2009.00535.x

Erlenmeyer-Kimling, L. (1972). Gene-environment interactions and the variability of behavior. In L. Ehrman, G. S. Omenn, & E. Caspari (Eds.), *Genetics, environment, and behavior* (pp. 181–208). San Diego: Academic Press.

Erlenmeyer-Kimling, L., & Jarvik, L. F. (1963). Genetics and intelligence: A review. *Science, 142,* 1477–1479. DOI: 10.1126/science.142.3598.1477

Erlenmeyer-Kimling, L., Squires-Wheeler, E., Adamo, U. H., Bassett, A. S., Cornblatt, B. A., Kestenbaum, C. J., . . . Gottesman, I. I. (1995). The New York High-Risk Project: Psychoses and cluster A personality disorders in offspring of schizophrenic parents at 23 years of follow-up. *Archives of General Psychiatry, 52,* 857–865. DOI: 10.1001/archpsyc.1995.03950220067013

Estourgie-van Burk, G. F., Bartels, M., van Beijsterveldt, T. C., Delemarre-van de Waal, H. A., & Boomsma, D. I. (2006). Body size in five-year-old twins: Heritability and comparison to singleton standards. *Twin Research and Human Genetics, 9,* 646–655. DOI: 10.1375/twin.9.5.646

Etcoff, N. (1999). *Survival of the prettiest: The science of beauty.* London: Little, Brown & Co.

Evans, W. E., & Relling, M. V. (2004). Moving towards individualized medicine with pharmacogenomics. *Nature, 429,* 464–468. DOI: 10.1038/nature02626

Eyler, L. T., Prom-Wormley, E., Panizzon, M. S., Kaup, A. R., Fennema-Notestine, C., Neale, M. C., . . . Kremen, W. S. (2011). Genetic and environmental contributions to regional cortical surface area in humans: A magnetic resonance imaging twin study. *Cerebral Cortex, 21,* 2313–2321. DOI: 10.1093/cercor/bhr013

Eysenck, H. J. (1952). *The scientific study of personality.* London: Routledge & Kegan Paul.

Fagard, R., Bielen, E., & Amery, A. (1991). Heritability of aerobic power and anaerobic energy generation during exercise. *Journal of Applied Physiology, 70,* 357–362.

Fagard, R. H., Loos, R. J., Beunen, G., Derom, C., & Vlietinck, R. (2003). Influence of chorionicity on the heritability estimates of blood pressure:

A study in twins. *Journal of Hypertension, 21,* 1313–1318. DOI: 10.1097/00004872-200307000-00019

Fagnani, C., Fibiger, S., Skytthe, A., & Hjelmborg, J. V. B. (2011). Heritability and environmental effects for self-reported periods with stuttering: A twin study from Denmark. *Logopedics Phoniatrics Vocology, 36,* 114–120. DOI: 10.3109/14015439.2010.534503

Falconer, D. S. (1965). The inheritance of liability to certain diseases estimated from the incidence among relatives. *Annals of Human Genetics, 29,* 51–76. DOI: 10.1111/j.1469-1809.1965.tb00500.x

Falconer, D. S., & MacKay, T. F. C. (1996). *Introduction to quantitative genetics* (4th ed.) Harlow, UK: Longman.

Faraone, S. V. (2004). Genetics of adult attention-deficit/hyperactivity disorder. *Psychiatric Clinics of North America, 27,* 303–321. DOI: 10.1016/S0193-953X(03)00090-X

Faraone, S. V., Biederman, J., & Mick, E. (2006). The age-dependent decline of attention deficit hyperactivity disorder: A meta-analysis of follow-up studies. *Psychological Medicine, 36,* 159–165. DOI: 10.1017/S003329170500471X

Faraone, S. V., Biederman, J., & Monuteaux, M. C. (2000). Attention-deficit disorder and conduct disorder in girls: Evidence for a familial subtype. *Biological Psychiatry, 48,* 21–29. DOI: 10.1016/S0006-3223(00)00230-4

Faraone, S. V., Perlis, R. H., Doyle, A. E., Smoller, J. W., Goralnick, J. J., Holmgren, M. A., & Sklar, P. (2005). Molecular genetics of attention-deficit/hyperactivity disorder. *Biological Psychiatry, 57,* 1313–1323. DOI: 10.1016/j.biopsych.2004.11.024

Faraone, S. V., Sergeant, J., Gillberg, C., & Biederman, J. (2003). The worldwide prevalence of ADHD: Is it an American condition? *World Psychiatry, 2,* 104–113.

Faraone, S. V., Tsuang, M. T., & Tsuang, D. W. (2002). *Genetics of mental disorders: What practitioners and students need to know.* New York: Guilford Press.

Farmer, A., Elkin, A., & McGuffin, P. (2007). The genetics of bipolar affective disorder. *Current Opinion in Psychiatry, 20,* 8–12. DOI: 10.1097/YCO.0b013e3280117722

Farmer, A., Scourfield, J., Martin, N., Cardno, A., & McGuffin, P. (1999). Is disabling fatigue in childhood influenced by genes? *Psychological Medicine, 29,* 279–282. DOI: 10.1017/S0033291798008095

Farmer, A. E., McGuffin, P., & Gottesman, I. I. (1987). Twin concordance for DSM-III schizophrenia: Scrutinzing the validity of the definition. *Archives of General Psychiatry, 44,* 634–641. DOI: 10.1001/archpsyc.1987.01800190054009

Farooqi, I. S., Keogh, J. M., Yeo, G. S. H., Lank, E. J., Cheetham, T., & O'Rahilly, S. (2003). Clinical spectrum of obesity and mutations in the melanocortin 4 receptor gene. *New England Journal of Medicine, 348,* 1085–1095. DOI: 10.1056/NEJMoa022050

Farooqi, I. S., Wangensteen, T., Collins, S., Kimber, W., Matarese, G., Keogh, J. M., . . . O'Rahilly, S. (2007). Clinical and molecular genetic spectrum of congenital deficiency of the leptin receptor. *New England Journal of Medicine, 356,* 237–247. DOI: 10.1056/NEJMoa063988

Fazel, S., & Danesh, J. (2002). Serious mental disorder in 23000 prisoners: A systematic review of 62 surveys. *Lancet, 359,* 545–550. DOI: 10.1016/S0140-6736(02)07740-1

Fearon, R. M., van Ijzendoorn, M. H., Fonagy, P., Bakermans-Kranenburg, M. J., Schuengel, C., & Bokhorst, C. L. (2006). In search of shared and nonshared environmental factors in security of attachment: A behavior-genetic study of the association between sensitivity and attachment security. *Developmental Psychology, 42,* 1026–1040. DOI: 10.1037/0012-1649.42.6.1026

Federenko, I. S., Schlotz, W., Kirschbaum, C., Bartels, M., Hellhammer, D. H., & Wust, S. (2006). The heritability of perceived stress. *Psychological Medicine, 36,* 375–385. DOI: 10.1017/S0033291705006616

Fehr, E., & Fischbacher, U. (2003). The nature of human altruism. *Nature, 425,* 785–791. DOI: 10.1038/nature02043

Feigon, S. A., Waldman, I. D., Levy, F., & Hay, D. A. (2001). Genetic and environmental influences on separation anxiety disorder symptoms and their moderation by age and sex. *Behavior Genetics, 31,* 403–411. DOI: 10.1023/A:1012738304233

Feinberg, M. E., Button, T. M. M., Neiderhiser, J. M., Reiss, D., & Hetherington, E. M. (2007). Parenting and adolescent antisocial behavior and depression—evidence of genotype X parenting environment interaction. *Archives of General Psychiatry, 64,* 457–465. DOI: 10.1001/archpsyc.64.4.457

Felsenfeld, S. (1994). Developmental speech and language disorders. In J. C. DeFries, R. Plomin & D. W. Fulker (Eds.), *Nature and nurture during middle childhood* (pp. 102–119). Oxford: Blackwell.

Felsenfeld, S., & Plomin, R. (1997). Epidemiological and offspring analyses of developmental speech disorders using data from the Colorado Adoption Project. *Journal of Speech, Language, and Hearing Research, 40,* 778–791.

Ferguson, C. J. (2010). Genetic contributions to antisocial personality and behavior: A meta-analytic review from an evolutionary perspective. *Journal of Social Psychology, 150,* 160–180. DOI: 10.1080/00224540903366503

Filiou, M. D., Turck, C. W., & Martins-de-Souza, D. (2011). Quantitative proteomics for investigating psychiatric disorders. *Proteomics Clinical Applications, 5,* 38–49. DOI: 10.1002/prca.201000060

Finkel, D., & McGue, M. (2007). Genetic and environmental influences on intraindividual variability in reaction time. *Experimental Aging Research, 33,* 13–35. DOI: 10.1080/03610730601006222

Finkel, D., & Pedersen, N. L. (2000). Contribution of age, genes, and environment to the relationship between perceptual speed and cognitive ability. *Psychology and Aging, 15,* 56–64. DOI: 10.1037/0882-7974.15.1.56

Finkel, D., & Reynolds, C. A. (2010). Behavioral genetic investigations of cognitive aging. In Y.-K. Kim (Ed.), *Handbook of behavior genetics* (pp. 101–112). New York: Springer. DOI: 10.1007/978-0-387-76727-7_7

Finkel, D., Wille, D. E., & Matheny, A. P., Jr. (1998). Preliminary results from a twin study of infant-caregiver attachment. *Behavior Genetics, 28,* 1–8. DOI: 10.1023/A:1021448429653

Finn, C. T., & Smoller, J. W. (2006). Genetic counseling in psychiatry. *Harvard Review of Psychiatry, 14,* 109–121. DOI: 10.1080/10673220600655723

Fischer, P. J., & Breakey, W. R. (1991). The epidemiology of alcohol, drug and mental disorders among homeless persons. *American Psychologist, 46,* 1115–1128. DOI: 10.1037/0003-066X.46.11.1115

Fisher, R. A. (1918). The correlation between relatives on the supposition of Mendelian inheritance. *Transactions of the Royal Society of Edinburgh, 52,* 399–433.

Fisher, R. A. (1922). On the mathematical foundations of theoretical statistics. *Philosophical Transactions of the Royal Society of London, 222,* 309–368. DOI: 10.1098/rsta.1922.0009

Fisher, R. A. (1930). *The genetical theory of natural selection.* Oxford: Clarendon Press.

Fisher, R. C. (2004). *Why men won't ask for directions: The seductions of sociobiology.* Princeton, NJ: Princeton University Press.

Fisher, S. E. (2010). Genetic susceptibility to stuttering. *New England Journal of Medicine, 362,* 750–752. DOI: 10.1056/NEJMe0912594

Fisher, S. E., & DeFries, J. C. (2002). Developmental dyslexia: Genetic dissection of a complex cognitive trait. *Nature Reviews Neuroscience, 3,* 767–780. DOI: 10.1038/nrn936

Flachsbart, F., Caliebeb, A., Kleindorp, R., Blanche, H., von Eller-Eberstein, H., Nikolaus, S., . . . Nebel, A. (2009). Association of FOXO3A variation with human longevity confirmed in German centenarians. *Proceedings of the National Academy of Sciences (USA), 106,* 2700–2705. DOI: 10.1073/pnas.0809594106

Flaxman, S. M., & Sherman, P. W. (2000). Morning sickness: A mechanism for protecting mother and embryo. *Quarterly Review of Biology, 75,* 113–148. DOI: 10.1086/393377

Flegal, K. M., Carroll, M. D., Kit, B. K., & Ogden, C. L. (2012). Prevalence of obesity and trends in the distribution of body mass index among US adults, 1999–2010. *Journal of the American Medical Association, 307,* 491–497. DOI: 10.1001/jama.2012.39

Flegal, K. M., Graubard, B. I., Williamson, D. F., & Gail, M. H. (2007). Cause-specific excess deaths associated with underweight, overweight, and obesity. *Journal of the American Medical Association, 298,* 2028–2037. DOI: 10.1001/jama.298.17.2028

Fletcher, R. (1990). *The Cyril Burt scandal: Case for the defense.* New York: Macmillan.

Flint, J. (2004). The genetic basis of neuroticism. *Neurosciences and Biobehavioral Reviews, 28,* 307–316. DOI: 10.1016/j.neubiorev.2004.01.004

Flint, J. (2011). Mapping quantitative traits and strategies to find quantitative trait genes. *Methods, 53,* 163–174. DOI: 10.1016/j.ymeth.2010.07.007

Flint, J., Corley, R., DeFries, J. C., Fulker, D. W., Gray, J. A., Miller, S., & Collins, A. C. (1995). A simple genetic basis for a complex psychological trait in laboratory mice. *Science, 269,* 1432–1435. DOI: 10.1126/science.7660127

Flint, J., & Munafo, M. R. (2007). The endophenotype concept in psychiatric genetics. *Psychological Medicine, 37,* 163–180. DOI: 10.1017/S0033291706008750

Flint, J., Valdar, W., Shifman, S., & Mott, R. (2005). Strategies for mapping and cloning quantitative trait genes in rodents. *Nature Reviews Genetics, 6,* 271–286. DOI: 10.1038/nrg1576

Flint, J., & Willis-Owen, S. (2010). The genetics of personality. In M. R. Speicher, S. E. Antonarakis, & A. G. Motulsky (Eds.), *Vogel and Motulsky's human genetics: Problems and approaches* (4th ed.) (pp. 651–661). New York: Springer.

Focking, M., Dicker, P., English, J. A., Schubert, K. O., Dunn, M. J., & Cotter, D. R. (2011). Common proteomic changes in the hippocampus in schizophrenia and bipolar disorder and particular evidence for involvement of cornu ammonis regions 2 and 3. *Archives of General Psychiatry, 68,* 477-488. DOI: 10.1001/archgenpsychiatry.2011.43

Folstein, S., & Rutter, M. (1977). Infantile autism: A genetic study of 21 twin pairs. *Journal of Child Psychology and Psychiatry, 18,* 297–321. DOI: 10.1111/j.1469-7610.1977.tb00443.x

Fontaine, N. M. G., McCrory, E. J. P., Boivin, M., Moffitt, T. E., & Viding, E. (2011). Predictors and outcomes of joint trajectories of callous-unemotional traits and conduct problems in childhood. *Journal of Abnormal Psychology, 120,* 730–742. DOI: 10.1037/a0022620

Foroud, T., Edenberg, H. J., Goate, A., Rice, J., Flury, L., Koller, D. L., . . . Reich, T. (2000). Alcoholism susceptibility loci: Confirmation studies in a replicate sample and further mapping. *Alcoholism: Clinical and Experimental Research, 24,* 933–945. DOI: 10.1111/j.1530-0277.2000.tb04634.x

Forsman, M., Lichtenstein, P., Andershed, H., & Larsson, H. (2010). A longitudinal twin study of the direction of effects between psychopathic personality and antisocial behaviour. *Journal of Child Psychology and Psychiatry, 51,* 39–47. DOI: 10.1111/j.1469-7610.2009.02141.x

Fountoulakis, M., & Kossida, S. (2006). Proteomics-driven progress in neurodegeneration research. *Electrophoresis, 27,* 1556–1573. DOI: 10.1002/elps.200500738

Fountoulakis, M., Tsangaris, G., Maris, A., & Lubec, G. (2005). The rat brain hippocampus proteome. *Journal of Chromatography B, 819,* 115–129. DOI: 10.1016/j.jchromb.2005.01.037

Fowler, J. H., Baker, L. A., & Dawes, C. T. (2008). Genetic variation in political participation. *American Political Science Review, 102,* 233–248. DOI: 10.1017/S0003055408080209

Fowler, J. H., & Schreiber, D. (2008). Biology, politics, and the emerging science of human nature. *Science, 322,* 912–914. DOI: 10.1126/science.1158188

Fowler, T., Lifford, K., Shelton, K., Rice, F., Thapar, A., Neale, M. C., . . . Van Den Bree, M. B. M. (2007). Exploring the relationship between genetic and environmental influences on initiation and progression of substance use. *Addiction, 102,* 413–422. DOI: 10.1111/j.1360-0443.2006.01694.x

Fraga, M. F., Ballestar, E., Paz, M. F., Ropero, S., Setien, F., Ballestar, M. L., . . . Esteller, M. (2005). Epigenetic differences arise during the lifetime of monozygotic twins. *Proceedings of the National Academy of Sciences (USA), 102,* 10604–10609. DOI: 10.1073/pnas.0500398102

Franic, S., Middeldorp, C. M., Dolan, C. V., Ligthart, L., & Boomsma, D. I. (2010). Childhood and adolescent anxiety and depression: Beyond heritability. *Journal of the American Academy of Child and Adolescent Psychiatry, 49,* 820–829. DOI: 10.1016/j.jaac.2010.05.013

Frayling, T. M., Timpson, N. J., Weedon, M. N., Zeggini, E., Freathy, R. M., Lindgren, C. M., . . . McCarthy, M. I. (2007). A common variant in the FTO gene is associated with body mass index and predisposes to childhood and adult obesity. *Science, 316,* 889–894. DOI: 10.1126/science.1141634

Frazer, K. A., Ballinger, D. G., Cox, D. R., Hinds, D. A., Stuve, L. L., Gibbs, R. A., . . . Stewart, J. (2007). A second generation human haplotype map of over 3.1 million SNPs. *Nature, 449,* 851–853. DOI: 10.1038/nature06258

Freeman, F. N., Holzinger, K. J., & Mitchell, B. (1928). The influence of environment on the intelligence, school achievement, and conduct of foster children. *Yearbook of the National Society for the Study of Education (Part 1), 27,* 103–217.

Freitag, C. M. (2007). The genetics of autistic disorders and its clinical relevance: A review of the literature. *Molecular Psychiatry, 12,* 2–22. DOI: 10.1038/sj.mp.4001896

Friedman, N. P., Miyake, A., Corley, R. P., Young, S. E., DeFries, J. C., & Hewitt, J. K. (2006). Not all executive functions are related to intelligence. *Psychological Science, 17,* 172–179. DOI: 10.1111/j.1467-9280.2006.01681.x

Friedman, N. P., Miyake, A., Young, S. E., DeFries, J. C., Corley, R. P., & Hewitt, J. K. (2008). Individual differences in executive functions are almost entirely genetic in origin. *Journal of Experimental Psychology: General, 137,* 201–225. DOI: 10.1037/0096-3445.137.2.201

Fu, Q. A., Heath, A. C., Bucholz, K. K., Nelson, E., Goldberg, J., Lyons, M. J., . . . Eisen, S. A. (2002). Shared genetic risk of major depression, alcohol dependence, and marijuana dependence—contribution of antisocial personality disorder in men. *Archives of General Psychiatry, 59,* 1125–1132. DOI: 10.1001/archpsyc.59.12.1125

Fulker, D. W. (1979). Nature and nurture: Heredity. In H. J. Eysenck (Ed.), *The structure and measurement of intelligence* (pp. 102–132). New York: Springer-Verlag.

Fulker, D. W., Cherny, S. S., & Cardon, L. R. (1993). Continuity and change in cognitive development. In R. Plomin & G. E. McClearn (Eds.), *Nature, nurture, and psychology* (pp. 77–97). Washington, DC: American Psychological Association. DOI: 10.1037/10131-004

Fulker, D. W., DeFries, J. C., & Plomin, R. (1988). Genetic influence on general mental ability increases between infancy and middle childhood. *Nature, 336,* 767–769. DOI: 10.1038/336767a0

Fulker, D. W., Eysenck, S. B. G., & Zuckerman, M. (1980). A genetic and environmental analysis of sensation seeking. *Journal of Research in Personality, 14,* 261–281. DOI: 10.1016/0092-6566(80)90033-1

Fuller, J. L., & Thompson, W. R. (1960). *Behavior genetics.* New York: Wiley.

Fuller, J. L., & Thompson, W. R. (1978). *Foundations of behavior genetics.* St Louis, MO: Mosby.

Fullerton, J. (2006). New approaches to the genetic analysis of neuroticism and anxiety. *Behavior Genetics, 36,* 147–161. DOI: 10.1007/s10519-005-9000-4

Fullerton, J., Cubin, M., Tiwari, H., Wang, C., Bomhra, A., Davidson, S., . . . Flint, J. (2003). Linkage analysis of extremely discordant and concordant sibling pairs identifies quantitative-trait loci that influence variation in the human personality trait neuroticism. *American Journal of Human Genetics, 72,* 879–890. DOI: 10.1086/374178

Fyer, A. J., Hamilton, S. P., Durner, M., Haghighi, F., Heiman, G. A., Costa, R., . . . Knowles, J. A. (2006). A third-pass genome scan in panic disorder: Evidence for multiple susceptibility loci. *Biological Psychiatry, 60,* 388–401. DOI: 10.1016/j.biopsych.2006.04.018

Fyer, A. J., Mannuzza, S., Chapman, T. F., Martin, L. Y., & Klein, D. F. (1995). Specificity in familial aggregation of phobic disorders. *Archives of General Psychiatry, 52,* 564–573. DOI: 10.1001/archpsyc.1995.03950190046007

Galsworthy, M. J., Paya-Cano, J. L., Liu, L., Monleon, S., Gregoryan, G., Fernandes, C., . . . Plomin, R. (2005). Assessing reliability, heritability and general cognitive ability in a battery of cognitive tasks for laboratory mice. *Behavior Genetics, 35,* 675–692. DOI: 10.1007/s10519-005-3423-9

Galton, F. (1865). Hereditary talent and character. *Macmillan's Magazine, 12,* 157–166, 318–327.

Galton, F. (1869). *Hereditary genius: An enquiry into its laws and consequences.* Cleveland, OH: World Publishing Co.

Galton, F. (1876). The history of twins as a criterion of the relative powers of nature and nurture. *Royal Anthropological Institute of Great Britain and Ireland Journal, 6,* 391–406. DOI: 10.2307/2840900

Galton, F. (1883). *Inquiries into human faculty and its development.* London: Macmillan.

Galton, F. (1889). *Natural inheritance.* London: Macmillan. DOI: 10.5962/bhl.title.32181

Gangestad, S. W., Haselton, M. G., & Buss, D. M. (2006). Evolutionary foundations of cultural variation: Evoked culture and mate preferences. *Psychological Inquiry, 17,* 75–95. DOI: 10.1207/s15327965pli1702_1

Gangestad, S. W., & Simpson, J. A. (2007). *The evolution of mind: Fundamental questions and controversies.* New York: Guilford Publications.

Gangestad, S. W., Thornhill, R., & Garver-Apgar, C. E. (2010a). Fertility in the cycle predicts women's interest in sexual opportunism. *Evolution and Human Behavior, 31,* 400–411. DOI: 10.1016/j.evolhumbehav.2010.05.003

Gangestad, S. W., Thornhill, R., & Garver-Apgar, C. E. (2010b). Men's facial masculinity predicts changes in their female partners' sexual interests across the ovulatory cycle, whereas men's intelligence does not. *Evolution and Human Behavior, 31,* 412–424. DOI: 10.1016/j.evolhumbehav.2010.06.001

Ganiban, J. M., Chou, C., Haddad, S., Lichtenstein, P., Reiss, D., Spotts, E. L., & Neiderhiser, J. M. (2009). Using behavior genetics methods to understand the structure of personality. *European Journal of Developmental Science, 3,* 195–214.

Ganiban, J. M., Saudino, K. J., Ulbricht, J., Neiderhiser, J. M., & Reiss, D. (2008). Stability and change in temperament during adolescence. *Journal of Personality and Social Psychology, 95,* 222–236. DOI: 10.1037/0022-3514.95.1.222

Ganiban, J. M., Ulbricht, J. A., Spotts, E. L., Lichtenstein, P., Reiss, D., Hansson, K., & Neiderhiser, J. M. (2009). Understanding the role of personality in explaining associations between marital quality and parenting. *Journal of Family Psychology, 23,* 646–660. DOI: 10.1037/a0016091

Gao, W., Li, L., Cao, W., Zhan, S., Lv, J., Qin, Y., . . . Hu, Y. (2006). Determination of zygosity by questionnaire and physical features comparison in Chinese adult twins. *Twin Research and Human Genetics, 9,* 266–271. DOI: 10.1375/twin.9.2.266

Gardenfors, P. (2006). *How homo became sapiens: On the evolution of thinking.* Oxford: Oxford University Press. DOI: 10.1093/acprof:oso/9780198528517.001.0001

Gardner, H. (2006). *Multiple intelligences: New horizons in theory and practice.* New York: Basic Books.

Garver-Apgar, C. E., Gangestad, S. W., Thornhill, R., Miller, R. D., & Olp, J. J. (2006). Major histocompatibility complex alleles, sexual responsivity, and unfaithfulness in romantic couples. *Psychological Science, 17,* 830–835. DOI: 10.1111/j.1467-9280.2006.01789.x

Gatz, M., Pedersen, N. L., Berg, S., Johansson, B., Johansson, K., Mortimer, J. A., . . . Ahlbom, A. (1997). Heritability for Alzheimer's disease: The study of dementia in Swedish twins. *Journals of Gerontology. A: Biological Science and Medical Science, 52,* M117–M125. DOI: 10.1093/gerona/52A.2.M117

Gatz, M., Pedersen, N. L., Plomin, R., Nesselroade, J. R., & McClearn, G. E. (1992). Importance of shared genes and shared environments for symptoms of depression in older adults. *Journal of Abnormal Psychology, 101,* 701–708. DOI: 10.1037/0021-843X.101.4.701

Gatz, M., Reynolds, C. A., Fratiglioni, L., Johansson, B., Mortimer, J. A., Berg, S., . . . Pedersen, N. L. (2006). Role of genes and environments for explaining Alzheimer disease. *Archives of General Psychiatry, 63,* 168–174. DOI: 10.1001/archpsyc.63.2.168

Gayán, J., & Olson, R. K. (2003). Genetic and environmental influences on individual differences in printed word recognition. *Journal of Experimental Child Psychology, 84,* 97–123.DOI: 10.1016/S0022-0965(02)00181-9

Gayán, J., Willcutt, E. G., Fisher, S. E., Francks, C., Cardon, L. R., Olson, R. K., . . . DeFries, J. C. (2005). Bivariate linkage scan for reading disability and attention-deficit/hyperactivity disorder localizes pleiotropic loci. *Journal of Child Psychology and Psychiatry, 46,* 1045–1056. DOI: 10.1111/j.1469-7610.2005.01447.x

Ge, X., Conger, R. D., Cadoret, R. J., Neiderhiser, J. M., Yates, W., Troughton, E., & Stewart, M. A. (1996). The developmental interface between nature and nurture: A mutual influence model of child antisocial behaviour and parenting. *Developmental Psychology, 32,* 574–589. DOI: 10.1037/0012-1649.32.4.574

Ge, X., Natsuaki, M. N., Martin, D. M., Leve, L. D., Neiderhiser, J. M., Shaw, D. S., . . . Reiss, D. (2008). Bridging the divide: Openness in adoption and postadoption psychosocial adjustment among birth and adoptive parents. *Journal of Family Psychology, 22,* 529–540. DOI: 10.1037/a0012817

Geary, D. C. (2005). *The origin of mind: Evolution of brain, cognition, and general intelligence.* Washington, DC: American Psychological Association. DOI: 10.1037/10871-000

Gecz, J., Shoubridge, C., & Corbett, M. (2009). The genetic landscape of intellectual disability arising from chromosome X. *Trends in Genetics, 25,* 308–316. DOI: 10.1016/j.tig.2009.05.002

Gejman, P. V., Sanders, A. R., & Kendler, K. S. (2011). Genetics of schizophrenia: New findings and challenges. *Annual Review of Genomics and Human Genetics, 12,* 121–144. DOI: 10.1146/annurev-genom-082410-101459

Gelernter, J., & Kranzler, H. R. (2010). Genetics of drug dependence. *Dialogues in Clinical Neuroscience, 12,* 77–84.

Gelernter, J., Kranzler, H. R., Panhuysen, C., Weiss, R. D., Brady, K., Poling, J., & Farrer, L. (2009). Dense genomewide linkage scan for alcohol dependence in African Americans: Significant linkage on chromosome 10. *Biological Psychiatry, 65,* 111–115. DOI: 10.1016/j.biopsych.2008.08.036

Gelhorn, H., Stallings, M., Young, S., Corley, R., Rhee, S. H., Christian, H., & Hewitt, J. (2006). Common and specific genetic influences on aggressive and nonaggressive conduct disorder domains. *Journal of the American Academy of Child and Adolescent Psychiatry, 45,* 570–577. DOI: 10.1097/01.chi.0000198596.76443.b0

Geschwind, D. H. (2011). Genetics of autism spectrum disorders. *Trends in Cognitive Sciences, 15,* 409–416. DOI: 10.1016/j.tics.2011.07.003

Ghazalpour, A., Doss, S., Zhang, B., Wang, S., Plaisier, C., Castellanos, R., . . . Horvath, S. (2006). Integrating genetic and network analysis to characterize genes related to mouse weight. *PLoS Genetics, 2:* e130. DOI: 10.1371/journal.pgen.0020130

Gibb, B. E., Johnson, A. L., Benas, J. S., Uhrlass, D. J., Knopik, V. S., & McGeary, J. E. (2011). Children's 5-HTTLPR genotype moderates the link between maternal criticism and attentional biases specifically for facial displays of anger. *Cognition & Emotion, 25,* 1104–1120. DOI: 10.1080/02699931.2010.508267

Gibbs, R. A., Weinstock, G. M., Metzker, M. L., Muzny, D. M., Sodergren, E. J., Scherer, S., . . . Collins, F. (2004). Genome sequence of the brown Norway rat yields insights into mammalian evolution. *Nature, 428,* 493–521. DOI: 10.1038/nature02426

Giedd, J. N., Clasen, L. S., Wallace, G. L., Lenroot, R. K., Lerch, J. P., Wells, E. M., . . . Samango-Sprouse, C. A. (2007). XXY (Klinefelter syndrome): A pediatric quantitative brain magnetic resonance imaging case-control study. *Pediatrics, 119,* e232–e240. DOI: 10.1542/peds.2005-2969

Gill, S. C., Butterworth, P., Rodgers, B., Anstey, K. J., Villamil, E., & Melzer, D. (2006). Mental health and the timing of men's retirement. *Social Psychiatry and Psychiatric Epidemiology, 41,* 515–522. DOI: 10.1007/s00127-006-0064-0

Gillespie, N. A., Whitfield, J. B., Williams, B., Heath, A. C., & Martin, N. G. (2005). The relationship between stressful life events, the serotonin transporter (5-HTTLPR) genotype and major depression. *Psychological Medicine, 35,* 101–111. DOI: 10.1017/S0033291704002727

Gillespie, N. A., Zhu, G., Heath, A. C., Hickie, I. B., & Martin, N. G. (2000). The genetic aetiology of somatic distress. *Psychological Medicine, 30,* 1051–1061. DOI: 10.1017/S0033291799002640

Gillham, N. W. (2001). *A life of Sir Francis Galton: From African exploration to the birth of eugenics.* Oxford: Oxford University Press.

Gillis, J. J., Gilger, J. W., Pennington, B. F., & DeFries, J. C. (1992). Attention deficit disorder in reading-disabled twins: Evidence for a genetic etiology. *Journal of Abnormal Child Psychology, 20,* 303–315. DOI: 10.1007/BF00916694

Giot, L., Bader, J. S., Brouwer, C., Chaudhuri, A., Kuang, B., Li, Y., . . . Rothberg, J. M. (2003). A protein interaction map of Drosophila melanogaster. *Science, 302,* 1727–1736. DOI: 10.1126/science.1090289

Giros, B., Jaber, M., Jones, S. R., Wightman, R. M., & Caron, M. G. (1996). Hyperlocomotion and indifference to cocaine and amphetamine in mice lacking the dopamine transporter. *Nature, 379,* 606–612. DOI: 10.1038/379606a0

Godinho, S. I., & Nolan, P. M. (2006). The role of mutagenesis in defining genes in behaviour. *European Journal of Human Genetics, 14,* 651–659. DOI: 10.1038/sj.ejhg.5201545

Goedert, M., & Spillantini, M. G. (2006). A century of Alzheimer's disease. *Science, 314,* 777–781. DOI: 10.1126/science.1132814

Goetz, A. T., & Shackelford, T. K. (2006). Modern application of evolutionary theory to psychology: Key concepts and clarifications. *American Journal of Psychology, 119,* 567–584. DOI: 10.2307/20445364

Goldberg, L. R. (1990). An alternative description of personality: The Big-Five factor structure. *Journal of Personality and Social Psychology, 59,* 1216–1229. DOI: 10.1037/0022-3514.59.6.1216

Goldberg, T. E., & Weinberger, D. R. (2004). Genes and the parsing of cognitive processes. *Trends in Cognitive Sciences, 8,* 325–335. DOI: 10.1016/j.tics.2004.05.011

Goldman, D., Oroszi, G., & Ducci, F. (2005). The genetics of addictions: Uncovering the genes. *Nature Reviews Genetics, 6,* 521–532. DOI: 10.1038/nrg1635

Goldsmith, H. H. (1983). Genetic influences on personality from infancy to adulthood. *Child Development, 54,* 331–355. DOI: 10.2307/1129695

Goldsmith, H. H., Buss, A. H., Plomin, R., Rothbart, M. K., Chess, S., Hinde, R. A., & McCall, R. B. (1987). Roundtable: What is temperament? Four approaches. *Child Development, 58,* 505–529. DOI: 10.2307/1130527

Goldsmith, H. H., & Campos, J. J. (1986). Fundamental issues in the study of early development: The Denver Twin Temperament Study. In M. E. Lamb, A. L. Brown, & B. Rogoff (Eds.), *Advances in developmental psychology* (pp. 231–283). Hillsdale, NJ: Erlbaum.

Goldstein, D. B., Tate, S. K., & Sisodiya, S. M. (2003). Pharmacogenetics goes genomic. *Nature Reviews Genetics, 4,* 937–947. DOI: 10.1038/nrg1229

Goleman, D. (2005). *Emotional intelligence.* New York: Bantam Books.

Goodman, R., & Stevenson, J. (1989). A twin study of hyperactivity. II. The aetiological role of genes, family relationships and perinatal adversity. *Journal of Child Psychology and Psychiatry, 30,* 691–709. DOI: 10.1111/j.1469-7610.1989.tb00782.x

Goodwin, D. W., Schulsinger, F., Hermansen, L., Guze, S. B., & Winokur, G. (1973). Alcohol problems in adoptees raised apart from alcoholic biological parents. *Archives of General Psychiatry, 28,* 238–243. DOI: 10.1001/archpsyc.1973.01750320068011

Goodwin, D. W., Schulsinger, F., Knop, J., Mednick, S., & Guze, S. B. (1977). Alcoholism and de-pression in adopted-out daughters of alcoholics. *Archives of General Psychiatry, 34,* 751–755. DOI: 10.1001/archpsyc.1977.01770190013001

Goodwin, F. K., & Jamison, K. R. (1990). *Manic-depressive illness.* New York: Oxford University Press.

Gornick, M. C., Addington, A. M., Sporn, A., Gogtay, N., Greenstein, D., Lenane, M., . . . Straub, R. E. (2005). Dysbindin (DTNBP1, 6p22.3) is associated with childhood-onset psychosis and endophenotypes measured by the Premorbid Adjustment Scale (PAS). *Journal of Autism and Developmental Disorders, 35,* 831–838. DOI: 10.1007/s10803-005-0028-3

Gottesman, I. I. (1991). *Schizophrenia genesis: The origins of madness.* New York: Freeman.

Gottesman, I. I., & Bertelsen, A. (1989). Confirming unexpressed genotypes for schizophrenia. *Archives of General Psychiatry, 46,* 867–872. DOI: 10.1001/archpsyc.1989.01810100009002

Gottesman, I. I., & Gould, T. D. (2003). The endophenotype concept in psychiatry: Etymology and strategic intentions. *American Journal of Psychiatry, 160,* 636–645. DOI: 10.1176/appi.ajp.160.4.636

Gottfredson, L. S. (1997). Why *g* matters: The complexity of everyday life. *Intelligence, 24,* 79–132. DOI: 10.1016/S0160-2896(97)90014-3

Gottschall, J., & Wilson, D. X. (2005). *The literary animal evolution and the nature of narrative.* Evanston, IL: Northwestern Unversity Press.

Gould, S. J. (2011). *Rocks of ages: Science and religion in the fullness of life.* New York: Ballantine.

Gould, T. D., & Gottesman, I. I. (2006). Psychiatric endophenotypes and the development of valid animal models. *Genes, Brain and Behavior, 5,* 113–119. DOI: 10.1111/j.1601-183X.2005.00186.x

Grados, M. A., & Walkup, J. T. (2006). A new gene for Tourette's syndrome: A window into causal mechanisms? *Trends in Genetics, 22,* 291–293. DOI: 10.1016/j.tig.2006.04.003

Granon, S., Faure, P., & Changeux, J. P. (2003). Executive and social behaviors under nicotinic receptor regulation. *Proceedings of the National Academy of Sciences (USA), 100,* 9596–9601. DOI: 10.1073/pnas.1533498100

Grant, B. F., Hasin, D. S., Chou, S. P., Stinson, F. S., & Dawson, D. A. (2004). Nicotine dependence and psychiatric disorders in the United States—results from the National Epidemiologic Survey on Alcohol and Related Conditions. *Archives of General Psychiatry, 61,* 1107–1115. DOI: 10.1001/archpsyc.61.11.1107

Grant, J. D., Heath, A. C., Bucholz, K. K., Madden, P. A. F., Agrawal, A., Statham, D. J., & Martin, N. G. (2007). Spousal concordance for alcohol dependence: Evidence for assortative mating or spousal interaction effects? *Alcoholism: Clinical and Experimental Research, 31,* 717–728. DOI: 10.1111/j.1530-0277.2007.00356.x

Grant, S. G. N., Marshall, M. C., Page, K. L., Cumiskey, M. A., & Armstrong, J. D. (2005). Synapse proteomics of multiprotein complexes: En route from genes to nervous system diseases. *Human Molecular Genetics, 14,* R225–R234. DOI: 10.1093/hmg/ddi330

Green, A. S., & Grahame, N. J. (2008). Ethanol drinking in rodents: Is free-choice drinking related to the reinforcing effects of ethanol? *Alcohol, 42,* 1–11. DOI: 10.1016/j.alcohol.2007.10.005

Green, E. D., & Guyer, M. S. (2011). Charting a course for genomic medicine from base pairs to bedside. *Nature, 470,* 204–213. DOI: 10.1038/nature09764

Green, R. E., Krause, J., Briggs, A. W., Maricic, T., Stenzel, U., Kircher, M., . . . Pääbo, S. (2010). A draft sequence of the Neandertal genome. *Science, 328,* 710–722. DOI: 10.1126/science.1188021

Greengross, G., & Miller, G. (2011). Humor ability reveals intelligence, predicts mating success, and is higher in males. *Intelligence, 39,* 188–192. DOI: 10.1016/j.intell.2011.03.006

Greenspan, R. J. (1995). Understanding the genetic construction of behavior. *Scientific American, 272(4),* 72–78. DOI: 10.1038/scientificamerican0495-72

Gregory, S. G., Barlow, K. F., McLay, K. E., Kaul, R., Swarbreck, D., Dunham, A., . . . Prigmore, E. (2006). The DNA sequence and biological annotation of human chromosome 1. *Nature, 441,* 315–321. DOI: 10.1038/nature04727

Greven, C., Rijsdijk, F., & Plomin, R. (2011). A twin study of ADHD symptoms in early adolescence: Hyperactivity-impulsivity and inattentiveness show substantial genetic overlap but also genetic specificity. *Journal of Abnormal Child Psychology, 39,* 265–275. DOI: 10.1007/s10802-010-9451-9

Greven, C. U., Asherson, P., Rijsdijk, F. V., & Plomin, R. (2011). A longitudinal twin study on the association between inattentive and hyperactive-impulsive ADHD symptoms. *Journal of Abnormal Child Psychology, 39,* 623–632. DOI: 10.1007/s10802-011-9513-7

Grilo, C. M., & Pogue-Geile, M. F. (1991). The nature of environmental influences on weight and obesity: A behavior genetic analysis. *Psychological Bulletin, 10,* 520–537. DOI: 10.1037/0033-2909.110.3.520

Grof, P., Duffy, A., Cavazzoni, P., Grof, E., Garnham, J., MacDougall, M., . . . Alda, M. (2002). Is response to prophylactic lithium a familial trait? *Journal of Clinical Psychiatry, 63,* 942–947. DOI: 10.4088/JCP.v63n1013

Gunderson, E. P., Tsai, A. L., Selby, J. V., Caan, B., Mayer-Davis, E. J., & Risch, N. (2006). Twins of mistaken zygosity (TOMZ): Evidence for genetic contributions to dietary patterns and physiologic traits. *Twin Research and Human Genetics, 9,* 540–549. DOI: 10.1375/twin.9.4.540

Gunter, T. D., Vaughn, M. G., & Philibert, R. A. (2010). Behavioral genetics in antisocial spectrum disorders and psychopathy: A review of the recent literature. *Behavioral Sciences and the Law, 28,* 148–173. DOI: 10.1002/bsl.923

Guo, G. (2006). Genetic similarity shared by best friends among adolescents. *Twin Research and Human Genetics, 9,* 113–121. DOI: 10.1375/twin.9.1.113

Guo, S. (2004). Linking genes to brain, behavior and neurological diseases: What can we learn from zebrafish? *Genes, Brain and Behavior, 3,* 63–74. DOI: 10.1046/j.1601-183X.2003.00053.x

Gusella, J. F., Tanzi, R. E., Anderson, M. A., Hobbs, W., Gibbons, K., Raschtchian, R., . . . Conneally, P. M. (1984). DNA markers for nervous-system diseases. *Science, 225,* 1320–1326. DOI: 10.1126/science.6089346

Gusella, J. F., Wexler, N. S., Conneally, P. M., Naylor, S. L., Anderson, M. A., & Tanzi, R. E. (1983). A polymorphic DNA marker genetically linked to Huntington's disease. *Nature, 306,* 234–238. DOI: 10.1038/306234a0

Guthrie, R. (1996). The introduction of newborn screening for phenylketonuria: A personal history. *European Journal of Pediatrics, 155,* S4–S5. DOI: 10.1007/PL00014247

Gutknecht, L., Spitz, E., & Carlier, M. (1999). Long-term effect of placental type on anthropometrical and psychological traits among monozygotic twins: A follow up study. *Twin Research, 2,* 212–217. DOI: 10.1375/136905299320565889

Guze, S. B. (1993). Genetics of Briquet's syndrome and somatization disorder: A review of family, adoption and twin studies. *Annals of Clinical Psychiatry, 5,* 225–230. DOI: 10.3109/10401239309148821

Guze, S. B., Cloninger, C. R., Martin, R. L., & Clayton, P. J. (1986). A follow-up and family study of Briquet's syndrome. *British Journal of Psychiatry, 149,* 17–23. DOI: 10.1192/bjp.149.1.17

Haberstick, B. C., Ehringer, M. A., Lessem, J. M., Hopfer, C. J., & Hewitt, J. K. (2011). Dizziness and the genetic influences on subjective experiences to initial cigarette use. *Addiction, 106,* 391–399. DOI: 10.1111/j.1360-0443.2010.03133.x

Haberstick, B. C., Lessem, J. M., McQueen, M. B., Boardman, J. D., Hopfer, C. J., Smolen, A., & Hewitt, J. K. (2010). Stable genes and changing environments: Body mass index across adolescence and young adulthood. *Behavior Genetics, 40,* 495–504. DOI: 10.1007/s10519-009-9327-3

Haberstick, B. C., Timberlake, D., Ehringer, M. A., Lessem, J. M., Hopfer, C. J., Smolen, A., & Hewitt, J. K. (2007). Genes, time to first cigarette and nicotine dependence in a general population sample of young adults. *Addiction, 102,* 655–665. DOI: 10.1111/j.1360-0443.2007.01746.x

Haig, D. (2000). Genomic imprinting, sex-biased dispersal, and social behavior. *Annals of the New York Academy of Sciences, 907,* 149–163. DOI: 10.1111/j.1749-6632.2000.tb06621.x

Halaas, J. L., Gajiwala, K. S., Maffei, M., Cohen, S. L., Chait, B. T., Rabinowitz, D., . . . Friedman, J. M. (1995). Weight-reducing effects of the plasma protein encoded by the obese gene. *Science, 269,* 543–546. DOI: 10.1126/science.7624777

Hall, B., Limaye, A., & Kulkarni, A. B. (2009). Overview: Generation of gene knockout mice. *Current Protocols in Cell Biology, 44,* Unit 19.12.1–17.

Hallett, V., Ronald, A., Rijsdijk, F., & Eley, T. C. (2009). Phenotypic and genetic differentiation of anxiety-related behaviors in middle childhood. *Depression and Anxiety, 26,* 316–324. DOI: 10.1002/da.20539

Hallgren, B. (1957). Enuresis, a clinical and genetic study. *Acta Psychiatrica et Neurologica Scandinavica, Supplementum 114,* 1–159.

Hallmayer, J., Cleveland, S., Torres, A., Phillips, J., Cohen, B., Torigoe, T., . . . Risch, N. (2011). Genetic heritability and shared environmental factors among twin pairs with autism. *Archives of General Psychiatry, 68,* 1095–1102. DOI: 10.1001/archgenpsychiatry.2011.76

Hamer, D., & Copeland, P. (1998). *Living with our genes.* New York: Doubleday.

Hamer, D. H., Hu, S., Magnuson, V. L., Hu, N., & Pattatucci, A. M. L. (1993). A linkage between DNA markers on the X chromosome and male sexual orientation. *Science, 2,* 321–327. DOI: 10.1126/science.8332896

Hamilton, A. S., Lessov-Schlaggar, C. N., Cockburn, M. G., Unger, J. B., Cozen, W., & Mack, T. M. (2006). Gender differences in determinants of smoking initiation and persistence in California twins. *Cancer Epidemiology, Biomarkers and Prevention, 15,* 1189–1197. DOI: 10.1158/1055-9965.EPI-05-0675

Hamilton, W. D. (1964). Genetical evolution of social behaviour: II. *Journal of Theoretical Biology, 7,* 17–52. DOI: 10.1016/0022-5193(64)90039-6

Hammen, C., Brennan, P. A., Keenan-Miller, D., Hazel, N. A., & Najman, J. M. (2010). Chronic and acute stress, gender, and serotonin transporter gene-environment interactions predicting depression symptoms in youth. *Journal of Child Psychology*

and Psychiatry, 51, 180–187. DOI: 10.1111/j.1469-7610.2009.02177.x

Hannon, G. J. (2002). RNA interference. *Nature, 418,* 244–251. DOI: 10.1038/418244a

Hanscombe, K. B., Trzaskowski, M., Haworth, C. M. A., Davis, O. S. P., Dale, P. S., & Plomin, R. (2012). Socioeconomic status (SES) and children's intelligence (IQ): In a UK-representative sample SES moderates the environmental, not genetic, effect on IQ. *PLoS One, 7:* e30320. DOI: 10.1371/journal.pone.0030320

Hansell, N. K., Wright, M. J., Luciano, M., Geffen, G. M., Geffen, L. B., & Martin, N. G. (2005). Genetic covariation between event-related potential (ERP) and behavioral non-ERP measures of working-memory, processing speed, and IQ. *Behavior Genetics, 35,* 695–706. DOI: 10.1007/s10519-005-6188-2

Hansen, E. E., Lozupone, C. A., Rey, F. E., Wu, M., Guruge, J. L., Narra, A., . . . Gordon, J. I. (2011). Pan-genome of the dominant human gut-associated archaeon, Methanobrevibacter smithii, studied in twins. *Proceedings of the National Academy of Sciences (USA), 108,* 4599–4606. DOI: 10.1073/pnas.1000071108

Hanson, D. R., & Gottesman, I. I. (1976). The genetics, if any, of infantile autism and childhood schizophrenia. *Journal of Autism and Developmental Disorders, 6,* 209–234. DOI: 10.1007/BF01543463

Happé, F., & Frith, U. (2006). The weak coherence account: Detail-focused cognitive style in autism spectrum disorders. *Journal of Autism and Developmental Disorders, 36,* 5–25. DOI: 10.1007/s10803-005-0039-0

Happé, F., & Ronald, A. (2008). The 'fractionable autism triad': A review of evidence from behavioural, genetic, cognitive and neural research. *Neuropsychology Review, 18,* 287–304. DOI: 10.1007/s11065-008-9076-8

Happé, F., Ronald, A., & Plomin, R. (2006). Time to give up on a single explanation for autism. *Nature Neuroscience, 9,* 1218–1220. DOI: 10.1038/nn1770

Harakeh, Z., Neiderhiser, J. M., Spotts, E. L., Engels, R. C. M. E., Scholte, R. H. J., & Reiss, D. (2008). Genetic factors contribute to the association between peers and young adults smoking: Univariate and multivariate behavioral genetic analyses. *Addictive Behaviors, 33,* 1113–1122. DOI: 10.1016/j.addbeh.2008.02.017

Harden, K. P., Turkheimer, E., & Loehlin, J. C. (2007). Genotype by environment interaction in adolescents' cognitive aptitude. *Behavior Genetics, 37,* 273–283. DOI: 10.1007/s10519-006-9113-4

Hardy, J. (1997). Amyloid, the presenilins and Alzheimer's disease. *Trends in Neuroscience, 20,* 154–159. DOI: 10.1016/S0166-2236(96)01030-2

Hare, R. D. (1993). *Without conscience: The disturbing world of psychopaths among us.* New York: Pocket Books.

Hariri, A. R., Drabant, E. M., Munoz, K. E., Kolachana, L. S., Mattay, V. S., Egan, M. F., & Weinberger, D. R. (2005). A susceptibility gene for affective disorders and the response of the human amygdala. *Archives of General Psychiatry, 62,* 146–152. DOI: 10.1001/archpsyc.62.2.146

Hariri, A. R., & Holmes, A. (2006). Genetics of emotional regulation: The role of the serotonin transporter in neural function. *Trends in Cognitive Sciences, 10,* 182–191. DOI: 10.1016/j.tics.2006.02.011

Hariri, A. R., Mattay, V. S., Tessitore, A., Kolachana, B., Fera, F., Goldman, D., . . . Weinberger, D. R. (2002). Serotonin transporter genetic variation and the response of the human amygdala. *Science, 297,* 400–403. DOI: 10.1126/science.1071829

Hariri, A. R., & Whalen, P. J. (2011). The amygdala: Inside and out. *F1000 Biology Reports, 3,* 2. DOI: 10.3410/B3-2

Harlaar, N. (2006). *Individual differences in early reading achievement: Developmental insights from a twin study.* Unpublished doctoral dissertation, King's College London, London, UK.

Harlaar, N., Butcher, L. M., Meaburn, E., Sham, P., Craig, I. W., & Plomin, R. (2005). A behavioural genomic analysis of DNA markers associated with general cognitive ability in 7-year-olds. *Journal of Child Psychology and Psychiatry, 46,* 1097–1107. DOI: 10.1111/j.1469-7610.2005.01515.x

Harlaar, N., Cutting, L., Deater-Deckard, K., DeThorne, L. S., Justice, L. M., Schatschneider, C., . . . Petrill, S. A. (2010). Predicting individual differences in reading comprehension: A twin study. *Annals of Dyslexia, 60,* 265–288. DOI: 10.1007/s11881-010-0044-7

Harlaar, N., Hayiou-Thomas, M. E., & Plomin, R. (2005). Reading and general cognitive ability: A multivariate analysis of 7-year-old twins. *Scientific Studies of Reading, 9,* 197–218. DOI: 10.1207/s1532799xssr0903_2

Harper, J. M., Leathers, C. W., & Austad, S. N. (2006). Does caloric restriction extend life in wild mice? *Aging Cell, 5,* 441–449. DOI: 10.1111/j.1474-9726.2006.00236.x

Harris, C. R. (2003). A review of sex differences in sexual jealousy, including self-report data, psychophysiological responses, interpersonal violence, and morbid jealousy. *Personality and Social Psychology Review, 7,* 102–128. DOI: 10.1207/S15327957PSPR0702_102-128

Harris, J. R. (1998). *The nurture assumption: Why children turn out the way they do.* New York: The Free Press.

Harris, J. R., Pedersen, N. L., Stacey, C., McClearn, G. E., & Nesselroade, J. R. (1992). Age differences in the etiology of the relationship between life satisfaction and self rated health. *Journal of Aging and Health, 4,* 349–-368. DOI: 10.1177/089826439200400302

Harris, K. M., Halpern, C. T., Whitsel, E., Hussy, J., Tabor, J., Entzel, P., & Udry, J. R. (2009). The National Longitudinal Study of Adolescent Health: Research design. Retrieved May 25, 2012, from http://www.cpc.unc.edu/projects/addhealth/design

Harris, T. W., Antoshechkin, I., Bieri, T., Blasiar, D., Chan, J., Chen, W. J., . . . Sternberg, P. W. (2010). Wormbase: A comprehensive resource for nematode research. *Nucleic Acids Research, 38,* D463-D467. DOI: 10.1093/nar/gkp952

Hart, S. A., Petrill, S. A., Thompson, L. A., & Plomin, R. (2009). The ABC's of math: A genetic analysis of mathematics and its links with reading ability and general cognitive ability. *Journal of Educational Psychology, 101,* 388–402. DOI: 10.1037/a0015115

Harter, S. (1983). Developmental perspectives on the self-system. In E. M. Hetherington (Ed.), *Handbook of child psychology: Socialization, personality, and social development* (4th ed.) (pp. 275–385). New York: Wiley.

Hartl, D. (2004). *A primer of population genetics* (3rd ed.) Sunderland, MA: Sinauer Associates.

Hartl, D., & Clark, H. G. (2006). *Principles of population genetics* (4th ed.) Sunderland, MA: Sinauer Associates.

Hartl, D. L., & Ruvolo, M. (2011). *Genetics, analysis of genes and genomes* (8th ed.) Burlington, MA: Jones and Bartlett Learning.

Hatemi, P. K., Alford, J. R., Hibbing, J. R., Martin, N. G., & Eaves, L. J. (2009). Is there a "party" in your genes? *Political Research Quarterly, 62,* 584–600. DOI: 10.1177/1065912908327606

Hatemi, P. K., & McDermot, R. (Eds.) (2011). *Man is by nature a political animal: Evolution, biology, and politics.* Chicago: University of Chicago Press.

Havlicek, J., & Roberts, S. C. (2009). MHC-correlated mate choice in humans: A review. *Psychoneuroendocrinology, 34,* 497–512. DOI: 10.1016/j.psyneuen.2008.10.007

Haworth, C. M. A., Asbury, K., Dale, P. S., & Plomin, R. (2011). Added value measures in education show genetic as well as environmental influence. *PLoS One, 6:* e16006. DOI: 10.1371/journal.pone.0016006

Haworth, C. M. A., Kovas, Y., Harlaar, N., Hayiou-Thomas, E. M., Petrill, S. A., Dale, P. S., & Plomin, R. (2009). Generalist genes and learning disabilities: A multivariate genetic analysis of low performance in reading, mathematics, language and general cognitive ability in a sample of 8000 12-year-old twins. *Journal of Child Psychology and Psychiatry, 50,* 1318–1325. DOI: 10.1111/j.1469-7610.2009.02114.x

Haworth, C. M. A., & Plomin, R. (2011). Genetics and education: Towards a genetically sensitive classroom. In K. R. Harris, S. Graham, & T. Urdan (Eds.), *The American Psychological Association handbook of educational psychology* (pp. 529–559). Washington, DC: American Psychological Association. DOI: 10.1037/13273-018

Haworth, C. M. A., Wright, M. J., Luciano, M., Martin, N. G., de Geus, E. J. C., van Beijsterveldt, C. E. M., . . . Plomin, R. (2010). The heritability of general cognitive ability increases linearly from childhood to young adulthood. *Molecular Psychiatry, 15,* 1112–1120. DOI: 10.1038/mp.2009.55

Haworth, C. M. A., Wright, M. J., Martin, N. W., Martin, N. G., Boomsma, D. I., Bartels, M., . . . Plomin, R. (2009). A twin study of the genetics of high cognitive ability selected from 11,000 twin pairs in six studies from four countries. *Behavior Genetics, 39,* 359–370. DOI: 10.1007/s10519-009-9262-3

Hayiou-Thomas, M. E., Harlaar, N., Dale, P. S., & Plomin, R. (2010). Preschool speech, language skills, and reading at 7, 9, and 10 years: Etiology of the relationship. *Journal of Speech, Language, and Hearing Research, 53,* 311–332. DOI: 10.1044/1092-4388(2009/07-0145)

Hearnshaw, L. S. (1979). *Cyril Burt, psychologist.* Ithaca, NY: Cornell University Press.

Heath, A. C., Jardine, R., & Martin, N. G. (1989). Interactive effects of genotype and social environment on alcohol consumption. *Journal of Studies on Alcohol, 50,* 38–48.

Heath, A. C., Madden, P. A. F., Bucholz, K. K., Nelson, E. C., Todorov, A., Price, R. K., . . . Martin, N. G. (2003). Genetic and environmental risks of dependence on alcohol, tobacco, and other drugs. In R. Plomin, J. C. DeFries, I. W. Craig, & P. McGuffin (Eds.), *Behavioral genetics in the postgenomic era* (pp. 309–334). Washington, DC: American Psychological Association. DOI: 10.1037/10480-017

Heath, A. C., Martin, N. G., Lynskey, M. T., Todorov, A. A., & Madden, P. A. F. (2002). Estimating two-stage models for genetic influences on alcohol, tobacco or drug use initiation and dependence vulnerability in twin and family data. *Twin Research, 5,* 113–124. DOI: 10.1375/1369052022983

Heath, A. C., Neale, M. C., Kessler, R. C., Eaves, L. J., & Kendler, K. S. (1992). Evidence for genetic influences on personality from self-reports and informant ratings. *Journal of Social and Personality Psychology, 63,* 85–96. DOI: 10.1037/0022-3514.63.1.85

Heath, A. C., Whitfield, J. B., Martin, N. G., Pergadia, M. L., Goate, A. M., Lind, P. A., . . . Montgomery, G. W. (2011). A quantitative-trait genome-wide association study of alcoholism risk in the community: Findings and implications. *Biological Psychiatry, 70,* 513–518. DOI: 10.1016/j.biopsych.2011.02.028

Hebb, D. O. (1949) *The organization of behavior.* New York: Wiley.

Hebebrand, J. (1992). A critical appraisal of X-linked bipolar ilness: Evidence for the assumed mode of inheritance is lacking. *British Journal of Psychiatry, 160,* 7–11. DOI: 10.1192/bjp.160.1.7

Heiman, N., Stallings, M. C., Young, S. E., & Hewitt, J. K. (2004). Investigating the genetic and environmental structure of Cloninger's personality dimensions in adolescence. *Twin Research, 7,* 462–470. DOI: 10.1375/1369052042335313

Heisenberg, M. (2003). Mushroom body memoir: From maps to models. *Nature Reviews Neuroscience, 4,* 266–275. DOI: 10.1038/nrn1074

Heitmann, B. L., Kaprio, J., Harris, J. R., Rissanen, A., Korkeila, M., & Koskenvuo, M. (1997). Are genetic determinants of weight gain modified by leisure-time physical activity? A prospective study of Finnish twins. *American Journal of Clinical Nutrition, 66,* 672–678.

Hemmings, S. M., & Stein, D. J. (2006). The current status of association studies in obsessive-compulsive disorder. *Psychiatric Clinics of North America, 29,* 411–444. DOI: 10.1016/j.psc.2006.02.011

Henderson, N. D. (1967). Prior treatment effects on open field behaviour of mice—a genetic analysis. *Animal Behavior, 15,* 365–376. DOI: 10.1016/0003-3472(67)90023-1

Henderson, N. D. (1972). Relative effects of early rearing environment on discrimination learning in housemice. *Journal of Comparative and Psychological Psychology, 72,* 505–511. DOI: 10.1037/h0029743

Henderson, N. D., Turri, M. G., DeFries, J. C., & Flint, J. (2004). QTL analysis of multiple behavioral measures of anxiety in mice. *Behavior Genetics, 34,* 267–293. DOI: 10.1023/B:BEGE.0000017872.25069.44

Hensler, B. S., Schatschneider, C., Taylor, J., & Wagner, R. K. (2010). Behavioral genetic approach to the study of dyslexia. *Journal of Developmental and Behavioral Pediatrics, 31,* 525–532. DOI: 10.1097/DBP.0b013e3181ee4b70

Herrera, B. M., Keildson, S., & Lindgren, C. M. (2011). Genetics and epigenetics of obesity. *Maturitas, 69,* 41–49. DOI: 10.1016/j.maturitas.2011.02.018

Herrnstein, R. J., & Murray, C. (1994). *The bell curve: Intelligence and class structure in American life.* New York: Free Press.

Hershberger, S. L., Lichtenstein, P., & Knox, S. S. (1994). Genetic and environmental influences on perceptions of organizational climate. *Journal* of *Applied Psychology, 79,* 24–33. DOI: 10.1037/0021-9010.79.1.24

Hershberger, S. L., Plomin, R., & Pedersen, N. L. (1995). Traits and metatraits: Their reliability, stability, and shared genetic influence. *Journal of Personality and Social Psychology, 69,* 673–684. DOI: 10.1037/0022-3514.69.4.673

Heston, L. L. (1966). Psychiatric disorders in foster home reared children of schizophrenic mothers. *British Journal of Psychiatry, 112,* 819–825. DOI: 10.1192/bjp.112.489.819

Hetherington, E. M., & Clingempeel, W. G. (1992). Coping with marital transitions: A family systems perspective. *Monographs of the Society for Research in Child Development, 57 (2-3),* 1–238. DOI: 10.1111/j.1540-5834.1992.tb00298.x

Hettema, J. M. (2010). Genetics of depression. *Focus, 8,* 316–322.

Hettema, J. M., Annas, P., Neale, M. C., Kendler, K. S., & Fredrikson, M. (2003). A twin study of the genetics of fear conditioning. *Archives of General Psychiatry, 60,* 702–708. DOI: 10.1001/archpsyc.60.7.702

Hettema, J. M., Neale, M. C., & Kendler, K. S. (2001). A review and meta-analysis of the genetic epidemiology of anxiety disorders. *American Journal of Psychiatry, 158,* 1568–1578. DOI: 10.1176/appi.ajp.158.10.1568

Hettema, J. M., Neale, M. C., Myers, J. M., Prescott, C. A., Kendler, K. S., Hettema, J. M., . . . Kendler, K. S. (2006). A population-based twin study of the relationship between neuroticism and internalizing disorders. *American Journal of Psychiatry, 163,* 857–864. DOI: 10.1176/appi.ajp.163.5.857

Hettema, J. M., Prescott, C. A., & Kendler, K. S. (2001). A population-based twin study of generalized anxiety disorder in men and women. *Journal of Nervous and Mental Disease, 189,* 413–420. DOI: 10.1097/00005053-200107000-00001

Hettema, J. M., Prescott, C. A., Myers, J. M., Neale, M. C., & Kendler, K. S. (2005). The structure of genetic and environmental risk factors for anxiety disorders in men and women. *Archives of General Psychiatry, 62,* 182–189. DOI: 10.1001/archpsyc.62.2.182

Hicks, B. M., Krueger, R. F., Iacono, W. G., McGue, M., & Patrick, C. J. (2004). Family transmission and heritability of externalizing disorders—a twin-family study. *Archives of General Psychiatry, 61,* 922–928. DOI: 10.1001/archpsyc.61.9.922

Hicks, B. M., South, S. C., DiRago, A. C., Iacono, W. G., & McGue, M. (2009). Environmental adversity and increasing genetic risk for externalizing disorders. *Archives of General Psychiatry, 66,* 640–648. DOI: 10.1001/archgenpsychiatry.2008.554

Higuchi, S., Matsushita, S., Masaki, T., Yokoyama, A., Kimura, M., Suzuki, G., & Mochizuki, H. (2004). Influence of genetic variations of ethanol-metabolizing enzymes on phenotypes of alcohol-related disorders. *Annals of the New York Academy of Sciences, 1025,* 472–480. DOI: 10.1196/annals.1316.058

Hill, S. Y. (2010). Neural plasticity, human genetics, and risk for alcohol dependence. *International Review of Neurobiology, 91,* 53–94. DOI: 10.1016/S0074-7742(10)91003-9

Hirschhorn, J. N., & Daly, M. J. (2005). Genome-wide association studies for common diseases and complex traits. *Nature Reviews Genetics, 6,* 95–108. DOI: 10.1038/nrg1521

Hirschhorn, J. N., Lohmueller, K., Byrne, E., & Hirschhorn, K. (2002). A comprehensive review of genetic association studies. *Genetics in Medicine, 4,* 45–61. DOI: 10.1097/00125817-200203000-00002

Hitzemann, R., Edmunds, S., Wu, W., Malmanger, B., Walter, N., Belknap, J., . . . McWeeney, S. (2009). Detection of reciprocal quantitative trait loci for acute ethanol withdrawal and ethanol consumption in heterogeneous stock mice. *Psychopharmacology, 203,* 713–722. DOI: 10.1007/s00213-008-1418-y

Hjelmborg, J. V., Iachine, I., Skythe, A., Vaupel, J. W., McGue, M., Koskenvuo, M., . . . Christensen, K. (2006). Genetic influence on human lifespan and longevity. *Human Genetics, 119,* 312–321. DOI: 10.1007/s00439-006-0144-y

Ho, H., Baker, L., & Decker, S. N. (1988). Covariation between intelligence and speed of cognitive processing: Genetic and environmental influences. *Behavior Genetics, 18,* 247–261. DOI: 10.1007/BF01067845

Ho, M. K., & Tyndale, R. F. (2007). Overview of the pharmacogenomics of cigarette smoking. *Pharmacogenomics Journal, 7,* 81–98. DOI: 10.1038/sj.tpj.6500436

Hobcraft, J. (2006). The ABC of demographic behaviour: How the interplays of alleles, brains, and contexts over the life course should shape research aimed at understanding population processes. *Population Studies, 60,* 153–187. DOI: 10.1080/00324720600646410

Hobert, O. (2003). Behavioral plasticity in C. elegans: Paradigms, circuits, genes. *Journal of Neurobiology, 54,* 203–223. DOI: 10.1002/neu.10168

Hollingworth, P., Harold, D., Jones, L., Owen, M. J., & Williams, J. (2011). Alzheimer's disease genetics: Current knowledge and future challenges. *International Journal of Geriatric Psychiatry, 26,* 793–802. DOI: 10.1002/gps.2628

Hollister, J. M., Mednick, S. A., Brennan, P., & Cannon, T. D. (1994). Impaired autonomic nervous system habituation in those at genetic risk for schizophrenia. *Archives of General Psychiatry, 51,* 552–558. DOI: 10.1001/archpsyc.1994.03950070044009

Holmans, P., Weissman, M. M., Zubenko, G. S., Scheftner, W. A., Crowe, R. R., Depaulo, J. R., Jr., . . . Levinson, D. F. (2007). Genetics of recurrent early-onset major depression (GenRED): Final genome scan report. *American Journal of Psychiatry, 164,* 248–258. DOI: 10.1176/appi.ajp.164.2.248

Hong, D. S., Dunkin, B., & Reiss, A. L. (2011). Psychosocial functioning and social cognitive processing in girls with Turner syndrome. *Journal of Developmental and Behavioral Pediatrics, 32,* 512–520. DOI: 10.1097/DBP.0b013e3182255301

Hopwood, C. J., Donnellan, M. B., Blonigen, D. M., Krueger, R. F., McGue, M., Iacono, W. G., & Burt, S. A. (2011). Genetic and environmental influences on personality trait stability and growth during the transition to adulthood: A three-wave longitudinal study. *Journal of Personality and Social Psychology, 100,* 545–556. DOI: 10.1037/a0022409

Horwitz, B. N., Ganiban, J. M., Spotts, E. L., Lichtenstein, P., Reiss, D., & Neiderhiser, J. M. (2011). The role of aggressive personality and family relationships in explaining family conflict. *Journal of Family Psychology, 25,* 174–183. DOI: 10.1037/a0023049

Hotta, Y., & Benzer, S. (1970). Genetic dissection of the Drosophila nervous sytem by means of mosaics. *Proceedings of the National Academy of Sciences (USA), 67,* 1156–1163. DOI: 10.1073/pnas.67.3.1156

Houts, R. M., Caspi, A., Pianta, R. C., Arseneault, L., & Moffitt, T. E. (2010). The challenging pupil in the classroom: The effect of the child on the teacher. *Psychological Science, 21,* 1802–1810. DOI: 10.1177/0956797610388047

Hrdy, S. B. (1999). *Mother nature: A history of mothers, infants and natural selection.* London: Chatto & Windus.

Hu, S., Pattatucci, A. M. L., Patterson, C., Li, L., Fulker, D. W., Cherny, S. S., . . . Hamer, D. H. (1995). Linkage between sexual orientation and chromosome Xq28 in males but not in females. *Nature Genetics, 11,* 248–256. DOI: 10.1038/ng1195-248

Hublin, C., Kaprio, J., Partinen, M., & Koskenvuo, M. (1998). Nocturnal enuresis in a nationwide twin cohort. *Sleep, 21,* 579–585.

Hudziak, J. J. (2008). *Developmental psychopathology and wellness: Genetic and environmental influences.* Arlington, VA: American Psychiatric Publishing, Inc.

Hudziak, J. J., Derks, E. M., Althoff, R. R., Rettew, D. C., & Boomsma, D. I. (2005). The genetic and environmental contributions to attention deficit hyperactivity disorder as measured by the Conners' Rating Scales–Revised. *American Journal of Psychiatry, 162,* 1614–1620. DOI: 10.1176/appi.ajp.162.9.1614

Hudziak, J. J., Van Beijsterveldt, C. E., Althoff, R. R., Stanger, C., Rettew, D. C., Nelson, E. C., . . . Boomsma, D. I. (2004). Genetic and environmental contributions to the Child Behavior Checklist Obsessive-Compulsive Scale: A cross-cultural twin study. *Archives of General Psychiatry, 61,* 608–616. DOI: 10.1001/archpsyc.61.6.608

Hulshoff Pol, H. E., Schnack, H. G., Posthuma, D., Mandl, R. C. W., Baare, W. F., van Oel, C., . . . Kahn, R. S. (2006). Genetic contributions to human brain morphology and intelligence. *Journal of Neuroscience, 26,* 10235–10242. DOI: 10.1523/JNEUROSCI.1312-06.2006

Hunt, E. B. (2011). *Human intelligence.* Cambridge: Cambridge University Press.

HUPO Views. (2010). A gene-centric human proteome project: HUPO—the human proteome organization. *Molecular and Cellular Proteomics, 9,* 427–429. DOI: 10.1074/mcp.H900001-MCP200

Hur, Y.-M., & Shin, J.-S. (2008). Effects of chorion type on genetic and environmental influences on height, weight, and body mass index in South Korean young twins. *Twin Research and Human Genetics, 11,* 63–69. DOI: 10.1375/twin.11.1.63

Husén, T. (1959). *Psychological twin research.* Stockholm: Almqvist & Wiksell.

Huszar, D., Lynch, C. A., Fairchild-Huntress, V., Dunmore, J. H., Fang, Q., Berkemeier, L. R., . . . Lee, F. (1997). Targeted disruption of the melanocortin-4 receptor results in obesity in mice. *Cell, 88,* 131–141. DOI: 10.1016/S0092-8674(00)81865-6

Iervolino, A. C., Pike, A., Manke, B., Reiss, D., Hetherington, E. M., & Plomin, R. (2002). Genetic and environmental influences in adolescent peer socialization: Evidence from two genetically sensitive designs. *Child Development, 73,* 162–175. DOI: 10.1111/1467-8624.00398

Inlow, J. K., & Restifo, L. L. (2004). Molecular and comparative genetics of mental retardation. *Genetics, 166,* 835–881. DOI: 10.1534/genetics.166.2.835

International Human Genome Sequencing Consortium. (2001). Initial sequencing and analysis of the human genome. *Nature, 409,* 860–921. DOI: 10.1038/35057062

International Molecular Genetic Study of Autism Consortium. (1998). A full genome screen for autism with evidence for linkage to a region on chromosome 7q. *Human Molecular Genetics, 7,* 571–578. DOI: 10.1093/hmg/7.3.571

Ioannides, A. A. (2006). Magnetoencephalography as a research tool in neuroscience:

State of the art. *Neuroscientist, 12,* 524–544. DOI: 10.1177/1073858406293696

Ioannidis, J. P., Ntzani, E. E., Trikalinos, T. A., & Contopoulos-Ioannidis, D. G. (2001). Replication validity of genetic association studies. *Nature Genetics, 29,* 306–309. DOI: 10.1038/ng749

Irons, D. E., McGue, M., Iacono, W. G., & Oetting, W. S. (2007). Mendelian randomization: A novel test of the gateway hypothesis and models of gene-environment interplay. *Development and Psychopathology, 19,* 1181–1195. DOI: 10.1017/S0954579407000612

Jacob, H. J., & Kwitek, A. E. (2002). Rat genetics: Attaching physiology and pharmacology to the genome. *Nature Reviews Genetics, 3,* 33–42. DOI: 10.1038/nrg702

Jacobs, N., Van Gestel, S., Derom, C., Thiery, E., Vernon, P., Derom, R., & Vlietinck, R. (2001). Heritability estimates of intelligence in twins: Effect of chorion type. *Behavior Genetics, 31,* 209–217. DOI: 10.1023/A:1010257512183

Jacobson, K. C., Beseler, C. L., Lasky-Su, J., Faraone, S. V., Glatt, S. J., Kremen, W. S., . . . Tsuang, M. T. (2008). Ordered subsets linkage analysis of antisocial behavior in substance use disorder among participants in the collaborative study on the genetics of alcoholism. *American Journal of Medical Genetics. B: Neuropsychiatric Genetics, 147B,* 1258–1269. DOI: 10.1002/ajmg.b.30771

Jacobson, K. C., & Rowe, D. C. (1999). Genetic and environmental influences on the relationships between family connectedness, school connectedness, and adolescent depressed mood: Sex differences. *Developmental Psychology, 35,* 926–939. DOI: 10.1037/0012-1649.35.4.926

Jacobson, P., Torgerson, J. S., Sjostrom, L., & Bouchard, C. (2007). Spouse resemblance in body mass index: Effects on adult obesity prevalence in the offspring generation. *American Journal of Epidemiology, 165,* 101–108. DOI: 10.1093/aje/kwj342

Jaffee, S. R., Caspi, A., Moffitt, T. E., Dodge, K. A., Rutter, M., Taylor, A., & Tully, L. A. (2005). Nature X nurture: Genetic vulnerabilities interact with physical maltreatment to promote conduct problems. *Development and Psychopathology, 17,* 67–84. DOI: 10.1017/S0954579405050042

Jaffee, S. R., Caspi, A., Moffitt, T. E., Polo-Tomas, M., Price, T. S., & Taylor, A. (2004). The limits of child effects: Evidence for genetically mediated child effects on corporal punishment but not on physical maltreatment. *Developmental Psychology, 40,* 1047–1058. DOI: 10.1037/0012-1649.40.6.1047

Jaffee, S. R., & Price, T. S. (2007). Gene-environment correlations: A review of the evidence and implications for prevention of mental illness. *Molecular Psychiatry, 12,* 432–442. DOI: 10.1038/sj.mp.4001950

Jaffee, S. R., & Price, T. S. (in press). The implications of genotype-environment correlation for establishing causal processes in psychopathology. *Development and Psychopathology.*

Jaffee, S. R., Strait, L. B., & Odgers, C. L. (2012). From correlates to causes: Can quasi-experimental studies and statistical innovations bring us closer to identifying the causes of antisocial behavior? *Psychological Bulletin, 138,* 272–295. DOI: 10.1037/a0026020

Jahanshad, N., Lee, A. D., Barysheva, M., McMahon, K. L., de Zubicaray, G. I., Martin, N. G., . . . Thompson, P. M. (2010). Genetic influences on brain asymmetry: A DTI study of 374 twins and siblings. *NeuroImage, 52,* 455–469. DOI: 10.1016/j.neuroimage.2010.04.236

James, W. (1890). *Principles of psychology.* New York: Holt. DOI: 10.1037/11059-000

Jang, K. L. (2005). *The behavioral genetics of psychopathology: A clinical guide.* Mahwah, NJ: Lawrence Erlbaum Associates.

Jang, K. L., Lam, R. W., Livesley, W. J., & Vernon, P. A. (1997). The relationship between seasonal mood change and personality: More apparent than real? *Acta Psychiatrica Scandinavica, 95,* 539–543. DOI: 10.1111/j.1600-0447.1997.tb10143.x

Jang, K. L., Livesley, W. J., Ando, J., Yamagata, S., Suzuki, A., Angleitner, A., . . . Spinath, F. (2006). Behavioral genetics of the higher-order factors of the Big Five. *Personality and Individual Differences, 41,* 261–272. DOI: 10.1016/j.paid.2005.11.033

Jang, K. L., Livesley, W. J., & Vernon, P. A. (1996). Heritability of the Big Five dimensions and their facets: A twin study. *Journal of Personality, 64,* 577–591. DOI: 10.1111/j.1467-6494.1996.tb00522.x

Jang, K. L., McCrae, R. R., Angleitner, A., Riemann, R., & Livesley, W. J. (1998). Heritability of facet-level traits in a cross-cultural twin sample: Support for a hierarchical model of personality. *Journal of Personality and Social Psychology, 74,* 1556–1565. DOI: 10.1037/0022-3514.74.6.1556

Jang, K. L., Woodward, T. S., Lang, D., Honer, W. G., & Livesley, W. J. (2005). The genetic and environmental basis of the relationship between schizotypy and personality: A twin study. *Journal of Nervous and Mental Disease, 193,* 153–159. DOI: 10.1097/01.nmd.0000154842.26600.bd

Jansen, R. C., & Nap, J. P. (2001). Genetical genomics: The added value from segregation. *Trends in Genetics, 17,* 388–391. DOI: 10.1016/S0168-9525(01)02310-1

Janzing, J. G. E., de Graaf, R., ten Have, M., Vollebergh, W. A., Verhagen, M., & Buitelaar, J. K. (2009). Familiality of depression in the community; associations with gender and phenotype of major depressive disorder. *Social Psychiatry* and Psychiatric Epidemiology, 44, 1067–1074. DOI: 10.1007/s00127-009-0026-4

Jensen, A. R. (1969). How much can we boost IQ and scholastic achievement? Harvard Educational Review, 39, 1–123.

Jensen, A. R. (1978). Genetic and behavioral effects of nonrandom mating. In R. T. Osbourne, C. E. Noble, & N. Weyl (Eds.), Human variation: The biopsychology of age, race, and sex (pp. 51–105). New York: Academic Press.

Jensen, A. R. (1998). The g factor: The science of mental ability. Westport, CT: Praeger.

Jinks, J. L., & Fulker, D. W. (1970). Comparison of the biometrical genetical, MAVA, and classical approaches to the analysis of human behavior. Psychological Bulletin, 73, 311–349. DOI: 10.1037/h0029135

John, O. P., Robins, R. W., & Pervin, L. A. (Eds.) (2008). Handbook of personality: Theory and research (3rd ed.) New York: Guilford Press.

Johnson, W., & Krueger, R. F. (2005). Genetic effects on physical health: Lower at higher income levels. Behavior Genetics, 35, 579–590. DOI: 10.1007/s10519-005-3598-0

Johnson, W., Krueger, R. F., Bouchard, T. J., Jr., & McGue, M. (2002). The personalities of twins: Just ordinary folks. Twin Research, 5, 125–131. DOI: 10.1375/twin.5.2.125

Johnson, W., McGue, M., Gaist, D., Vaupel, J. W., & Christensen, K. (2002). Frequency and heritability of depression symptomatology in the second half of life: Evidence from Danish twins over 45. Psychological Medicine, 32, 1175–1185. DOI: 10.1017/S0033291702006207

Johnson, W., McGue, M., & Krueger, R. F. (2005). Personality stability in late adulthood: A behavioral genetic analysis. Journal of Personality, 73, 523–552. DOI: 10.1111/j.1467-6494.2005.00319.x

Johnson, W., McGue, M., Krueger, R. F., & Bouchard, T. J., Jr. (2004). Marriage and personality: A genetic analysis. Journal of Personality and Social Psychology, 86, 285–294. DOI: 10.1037/0022-3514.86.2.285

Jones, P. B., & Murray, R. M. (1991). Aberrant neurodevelopment as the expression of schizophrenia genotype. In P. McGuffin & R. Murray (Eds.), The new genetics of mental illness (pp. 112–129). Oxford: Butterworth-Heinemann.

Jones, S. (1999). Almost like a whale: The origin of species, updated. New York: Doubleday.

Jones, S. R., Gainetdinov, R. R., Jaber, M., Giros, B., Wightman, R. M., & Caron, M. G. (1998). Profound neuronal plasticity in response to inactivation of the dopamine transporter. Proceedings of the National Academy of Sciences (USA), 95, 4029–4034. DOI: 10.1073/pnas.95.7.4029

Jordan, K. W., Morgan, T. J., & Mackay, T. F. (2006). Quantitative trait loci for locomotor behavior in Drosophila melanogaster. Genetics, 174, 271–284. DOI: 10.1534/genetics.106.058099

Joshi, A. A., Lepore, N., Joshi, S. H., Lee, A. D., Barysheva, M., Stein, J. L., . . . Thompson, P. M. (2011). The contribution of genes to cortical thickness and volume. NeuroReport, 22, 101–105. DOI: 10.1097/WNR.0b013e3283424c84

Joynson, R. B. (1989). The Burt affair. London: Routledge.

Kafkafi, N., Benjamini, Y., Sakov, A., Elmer, G. I., & Golani, I. (2005). Genotype-environment interactions in mouse behavior: A way out of the problem. Proceedings of the National Academy of Sciences (USA), 102, 4619–4624. DOI: 10.1073/pnas.0409554102

Kaitz, M., Shalev, I., Sapir, N., Devor, N., Samet, Y., Mankuta, D., & Ebstein, R. P. (2010). Mothers' dopamine receptor polymorphism modulates the relation between infant fussiness and sensitive parenting. Developmental Psychobiology, 52, 149–157. DOI: 10.1002/dev.20423

Kallmann, F. J. (1952). Twin and sibship study of overt male homosexuality. Journal of Human Genetics, 4, 136–146.

Kallmann, F. J., & Kaplan, O. J. (1955). Genetic aspects of mental disorders in later life. In O. J. Kaplan (Ed.), Mental disorders in later life (pp. 26–46). Stanford, CA: Stanford University Press.

Kallmann, F. J., & Roth, B. (1956). Genetic aspects of preadolescent schizophrenia. American Journal of Psychiatry, 112, 599–606.

Kamakura, T., Ando, J., & Ono, Y. (2007). Genetic and environmental effects of stability and change in self-esteem during adolescence. Personality and Individual Differences, 42, 181–190. DOI: 10.1016/j.paid.2006.07.005

Kamin, L. J. (1974). The science and politics of IQ. Potomac, MD: Erlbaum.

Kang, C., & Drayna, D. (2011). Genetics of speech and language disorders. Annual Review of Genomics and Human Genetics, 12, 145–164. DOI: 10.1146/annurev-genom-090810-183119

Kanner, L. (1943). Autistic disturbances of affective contact. Nervous Child, 2, 217–250.

Karanjawala, Z. E., & Collins, F. S. (1998). Genetics in the context of medical practice. Journal of the American Medical Association, 280, 1533–1544. DOI: 10.1001/jama.280.17.1533

Karg, K., Burmeister, M., Shedden, K., & Sen, S. (2011). The serotonin transporter promoter variant (5-HTTLPR), stress, and depression meta-analysis revisited: Evidence of genetic moderation. Archives of General Psychiatry, 68, 444–454. DOI: 10.1001/archgenpsychiatry.2010.189

Karlsgodt, K. H., Bachman, P., Winkler, A. M., Bearden, C. E., & Glahn, D. C. (2011). Genetic influence on the working memory circuitry: Behavior, structure, function and extensions to illness. *Behavioural Brain Research, 225,* 610–622. DOI: 10.1016/j.bbr.2011.08.016

Kato, K., & Pedersen, N. L. (2005). Personality and coping: A study of twins reared apart and twins reared together. *Behavior Genetics, 35,* 147–158. DOI: 10.1007/s10519-004-1015-8

Kato, K., Sullivan, P. F., Evengard, B., & Pedersen, N. L. (2009). A population-based twin study of functional somatic syndromes. *Psychological Medicine, 39,* 497–505. DOI: 10.1017/S0033291708003784

Kato, K., Sullivan, P. F., & Pedersen, N. L. (2010). Latent class analysis of functional somatic symptoms in a population-based sample of twins. *Journal of Psychosomatic Research, 68,* 447–453. DOI: 10.1016/j.jpsychores.2010.01.010

Kato, T. (2007). Molecular genetics of bipolar disorder and depression. *Psychiatry and Clinical Neuroscience, 61,* 3–19. DOI: 10.1111/j.1440-1819.2007.01604.x

Katon, W. J., Lin, E. H. B., Russo, J., Von Korff, M., Ciechonowski, P., Simon, G., . . . Young, B. (2004). Cardiac risk factors in patients with diabetes mellitus and major depression. *Journal of General Internal Medicine, 19,* 1192–1199. DOI: 10.1111/j.1525-1497.2004.30405.x

Kedes, L., & Campany, G. (2011). The new date, new format, new goals and new sponsor of the Archon Genomics X PRIZE competition. *Nature Genetics, 43,* 1055–1058. DOI: 10.1038/ng.988

Keen-Kim, D., Mathews, C. A., Reus, V. I., Lowe, T. L., Diego Herrera, L., Budman, C. L., . . . Freimer, N. B. (2006). Overrepresentation of rare variants in a specific ethnic group may confuse interpretation of association analyses. *Human Molecular Genetics, 15,* 3324–3328. DOI: 10.1093/hmg/ddl408

Keller, L. M., Bouchard, T. J., Jr., Segal, N. L., & Dawes, R. V. (1992). Work values: Genetic and environmental influences. *Journal of Applied Psychology, 77,* 79–88. DOI: 10.1037/0021-9010.77.1.79

Keller, M. A., & Miller, G. (2006). An evolutionary framework for mental disorders: Integrating adaptationist and evolutionary genetic models. *Behavioral and Brain Sciences, 29,* 429–452. DOI: 10.1017/S0140525X06459094

Keller, M. C., Coventry, W. L., Heath, A. C., & Martin, N. G. (2005). Widespread evidence for non-additive genetic variation in Cloninger's and Eysenck's personality dimensions using a twin plus sibling design. *Behavior Genetics, 35,* 707–721. DOI: 10.1007/s10519-005-6041-7

Keller, M. C., Visscher, P. M., & Goddard, M. E. (2011). Quantification of inbreeding due to distant ancestors and its detection using dense single nucleotide polymorphism data. *Genetics, 189,* 237–249. DOI: 10.1534/genetics.111.130922

Kelly, T., Yang, W., Chen, C. S., Reynolds, K., & He, J. (2008). Global burden of obesity in 2005 and projections to 2030. *International Journal of Obesity, 32,* 1431–1437. DOI: 10.1038/ijo.2008.102

Kelsoe, J. R., Ginns, E. I., Egeland, J. A., Gerhard, D. S., Goldstein, A. M., Bale, S. J., . . . Paul, S. M. (1989). Re-evaluation of the linkage relationship between chromosome 11p loci and the gene for bipolar affective disorder in the old order Amish. *Nature, 342,* 238–242. DOI: 10.1038/342238a0

Kendler, K. S. (1996). Major depression and generalised anxiety disorder. Same genes, (partly) different environments—revisited. *British Journal of Psychiatry, Supplement 168,* 68–75.

Kendler, K. S. (2001). Twin studies of psychiatric illness: An update. *Archives of General Psychiatry, 58,* 1005–1014. DOI: 10.1001/archpsyc.58.11.1005

Kendler, K. S. (2005). Toward a philosophical structure for psychiatry. *American Journal of Psychiatry, 162,* 433–440. DOI: 10.1176/appi.ajp.162.3.433

Kendler, K. S., Aggen, S. H., Czajkowski, N., Roysamb, E., Tambs, K., Torgersen, S., . . . Reichborn-Kjennerud, T. (2008). The structure of genetic and environmental risk factors for DSM-IV personality disorders: A multivariate twin study. *Archives of General Psychiatry, 65,* 1438–1446. DOI: 10.1001/archpsyc.65.12.1438

Kendler, K. S., Aggen, S. H., & Patrick, C. J. (2012). A multivariate twin study of the DSM-IV criteria for antisocial personality disorder. *Biological Psychiatry, 71,* 247–253. DOI: 10.1016/j.biopsych.2011.05.019

Kendler, K. S., Aggen, S. H., Tambs, K., & Reichborn-Kjennerud, T. (2006). Illicit psychoactive substance use, abuse and dependence in a population-based sample of Norwegian twins. *Psychological Medicine, 36,* 955–962. DOI: 10.1017/S0033291706007720

Kendler, K. S., & Baker, J. H. (2007). Genetic influences on measures of the environment: A systematic review. *Psychological Medicine, 37,* 615–626. DOI: 10.1017/S0033291706009524

Kendler, K. S., Czajkowski, N., Tambs, K., Torgersen, S., Aggen, S. H., Neale, M. C., & Reichborn-Kjennerud, T. (2006). Dimensional representations of DSM-IV Cluster A personality disorders in a population-based sample of Norwegian twins: A multivariate study. *Psychological Medicine, 36,* 1583–1591. DOI: 10.1017/S0033291706008609

Kendler, K. S., & Eaves, L. (2005). *Psychiatric genetics.* Washington, DC: American Psychiatric Publishing.

Kendler, K. S., & Eaves, L. J. (1986). Models for the joint effects of genotype and environment on liability to psychiatric illness. *American Journal of Psychiatry, 143,* 279–289.

Kendler, K. S., Gardner, C., & Dick, D. M. (2011). Predicting alcohol consumption in adolescence from alcohol-specific and general externalizing genetic risk factors, key environmental exposures and their interaction. *Psychological Medicine, 41,* 1507–1516. DOI: 10.1017/S003329171000190X

Kendler, K. S., & Gardner, C. O. (2011). A longitudinal etiologic model for symptoms of anxiety and depression in women. *Psychological Medicine, 41,* 2035–2045. DOI: 10.1017/S0033291711000225

Kendler, K. S., Gardner, C. O., Annas, P., Neale, M. C., Eaves, L. J., & Lichtenstein, P. (2008). A longitudinal twin study of fears from middle childhood to early adulthood: Evidence for a developmentally dynamic genome. *Archives of General Psychiatry, 65,* 421–429. DOI: 10.1001/archpsyc.65.4.421

Kendler, K. S., Gardner, C. O., & Lichtenstein, P. (2008). A developmental twin study of symptoms of anxiety and depression: Evidence for genetic innovation and attenuation. *Psychological Medicine, 38,* 1567–1575. DOI: 10.1017/S003329170800384X

Kendler, K. S., Gardner, C. O., & Prescott, C. A. (2001). Panic syndromes in a population-based sample of male and female twins. *Psychological Medicine, 31,* 989–1000. DOI: 10.1017/S0033291701004226

Kendler, K. S., Gardner, C. O., & Prescott, C. A. (2003). Personality and the experience of environmental adversity. *Psychological Medicine, 33,* 1193–1202. DOI: 10.1017/S0033291703008298

Kendler, K. S., Gatz, M., Gardner, C. O., & Pedersen, N. L. (2006a). Personality and major depression: A Swedish longitudinal, population-based twin study. *Archives of General Psychiatry, 63,* 1113–1120. DOI: 10.1001/archpsyc.63.10.1113

Kendler, K. S., Gatz, M., Gardner, C. O., & Pedersen, N. L. (2006b). A Swedish national twin study of lifetime major depression. *American Journal of Psychiatry, 163,* 109–114. DOI: 10.1176/appi.ajp.163.1.109

Kendler, K. S., & Greenspan, R. J. (2006). The nature of genetic influences on behavior: Lessons from "simpler" organisms. *American Journal of Psychiatry, 163,* 1683–1694. DOI: 10.1176/appi.ajp.163.10.1683

Kendler, K. S., Gruenberg, A. M., & Kinney, D. K. (1994). Independent diagnoses of adoptees and relatives, as defined by DSM-II, in the provin-cial and national samples of the Danish Adoption Study of Schizophrenia. *Archives of General Psychiatry, 51,* 456–468. DOI: 10.1001/archpsyc.1994.03950060020002

Kendler, K. S., Jacobson, K. C., Gardner, C. O., Gillespie, N., Aggen, S. A., & Prescott, C. A. (2007). Creating a social world—a developmental twin study of peer-group deviance. *Archives of General Psychiatry, 64,* 958–965. DOI: 10.1001/archpsyc.64.8.958

Kendler, K. S., & Karkowski-Shuman, L. (1997). Stressful life events and genetic liability to major depression: Genetic control of exposure to the environment. *Psychological Medicine, 27,* 539–547. DOI: 10.1017/S0033291797004716

Kendler, K. S., Kessler, R. C., Walters, E. E., MacLean, C. J., Neale, M. C., Heath, A. C., & Eaves, L. J. (1995). Stressful life events, genetic liability, and onset of an episode of major depression in women. *American Journal of Psychiatry, 152,* 833–842.

Kendler, K. S., Kuhn, J. W., Vittum, J., Prescott, C. A., & Riley, B. (2005). The interaction of stressful life events and a serotonin transporter polymorphism in the prediction of episodes of major depression: A replication. *Archives of General Psychiatry, 62,* 529–535. DOI: 10.1001/archpsyc.62.5.529

Kendler, K. S., Myers, J., & Prescott, C. A. (2007). Specificity of genetic and environmental risk factors for symptoms of cannabis, cocaine, alcohol, caffeine, and nicotine dependence. *Archives of General Psychiatry, 64,* 1313–1320. DOI: 10.1001/archpsyc.64.11.1313

Kendler, K. S., Myers, J., Prescott, C. A., & Neale, M. C. (2001). The genetic epidemiology of irrational fears and phobias in men. *Archives of General Psychiatry, 58,* 257–265. DOI: 10.1001/archpsyc.58.3.257

Kendler, K. S., Neale, M., Kessler, R., Heath, A., & Eaves, L. (1993a). A twin study of recent life events and difficulties. *Archives of General Psychiatry, 50,* 789–796. DOI: 10.1001/archpsyc.1993.01820220041005

Kendler, K. S., & Neale, M. C. (2010). Endophenotype: A conceptual analysis. *Molecular Psychiatry, 15,* 789–797. DOI: 10.1038/mp.2010.8

Kendler, K. S., Neale, M. C., Kessler, R. C., Heath, A. C., & Eaves, L. J. (1992). Major depression and generalized anxiety disorder. Same genes, (partly) different environments? *Archives of General Psychiatry, 49,* 716–722. DOI: 10.1001/archpsyc.1992.01820090044008

Kendler, K. S., Neale, M. C., Kessler, R. C., Heath, A. C., & Eaves, L. J. (1993b). A test of the equal-environment assumption in twin studies of psychiatric illness. *Behavior Genetics, 23,* 21–27. DOI: 10.1007/BF01067551

Kendler, K. S., Neale, M. C., Kessler, R. C., Heath, A. C., & Eaves, L. J. (1994). Parental treatment and the equal environment assumption in twin studies of psychiatric illness. *Psychological Medicine, 24,* 579–590. DOI: 10.1017/S0033291700027732

Kendler, K. S., Neale, M. C., Sullivan, P., Corey, L. A., Gardner, C. O., & Prescott, C. A. (1999). A population-based twin study in women of smoking initiation and nicotine dependence. *Psychological Medicine, 29,* 299–308. DOI: 10.1017/S0033291798008022

Kendler, K. S., & Prescott, C. A. (1998). Cannabis use, abuse, and dependence in a population-based sample of female twins. *American Journal of Psychiatry, 155,* 1016–1022.

Kendler, K. S., & Prescott, C. A. (2007). *Genes, environment, and psychopathology: Understanding the causes of psychiatric and substance use disorders.* New York: Guilford Press.

Kendler, K. S., Prescott, C. A., Myers, J., & Neale, M. C. (2003). The structure of genetic and environmental risk factors for common psychiatric and substance use disorders in men and women. *Archives of General Psychiatry, 60,* 929–937. DOI: 10.1001/archpsyc.60.9.929

Kendler, K. S., Thornton, L. M., Gilman, S. E., & Kessler, R. C. (2000). Sexual orientation in a U.S. national sample of twin and nontwin sibling pairs. *American Journal of Psychiatry, 157,* 1843–1846. DOI: 10.1176/appi.ajp.157.11.1843

Kenrick, D. T., & Funder, D. C. (1988). Profiting from controversy: Lessons from the person-situation debate. *American Psychologist, 43,* 23–34. DOI: 10.1037/0003-066X.43.1.23

Kenyon, C. J. (2010). The genetics of ageing. *Nature, 464,* 504–512. DOI: 10.1038/nature08980

Kessler, R., McGonagle, K. A., Zhao, C. B., Nelson, C. B., Hughes, M., Eshleman, S., . . . Kendler, K. S. (1994). Lifetime and 12-month prevalence of DSM-III-R psychiatric disorders in the United States: Results from the National Comorbidity Study. *Archives of General Psychiatry, 51,* 8–19. DOI: 10.1001/archpsyc.1994.03950010008002

Kessler, R. C., Angermeyer, M., Anthony, J. C., de Graaf, R., Demyttenaere, K., Gasquet, I., . . . Üstün, T. B. (2007). Lifetime prevalence and age-of-onset distributions of mental disorders in the World Health Organization's World Mental Health Survey Initiative. *World Psychiatry, 6,* 168–176.

Kessler, R. C., Berglund, P., Demler, O., Jin, R., Merikangas, K. R., & Walters, E. E. (2005). Lifetime prevalence and age-of-onset distributions of DSM-IV disorders in the National Comorbidity Survey Replication. *Archives of General Psychiatry, 62,* 593–602. DOI: 10.1001/archpsyc.62.6.593

Kessler, R. C., Chiu, W. T., Demler, O., Merikangas, K. R., & Walters, E. E. (2005). Prevalence, severity, and comorbidity of 12-month DSM-IV disorders in the National Comorbidity Survey Replication. *Archives of General Psychiatry, 62,* 617–627. DOI: 10.1001/archpsyc.62.6.617

Kessler, R. C., Kendler, K. S., Heath, A. C., Neale, M. C., & Eaves, L. J. (1992). Social support, depressed mood, and adjustment to stress: A genetic epidemiological investigation. *Journal of Personality and Social Psychology, 62,* 257–272. DOI: 10.1037/0022-3514.62.2.257

Kety, S. S. (1987). The significance of genetic factors in the etiology of schizophrenia: Results from the national study of adoptees in Denmark. *Journal of Psychiatric Research, 21,* 423–430. DOI: 10.1016/0022-3956(87)90089-6

Kety, S. S., Wender, P. H., Jacobsen, B., Ingraham, L. J., Jansson, L., Faber, B., & Kinney, D. K. (1994). Mental illness in the biological and adoptive relatives of schizophrenic adoptees: Replication of the Copenhagen Study in the rest of Denmark. *Archives of General Psychiatry, 51,* 442–455. DOI: 10.1001/archpsyc.1994.03950060006001

Khan, A. A., Jacobson, K. C., Gardner, C. O., Prescott, C. A., & Kendler, K. S. (2005). Personality and comorbidity of common psychiatric disorders. *British Journal of Psychiatry, 186,* 190–196. DOI: 10.1192/bjp.186.3.190

Khoury, M. J., Yang, Q. H., Gwinn, M., Little, J. L., & Flanders, W. D. (2004). An epidemiologic assessment of genomic profiling for measuring susceptibility to common diseases and targeting interventions. *Genetics in Medicine, 6,* 38–47. DOI: 10.1097/01.GIM.0000105751.71430.79

Kidd, K. (1983). Recent progress on the genetics of stuttering. In C. Ludlow & J. Cooper (Eds.), *Genetic aspects of speech and language disorders* (pp. 197–213). New York: Academic Press.

Kieseppa, T., Partonen, T., Haukka, J., Kaprio, J., & Lonnqvist, J. (2004). High concordance of bipolar I disorder in a nationwide sample of twins. *American Journal of Psychiatry, 161,* 1814–1821. DOI: 10.1176/appi.ajp.161.10.1814

Kile, B. T., & Hilton, D. J. (2005). The art and design of genetic screens: Mouse. *Nature Reviews Genetics, 6,* 557–567. DOI: 10.1038/nrg1636

Kim, D. H., & Rossi, J. J. (2007). Strategies for silencing human disease using RNA interference. *Nature Reviews Genetics, 8,* 173–184. DOI: 10.1038/nrg2006

Kim, Y. S., Leventhal, B. L., Koh, Y.-J., Fombonne, E., Laska, E., Lim, E.-C., . . . Grinker, R. R. (2011). Prevalence of autism spectrum disorders in a total population sample. *American Journal of Psychiatry, 168,* 904–912. DOI: 10.1176/appi.ajp.2011.10101532

Kim-Cohen, J., Caspi, A., Taylor, A., Williams, B., Newcombe, R., Craig, I. W., & Moffitt, T. E. (2006). MAOA, maltreatment, and gene-environment interaction predicting children's mental health: New evidence and a meta-analysis. *Molecular Psychiatry, 11,* 903–913. DOI: 10.1038/sj.mp.4001851

Kimura, M., & Higuchi, S. (2011). Genetics of alcohol dependence. *Psychiatry and Clinical Neurosciences, 65,* 213–225. DOI: 10.1111/j.1440-1819.2011.02190.x

Kinsella, K., & He, W. (2009). *An aging world: 2008* (U.S. Census Bureau—International Population Reports, P95/09-1). Washington, DC: U.S. Government Printing Office.

Klein, R. G., & Mannuzza, S. (1991). Long-term outcome of hyperactive-children—a review. *Journal of the American Academy of Child and Adolescent Psychiatry, 30,* 383–387. DOI: 10.1097/00004583-199105000-00005

Kleinman, J. E., Law, A. J., Lipska, B. K., Hyde, T. M., Ellis, J. K., Harrison, P. J., & Weinberger, D. R. (2011). Genetic neuropathology of schizophrenia: New approaches to an old question and new uses for postmortem human brains. *Biological Psychiatry, 69,* 140–145. DOI: 10.1016/j.biopsych.2010.10.032

Klose, J., Nock, C., Herrmann, M., Stuhler, K., Marcus, K., Bluggel, M., . . . Lehrach, H. (2002). Genetic analysis of the mouse brain proteome. *Nature Genetics, 30,* 385–393. DOI: 10.1038/ng861

Klump, K. L., Keel, P. K., Sisk, C., & Burt, S. A. (2010). Preliminary evidence that estradiol moderates genetic influences on disordered eating attitudes and behaviors during puberty. *Psychological Medicine, 40,* 1745–1753. DOI: 10.1017/S0033291709992236

Klump, K. L., Suisman, J. L., Burt, S. A., McGue, M., & Iacono, W. G. (2009). Genetic and environmental influences on disordered eating: An adoption study. *Journal of Abnormal Psychology, 118,* 797–805. DOI: 10.1037/a0017204

Knafo, A., & Plomin, R. (2006a). Parental discipline and affection, and children's prosocial behavior: Genetic and environmental links. *Journal of Personality and Social Psychology, 90,* 147–164. DOI: 10.1037/0022-3514.90.1.147

Knafo, A., & Plomin, R. (2006b). Prosocial behavior from early to middle childhood: Genetic and environmental influences on stability and change. *Developmental Psychology, 42,* 771–786. DOI: 10.1037/0012-1649.42.5.771

Knafo, A., Zahn-Waxler, C., Van Hulle, C., Robinson, J. L., & Rhee, S. H. (2008). The developmental origins of a disposition toward empathy: Genetic and environmental contributions. *Emotion, 8,* 737–752. DOI: 10.1037/a0014179

Knickmeyer, R. C., Kang, C., Woolson, S., Smith, J. K., Hamer, R. M., Lin, W., . . . Gilmore, J. H. (2011). Twin-singleton differences in neonatal brain structure. *Twin Research and Human Genetics, 14,* 268–276. DOI: 10.1375/twin.14.3.268

Knopik, V. S., Alarcón, M., & DeFries, J. C. (1997). Comorbidity of mathematics and reading deficits: Evidence for a genetic etiology. *Behavior Genetics, 27,* 447–453. DOI: 10.1023/A:1025622400239

Knopik, V. S., Heath, A. C., Bucholz, K. K., Madden, P. A. F., & Waldron, M. (2009). Genetic and environmental influences on externalizing behavior and alcohol problems in adolescence: A female twin study. *Pharmacology, Biochemistry and Behavior, 93,* 313–321. DOI: 10.1016/j.pbb.2009.03.011

Knopik, V. S., Heath, A. C., Jacob, T., Slutske, W. S., Bucholz, K. K., Madden, P. A. F., . . . Martin, N. G. (2006). Maternal alcohol use disorder and offspring ADHD: Disentangling genetic and environmental effects using a children-of-twins design. *Psychological Medicine, 36,* 1461–1471. DOI: 10.1017/S0033291706007884

Knopik, V. S., Heath, A. C., Madden, P. A. F., Bucholz, K. K., Slutske, W. S., Nelson, E. C., . . . Martin, N. G. (2004). Genetic effects on alcohol dependence risk: Re-evaluating the importance of psychiatric and other heritable risk factors. *Psychological Medicine, 34,* 1519–1530. DOI: 10.1017/S0033291704002922

Knopik, V. S., Jacob, T., Haber, J. R., Swenson, L. P., & Howell, D. N. (2009). Paternal alcoholism and offspring ADHD problems: A children of twins design. *Twin Research and Human Genetics, 12,* 53–62. DOI: 10.1375/twin.12.1.53

Knoppien, P. (1985). Rare male mating advantage—a review. *Biological Reviews of the Cambridge Philosophical Society, 60,* 81–117. DOI: 10.1111/j.1469-185X.1985.tb00418.x

Koellinger, P., van der Loos, M., Groenen, P., Thurik, A., Rivadeneira, F., van Rooij, F., . . . Hofman, A. (2010). Genome-wide association studies in economics and entrepreneurship research: Promises and limitations. *Small Business Economics, 35,* 1–18. DOI: 10.1007/s11187-010-9286-3

Koenig, L. B., McGue, M., Krueger, R. F., & Bouchard, T. J., Jr. (2005). Genetic and environmental influences on religiousness: Findings for retrospective and current religiousness ratings. *Journal of Personality, 73,* 471–488. DOI: 10.1111/j.1467-6494.2005.00316.x

Koeppen-Schomerus, G., Spinath, F. M., & Plomin, R. (2003). Twins and non-twin siblings: Different estimates of shared environmental influence in early childhood. *Twin Research, 6,* 97–105. DOI: 10.1375/136905203321536227

Kohnstamm, G. A., Bates, J. E., & Rothbart, M. K. (1989). *Temperament in childhood.* New York: Wiley.

Koivumaa-Honkanen, H., Kaprio, J., Honkanen, R., Viinamaki, H., & Koskenvuo, M. (2004). Life satisfaction and depression in a 15-year follow-up of healthy adults. *Social Psychiatry and Psychiatric Epidemiology, 39*, 994–999. DOI: 10.1007/s00127-004-0833-6

Kolevzon, A., Smith, C. J., Schmeidler, J., Buxbaum, J. D., & Silverman, J. M. (2004). Familial symptom domains in monozygotic siblings with autism. *American Journal of Medical Genetics. B: Neuropsychiatriatric Genetics, 129*, 76–81. DOI: 10.1002/ajmg.b.30011

Komaromy, A. M., Alexander, J. J., Rowlan, J. S., Garcia, M. M., Chiodo, V. A., Kaya, A., . . . Aguirre, G. D. (2010). Gene therapy rescues cone function in congenital achromatopsia. *Human Molecular Genetics, 19*, 2581–2593. DOI: 10.1093/hmg/ddq136

Konradi, C. (2005). Gene expression microarray studies in polygenic psychiatric disorders: Applications and data analysis. *Brain Research Reviews, 50*, 142–155. DOI: 10.1016/j.brainres rev.2005.05.004

Koopmans, J. R., Boomsma, D. I., Heath, A. C., & van Doornen, L. J. P. (1995). A multivariate genetic analysis of sensation seeking. *Behavior Genetics, 25*, 349–356. DOI: 10.1007/BF02197284

Koopmans, J. R., Slutske, W. S., van Baal, G. C., & Boomsma, D. I. (1999). The influence of religion on alcohol use initiation: Evidence for genotype × environment interaction. *Behavior Genetics, 29*, 445–453. DOI: 10.1023/A:1021679005623

Koten, J. W., Jr., Wood, G., Hagoort, P., Goebel, R., Propping, P., Willmes, K., & Boomsma, D. I. (2009). Genetic contribution to variation in cognitive function: An fMRI study in twins. *Science, 323*, 1737–1740. DOI: 10.1126/science.1167371

Kovas, Y., Haworth, C. M., Harlaar, N., Petrill, S. A., Dale, P. S., & Plomin, R. (2007). Overlap and specificity of genetic and environmental influences on mathematics and reading disability in 10-year-old twins. *Journal of Child Psychology and Psychiatry, 48*, 914–922. DOI: 10.1111/j.1469-7610.2007.01748.x

Kovas, Y., Haworth, C. M. A., Dale, P. S., & Plomin, R. (2007). The genetic and environmental origins of learning abilities and disabilities in the early school years. *Monographs of the Society for Research in Child Development, 72*, 1–144. DOI: 10.1111/j.1540-5834.2007.00453..x

Kovas, Y., Haworth, C. M. A., Petrill, S., & Plomin, R. (2007). Mathematical ability of 10-year-old boys and girls: Genetic and environmental etiology of normal and low performance. *Journal of Learning Disabilities, 40*, 554–567. DOI: 10.1177/00222194070400060601

Kovas, Y., & Plomin, R. (2006). Generalist genes: Implications for the cognitive sciences.

Trends in Cognitive Sciences, 10, 198–203. DOI: 10.1016/j.tics.2006.03.001

Kozell, L., Belknap, J. K., Hofstetter, J. R., Mayeda, A., & Buck, K. J. (2008). Mapping a locus for alcohol physical dependence and associated withdrawal to a 1.1 Mb interval of mouse chromosome 1 syntenic with human chromosome 1q23.2-23.3. *Genes, Brain and Behavior, 7*, 560–567. DOI: 10.1111/j.1601-183X.2008.00391.x

Krebs, N. F., Himes, J. H., Jacobson, D., Nicklas, T. A., Guilday, P., & Styne, D. (2007). Assessment of child and adolescent overweight and obesity. *Pediatrics, 120*, S193–S228. DOI: 10.1542/peds.2007-2329D

Kreek, M. J., Zhou, Y., Butelman, E. R., & Levran, O. (2009). Opiate and cocaine addiction: From bench to clinic and back to the bench. *Current Opinion in Pharmacology, 9*, 74–80. DOI: 10.1016/j.coph.2008.12.016

Kremen, W. S., Jacobson, K. C., Xian, H., Eisen, S. A., Waterman, B., Toomey, R., . . . Lyons, M. J. (2005). Heritability of word recognition in middle-aged men varies as a function of parental education. *Behavior Genetics, 35*, 417–433. DOI: 10.1007/s10519-004-3876-2

Kringlen, E., & Cramer, G. (1989). Offspring of monozygotic twins discordant for schizophrenia. *Archives of General Psychiatry, 46*, 873–877. DOI: 10.1001/archpsyc.1989.01810100015003

Krueger, R. F. (1999). The structure of common mental disorders. *Archives of General Psychiatry, 56*, 921–926. DOI: 10.1001/archpsyc.56.10.921

Krueger, R. F., Caspi, A., Moffitt, T. E., Silva, A., & McGee, R. (1996). Personality traits are differentially linked to mental disorders: A multitrait-multidiagnosis study of an adolescent birth cohort. *Journal of Abnormal Psychology, 105*, 299–312. DOI: 10.1037/0021-843X.105.3.299

Krueger, R. F., Hicks, B. M., Patrick, C. J., Carlson, S. R., Iacono, W. G., & McGue, M. (2002). Etiologic connections among substance dependence, antisocial behavior, and personality: Modeling the externalizing spectrum. *Journal of Abnormal Psychology, 111*, 411–424. DOI: 10.1037/0021-843X.111.3.411

Krueger, R. F., Markon, K. E., & Bouchard, T. J., Jr. (2003). The extended genotype: The heritability of personality accounts for the heritability of recalled family environments in twins reared apart. *Journal of Personality, 71*, 809–833. DOI: 10.1111/1467-6494.7105005

Krueger, R. F., South, S., Johnson, W., & Iacono, W. (2008). The heritability of personality is not always 50%: Gene-environment interactions and correlations between personality and parenting. *Journal of Personality, 76*, 1485–1521. DOI: 10.1111/j.1467-6494.2008.00529.x

Kruger, D. J. (2006). Male facial masculinity influences attributions of personality and reproductive strategy. *Personal Relationships, 13,* 451–463. DOI: 10.1111/j.1475-6811.2006.00129.x

Ku, C.-S., Naidoo, N., & Pawitan, Y. (2011). Revisiting Mendelian disorders through exome sequencing. *Human Genetics, 129,* 351–370. DOI: 10.1007/s00439-011-0964-2

Kukekova, A. V., Trut, L. N., Chase, K., Kharlamova, A. V., Johnson, J. L., Temnykh, S. V., . . . Lark, K. G. (2011). Mapping loci for fox domestication: Deconstruction/reconstruction of a behavioral phenotype. *Behavior Genetics, 41,* 593–606. DOI: 10.1007/s10519-010-9418-1

Kumar, V., Kim, K., Joseph, C., Thomas, L. C., Hong, H., & Takahashi, J. S. (2011). Second-generation high-throughput forward genetic screen in mice to isolate subtle behavioral mutants. *Proceedings of the National Academy of Sciences (USA), 108,* 15557–15564. DOI: 10.1073/pnas.1107726108

Kuntsi, J., Rijsdijk, F., Ronald, A., Asherson, P., & Plomin, R. (2005). Genetic influences on the stability of attention-deficit/hyperactivity disorder symptoms from early to middle childhood. *Biological Psychiatry, 57,* 647–654. DOI: 10.1016/j.biopsych.2004.12.032

Kuo, P.-H., Neale, M. C., Walsh, D., Patterson, D. G., Riley, B., Prescott, C. A., & Kendler, K. S. (2010). Genome-wide linkage scans for major depression in individuals with alcohol dependence. *Journal of Psychiatric Research, 44,* 616–619. DOI: 10.1016/j.jpsychires.2009.12.005

Kupper, N., Boomsma, D. I., de Geus, E. J. C., Denollet, J., & Willemsen, G. (2011). Nine-year stability of Type D personality: Contributions of genes and environment. *Psychosomatic Medicine, 73,* 75–82. DOI: 10.1097/PSY.0b013e3181fdce54

Labuda, M. C., Defries, J. C., & Fulker, D. W. (1987). Genetic and environmental covariance-structures among WISC-R subtests—a twin study. *Intelligence, 11,* 233–244. DOI: 10.1016/0160-2896(87)90008-0

Lachance, J. (2009). Detecting selection-induced departures from Hardy-Weinberg proportions. *Genetics Selection Evolution, 41:* 15. DOI: 10.1186/1297-9686-41-15

Lack, D. (1953a). *Darwin's finches.* Cambridge: Cambridge University Press.

Lack, D. (1953b). Darwin's finches. *Scientific American, 188,* 67.

Lai, C. S., Fisher, S. E., Hurst, J. A., Vargha-Khadem, F., & Monaco, A. P. (2001). A forkhead-domain gene is mutated in a severe speech and language disorder. *Nature, 413,* 519–523. DOI: 10.1038/35097076

Lajunen, H.-R., Kaprio, J., Rose, R. J., Pulkkinen, L., & Silventoinen, K. (2012). Genetic and environmental influences on BMI from late childhood to adolescence are modified by parental education. *Obesity, 20,* 583–589. DOI: 10.1038/oby.2011.304

Lamb, D. J., Middeldorp, C. M., van Beijsterveldt, C. E. M., Bartels, M., van der Aa, N., Polderman, T. J. C., & Boomsma, D. I. (2010). Heritability of anxious-depressive and withdrawn behavior: Age-related changes during adolescence. *Journal of the American Academy of Child and Adolescent Psychiatry, 49,* 248–255. DOI: 10.1016/j.jaac.2009.11.014

Lana-Elola, E., Watson-Scales, S. D., Fisher, E. M. C., & Tybulewicz, V. L. J. (2011). Down syndrome: Searching for the genetic culprits. *Disease Models and Mechanisms, 4,* 586–595. DOI: 10.1242/dmm.008078

Lander, E. S. (2011). Initial impact of the sequencing of the human genome. *Nature, 470,* 187–197. DOI: 10.1038/nature09792

Lanfranco, F., Kamischke, A., Zitzmann, M., & Nieschlag, E. (2004). Klinefelter's syndrome. *Lancet, 364,* 273–283. DOI: 10.1016/S0140-6736(04)16678-6

Langstrom, N., Rahman, Q., Carlstrom, E., & Lichtenstein, P. (2010). Genetic and environmental effects on same-sex sexual behavior: A population study of twins in Sweden. *Archives of Sexual Behavior, 39,* 75–80. DOI: 10.1007/s10508-008-9386-1

Larsson, H., Anckarsater, H., Rastam, M., Chang, Z., & Lichtenstein, P. (2012). Childhood attention-deficit hyperactivity disorder as an extreme of a continuous trait: A quantitative genetic study of 8,500 twin pairs. *Journal of Child Psychology and Psychiatry, 53,* 73–80. DOI: 10.1111/j.1469-7610.2011.02467.x

Larsson, H., Andershed, H., & Lichtenstein, P. (2006). A genetic factor explains most of the variation in the psychopathic personality. *Journal of Abnormal Psychology, 115,* 221–230. DOI: 10.1037/0021-843X.115.2.221

Larsson, H., Dilshad, R., Lichtenstein, P., & Barker, E. D. (2011). Developmental trajectories of DSM-IV symptoms of attention-deficit/hyperactivity disorder: Genetic effects, family risk and associated psychopathology. *Journal of Child Psychology and Psychiatry, 52,* 954–963. DOI: 10.1111/j.1469-7610.2011.02379.x

Larsson, H., Lichtenstein, P., & Larsson, J. O. (2006). Genetic contributions to the development of ADHD subtypes from childhood to adolescence. *Journal of the American Academy of Child and Adolescent Psychiatry, 45,* 973–981. DOI: 10.1097/01.chi.0000222787.57100.d8

Larsson, H., Tuvblad, C., Rijsdijk, F. V., Andershed, H., Grann, M., & Lichtenstein, P. (2007).

A common genetic factor explains the association between psychopathic personality and antisocial behavior. *Psychological Medicine, 37,* 15–26. DOI: 10.1017/S003329170600907X

Lataster, T., Myin-Germeys, I., Derom, C., Thiery, E., & van Os, J. (2009). Evidence that self-reported psychotic experiences represent the transitory developmental expression of genetic liability to psychosis in the general population. *American Journal of Medical Genetics. B: Neuropsychiatric Genetics, 150B,* 1078–1084. DOI: 10.1002/ajmg.b.30933

Lau, B., Bretaud, S., Huang, Y., Lin, E., & Guo, S. (2006). Dissociation of food and opiate preference by a genetic mutation in zebrafish. *Genes, Brain, and Behavior, 5,* 497–505. DOI: 10.1111/j.1601-183X.2005.00185.x

Laucht, M., Blomeyer, D., Buchmann, A. F., Treutlein, J., Schmidt, M. H., Esser, G., . . . Banaschewski, T. (2012). Catechol-O-methyltransferase Val158Met genotype, parenting practices and adolescent alcohol use: Testing the differential susceptibility hypothesis. *Journal of Child Psychology and Psychiatry, 53,* 351–359. DOI: 10.1111/j.1469-7610.2011.02408.x

Laursen, T. M., Agerbo, E., & Pedersen, C. B. (2009). Bipolar disorder, schizoaffective disorder, and schizophrenia overlap: A new comorbidity index. *Journal of Clinical Psychiatry, 70,* 1432–1438. DOI: 10.4088/JCP.08m04807

Le, A. T., Miller, P. W., Slutske, W. S., & Martin, N. G. (2010). Are attitudes towards economic risk heritable? Analyses using the Australian Twin Study of Gambling. *Twin Research and Human Genetics, 13,* 330–339. DOI: 10.1375/twin.13.4.330

Le Couteur, A., Bailey, A., Goode, S., Pickles, A., Robertson, S., Gottesman, I. I., & Rutter, M. (1996). A broader phenotype of autism: The clinical spectrum in twins. *Journal of Child Psychology and Psychiatry, 37,* 785–801. DOI: 10.1111/j.1469-7610.1996.tb01475.x

Leahy, A. M. (1935). Nature-nurture and intelligence. *Genetic Psychology Monographs, 17,* 236–308.

LeDoux, J. E. (2000). Emotion circuits in the brain. *Annual Review of Neuroscience, 23,* 155–184. DOI: 10.1146/annurev.neuro.23.1.155

Lee, S. H., Wray, N. R., Goddard, M. E., & Visscher, P. M. (2011). Estimating missing heritability for disease from genome-wide association studies. *American Journal of Human Genetics, 88,* 294–305. DOI: 10.1016/j.ajhg.2011.02.002

Lee, S. S. (2011). Deviant peer affiliation and antisocial behavior: Interaction with monoamine oxidase A (MAOA) genotype. *Journal of Abnormal Child Psychology, 39,* 321–332. DOI: 10.1007/s10802-010-9474-2

Lee, S. S., Chronis-Tuscano, A., Keenan, K., Pelham, W. E., Loney, J., Van Hulle, C. A., . . . Lahey, B. B. (2010). Association of maternal dopamine transporter genotype with negative parenting: Evidence for gene × environment interaction with child disruptive behavior. *Molecular Psychiatry, 15,* 548–558. DOI: 10.1038/mp.2008.102

Lee, T., Henry, J. D., Trollor, J. N., & Sachdev, P. S. (2010). Genetic influences on cognitive functions in the elderly: A selective review of twin studies. *Brain Research Reviews, 64,* 1–13. DOI: 10.1016/j.brainresrev.2010.02.001

Lee, T., Mosing, M. A., Henry, J. D., Trollor, J. N., Lammel, A., Ames, D., . . . Sachdev, P. S. (2012). Genetic influences on five measures of processing speed and their covariation with general cognitive ability in the elderly: The Older Australian Twins Study. *Behavior Genetics, 42,* 96–106. DOI: 10.1007/s10519-011-9474-1

Leggett, V., Jacobs, P., Nation, K., Scerif, G., & Bishop, D. V. M. (2010). Neurocognitive outcomes of individuals with a sex chromosome trisomy: XXX, XYY, or XXY: A systematic review. *Developmental Medicine and Child Neurology, 52,* 119–129. DOI: 10.1111/j.1469-8749.2009.03545.x

Legrand, L. N., McGue, M., & Iacono, W. G. (1999). A twin study of state and trait anxiety in childhood and adolescence. *Journal of Child Psychology and Psychiatry, 40,* 953–958. DOI: 10.1111/1469-7610.00512

Lemery-Chalfant, K., Doelger, L., & Goldsmith, H. H. (2008). Genetic relations between effortful and attentional control and symptoms of psychopathology in middle childhood. *Infant and Child Development, 17,* 365–385. DOI: 10.1002/icd.581

Lerner, I. M. (1968). *Heredity, evolution and society.* San Francisco: Freeman.

Lesch, K. P., Bengel, D., Heils, A., Zhang Sabol, S., Greenburg, B. D., Petri, S., . . . Murphy, D. L. (1996). Association of anxiety-related traits with a polymorphism in the serotonin transporter gene regulatory region. *Science, 274,* 1527–1531. DOI: 10.1126/science.274.5292.1527

Letwin, N. E., Kafkafi, N., Benjamini, Y., Mayo, C., Frank, B. C., Luu, T., . . . Elmer, G. I. (2006). Combined application of behavior genetics and microarray analysis to identify regional expression themes and gene-behavior associations. *Journal of Neuroscience, 26,* 5277–5287. DOI: 10.1523/JNEUROSCI.4602-05.2006

Leve, L. D., Harold, G. T., Ge, X., Neiderhiser, J. M., Shaw, D., Scaramella, L. V., & Reiss, D. (2009). Structured parenting of toddlers at high versus low genetic risk: Two pathways to child problems. *Journal of the American Academy of Child and*

Adolescent Psychiatry, 48, 1102–1109. DOI: 10.1097/ CHI.0b013e3181b8bfc0

Leve, L. D., Kerr, D. C. R., Shaw, D., Ge, X., Neiderhiser, J. M., Scaramella, L. V., . . . Reiss, D. (2010). Infant pathways to externalizing behavior: Evidence of genotype X environment interaction. *Child Development, 81,* 340–356. DOI: 10.1111/j.1467-8624.2009.01398.x

Leve, L. D., Neiderhiser, J. M., Scaramella, L. V., & Reiss, D. (2010). The early growth and development study: Using the prospective adoption design to examine genotype-environment interplay. *Behavior Genetics, 40,* 306–314. DOI: 10.1007/ s10519-010-9353-1

Levinson, D. F., Evgrafov, O. V., Knowles, J. A., Potash, J. B., Weissman, M. M., Scheftner, W. A., . . . Holmans, P. (2007). Genetics of recurrent early-onset major depression (GenRED): Significant linkage on chromosome 15q25-q26 after fine mapping with single nucleotide polymorphism markers. *American Journal of Psychiatry, 164,* 259–264. DOI: 10.1176/appi.ajp.164.2.259

Levy, D., Ronemus, M., Yamrom, B., Lee, Y.-H., Leotta, A., Kendall, J., . . . Wigler, M. (2011). Rare de novo and transmitted copy-number variation in autistic spectrum disorders. *Neuron, 70,* 886–897. DOI: 10.1016/j.neuron.2011.05.015

Levy, D. L., Holzman, P. S., Matthysse, S., & Mendell, N. R. (1993). Eye tracking disfunction and schizophrenia: A critical perspective. *Schizophrenia Bulletin, 19,* 461–536. DOI: 10.1093/ schbul/19.3.461

Levy, R., Mirlesse, V., Jacquemard, F., & Daffos, F. (2002). Prenatal diagnosis of zygosity by fetal DNA analysis, a contribution to the management of multiple pregnancies. A series of 31 cases. *Fetal Diagnosis and Therapy, 17,* 339–342. DOI: 10.1159/000065381

Lewis, C. M., Levinson, D. F., Wise, L. H., DeLisi, L. E., Straub, R. E., Hovatta, I., . . . Helgason, T. (2003). Genome scan meta-analysis of schizophrenia and bipolar disorder, Part II: Schizophrenia. *American Journal of Human Genetics, 73,* 34–48. DOI: 10.1086/376549

Li, D., Sham, P. C., Owen, M. J., & He, L. (2006). Meta-analysis shows significant association between dopamine system genes and attention deficit hyperactivity disorder (ADHD). *Human Molecular Genetics, 15,* 2276–2284. DOI: 10.1093/hmg/ddl152

Li, J., & Burmeister, M. (2005). Genetical genomics: Combining genetics with gene expression analysis. *Human Molecular Genetics, 14,* R163–R169. DOI: 10.1093/hmg/ddi267

Li, M., Sloboda, D. M., & Vickers, M. H. (2011). Maternal obesity and developmental programming of metabolic disorders in offspring: Evidence

from animal models. *Experimental Diabetes Research, 2011:* 592408. DOI: 10.1155/2011/592408

Li, M. D., Cheng, R., Ma, J. Z., & Swan, G. E. (2003). A meta-analysis of estimated genetic and environmental effects on smoking behavior in male and female adult twins. *Addiction, 98,* 23–31. DOI: 10.1046/j.1360-0443.2003.00295.x

Li, Y., Breitling, R., & Jansen, R. C. (2008). Generalizing genetical genomics: Getting added value from environmental perturbation. *Trends in Genetics, 24,* 518–524. DOI: 10.1016/j.tig.2008.08.001

Li, Y., Wang, W.-J., Cao, H., Lu, J., Wu, C., Hu, F.-Y., . . . Tian, X.-L. (2009). Genetic association of FOXO1A and FOXO3A with longevity trait in Han Chinese populations. *Human Molecular Genetics, 18,* 4897–4904. DOI: 10.1093/hmg/ddp459

Liang, Z. S., Trang, N., Mattila, H. R., Rodriguez-Zas, S. L., Seeley, T. D., & Robinson, G. E. (2012). Molecular determinants of scouting behavior in honey bees. *Science, 335,* 1225–1228. DOI: 10.1126/science.1213962

Liao, C. Y., Rikke, B. A., Johnson, T. E., Diaz, V., & Nelson, J. F. (2010). Genetic variation in the murine lifespan response to dietary restriction: From life extension to life shortening. *Aging Cell, 9,* 92–95. DOI: 10.1111/j.1474-9726.2009.00533.x

Lichtenstein, P., Carlstrom, E., Rastam, M., Gillberg, C., & Anckarsater, H. (2010). The genetics of autism spectrum disorders and related neuropsychiatric disorders in childhood. *American Journal of Psychiatry, 167,* 1357–1363. DOI: 10.1176/ appi.ajp.2010.10020223

Lichtenstein, P., Harris, J. R., Pedersen, N. L., & McClearn, G. E. (1992). Socioeconomic status and physical health, how are they related? An empirical study based on twins reared apart and twins reared together. *Social Science and Medicine, 36,* 441–450. DOI: 10.1016/0277-9536(93)90406-T

Lichtenstein, P., Holm, N. V., Verkasalo, P. K., Iliadou, A., Kaprio, J., Koskenvuo, M., . . . Hemminki, K. (2000). Environmental and heritable factors in the causation of cancer—analysis of cohorts of twins from Sweden, Denmark, and Finland. *New England Journal of Medicine, 343,* 78–85. DOI: 10.1056/NEJM200007133430201

Lichtenstein, P., Pedersen, N. L., & McClearn, G. E. (1992). The origins of individual differences in occupational status and educational level. *Acta Sociologica, 35,* 13–31. DOI: 10.1177/000169939203500102

Lichtenstein, P., Yip, B. H., Bjork, C., Pawitan, Y., Cannon, T. D., Sullivan, P. F., & Hultman, C. M. (2009). Common genetic determinants of schizophrenia and bipolar disorder in Swedish families: A population-based study. *Lancet, 373,* 234–239. DOI: 10.1016/S0140-6736(09)60072-6

Lidsky, A. S., Robson, K., Chandra, T., Barker, P., Ruddle, F., & Woo, S. L. C. (1984). The PKU locus in Man is on chromosome 12. *American Journal of Human Genetics, 36,* 527–533.

Lie, H. C., Rhodes, G., & Simmons, L. W. (2008). Genetic diversity revealed in human faces. *Evolution, 62,* 2473–2486. DOI: 10.1111/j.1558-5646.2008.00478.x

Lilienfeld, S. O. (1992). The association between antisocial personality and somatization disorders: A review and integration of theorietical models. *Clinical Psychology Review, 12,* 641–662. DOI: 10.1016/0272-7358(92)90136-V

Lim, L. P., Lau, N. C., Garrett-Engele, P., Grimson, A., Schelter, J. M., Castle, J., . . . Johnson, J. M. (2005). Microarray analysis shows that some microRNAs downregulate large numbers of target mRNAs. *Nature, 433,* 769–773. DOI: 10.1038/nature03315

Lindblad-Toh, K., Wade, C. M., Mikkelsen, T. S., Karlsson, E. K., Jaffe, D. B., Kamal, M., . . . Lander, E. S. (2005). Genome sequence, comparative analysis and haplotype structure of the domestic dog. *Nature, 438,* 803–819. DOI: 10.1038/nature04338

Lindenberger, U. (2001). Lifespan theories of cognitive development. In N. J. Smelser & P. B. Bates (Eds.), *International encyclopedia of the social and behavior sciences* (pp. 8848–8854). Oxford: Elsevier. DOI: 10.1016/B0-08-043076-7/01572-2

Linney, Y. M., Murray, R. M., Peters, E. R., MacDonald, A. M., Rijsdijk, F., & Sham, P. C. (2003). A quantitative genetic analysis of schizotypal personality traits. *Psychological Medicine, 33,* 803–816. DOI: 10.1017/S0033291703007906

Little, A. C., Jones, B. C., Penton-Voak, I. S., Burt, D. M., & Perrett, D. I. (2002). Partnership status and the temporal context of relationships influence human female preferences for sexual dimorphism in male face shape. *Proceedings of the Royal Society of London. B: Biological Sciences, 269,* 1095–1100. DOI: 10.1098/rspb.2002.1984

Liu, N., Gong, K. Z., Cai, Y. B., & Li, Z. J. (2011). Identification of proteins responding to adrenergic receptor subtype-specific hypertrophy in cardiomyocytes by proteomic approaches. *Biochemistry (Moscow), 76,* 1140–1146. DOI: 10.1134/S0006297911100075

Liu, X., & Davis, R. L. (2006). Insect olfactory memory in time and space. *Current Opinion in Neurobiology, 16,* 679–685. DOI: 10.1016/j.conb.2006.09.003

Liu, X.-Q., Georgiades, S., Duku, E., Thompson, A., Devlin, B., Cook, E. H., . . . Szatmari, P. (2011). Identification of genetic loci underlying the phenotypic constructs of autism spectrum disorders. *Journal of the American Academy of Child and Adolescent Psychiatry, 50,* 687–696. DOI: 10.1016/j.jaac.2011.05.002

Livesley, W. J., Jang, K. L., & Vernon, P. A. (1998). Phenotypic and genetic structure of traits delineating personality disorder. *Archives of General Psychiatry, 55,* 941–948. DOI: 10.1001/archpsyc.55.10.941

Lize, A., Cortesero, A. M., Atlan, A., & Poinsot, D. (2007). Kin recognition in Aleochara bilineata could support the kinship theory of genomic imprinting. *Genetics, 175,* 1735–1740. DOI: 10.1534/genetics.106.070045

Llewellyn, C. H., van Jaarsveld, C. H., Johnson, L., Carnell, S., & Wardle, J. (2010). Nature and nurture in infant appetite: Analysis of the Gemini twin birth cohort. *American Journal of Clinical Nutrition, 91,* 1172–1179. DOI: 10.3945/ajcn.2009.28868

Loehlin, J. C. (1989). Partitioning environmental and genetic contributions to behavioral development. *American Psychologist, 44,* 1285–1292. DOI: 10.1037/0003-066X.44.10.1285

Loehlin, J. C. (1992). *Genes and environment in personality development.* Newbury Park, CA: Sage Publications, Inc.

Loehlin, J. C. (1997). Genes and environment. In D. Magnusson (Ed.), *The lifespan development of individuals: Behavioral, neurobiological, and psychosocial perspectives: A synthesis* (pp. 38–51). New York: Cambridge University Press.

Loehlin, J. C. (2010). Is there an active gene-environment correlation in adolescent drinking behavior? *Behavior Genetics, 40,* 447–451. DOI: 10.1007/s10519-010-9347-z

Loehlin, J. C., Horn, J. M., & Willerman, L. (1989). Modeling IQ change: Evidence from the Texas Adoption Project. *Child Development, 60,* 993–1004. DOI: 10.2307/1131039

Loehlin, J. C., Horn, J. M., & Willerman, L. (1990). Heredity, environment, and personality change: Evidence from the Texas Adoption Project. *Journal of Personality, 58,* 221–243. DOI: 10.1111/j.1467-6494.1990.tb00914.x

Loehlin, J. C., & Martin, N. G. (2001). Age changes in personality traits and their heritabilities during the adult years: Evidence from Australian twin registry samples. *Personality and Individual Differences, 30,* 1147–1174. DOI: 10.1016/S0191-8869(00)00099-4

Loehlin, J. C., Neiderhiser, J. M., & Reiss, D. (2003). The behavior genetics of personality and the NEAD study. *Journal of Research in Personality, 37,* 373–387. DOI: 10.1016/S0092-6566(03)00012-6

Loehlin, J. C., & Nichols, J. (1976). *Heredity, environment and personality.* Austin: University of Texas Press.

Loehlin, J. C., Willerman, L., & Horn, J. M. (1982). Personality resemblances between unwed mothers and their adopted-away offspring. *Journal of Personality and Social Psychology, 42,* 1089–1099. DOI: 10.1037/0022-3514.42.6.1089

Lohmueller, K. E., Pearce, C. L., Pike, M., Lander, E. S., & Hirschhorn, J. N. (2003). Meta-analysis of genetic association studies supports a contribution of common variants to susceptibility to common disease. *Nature Genetics, 33,* 177–182. DOI: 10.1038/ng1071

Long, J., Knowler, W., Hanson, R., Robin, R., Urbanek, M., Moore, E., ... Goldman, D. (1998). Evidence for genetic linkage to alcohol dependence on chromosomes 4 and 11 from an autosome-wide scan in an American Indian population. *American Journal of Medical Genetics, 81,* 216–221. DOI: 10.1002/(SICI)1096-8628(19980508)81:3<216::AID-AJMG2>3.0.CO;2-U

Losoya, S. H., Callor, S., Rowe, D. C., & Goldsmith, H. H. (1997). Origins of familial similarity in parenting: A study of twins and adoptive siblings. *Developmental Psychology, 33,* 1012–1023. DOI: 10.1037/0012-1649.33.6.1012

Lott, I. T., & Dierssen, M. (2010). Cognitive deficits and associated neurological complications in individuals with Down's syndrome. *Lancet Neurology, 9,* 623–633. DOI: 10.1016/S1474-4422(10)70112-5

Lovinger, D. M., & Crabbe, J. C. (2005). Laboratory models of alcoholism: Treatment target identification and insight into mechanisms. *Nature Neuroscience, 8,* 1471–1480. DOI: 10.1038/nn1581

Lucht, M., Barnow, S., Schroeder, W., Grabe, H. J., Finckh, U., John, U., ... Herrmann, F. H. (2006). Negative perceived paternal parenting is associated with dopamine D2 receptor exon 8 and GABA(A) alpha 6 receptor variants: An explorative study. *American Journal of Medical Genetics. B: Neuropsychiatriatric Genetics, 141,* 167–172. DOI: 10.1002/ajmg.b.30255

Luciano, M., Hansell, N. K., Lahti, J., Davies, G., Medland, S. E., Raikkonen, K., ... Deary, I. J. (2011). Whole genome association scan for genetic polymorphisms influencing information processing speed. *Biological Psychology, 86,* 193–202. DOI: 10.1016/j.biopsycho.2010.11.008

Luciano, M., Montgomery, G. W., Martin, N. G., Wright, M. J., & Bates, T. C. (2011). SNP sets and reading ability: Testing confirmation of a 10-SNP set in a population sample. *Twin Research and Human Genetics, 14,* 228–232. DOI: 10.1375/twin.14.3.228

Luciano, M., Wright, M., Smith, G. A., Geffen, G. M., Geffen, L. B., & Martin, N. G. (2001). Genetic covariance among measures of information processing speed, working memory, and IQ. *Behavior Genetics, 31,* 581–592. DOI: 10.1023/A:1013397428612

Luciano, M., Wright, M. J., Duffy, D. L., Wainwright, M. A., Zhu, G., Evans, D. M., ... Martin, N. G. (2006). Genome-wide scan of IQ finds significant linkage to a quantitative trait locus on 2q. *Behavior Genetics, 36,* 45–55. DOI: 10.1007/s10519-005-9003-1

Luciano, M., Wright, M. J., Geffen, G. M., Geffen, L. B., Smith, G. A., Evans, D. M., & Martin, N. G. (2003). A genetic two-factor model of the covariation among a subset of Multidimensional Aptitude Battery and Wechsler Adult Intelligence Scale—Revised subtests. *Intelligence, 31,* 589–605. DOI: 10.1016/S0160-2896(03)00057-6

Lundstrom, S., Chang, Z., Rastam, M., Gillberg, C., Larsson, H., Anckarsater, H., & Lichtenstein, P. (2012). Autism spectrum disorders and autisticlike traits: Similar etiology in the extreme end and the normal variation. *Archives of General Psychiatry, 69,* 46–52. DOI: 10.1001/archgenpsychiatry.2011.144

Luo, D., Petrill, S. A., & Thompson, L. A. (1994). An exploration of genetic g: Hierarchical factor analysis of cognitive data from the Western Reserve Twin Project. *Intelligence, 18,* 335–348. DOI: 10.1016/0160-2896(94)90033-7

Luo, X. G., Kranzler, H. R., Zuo, L. J., Wang, S., Blumberg, H. P., & Gelernter, J. (2005). CHRM2 gene predisposes to alcohol dependence, drug dependence and affective disorders: Results from an extended case-control structured association study. *Human Molecular Genetics, 14,* 2421–2434. DOI: 10.1093/hmg/ddi244

Lykken, D. T. (2006). The mechanism of emergenesis. *Genes, Brain and Behavior, 5,* 306–310. DOI: 10.1111/j.1601-183X.2006.00233.x

Lykken, D. T., & Tellegen, A. (1993). Is human mating adventitious or the result of lawful choice? A twin study of mate selection. *Journal of Personality and Social Psychology, 65,* 56–68. DOI: 10.1037/0022-3514.65.1.56

Lynch, M. A. (2004). Long-term potentiation and memory. *Physiological Reviews, 84,* 87–136. DOI: 10.1152/physrev.00014.2003

Lynch, S. K., Turkheimer, E., D'Onofrio, B. M., Mendle, J., Emery, R. E., Slutske, W. S., & Martin, N. G. (2006). A genetically informed study of the association between harsh punishment and offspring behavioral problems. *Journal of Family Psychology, 20,* 190–198. DOI: 10.1037/0893-3200.20.2.190

Lynskey, M. T., Agrawal, A., & Heath, A. C. (2010). Genetically informative research on adolescent substance use: Methods, findings, and challenges. *Journal of the American Academy of Child and Adolescent Psychiatry, 49,* 1202–1214. DOI: 10.1016/j.jaac.2010.09.004

Lyons, M. J. (1996). A twin study of self-reported criminal behaviour. In G. R. Bock & J. A. Goode (Eds.), *Ciba Foundation Symposium 194—Genetics of criminal and antisocial behaviour* (pp. 70–75). Chichester, UK: John Wiley & Sons. DOI: 10.1002/9780470514825.ch4

Lyons, M. J., Goldberg, J., Eisen, S. A., True, W., Tsuang, M. T., Meyer, J. M., & Henderson, W. G. (1993). Do genes influence exposure to trauma: A twin study of combat. *American Journal of Medical Genetics. B: Neuropsychiatriatric Genetics, 48,* 22–27. DOI: 10.1002/ajmg.1320480107

Lyons, M. J., True, W. R., Eisen, S. A., Goldberg, J., Meyer, J. M., Faraone, S. V., . . . Tsuang, M. T. (1995). Differential heritability of adult and juvenile antisocial traits. *Archives of General Psychiatry, 52,* 906–915. DOI: 10.1001/archpsyc.1995.03950230020005

Lyons, M. J., York, T. P., Franz, C. E., Grant, M. D., Eaves, L. J., Jacobson, K. C., . . . Kremen, W. S. (2009). Genes determine stability and the environment determines change in cognitive ability during 35 years of adulthood. *Psychological Science, 20,* 1146–1152. DOI: 10.1111/j.1467-9280.2009.02425.x

Lytton, H. (1977). Do parents create or respond to differences in twins? *Developmental Psychology, 13,* 456–459. DOI: 10.1037/0012-1649.13.5.456

Lytton, H. (1980). *Parent-child interaction: The socialization process observed in twin and singleton families.* New York: Plenum.

Lytton, H. (1991). Different parental practices—different sources of influence. *Behavioral and Brain Sciences, 14,* 399–400. DOI: 10.1017/S0140525X00070436

Ma, D. Q., Cuccaro, M. L., Jaworski, J. M., Haynes, C. S., Stephan, D. A., Parod, J., . . . Pericak-Vance, M. A. (2007). Dissecting the locus heterogeneity of autism: Significant linkage to chromosome 12q14. *Molecular Psychiatry, 12,* 376–384. DOI: 10.1038/sj.mp.4001927

Maat-Kievit, A., Vegter-van der Vlis, M., Zoeteweij, M., Losekoot, M., van Haeringen, A., & Roos, R. (2000). Paradox of a better test for Huntington's disease. *Journal of Neurology, Neurosurgery and Psychiatry, 69,* 579–583. DOI: 10.1136/jnnp.69.5.579

Mabb, A. M., Judson, M. C., Zylka, M. J., & Philpot, B. D. (2011). Angelman syndrome: Insights into genomic imprinting and neurodevelopmental phenotypes. *Trends in Neurosciences, 34,* 293–303. DOI: 10.1016/j.tins.2011.04.001

Maccani, M. A., Avissar-Whiting, M., Banister, C. E., McGonnigal, B., Padbury, J. F., & Marsit, C. J. (2010). Maternal cigarette smoking during pregnancy is associated with downregulation of miR-16, miR-21 and miR-146a in the placenta. *Epigenetics, 5,* 583–589. DOI: 10.4161/epi.5.7.12762

Maccani, M. A., & Marsit, C. J. (2009). Epigenetics in the placenta. *American Journal of Reproductive Immunology, 62,* 78–89. DOI: 10.1111/j.1600-0897.2009.00716.x

MacGillivray, I., Campbell, D. M., & Thompson, B. (1988). *Twinning and twins.* Chichester, UK: Wiley.

Mackay, T. F., & Anholt, R. R. (2006). Of flies and man: Drosophila as a model for human complex traits. *Annual Review of Genomics and Human Genetics, 7,* 339–367. DOI: 10.1146/annurev.genom.7.080505.115758

MacKillop, J., Amlung, M. T., Few, L. R., Ray, L. A., Sweet, L. H., & Munafo, M. R. (2011). Delayed reward discounting and addictive behavior: A meta-analysis. *Psychopharmacology, 216,* 305–321. DOI: 10.1007/s00213-011-2229-0

MacKillop, J., Obasi, E. M., Amlung, M. T., McGeary, J. E., & Knopik, V. S. (2010). The role of genetics in nicotine dependence: Mapping the pathways from genome to syndrome. *Current Cardiovascular Risk Reports, 4,* 446–453. DOI: 10.1007/s12170-010-0132-6

Mackintosh, M. A., Gatz, M., Wetherell, J. L., & Pedersen, N. L. (2006). A twin study of lifetime generalized anxiety disorder (GAD) in older adults: Genetic and environmental influences shared by neuroticism and GAD. *Twin Research and Human Genetics, 9,* 30–37. DOI: 10.1375/twin.9.1.30

Mackintosh, N. J. (1995). *Cyril Burt: Fraud or framed?* Oxford: Oxford University Press. DOI: 10.1093/acprof:oso/9780198523369.001.0001

Mackintosh, N. J. (1998). *IQ and human intelligence.* Oxford: Oxford University Press.

Maes, H. H., Neale, M. C., & Eaves, L. J. (1997). Genetic and environmental factors in relative body weight and human adiposity. *Behavior Genetics, 27,* 325–351. DOI: 10.1023/A:1025635913927

Maes, H. H., Sullivan, P. F., Bulik, C. M., Neale, M. C., Prescott, C. A., Eaves, L. J., & Kendler, K. S. (2004). A twin study of genetic and environmental influences on tobacco initiation, regular tobacco use and nicotine dependence. *Psychological Medicine, 34,* 1251–1261. DOI: 10.1017/S0033291704002405

Maguire, E. A., Gadian, D. G., Johnsrude, I. S., Good, C. D., Ashburner, J., Frackowiak, R. S. J., & Frith, C. D. (2000). Navigation-related structural change in the hippocampi of taxi drivers. *Proceedings of the National Academy of Sciences (USA), 97,* 4398–4403. DOI: 10.1073/pnas.070039597

Maguire, E. A., Woollett, K., & Spiers, H. J. (2006). London taxi drivers and bus drivers: A structural MRI and neuropsychological analysis. *Hippocampus, 16,* 1091–1101. DOI: 10.1002/hipo.20233

Maher, B. (2008). Personal genomes: The case of the missing heritability. *Nature, 456,* 18–21. DOI: 10.1038/456018a

Mahowald, M. B., Verp, M. S., & Anderson, R. R. (1998). Genetic counseling: Clinical and ethical challenges. *Annual Review of Genetics, 32,* 547–559. DOI: 10.1146/annurev.genet.32.1.547

Malykh, S. B., Iskoldsky, N. V., & Gindina, E. V. (2005). Genetic analysis of IQ in young adulthood: A Russian twin study. *Personality and Individual Differences, 38,* 1475–1485. DOI: 10.1016/j.paid.2003.06.014

Manakov, S. A., Grant, S. G. N., & Enright, A. J. (2009). Reciprocal regulation of microRNA and mRNA profiles in neuronal development and synapse formation. *BMC Genomics, 10:* 419. DOI: 10.1186/1471-2164-10-419

Mandoki, M. W., Sumner, G. S., Hoffman, R. P., & Riconda, D. L. (1991). A review of Klinefelter's syndrome in children and adolescents. *Journal of the American Academy of Child and Adolescent Psychiatry, 30,* 167–172. DOI: 10.1097/00004583-199103000-00001

Manke, B., McGuire, S., Reiss, D., Hetherington, E. M., & Plomin, R. (1995). Genetic contributions to adolescents' extrafamilial social interactions: Teachers, best friends, and peers. *Social Development, 4,* 238–256. DOI: 10.1111/j.1467-9507.1995.tb00064.x

Manolio, T. A., Collins, F. S., Cox, N. J., Goldstein, D. B., Hindorff, L. A., Hunter, D. J., . . . Visscher, P. M. (2009). Finding the missing heritability of complex diseases. *Nature, 461,* 747–753. DOI: 10.1038/nature08494

Margulies, C., Tully, T., & Dubnau, J. (2005). Deconstructing memory in Drosophila. *Current Biology, 15,* R700–R713. DOI: 10.1016/j.cub.2005.08.024

Marks, I. M., & Nesse, R. M. (1994). Fear and fitness: An evolutionary analysis of anxiety disorders. *Etiology and Sociobiology, 15,* 247–261. DOI: 10.1016/0162-3095(94)90002-7

Maron, E., Toru, I., Must, A., Tasa, G., Toover, E., Vasar, V., . . . Shlik, J. (2007). Association study of tryptophan hydroxylase 2 gene polymorphisms in panic disorder. *Neuroscience Letters, 411,* 180–184. DOI: 10.1016/j.neulet.2006.09.060

Martens, M. A., Wilson, S. J., & Reutens, D. C. (2008). Research review: Williams syndrome: A critical review of the cognitive, behavioral, and neuroanatomical phenotype. *Journal of Child Psychology and Psychiatry, 49,* 576–608. DOI: 10.1111/j.1469-7610.2008.01887.x

Martin, G. M. (2011). The biology of aging: 1985–2010 and beyond. *FASEB Journal, 25,* 3756–3762. DOI: 10.1096/fj.11-1102.ufm

Martin, N., Boomsma, D. I., & Machin, G. (1997). A twin-pronged attack on complex trait. *Nature Genetics, 17,* 387–392. DOI: 10.1038/ng1297-387

Martin, N. G., & Eaves, L. J. (1977). Genetic-analysis of covariance structure. *Heredity, 38,* 79–95. DOI: 10.1038/hdy.1977.9

Martinez, D., & Narendran, R. (2010). Imaging neurotransmitter release by drugs of abuse. In D. W. Self & J. K. Staley Gottschalk (Eds.), *Behavioral neuroscience of drug addiction* (3rd ed.) (pp. 219–245). New York: Springer. DOI: 10.1007/7854_2009_34

Martynyuk, A. E., van Spronsen, F. J., & Van der Zee, E. A. (2010). Animal models of brain dysfunction in phenylketonuria. *Molecular Genetics and Metabolism, 99,* S100–S105. DOI: 10.1016/j.ymgme.2009.10.181

Matheny, A. P., Jr. (1980). Bayley's Infant Behavioral Record: Behavioral components and twin analysis. *Child Development, 51,* 1157–1167. DOI: 10.2307/1129557

Matheny, A. P., Jr. (1989). Children's behavioral inhibition over age and across situations: Genetic similarity for a trait during change. *Journal of Personality, 57,* 215–235. DOI: 10.1111/j.1467-6494.1989.tb00481.x

Matheny, A. P., Jr. (1990). Developmental behavior genetics: Contributions from the Louisville Twin Study. In M. E. Hahn, J. K. Hewitt, N. D. Henderson & R. H. Benno (Eds.), *Developmental behavior genetics: Neural, biometrical, and evolutionary approaches* (pp. 25–39). New York: Chapman & Hall.

Matheny, A. P., Jr., & Dolan, A. B. (1975). Persons, situations, and time: A genetic view of behavioral change in children. *Journal of Personality and Social Psychology, 14,* 224–234.

Mather, K., & Jinks, J. L. (1982). *Biometrical genetics: The study of continuous variation* (3rd ed.) New York: Chapman & Hall.

Mathes, W. F., Kelly, S. A., & Pomp, D. (2011). Advances in comparative genetics: Influence of genetics on obesity. *British Journal of Nutrition, 106,* S1–S10. DOI: 10.1017/S0007114511001905

Matsumoto, J., Sugiura, Y., Yuki, D., Hayasaka, T., Goto-Inoue, N., Zaima, N., . . . Niwa, S. (2011). Abnormal phospholipids distribution in the prefrontal cortex from a patient with schizophrenia revealed by matrix-assisted laser desorption/ionization imaging mass spectrometry. *Analytical and Bioanalytical Chemistry, 400,* 1933–1943. DOI: 10.1007/s00216-011-4909-3

Mattay, V. S., & Goldberg, T. E. (2004). Imaging genetic influences in human brain function. *Current Opinion in Neurobiology, 14,* 239–247. DOI: 10.1016/j.conb.2004.03.014

Matthews, K. A., Kaufman, T. C., & Gelbart, W. M. (2005). Research resources for Drosophila: The expanding universe. *Nature Reviews Genetics, 6,* 179–193. DOI: 10.1038/nrg1554

Matzel, L. D., & Kolata, S. (2010). Selective attention, working memory, and animal intelligence. *Neuroscience and Biobehavioral Reviews, 34,* 23–30. DOI: 10.1016/j.neubiorev.2009.07.002

Maubourguet, N., Lesne, A., Changeux, J.-P., Maskos, U., & Faure, P. (2008). Behavioral sequence analysis reveals a novel role for β2* nicotinic receptors in exploration. *PLoS Computational Biology, 4:* e1000229. DOI: 10.1371/journal.pcbi.1000229

Maxson, S. C. (2009). The genetics of offensive aggression in mice. In Y.–K. Kim (Ed.), *Handbook of behavior genetics* (pp. 301–316). New York: Springer. DOI: 10.1007/978-0-387-76727-7_21

Mayford, M., & Kandel, E. R. (1999). Genetic approaches to memory storage. *Trends in Genetics, 15,* 463–470. DOI: 10.1016/S0168-9525(99)01846-6

Mazzeo, S. E., Mitchell, K. S., Bulik, C. M., Aggen, S. H., Kendler, K. S., & Neale, M. C. (2010). A twin study of specific bulimia nervosa symptoms. *Psychological Medicine, 40,* 1203–1213. DOI: 10.1017/S003329170999122X

McBride, C. M., Koehly, L. M., Sanderson, S. C., & Kaphingst, K. A. (2010). The behavioral response to personalized genetic information: Will genetic risk profiles motivate individuals and families to choose more healthful behaviors? *Annual Review of Public Health, 31,* 89–103. DOI: 10.1146/annurev.publhealth.012809.103532

McCartney, K., Harris, M. J., & Bernieri, F. (1990). Growing up and growing apart: A developmental meta-analysis of twin studies. *Psychological Bulletin, 107,* 226–237. DOI: 10.1037/0033-2909.107.2.226

McCay, C. M., Crowell, M. F., & Maynard, L. A. (1935). The effect of retarded growth upon the length of life span and upon the ultimate body size. *Journal of Nutrition, 10,* 63–79.

McClearn, G. E. (1963). The inheritance of behavior. In L. J. Postman (Ed.), *Psychology in the making* (pp. 144–252). New York: Alfred A. Knopf, Inc.

McClearn, G. E. (1976). Experimental behavioural genetics. In D. Barltrop (Ed.), *Aspects of genetics in paediatrics* (pp. 31–39). London: Fellowship of Postdoctorate Medicine.

McClearn, G. E., Johansson, B., Berg, S., Pedersen, N. L., Ahern, F., Petrill, S. A., & Plomin, R. (1997). Substantial genetic influence on cognitive abilities in twins 80 or more years old. *Science, 276,* 1560–1563. DOI: 10.1126/science.276.5318.1560

McClearn, G. E., & Rodgers, D. A. (1959). Differences in alcohol preference among inbred strains of mice. *Quarterly Journal of Studies on Alcohol, 52,* 62–67.

McCourt, K., Bouchard, T. J., Jr., Lykken, D. T., Tellegen, A., & Keyes, M. (1999). Authoritarianism revisted: Genetic and environmental influences examined in twins reared apart and together. *Personality and Individual Differences, 27,* 985–1014. DOI: 10.1016/S0191-8869(99)00048-3

McEvoy, J. P. (2007). The costs of schizophrenia. *Journal of Clinical Psychiatry, 68 (Suppl. 14),* 4–7.

McGeary, J. (2009). The DRD4 exon 3 VNTR polymorphism and addiction-related phenotypes: A review. *Pharmacology, Biochemistry and Behavior, 93,* 222–229. DOI: 10.1016/j.pbb.2009.03.010

McGough, J. J., Loo, S. K., McCracken, J. T., Dang, J., Clark, S., Nelson, S. F., & Smalley, S. L. (2008). CBCL Pediatric Bipolar Disorder profile and ADHD: Comorbidity and quantitative trait loci analysis. *Journal of the American Academy of Child and Adolescent Psychiatry, 47,* 1151–1157. DOI: 10.1097/CHI.0b013e3181825a68

McGrath, M., Kawachi, I., Ascherio, A., Colditz, G. A., Hunter, D. J., & DeVivo, V. I. (2004). Association between catechol-O-methyltransferase and phobic anxiety. *American Journal of Psychiatry, 161,* 1703–1705. DOI: 10.1176/appi.ajp.161.9.1703

McGue, M., Bacon, S., & Lykken, D. T. (1993). Personality stability and change in early adulthood: A behavioral genetic analysis. *Developmental Psychology, 29,* 96–109. DOI: 10.1037/0012-1649.29.1.96

McGue, M., & Bouchard, T. J., Jr. (1989). Genetic and environmental determinants of information processing and special mental abilities; a twin analysis. In R. J. Sternberg (Ed.), *Advances in the psychology of human intelligence* (5th ed.) (pp. 7–45). Hillsdale, NJ: Erlbaum.

McGue, M., Bouchard, T. J., Jr., Iacono, W. G., & Lykken, D. T. (1993). Behavioral genetics of cognitive ability: A life-span perspective. In R. Plomin & G. E. McClearn (Eds.), *Nature, nurture, and psychology* (pp. 59–76). Washington, DC: American Psychological Association. DOI: 10.1037/10131-003

McGue, M., Elkins, I., Walden, B., & Iacono, W. G. (2005). Perceptions of the parent-adolescent relationship: A longitudinal investigation. *Developmental Psychology, 41,* 971-984. DOI: 10.1037/0012-1649.41.6.971

McGue, M., & Gottesman, I. I. (1989). Genetic linkage in schizophrenia: Perspectives from genetic epidemiology. *Schizophrenia Bulletin, 15,* 453–464. DOI: 10.1093/schbul/15.3.453

McGue, M., Hirsch, B., & Lykken, D. T. (1993). Age and the self-perception of ability: A twin study analysis. *Psychology and Aging, 8,* 72–80. DOI: 10.1037/0882-7974.8.1.72

McGue, M., Keyes, M., Sharma, A., Elkins, I., Legrand, L., Johnson, W., & Iacono, W. G. (2007). The environments of adopted and non-adopted youth: Evidence on range restriction from the sibling interaction and behavior study (SIBS). *Behavior Genetics, 37,* 449–462. DOI: 10.1007/s10519-007-9142-7

McGue, M., & Lykken, D. T. (1992). Genetic influence on risk of divorce. *Psychological Science, 3,* 368–373. DOI: 10.1111/j.1467-9280.1992.tb00049.x

McGue, M., Sharma, S., & Benson, P. (1996). Parent and sibling influences on adolescent alcohol use and misuse: Evidence from a U.S. adoption court. *Journal of Studies on Alcohol, 57,* 8–18.

McGuffin, P., Cohen, S., & Knight, J. (2007). Homing in on depression genes. *American Journal of Psychiatry, 164,* 195–197. DOI: 10.1176/appi.ajp.164.2.195

McGuffin, P., Farmer, A. E., & Gottesman, I. I. (1987). Is there really a split in schizophrenia? The genetic evidence. *British Journal of Psychiatry, 50,* 581–592. DOI: 10.1192/bjp.150.5.581

McGuffin, P., & Gottesman, I. I. (1985). Genetic influences on normal and abnormal development. In M. Rutter & L. Hersov (Eds.), *Child and adolescent psychiatry: Modern approaches* (2nd ed.) (pp. 17–33). Oxford: Blackwell Scientific.

McGuffin, P., Gottesman, I. I., & Owen, M. J. (2002). *Psychiatric genetics and genomics.* Oxford: Oxford University Press.

McGuffin, P., & Katz, R. (1986). Nature, nurture, and affective disorder. In J. W. F. Deakin (Ed.), *The biology of depression* (pp. 26–51). London: Gaskell Press.

McGuffin, P., Katz, R., & Rutherford, J. (1991). Nature, nurture and depression: A twin study. *Psychological Medicine, 21,* 329–335. DOI: 10.1017/S0033291700020432

McGuffin, P., Katz, R., Watkins, S., & Rutherford, J. (1996). A hospital-based twin register of the heritability of DSM-IV unipolar depression. *Archives of General Psychiatry, 53,* 129–136. DOI: 10.1001/archpsyc.1996.01830020047006

McGuffin, P., Knight, J., Breen, G., Brewster, S., Boyd, P. R., Craddock, N., . . . Farmer, A. E. (2005). Whole genome linkage scan of recurrent depressive disorder from the depression network study. *Human Molecular Genetics, 14,* 3337–3345. DOI: 10.1093/hmg/ddi363

McGuffin, P., Owen, M. J., O'Donovan, M. C., Thapar, A., & Gottesman, I. I. (1994). *Seminars in psychiatric genetics.* London: Gaskell.

McGuffin, P., Rijsdijk, F., Andrew, M., Sham, P., Katz, R., & Cardno, A. (2003). The heritability of bipolar affective disorder and the genetic relationship to unipolar depression. *Archives of General Psychiatry, 60,* 497–502. DOI: 10.1001/archpsyc.60.5.497

McGuffin, P., Sargeant, M., Hetti, G., Tidmarsh, S., Whatley, S., & Marchbanks, R. M. (1990). Exclusion of a schizophrenia susceptibility gene from the chromosome 5q11-q13 region. New data and a reanalysis of previous reports. *American Journal of Human Genetics, 47,* 534–535.

McGuffin, P., & Sturt, E. (1986). Genetic markers in schizophrenia. *Human Heredity, 36,* 65–88. DOI: 10.1159/000153604

McGuire, M., & Troisi, A. (1998). *Darwinian psychiatry.* Oxford: Oxford University Press.

McGuire, S., Neiderhiser, J. M., Reiss, D., Hetherington, E. M., & Plomin, R. (1994). Genetic and environmental influences on perceptions of self-worth and competence in adolescence: A study of twins, full siblings, and step-siblings. *Child Development, 65,* 785–799. DOI: 10.2307/1131418

McGuire, S. E., Deshazer, M., & Davis, R. L. (2005). Thirty years of olfactory learning and memory research in Drosophila melanogaster. *Progress in Neurobiology, 76,* 328–347. DOI: 10.1016/j.pneurobio.2005.09.003

McKie, R. (2007). *Face of Britain: How our genes reveal the history of Britain.* London: Simon & Schuster.

McLoughlin, G., Rijsdijk, F., Asherson, P., & Kuntsi, J. (2011). Parents and teachers make different contributions to a shared perspective on hyperactive-impulsive and inattentive symptoms: A multivariate analysis of parent and teacher ratings on the symptom domains of ADHD. *Behavior Genetics, 41,* 668–679. DOI: 10.1007/s10519-011-9473-2

McLoughlin, G., Ronald, A., Kuntsi, J., Asherson, P., & Plomin, R. (2007). Genetic support for the dual nature of attention deficit hyperactivity disorder: Substantial genetic overlap between the inattentive and hyperactive-impulsive components. *Journal of Abnormal Child Psychology, 35,* 999–1008. DOI: 10.1007/s10802-007-9149-9

McMahon, R. C. (1980). Genetic etiology in the hyperactive child syndrome: A critical review. *American Journal of Orthopsychiatry, 50,* 145–150. DOI: 10.1111/j.1939-0025.1980.tb03270.x

McQueen, M. B., Devlin, B., Faraone, S. V., Nimgaonkar, V. L., Sklar, P., Smoller, J. W., . . . Laird, N. M. (2005). Combined analysis from eleven linkage studies of bipolar disorder provides strong evidence of susceptibility loci on chromosomes 6q and 8q. *American Journal of Human Genetics, 77,* 582–595. DOI: 10.1086/491603

McRae, A. F., Matigian, N. A., Vadlamudi, L., Mulley, J. C., Mowry, B., Martin, N. G., . . . Visscher, P. M. (2007). Replicated effects of sex and genotype on gene expression in human lymphoblastoid cell lines. *Human Molecular Genetics, 16,* 364–373. DOI: 10.1093/hmg/ddl456

Meaburn, E., Dale, P. S., Craig, I. W., & Plomin, R. (2002). Language-impaired children: No sign of the FOXP2 mutation. *NeuroReport, 13,* 1075–1077. DOI: 10.1097/00001756-200206120-00020

Meaburn, E. L., Harlaar, N., Craig, I. W., Schalkwyk, L. C., & Plomin, R. (2008). Quantitative trait locus association scan of early reading disability and ability using pooled DNA and 100k SNP microarrays in a sample of 5760 children. *Molecular Psychiatry, 13,* 729–740. DOI: 10.1038/sj.mp.4002063

Meaney, M. J. (2010). Epigenetics and the biological definition of gene × environment interactions. *Child Development, 81,* 41–79. DOI: 10.1111/j.1467-8624.2009.01381.x

Medlund, P., Cederlof, R., Floderus-Myrhed, B., Friberg, L., & Sorensen, S. (1977). A new Swedish twin registry. *Acta Medica Scandinavica Supplementum, 60,* 1–11.

Mednick, S. A., Gabrielli, W. F., & Hutchings, B. (1984). Genetic factors in criminal behavior: Evidence from an adoption cohort. *Science, 224,* 891–893. DOI: 10.1126/science.6719119

Mendel, G. J. (1866). Versuche ueber pflanzenhybriden. *Verhandlungen des Naturforschunden Vereines in Bruenn, 4,* 3–47.

Mendes Soares, L. M. M., & Valcárcel, J. (2006). The expanding transcriptome: The genome as the 'book of sand'. *EMBO Journal, 25,* 923–931. DOI: 10.1038/sj.emboj.7601023

Mendlewicz, J., & Rainer, J. D. (1977). Adoption study supporting genetic transmission in manic-depressive illness. *Nature, 268,* 327–329. DOI: 10.1038/268327a0

Merikangas, K. R., He, J.-P., Brody, D., Fisher, P. W., Bourdon, K., & Koretz, D. S. (2010). Prevalence and treatment of mental disorders among US children in the 2001–2004 NHANES. *Pediatrics, 125,* 75–81. DOI: 10.1542/peds.2008-2598

Merikangas, K. R., Stolar, M., Stevens, D. E., Goulet, J., Preisig, M. A., Fenton, B., . . . Rounsaville, B. J. (1998). Familial transmission of substance use disorders. *Archives of General Psychiatry, 55,* 973–979. DOI: 10.1001/archpsyc.55.11.973

Merriman, C. (1924). The intellectual resemblance of twins. *Psychological Monographs, 33,* 1–58. DOI: 10.1037/h0093212

Meyer, J. M. (1995). Genetic studies of obesity across the life span. In L. R. Cardon & J. K. Hewitt (Eds.), *Behavior genetic approaches to behavioral medicine* (pp. 145–166). New York: Plenum.

Middeldorp, C. M., Cath, D. C., Van Dyck, R., & Boomsma, D. I. (2005). The co-morbidity of anxiety and depression in the perspective of genetic epidemiology. A review of twin and family studies. *Psychological Medicine, 35,* 611–624. DOI: 10.1017/S003329170400412X

Middeldorp, C. M., Cath, D. C., Vink, J. M., & Boomsma, D. I. (2005). Twin and genetic effects on life events. *Twin Research and Human Genetics, 8,* 224–231. DOI: 10.1375/twin.8.3.224

Milagro, F. I., Campion, J., Garcia-Diaz, D. F., Goyenechea, E., Paternain, L., & Martinez, J. A. (2009). High fat diet-induced obesity modifies the methylation pattern of leptin promoter in rats. *Journal of Physiology and Biochemistry, 65,* 1–9. DOI: 10.1007/BF03165964

Miles, D. R., Silberg, J. L., Pickens, R. W., & Eaves, L. J. (2005). Familial influences on alcohol use in adolescent female twins: Testing for genetic and environmental interactions. *Journal of Studies of Alcohol, 66,* 445–451.

Miller, G. F. (2000). *The mating mind.* New York: Doubleday.

Miller, N., & Gerlai, R. (2007). Quantification of shoaling behaviour in zebrafish (Danio rerio). *Behavioural Brain Research, 184,* 157–166. DOI: 10.1016/j.bbr.2007.07.007

Mills, R. E., Walter, K., Stewart, C., Handsaker, R. E., Chen, K., Alkan, C., . . . 1000 Genomes Project. (2011). Mapping copy number variation by population-scale genome sequencing. *Nature, 470,* 59–65. DOI: 10.1038/nature09708

Milne, B. J., Caspi, A., Harrington, H., Poulton, R., Rutter, M., & Moffitt, T. E. (2009). Predictive value of family history on severity of illness: The case for depression, anxiety, alcohol dependence, and drug dependence. *Archives of General Psychiatry, 66,* 738–747. DOI: 10.1001/archgenpsychiatry.2009.55

Mitchell, J. J., Trakadis, Y. J., & Scriver, C. R. (2011). Phenylalanine hydroxylase deficiency. *Genetics in Medicine, 13,* 697–707. DOI: 10.1097/GIM.0b013e3182141b48

Mitchell, K. S., Neale, M. C., Bulik, C. M., Aggen, S. H., Kendler, K. S., & Mazzeo, S. E. (2010). Binge eating disorder: A symptom-level investigation of genetic and environmental influences on liability. *Psychological Medicine, 40,* 1899–1906. DOI: 10.1017/S0033291710000139

Moberg, T., Lichtenstein, P., Forsman, M., & Larsson, H. (2011). Internalizing behavior in adolescent girls affects parental emotional over-involvement: A cross-lagged twin study. *Behavior*

Genetics, 41, 223–233. DOI: 10.1007/s10519-010-9383-8

Moehring, A. J., & Mackay, T. F. (2004). The quantitative genetic basis of male mating behavior in Drosophila melanogaster. *Genetics, 167,* 1249–1263. DOI: 10.1534/genetics.103.024372

Moeller, F. G., & Dougherty, D. M. (2001). Antisocial personality disorder, alcohol, and aggression. *Alcohol Research and Health, 25,* 5–11.

Moffitt, T. E. (1993). Adolescence-limited and life-course-persistent antisocial behavior: A developmental taxonomy. *Psychological Review, 100,* 674–701. DOI: 10.1534/genetics.103.024372

Moffitt, T. E. (2005). The new look of behavioral genetics in developmental psychopathology: Gene-environment interplay in antisocial behaviors. *Psychological Bulletin, 131,* 533–554. DOI: 10.1037/0033-2909.131.4.533

Monks, S. A., Leonardson, A., Zhu, H., Cundiff, P., Pietrusiak, P., Edwards, S., . . . Schadt, E. E. (2004). Genetic inheritance of gene expression in human cell lines. *American Journal of Human Genetics, 75,* 1094–1105. DOI: 10.1086/426461

Montague, C. T., Farooqi, I. S., Whitehead, J. P., Soos, M. A., Rau, H., Wareham, N. J., . . . O'Rahilly, S. (1997). Congenital leptin deficiency is associated with severe early-onset obesity in humans. *Nature, 387,* 904–908.

Moore, T., & Haig, D. (1991). Genomic imprinting in mammalian development: A parental tug-of-war. *Trends in Genetics, 7,* 45–49.

Moressis, A., Friedrich, A. R., Pavlopoulos, E., Davis, R. L., & Skoulakis, E. M. C. (2009). A dual role for the adaptor protein DRK in Drosophila olfactory learning and memory. *Journal of Neuroscience, 29,* 2611–2625. DOI: 10.1523/JNEUROSCI.3670-08.2009

Morgan, T. H., Sturtevant, A. H., Muller, H. J., & Bridges, C. B. (1915). *The mechanism of Mendelian heredity.* New York: Holt. DOI: 10.5962/bhl.title.6001

Morison, I. M., Ramsay, J. P., & Spencer, H. G. (2005). A census of mammalian imprinting. *Trends in Genetics, 21,* 457–465. DOI: 10.1016/j.tig.2005.06.008

Morley, K. I., Lynskey, M. T., Madden, P. A. F., Treloar, S. A., Heath, A. C., & Martin, N. G. (2007). Exploring the inter-relationship of smoking age-at-onset, cigarette consumption and smoking persistence: Genes or environment? *Psychological Medicine, 37,* 1357–1367. DOI: 10.1017/S0033291707000748

Morley, M., Molony, C. M., Weber, T. M., Devlin, J. L., Ewens, K. G., Spielman, R. S., & Cheung, V. G. (2004). Genetic analysis of genome-wide variation in human gene expression. *Nature, 430,* 743–747. DOI: 10.1038/nature02797

Morris, J. A., Royall, J. J., Bertagnolli, D., Boe, A. F., Burnell, J. J., Byrnes, E. J., . . . Jones, A. R. (2010). Divergent and nonuniform gene expression patterns in mouse brain. *Proceedings of the National Academy of Sciences (USA), 107,* 19049–19054. DOI: 10.1073/pnas.1003732107

Morris-Yates, A., Andrews, G., Howie, P., & Henderson, S. (1990). Twins: A test of the equal environments assumption. *Acta Psychiatrica Scandinavica, 81,* 322–326. DOI: 10.1111/j.1600-0447.1990.tb05457.x

Morrow, E. M. (2010). Genomic copy number variation in disorders of cognitive development. *Journal of the American Academy of Child and Adolescent Psychiatry, 49,* 1091–1104. DOI: 10.1016/j.jaac.2010.08.009

Mosher, L. R., Polling, W., & Stabenau, J. R. (1971). Identical twins discordant for schizophrenia: Neurological findings. *Archives of General Psychiatry, 24,* 422–430. DOI: 10.1001/archpsyc.1971.01750110034006

Mosing, M. A., Gordon, S. D., Medland, S. E., Statham, D. J., Nelson, E. C., Heath, A. C., . . . Wray, N. R. (2009). Genetic and environmental influences on the co-morbidity between depression, panic disorder, agoraphobia, and social phobia: A twin study. *Depression and Anxiety, 26,* 1004–1011. DOI: 10.1002/da.20611

Mosing, M. A., Verweij, K. J., Medland, S. E., Painter, J., Gordon, S. D., Heath, A. C., . . . Martin, N. G. (2010). A genome-wide association study of self-rated health. *Twin Research and Human Genetics, 13,* 398–403. DOI: 10.1375/twin.13.4.398

Mosing, M. A., Zietsch, B. P., Shekar, S. N., Wright, M. J., & Martin, N. G. (2009). Genetic and environmental influences on optimism and its relationship to mental and self-rated health: A study of aging twins. *Behavior Genetics, 39,* 597–604. DOI: 10.1007/s10519-009-9287-7

Mrozek-Budzyn, D., Kieltyka, A., & Majewska, R. (2010). Lack of association between measles-mumps-rubella vaccination and autism in children: A case-control study. *Pediatric Infectious Disease Journal, 29,* 397–400. DOI: 10.1097/INF.0b013e3181c40a8a

Muhle, R., Trentacoste, S. V., & Rapin, I. (2004). The genetics of autism. *Pediatrics, 113,* e472–e486. DOI: 10.1542/peds.113.5.e472

Mullineaux, P. Y., Deater-Deckard, K., Petrill, S. A., Thompson, L. A., & DeThorne, L. S. (2009). Temperament in middle childhood: A behavioral genetic analysis of fathers' and mothers' reports. *Journal of Research in Personality, 43,* 737–746. DOI: 10.1016/j.jrp.2009.04.008

Munafo, M. R., Clark, T., & Flint, J. (2005). Does measurement instrument moderate the association between the serotonin transporter gene and anxiety-related personality traits? A meta-analysis. *Molecular Psychiatry, 10,* 415–419.DOI: 10.1038/sj.mp.4001627

Munafo, M. R., Clark, T. G., Moore, L. R., Payne, E., Walton, R., & Flint, J. (2003). Genetic polymorphisms and personality in healthy adults: A systematic review and meta-analysis. *Molecular Psychiatry, 8,* 471-484. DOI: 10.1038/sj.mp.4001326

Munafo, M. R., Durrant, C., Lewis, G., & Flint, J. (2009). Gene × environment interactions at the serotonin transporter locus. *Biological Psychiatry, 65,* 211–219. DOI: 10.1016/j.biopsych.2008.06.009

Munafo, M. R., & Flint, J. (2011). Dissecting the genetic architecture of human personality. *Trends in Cognitive Sciences, 15,* 395–400. DOI: 10.1016/j.tics.2011.07.007

Murray, R. M., Lewis, S. W., & Reveley, A. M. (1985). Towards an aetiological classification of schizophrenia. *Lancet, 1,* 1023–1026. DOI: 10.1016/S0140-6736(85)91623-X

Mustelin, L., Joutsi, J., Latvala, A., Pietilainen, K. H., Rissanen, A., & Kaprio, J. (2012). Genetic influences on physical activity in young adults: A twin study. *Medicine and Science in Sports and Exercise, 44,* 1293–1301. DOI: 10.1249/MSS.0b013e3182479747

Mustelin, L., Silventoinen, K., Pietilainen, K., Rissanen, A., & Kaprio, J. (2009). Physical activity reduces the influence of genetic effects on BMI and waist circumference: A study in young adult twins. *International Journal of Obesity, 33,* 29–36. DOI: 10.1038/ijo.2008.258

Nadder, T. S., Rutter, M., Silberg, J. L., Maes, H. H., & Eaves, L. J. (2002). Genetic effects on the variation and covariation of attention deficit-hyperactivity disorder (ADHD) and oppositional-defiant disorder/conduct disorder (ODD/CD) symptomatologies across informant and occasion of measurement. *Psychological Medicine, 32,* 39–53. DOI: 10.1017/S0033291701004792

Nadler, J. J., Zou, F., Huang, H., Moy, S. S., Lauder, J., Crawley, J. N., . . . Magnuson, T. R. (2006). Large-scale gene expression differences across brain regions and inbred strains correlate with a behavioral phenotype. *Genetics, 174,* 1229–1236. DOI: 10.1534/genetics.106.061481

Narusyte, J., Andershed, A. K., Neiderhiser, J. M., & Lichtenstein, P. (2007). Aggression as a mediator of genetic contributions to the association between negative parent-child relationships and adolescent antisocial behavior. *European Child and Adolescent Psychiatry, 16,* 128–137. DOI: 10.1007/s00787-006-0582-z

Narusyte, J., Neiderhiser, J. M., Andershed, A.-K., D'Onofrio, B. M., Reiss, D., Spotts, E., . . . Lichtenstein, P. (2011). Parental criticism and externalizing behavior problems in adolescents: The role of environment and genotype-environment correlation. *Journal of Abnormal Psychology, 120,* 365–376. DOI: 10.1037/a0021815

Narusyte, J., Neiderhiser, J. M., D'Onofrio, B. M., Reiss, D., Spotts, E. L., Ganiban, J., & Lichtenstein, P. (2008). Testing different types of genotype-environment correlation: An extended children-of-twins model. *Developmental Psychology, 44,* 1591–1603. DOI: 10.1037/a0013911

Nash, M. W., Huezo-Diaz, P., Williamson, R. J., Sterne, A., Purcell, S., Hoda, F., . . . Sham, P. C. (2004). Genome-wide linkage analysis of a composite index of neuroticism and mood-related scales in extreme selected sibships. *Human Molecular Genetics, 13,* 2173–2182. DOI: 10.1093/hmg/ddh239

National Human Genome Research Institute (2012). Genome.gov. Retrieved May 23, 2012, from http://www.genome.gov/

Natsuaki, M. N., Ge, X., Leve, L. D., Neiderhiser, J. M., Shaw, D. S., Conger, R. D., . . . Reiss, D. (2010). Genetic liability, environment, and the development of fussiness in toddlers: The roles of maternal depression and parental responsiveness. *Developmental Psychology, 46,* 1147–1158. DOI: 10.1037/a0019659

Neale, B. M., Rivas, M. A., Voight, B. F., Altshuler, D., Devlin, B., Orho-Melander, M., . . . Daly, M. J. (2011). Testing for an unusual distribution of rare variants. *PLoS Genetics, 7:* e1001322. DOI: 10.1371/journal.pgen.1001322

Neale, M. C., & Stevenson, J. (1989). Rater bias in the EASI temperament scales: A twin study. *Journal of Personality and Social Psychology, 56,* 446–455. DOI: 10.1037/0022-3514.56.3.446

Need, A. C., Attix, D. K., McEvoy, J. M., Cirulli, E. T., Linney, K. L., Hunt, P., . . . Goldstein, D. B. (2009). A genome-wide study of common SNPs and CNVs in cognitive performance in the CANTAB. *Human Molecular Genetics, 18,* 4650–4661. DOI: 10.1093/hmg/ddp413

Neher, A. (2006). Evolutionary psychology: Its programs, prospects, and pitfalls. *American Journal of Psychology, 119,* 517–566. DOI: 10.2307/20445363

Neiderhiser, J. M., & McGuire, S. (1994). Competence during middle childhood. In J. C. DeFries, R. Plomin, & D. W. Fulker (Eds.), *Nature and nurture during middle childhood* (pp. 141–151). Cambridge, MA: Blackwell.

Neiderhiser, J. M., Reiss, D., Hetherington, E. M., & Plomin, R. (1999). Relationships between parenting and adolescent adjustment over time: Genetic and environmental contributions. *Developmental Psychology, 35,* 680–692. DOI: 10.1037/0012-1649.35.3.680

Neiderhiser, J. M., Reiss, D., Lichtenstein, P., Spotts, E. L., & Ganiban, J. (2007). Father-adolescent relationships and the role of genotype-environment correlation. *Journal of Family Psychology, 21,* 560–571. DOI: 10.1037/0893-3200.21.4.560

Neiderhiser, J. M., Reiss, D., Pedersen, N. L., Lichtenstein, P., Spotts, E. L., & Hansson, K. (2004). Genetic and environmental influences on mothering of adolescents: A comparison of two samples. *Developmental Psychology, 40,* 335–351. DOI: 10.1037/0012-1649.40.3.335

Neiss, M. B., Sedikides, C., & Stevenson, J. (2006). Genetic influences on level and stability of self-esteem. *Self and Identity, 5,* 247–266. DOI: 10.1080/15298860600662106

Neiss, M. B., Stevenson, J., Legrand, L. N., Iacono, W. G., & Sedikides, C. (2009). Self-esteem, negative emotionality, and depression as a common temperamental core: A study of mid-adolescent twin girls. *Journal of Personality, 77,* 327–346. DOI: 10.1111/j.1467-6494.2008.00549.x

Neisser, U. (1997). Never a dull moment. *American Psychologist, 52,* 79–81. DOI: 10.1037/0003-066X.52.1.79

Neisser, U., Boodoo, G., Bouchard, T. J., Jr., Boykin, A. W., Brody, N., Ceci, S. J., . . . Urbina, S. (1996). Intelligence: Knowns and unknowns. *American Psychologist, 51,* 77–101. DOI: 10.1037/0003-066X.51.2.77

Nelson, R. J., Demas, G. E., Huang, P. L., Fishman, M. C., Dawson, V. L., Dawson, T. M., & Snyder, S. H. (1995). Behavioural abnormalities in male mice lacking neuronal nitric oxide synthase. *Nature, 378,* 383–386. DOI: 10.1038/378383a0

Nesse, R. M., & Williams, G. C. (1996). *Why we get sick.* New York: Times Books, Random House.

Nettle, D. (2006). The evolution of personality variation in humans and other animals. *American Psychologist, 61,* 622–631. DOI: 10.1037/0003-066X.61.6.622

Neubauer, A. C., & Fink, A. (2009). Intelligence and neural efficiency. *Neuroscience and Biobehavioral Reviews, 33,* 1004–1023. DOI: 10.1016/j.neubiorev.2009.04.001

Neubauer, A. C., Sange, G., & Pfurtscheller, G. (1999). Psychometric intelligence and event-related desynchronisation during performance of a letter matching task. In G. Pfurtscheller & S. Lopes da Silva (Eds.), *Event-related desynchronisation (ERD)—and related oscillatory EEG—phenomena of the awake brain.* Amsterdam, Netherlands: Elsevier.

Neubauer, A. C., Spinath, F. M., Riemann, R., Borkenau, P., & Angleitner, A. (2000). Genetic (and environmental) influence on two measures of speed of information processing and their relation to psychometric intelligence: Evidence from the German Observational Study of Adult Twins. *Intelligence, 28,* 267–289. DOI: 10.1016/S0160-2896(00)00036-2

Neul, J. L., Kaufmann, W. E., Glaze, D. G., Christodoulou, J., Clarke, A. J., Bahi Buisson, N., . . . Percy, A. K. for the RettSearch Consortium (2010). Rett syndrome: Revised diagnostic criteria and nomenclature. *Annals of Neurology, 68,* 944–950. DOI: 10.1002/ana.22124

Newbury, D. F., Bonora, E., Lamb, J. A., Fisher, S. E., Lai, C. S. L., Baird, G., . . . the International Molecular Genetic Study of Autism Consortium. (2002). FOXP2 is not a major susceptibility gene for autism or specific language impairment. *American Journal of Human Genetics, 70,* 1318–1327. DOI: 10.1086/339931

Newcomer, J. W., & Krystal, J. H. (2001). NMDA receptor regulation of memory and behavior in humans. *Hippocampus, 11,* 529–542. DOI: 10.1002/hipo.1069

Newson, A., & Williamson, R. (1999). Should we undertake genetic research on intelligence? *Bioethics, 13,* 327–342. DOI: 10.1111/1467-8519.00161

Ng, S. B., Buckingham, K. J., Lee, C., Bigham, A. W., Tabor, H. K., Dent, K. M., . . . Bamshad, M. J. (2010). Exome sequencing identifies the cause of a Mendelian disorder. *Nature Genetics, 42,* 30–35. DOI: 10.1038/ng.499

Nichols, P. L. (1984). Familial mental retardation. *Behavior Genetics, 14,* 161–170. DOI: 10.1007/BF01065538

Nichols, R. C. (1978). Twin studies of ability, personality, and interests. *Homo, 29,* 158–173.

Nicolson, R., Brookner, F. B., Lenane, M., Gochman, P., Ingraham, L. J., Egan, M. F., . . . Rapoport, J. L. (2003). Parental schizophrenia spectrum disorders in childhood-onset and adult-onset schizophrenia. *American Journal of Psychiatry, 160,* 490–495. DOI: 10.1176/appi.ajp.160.3.490

Nicolson, R., & Rapoport, J. L. (1999). Childhood-onset schizophrenia: Rare but worth studying. *Biological Psychiatry, 46,* 1418–1428. DOI: 10.1016/S0006-3223(99)00231-0

Nigg, J. T., & Goldsmith, H. H. (1994). Genetics of personality disorders: Perspectives from personality and psychopathology research. *Psychological Bulletin, 115,* 346–380. DOI: 10.1037/0033-2909.115.3.346

Nikolas, M. A., & Burt, S. A. (2010). Genetic and environmental influences on ADHD symptom dimensions of inattention and hyperactivity: A meta-analysis. *Journal of Abnormal Psychology, 119,* 1–17. DOI: 10.1037/a0018010

Nilsson, T., Mann, M., Aebersold, R., Yates, J. R., Bairoch, A., & Bergeron, J. J. M. (2010). Mass spectrometry in high-throughput proteomics:

Ready for the big time. *Nature Methods, 7,* 681–685. DOI: 10.1038/nmeth0910-681

Novelli, V., Anselmi, C. V., Roncarati, R., Guffanti, G., Malovini, A., Piluso, G., & Puca, A. A. (2008). Lack of replication of genetic associations with human longevity. *Biogerontology, 9,* 85–92. DOI: 10.1007/s10522-007-9116-4

Nuffield Council on Bioethics. (2002). *Genetics and human behaviour: The ethical context.* London: Nuffield Council on Bioethics.

Nurnberger, J. I., Wiegand, R., Bucholz, K., O'Connor, S., Meyer, E. T., Reich, T., . . . Porjesz, B. (2004). A family study of alcohol dependence—coaggregation of multiple disorders in relatives of alcohol-dependent probands. *Archives of General Psychiatry, 61,* 1246–1256. DOI: 10.1001/archpsyc.61.12.1246

O'Connor, T. G., & Croft, C. M. (2001). A twin study of attachment in preschool children. *Child Development, 72,* 1501–1511. DOI: 10.1111/1467-8624.00362

O'Connor, T. G., Deater-Deckard, K., Fulker, D. W., Rutter, M., & Plomin, R. (1998). Genotype-environment correlations in late childhood and early adolescence: Antisocial behavioural problems in the Colorado Adoption Project. *Developmental Psychology, 34,* 970–981. DOI: 10.1037/0012-1649.34.5.970

O'Connor, T. G., Hetherington, E. M., Reiss, D., & Plomin, R. (1995). A twin-sibling study of observed parent-adolescent interactions. *Child Development, 66,* 812–829. DOI: 10.2307/1131952

O'Roak, B. J., Morgan, T. M., Fishman, D. O., Saus, E., Alonso, P., Gratacos, M., . . . State, M. W. (2010). Additional support for the association of SLITRK1 var321 and Tourette syndrome. *Molecular Psychiatry, 15,* 447–450. DOI: 10.1038/mp.2009.105

Ogden, C. L., Carroll, M. D., Kit, B. K., & Flegal, K. M. (2012). Prevalence of obesity and trends in body mass index among US children and adolescents, 1999–2010. *Journal of the American Medical Association, 307,* 483–490. DOI: 10.1001/jama.2012.40

Ogdie, M. N., Fisher, S. E., Yang, M., Ishii, J., Francks, C., Loo, S. K., . . . Nelson, S. F. (2004). Attention deficit hyperactivity disorder: Fine mapping supports linkage to 5p13, 6q12, 16p13, and 17p11. *American Journal of Human Genetics, 75,* 661–668. DOI: 10.1086/424387

Ogliari, A., Spatola, C. A., Pesenti-Gritti, P., Medda, E., Penna, L., Stazi, M. A., . . . Fagnani, C. (2010). The role of genes and environment in shaping co-occurrence of DSM-IV defined anxiety dimensions among Italian twins aged 8–17. *Journal of Anxiety Disorders, 24,* 433–439. DOI: 10.1016/j.janxdis.2010.02.008

Ohman, A., & Mineka, S. (2001). Fears, phobias, and preparedness: Toward an evolved module of fear and fear learning. *Psychological Review, 108,* 483–522. DOI: 10.1037/0033-295X.108.3.483

Oliver, B. R., Harlaar, N., Hayiou-Thomas, M. E., Kovas, Y., Walker, S. O., Petrill, S. A., . . . Plomin, R. (2004). A twin study of teacher-reported mathematics performance and low performance in 7-year-olds. *Journal of Educational Psychology, 96,* 504–517. DOI: 10.1037/0022-0663.96.3.504

Oliver, B. R., Pike, A., & Plomin, R. (2008). Nonshared environmental influences on teacher-reported behaviour problems: Monozygotic twin differences in perceptions of the classroom. *Journal of Child Psychology and Psychiatry, 49,* 646–653. DOI: 10.1111/j.1469-7610.2008.01891.x

Olson, J. M., Vernon, P. A., Harris, J. A., & Jang, K. L. (2001). The heritability of attitudes: A study of twins. *Journal of Personality and Social Psychology, 80,* 845–860. DOI: 10.1037/0022-3514.80.6.845

Olson, R. K. (2007). Introduction to the special issue on genes, environment and reading. *Reading and Writing, 20,* 1–11. DOI: 10.1007/s11145-006-9015-0

Ooki, S. (2005). Genetic and environmental influences on stuttering and tics in Japanese twin children. *Twin Research and Human Genetics, 8,* 69–75. DOI: 10.1375/twin.8.1.69

Oppenheimer, S. (2006). *The origins of the British: A genetic detective story.* London: Constable & Robinson.

Ortega-Alonso, A., Pietilainen, K. H., Silventoinen, K., Saarni, S. E., & Kaprio, J. (2012). Genetic and environmental factors influencing BMI development from adolescence to young adulthood. *Behavior Genetics, 42,* 73–85. DOI: 10.1007/s10519-011-9492-z

Ostrer, H. (2011). Changing the game with whole exome sequencing. *Clinical Genetics, 80,* 101–103. DOI: 10.1111/j.1399-0004.2011.01712.x

Ott, J., Kamatani, Y., & Lathrop, M. (2011). Family-based designs for genome-wide association studies. *Nature Reviews Genetics, 12,* 465–474. DOI: 10.1038/nrg2989

Owen, M. J., Liddle, M. B., & McGuffin, P. (1994). Alzheimer's disease: An association with apolipoprotein E4 may help unlock the puzzle. *British Medical Journal, 308,* 672–673. DOI: 10.1136/bmj.308.6930.672

Pagan, J. L., Rose, R. J., Viken, R. J., Pulkkinen, L., Kaprio, J., & Dick, D. M. (2006). Genetic and environmental influences on stages of alcohol use across adolescence and into young adulthood. *Behavior Genetics, 36,* 483–497. DOI: 10.1007/s10519-006-9062-y

Palmer, R. H., McGeary, J. E., Francazio, S., Raphael, B.J., Lander, A., Heath, A. C., & Knopik, V. S. (in press). The genetics of alcohol dependence: Advancing towards a systems-based approach. *Drug and Alcohol Dependence.*

Palmer, R. H. C., Button, T. M., Rhee, S. H., Corley, R. P., Young, S. E., Stallings, M. C., . . . Hewitt, J. K. (2012). Genetic etiology of the common liability to drug dependence: Evidence of common and specific mechanisms for DSM-IV dependence symptoms. *Drug and Alcohol Dependence, 123,* S24–S32. DOI: 10.1016/j.drugalcdep.2011.12.015

Panizzon, M. S., Lyons, M. J., Jacobson, K. C., Franz, C. E., Grant, M. D., Eisen, S. A., . . . Kremen, W. S. (2011). Genetic architecture of learning and delayed recall: A twin study of episodic memory. *Neuropsychology, 25,* 488–498. DOI: 10.1037/a0022569

Papassotiropoulos, A., Stephan, D. A., Huentelman, M. J., Hoerndli, F. J., Craig, D. W., Pearson, J. V., . . . de Quervain, D. J. (2006). Common KIBRA alleles are associated with human memory performance. *Science, 314,* 475–478. DOI: 10.1126/science.1129837

Paris, J. (1999). *Genetics and psychopathology: Predisposition-stress interactions.* Washington, DC: American Psychiatric Press.

Park, J., Shedden, K., & Polk, T. A. (2012). Correlation and heritability in neuroimaging datasets: A spatial decomposition approach with application to an fMRI study of twins. *NeuroImage, 59,* 1132–1142. DOI: 10.1016/j.neuroimage.2011.06.066

Park, J. H., Wacholder, S., Gail, M. H., Peters, U., Jacobs, K. B., Chanock, S. J., & Chatterjee, N. (2010). Estimation of effect size distribution from genome-wide association studies and implications for future discoveries. *Nature Genetics, 42,* 570–575. DOI: 10.1038/ng.610

Parker, H. G., Kim, L. V., Sutter, N. B., Carlson, S., Lorentzen, T. D., Malek, T. B., . . . Kruglyak, L. (2004). Genetic structure of the purebred domestic dog. *Science, 304,* 1160–1164. DOI: 10.1126/science.1097406

Parnas, J., Cannon, T. D., Jacobsen, B., Schulsinger, H., Schulsinger, F., & Mednick, S. A. (1993). Lifetime DSM-II-R diagnostic outcomes in the offspring of schizophrenic mothers: Results from the Copenhagen High-Risk Study. *Archives of General Psychiatry, 50,* 707–714. DOI: 10.1001/archpsyc.1993.01820210041005

Patterson, D., & Costa, A. C. (2005). Down syndrome and genetics—a case of linked histories. *Nature Reviews Genetics, 6,* 137–147. DOI: 10.1038/nrg1525

Pauls, D. L. (1990). Genetic influences on child psychiatric conditions. In M. Lewis (Ed.), *Child and adolescent psychiatry: A comprehensive textbook* (pp. 351–353). Baltimore, MD: Williams & Wilkins.

Pauls, D. L. (2003). An update on the genetics of Gilles de la Tourette syndrome. *Journal of Psychosomatic Research, 55,* 7–12. DOI: 10.1016/S0022-3999(02)00586-X

Pauls, D. L., Leckman, J. F., & Cohen, D. J. (1993). Familial relationship between Gilles de la Tourette's syndrome, attention deficit disorder, learning difficulties, speech disorders, and stuttering. *Journal of the American Academy of Child and Adolescent Psychiatry, 32,* 1044–1050. DOI: 10.1097/00004583-199309000-00025

Pauls, D. L., Towbin, K. E., Leckman, J. F., Zahner, G. E. P., & Cohen, D. J. (1986). Gilles de la Tourette's syndrome and obsessive compulsive disorder. *Archives of General Psychiatry, 43,* 1180–1182. DOI: 10.1001/archpsyc.1986.01800120066013

Paunio, T., Korhonen, T., Hublin, C., Partinen, M., Kivimäki, M., Koskenvuo, M., & Kaprio, J. (2009). Longitudinal study on poor sleep and life dissatisfaction in a nationwide cohort of twins. *American Journal of Epidemiology, 169,* 206–213. DOI: 10.1093/aje/kwn305

Pavuluri, M. N., Birmaher, B., & Naylor, M. W. (2005). Pediatric bipolar disorder: A review of the past 10 years. *Journal of the American Academy of Child and Adolescent Psychiatry, 44,* 846–871. DOI: 10.1097/01.chi.0000170554.23422.c1

Payton, A. (2009). The impact of genetic research on our understanding of normal cognitive ageing: 1995 to 2009. *Neuropsychology Review, 19,* 451–477. DOI: 10.1007/s11065-009-9116-z

Pedersen, N. L. (1996). Gerontological behavioral genetics. In J. E. Birren & K. W. Schaie (Eds.), *Handbook of the psychology of aging* (4th ed.) (pp. 59-77). San Diego: Academic Press.

Pedersen, N. L., Gatz, M., Plomin, R., Nesselroade, J. R., & McClearn, G. E. (1989). Individual differences in locus of control during the second half of the life span for identical and fraternal twins reared apart and reared together. *Journal of Gerontology, 44,* 100–105.

Pedersen, N. L., Lichtenstein, P., Plomin, R., deFaire, U., McClearn, G. E., & Matthews, K. A. (1989). Genetic and environmental influences for Type A-like measures and related traits: A study of twins reared apart and twins reared together. *Psychosomatic Medicine, 51,* 428–440.

Pedersen, N. L., McClearn, G. E., Plomin, R., & Nesselroade, J. R. (1992). Effects of early rearing environment on twin similarity in the last half of the life span. *British Journal of Developmental Psychology, 10,* 255–267. DOI: 10.1111/j.2044-835X.1992.tb00576.x

Pedersen, N. L., Plomin, R., & McClearn, G. E. (1994). Is there *G* beyond *g*? (Is there genetic influence on specific cognitive abilities independent of genetic influence on general cognitive ability?). *Intelligence, 18,* 133–143. DOI: 10.1016/0160-2896(94)90024-8

Pedersen, N. L., Plomin, R., Nesselroade, J. R., & McClearn, G. E. (1992). A quantitative genetic analysis of cognitive abilities during the second half of the life span. *Psychological Science, 3,* 346–353. DOI: 10.1111/j.1467-9280.1992.tb00045.x

Peerbooms, O. L. J., van Os, J., Drukker, M., Kenis, G., Hoogveld, L., de Hert, M., . . . Rutten, B. P. F. (2011). Meta-analysis of MTHFR gene variants in schizophrenia, bipolar disorder and unipolar depressive disorder: Evidence for a common genetic vulnerability? *Brain, Behavior, and Immunity, 25,* 1530–1543. DOI: 10.1016/j.bbi.2010.12.006

Peirce, J. L., Li, H. Q., Wang, J. T., Manly, K. F., Hitzemann, R. J., Belknap, J. K., . . . Lu, L. (2006). How replicable are mRNA expression QTL? *Mammalian Genome, 17,* 643–656. DOI: 10.1007/s00335-005-0187-8

Pemberton, C. K., Neiderhiser, J. M., Leve, L. D., Natsuaki, M. N., Shaw, D. S., Reiss, D., & Ge, X. (2010). Influence of parental depressive symptoms on adopted toddler behaviors: An emerging developmental cascade of genetic and environmental effects. *Development and Psychopathology, 22,* 803–818. DOI: 10.1017/S0954579410000477

Pennington, B. F., & Bishop, D. V. M. (2009). Relations among speech, language, and reading disorders. *Annual Review of Psychology, 60,* 283–306. DOI: 10.1146/annurev.psych.60.110707.163548

Pennington, B. F., Filipek, P. A., Lefly, D., Chhabildas, N., Kennedy, D. N., Simon, J. H., . . . DeFries, J. C. (2000). A twin MRI study of size variations in the human brain. *Journal of Cognitive Neuroscience, 12,* 223–232. DOI: 10.1162/089892900561850

Peper, J. S., Brouwer, R. M., Boomsma, D. I., Kahn, R. S., & Hulshoff Pol, H. E. (2007). Genetic influences on human brain structure: A review of brain imaging studies in twins. *Human Brain Mapping, 28,* 464–473. DOI: 10.1002/hbm.20398

Pergadia, M. L., Agrawal, A., Heath, A. C., Martin, N. G., Bucholz, K. K., & Madden, P. A. F. (2010). Nicotine withdrawal symptoms in adolescent and adult twins. *Twin Research and Human Genetics, 13,* 359–369. DOI: 10.1375/twin.13.4.359

Pergadia, M. L., Heath, A. C., Martin, N. G., & Madden, P. A. (2006). Genetic analyses of DSM-IV nicotine withdrawal in adult twins. *Psychological Medicine, 36,* 963–972. DOI: 10.1017/S0033291706007495

Pergadia, M. L., Madden, P. A., Lessov, C. N., Todorov, A. A., Bucholz, K. K., Martin, N. G., &

Heath, A. C. (2006). Genetic and environmental influences on extreme personality dispositions in adolescent female twins. *Journal of Child Psychology & Psychiatry, 47,* 902–909. DOI: 10.1111/j.1469-7610.2005.01568.x

Petersen, I. T., Bates, J. E., Goodnight, J. A., Dodge, K. A., Lansford, J. E., Pettit, G. S., . . . Dick, D. M. (2012). Interaction between serotonin transporter polymorphism (5-HTTLPR) and stressful life events in adolescents' trajectories of anxious/depressed symptoms. *Developmental Psychology.* Advance online publication. DOI: 10.1037/a0027471

Petretto, E., Mangion, J., Dickens, N. J., Cook, S. A., Kumaran, M. K., Lu, H., . . . Aitman, T. J. (2006). Heritability and tissue specificity of expression quantitative trait loci. *PLoS Genetics, 2:* e172. DOI: 10.1371/journal.pgen.0020172

Petrill, S.A. (1997). Molarity versus modularity of cognitive functioning? A behavioral genetic perspective. *Current Directions in Psychological Science, 6,* 95–99.

Petrill, S. A., Deater-Deckard, K., Thompson, L. A., Schatschneider, C., DeThorne, L. S., & Vandenbergh, D. J. (2007). Longitudinal genetic analysis of early reading: The Western Reserve Reading Project. *Reading and Writing, 20,* 127–146. DOI: 10.1007/s11145-006-9021-2

Petrill, S. A., Lipton, P. A., Hewitt, J. K., Plomin, R., Cherny, S. S., Corley, R., & DeFries, J. C. (2004). Genetic and environmental contributions to general cognitive ability through the first 16 years of life. *Developmental Psychology, 40,* 805–812. DOI: 10.1037/0012-1649.40.5.805

Petrill, S. A., Luo, D., Thompson, L. A., & Detterman, D. K. (1996). The independent prediction of general intelligence by elementary cognitive tasks: Genetic and environmental influences. *Behavior Genetics, 26,* 135–147. DOI: 10.1007/BF02359891

Petrill, S. A., Plomin, R., Berg, S., Johansson, B., Pedersen, N. L., Ahern, F., & McClearn, G. E. (1998). The genetic and environmental relationship between general and specific cognitive abilities in twins age 80 and older. *Psychological Science, 9,* 183–189. DOI: 10.1111/1467-9280.00035

Petrill, S. A., Plomin, R., DeFries, J. C., & Hewitt, J. K. (2003). *Nature, nurture, and the transition to early adolescence.* Oxford: Oxford University Press. DOI: 10.1093/acprof:oso/9780195157475.001.0001

Petrill, S. A., Saudino, K. J., Cherny, S. S., Emde, R. N., Hewitt, J. K., Fulker, D. W., & Plomin, R. (1997). Exploring the genetic etiology of low general cognitive ability from 14 to 36 months. *Developmental Psychology, 33,* 544–548. DOI: 10.1037/0012-1649.33.3.544

Petrill, S. A., Thompson, L. A., & Detterman, D. K. (1995). The genetic and environmental variance underlying elementary cognitive tasks. *Behavior Genetics, 25,* 199–209. DOI: 10.1007/BF02197178

Petronis, A. (2006). Epigenetics and twins: Three variations on the theme. *Trends in Genetics, 22,* 347–350. DOI: 10.1016/j.tig.2006.04.010

Petronis, A. (2010). Epigenetics as a unifying principle in the aetiology of complex traits and diseases. *Nature, 465,* 721–727. DOI: 10.1038/nature09230

Pharoah, P. D. P., Antoniou, A., Bobrow, M., Zimmern, R. L., Easton, D. F., & Ponder, B. A. J. (2002). Polygenic susceptibility to breast cancer and implications for prevention. *Nature Genetics, 31,* 33–36. DOI: 10.1038/ng853

Phelps, E. A., & LeDoux, J. E. (2005). Contributions of the amygdala to emotion processing: From animal models to human behavior. *Neuron, 48,* 175–187. DOI: 10.1016/j.neuron.2005.09.025

Phelps, J. A., Davis, O. J., & Schwartz, K. M. (1997). Nature, nurture and twin research strategies. *Current Directions in Psychological Science, 6,* 117–121. DOI: 10.1111/1467-8721.ep10772877

Phillips, D. I. W. (1993). Twin studies in medical research: Can they tell us whether diseases are genetically determined? *Lancet, 341,* 1008–1009. DOI: 10.1016/0140-6736(93)91086-2

Phillips, K., & Matheny, A. P., Jr. (1995). Quantitative genetic analysis of injury liability in infants and toddlers. *American Journal of Medical Genetics. B: Neuropsychiatric Genetics, 60,* 64–71. DOI: 10.1002/ajmg.1320600112

Phillips, K., & Matheny, A. P., Jr. (1997). Evidence for genetic influence on both cross-situation and situation-specific components of behavior. *Journal of Personality and Social Psychology, 73,* 129–138. DOI: 10.1037/0022-3514.73.1.129

Phillips, T. J., Belknap, J. K., Buck, K. J., & Cunningham, C. L. (1998). Genes on mouse chromosomes 2 and 9 determine variation in ethanol consumption. *Mammalian Genome, 9,* 936–941. DOI: 10.1007/s003359900903

Phillips, T. J., Reed, C., Burkhart-Kasch, S., Li, N., Hitzemann, R., Yu, C.-H., . . . Belknap, J. K. (2010). A method for mapping intralocus interactions influencing excessive alcohol drinking. *Mammalian Genome, 21,* 39–51. DOI: 10.1007/s00335-009-9239-9

Picchioni, M. M., Walshe, M., Toulopoulou, T., McDonald, C., Taylor, M., Waters-Metenier, S., . . . Rijsdijk, F. (2010). Genetic modelling of childhood social development and personality in twins and siblings with schizophrenia. *Psychological Medicine, 40,* 1305–1316. DOI: 10.1017/S0033291709991425

Pietilainen, K. H., Kaprio, J., Rissanen, A., Winter, T., Rimpela, A., Viken, R. J., & Rose, R. J. (1999). Distribution and heritability of BMI in Finnish adolescents aged 16y and 17y: A study of 4884 twins and 2509 singletons. *International Journal of Obesity and Related Metabolic Disorders, 23,* 107–115.

Pike, A., & Atzaba-Poria, N. (2003). Do sibling and friend relationships share the same temperamental origins? A twin study. *Journal of Child Psychology and Psychiatry, 44,* 598–611. DOI: 10.1111/1469-7610.00148

Pike, A., McGuire, S., Hetherington, E. M., Reiss, D., & Plomin, R. (1996). Family environment and adolescent depressive symptoms and antisocial behavior: A multivariate genetic analysis. *Developmental Psychology, 32,* 590–603. DOI: 10.1037/0012-1649.32.4.590

Pike, A., Reiss, D., Hetherington, E. M., & Plomin, R. (1996). Using MZ differences in the search for nonshared environmental effects. *Journal of Child Psychology and Psychiatry, 37,* 695–704. DOI: 10.1111/j.1469-7610.1996.tb01461.x

Pillard, R. C., & Bailey, J. M. (1998). Human sexual orientation has a heritable component. *Human Biology, 70,* 347–365.

Pinker, S. (1994). *The language instinct.* New York: William Morrow & Co, Inc.

Pinker, S. (2002). *The blank slate: The modern denial of human nature.* New York: Penguin.

Pinker, S. (2010). The cognitive niche: Coevolution of intelligence, sociality, and language. *Proceedings of the National Academy of Sciences (USA), 107,* 8993–8999. DOI: 10.1073/pnas.0914630107

Pischon, T., Boeing, H., Hoffmann, K., Bergmann, M., Schulze, M. B., Overvad, K., . . . Riboli, E. (2008). General and abdominal adiposity and risk of death in Europe. *New England Journal of Medicine, 359,* 2105–2120. DOI: 10.1056/NEJMoa0801891

Platek, S., & Shackelford, T. (2006). *Female infidelity and paternal uncertainty: Evolutionary perspectives on male anti-cuckoldry tactics.* Cambridge: Cambridge University Press. DOI: 10.1017/CBO9780511617812

Plomin, R. (1986). *Development, genetics, and psychology.* Hillsdale, NJ: Erlbaum.

Plomin, R. (1987). Developmental behavioral genetics and infancy. In J. Osofsky (Ed.), *Handbook of infant development* (2nd ed.) (pp. 363–417). New York: Interscience.

Plomin, R. (1988). The nature and nurture of cognitive abilities. In R. J. Sternberg (Ed.), *Advances in the psychology of human intelligence* (4th ed.) (pp. 1–33). Hillsdale, NJ: Lawrence Erlbaum Associates.

Plomin, R. (1991). Genetic risk and psychosocial disorders: Links between the normal and abnormal. In M. Rutter & P. Casaer (Eds.), *Biological risk factors for psychosocial disorders* (pp. 101–138). Cambridge: Cambridge University Press.

Plomin, R. (1994). *Genetics and experience: The interplay between nature and nurture.* Thousand Oaks, CA: Sage Publications.

Plomin, R. (1999). Genetics and general cognitive ability. *Nature, 402,* C25–C29. DOI: 10.1038/35011520

Plomin, R. (2001). The genetics of *g* in human and mouse. *Nature Reviews Neuroscience, 2,* 136–141. DOI: 10.1038/35053584

Plomin, R. (2011). Commentary: Why are children in the same family so different? Non-shared environment three decades later. *International Journal of Epidemiology, 40,* 582–592. DOI: 10.1093/ije/dyq144

Plomin, R., Asbury, K., & Dunn, J. (2001). Why are children in the same family so different? Non-shared environment a decade later. *Canadian Journal of Psychiatry, 46,* 225–233.

Plomin, R., & Bergeman, C. S. (1991). The nature of nurture: Genetic influence on "environmental" measures. *Behavioral and Brain Sciences, 14,* 373–414. DOI: 10.1017/S0140525X00070278

Plomin, R., & Caspi, A. (1999). Behavioral genetics and personality. In L. A. Pervin & O. P. John (Eds.), *Handbook of personality: Theory and research* (2nd ed.) (pp. 251–276). New York: Guilford Press.

Plomin, R., Chipuer, H. M., & Loehlin, J. C. (1990). Behavioral genetics and personality. In L. A. Pervin (Ed.), *Handbook of personality: Theory and research* (pp. 225–243). New York: Guilford.

Plomin, R., Coon, H., Carey, G., DeFries, J. C., & Fulker, D. W. (1991). Parent-offspring and sibling adoption analyses of parental ratings of temperament in infancy and childhood. *Journal of Personality, 59,* 705–732. DOI: 10.1111/j.1467-6494.1991.tb00928.x

Plomin, R., Corley, R., Caspi, A., Fulker, D. W., & DeFries, J. C. (1998). Adoption results for self-reported personality: Evidence for nonadditive genetic effects? *Journal of Personality and Social Psychology, 75,* 211–218. DOI: 10.1037/0022-3514.75.1.211

Plomin, R., & Crabbe, J. C. (2000). DNA. *Psychological Bulletin, 126,* 806–828. DOI: 10.1037/0033-2909.126.6.806

Plomin, R., & Craig, I. W. (1997). Human behavioral genetics of cognitive abilities and disabilities. *BioEssays, 19,* 1117–1124. DOI: 10.1002/bies.950191211

Plomin, R., & Daniels, D. (1987). Why are children in the same family so different from each other? *Behavioral and Brain Sciences, 10,* 1–16. DOI: 10.1017/S0140525X00055941

Plomin, R., & Davis, O. S. P. (2006). Gene-environment interactions and correlations in the development of cognitive abilities and disabilities. In J. MacCabe, O. O'Daly, R. M. Murray, P. McGuffin,

& P. Wright (Eds.), *Beyond nature and nurture: Genes, environment and their interplay in psychiatry* (pp. 35–45). Andover, UK: Thomson Publishing Services.

Plomin, R., & DeFries, J. C. (1985). A parent-offspring adoption study of cognitive abilities in early childhood. *Intelligence, 9,* 341–356. DOI: 10.1016/0160-2896(85)90019-4

Plomin, R., & DeFries, J. C. (1998). The genetics of cognitive abilities and disabilities. *Scientific American, 278(5),* 62–69. DOI: 10.1038/scientificamerican0598-62

Plomin, R., DeFries, J. C., & Fulker, D. W. (1988). *Nature and nurture during infancy and early childhood.* Cambridge, UK: Cambridge University Press. DOI: 10.1017/CBO9780511527654

Plomin, R., DeFries, J. C., & Loehlin, J. C. (1977a). Assortative mating by unwed biological parents of adopted children. *Science, 196,* 449–450. DOI: 10.1126/science.850790

Plomin, R., DeFries, J. C., & Loehlin, J. C. (1977b). Genotype-environment interaction and correlation in the analysis of human behavior. *Psychological Bulletin, 84,* 309–322. DOI: 10.1037/0033-2909.84.2.309

Plomin, R., Emde, R. N., Braungart, J. M., Campos, J., Corley, R., Fulker, D. W., . . . DeFries, J. C. (1993). Genetic change and continuity from fourteen to twenty months: The MacArthur Longitudinal Twin Study. *Child Development, 64,* 1354–1376. DOI: 10.2307/1131539

Plomin, R., & Foch, T. T. (1980). A twin study of objectively assessed personality in childhood. *Journal of Personality and Social Psychology, 38,* 680–688. DOI: 10.1037/0022-3514.39.4.680

Plomin, R., Foch, T. T., & Rowe, D. C. (1981). Bobo clown aggression in childhood: Environment not genes. *Journal of Research in Personality, 14,* 331–342. DOI: 10.1016/0092-6566(81)90031-3

Plomin, R., Fulker, D. W., Corley, R., & DeFries, J. C. (1997). Nature, nurture and cognitive development from 1 to 16 years: A parent-offspring adoption study. *Psychological Science, 8,* 442–447. DOI: 10.1111/j.1467-9280.1997.tb00458.x

Plomin, R., Haworth, C. M. A., & Davis, O. S. P. (2009). Common disorders are quantitative traits. *Nature Reviews Genetics, 10,* 872–878. DOI: 10.1038/nrg2670

Plomin, R., Hill, L., Craig, I., McGuffin, P., Purcell, S., Sham, P., . . . Owen, M. J. (2001). A genome-wide scan of 1842 DNA markers for allelic associations with general cognitive ability: A five-stage design using DNA pooling and extreme selected groups. *Behavior Genetics, 31,* 497–509. DOI: 10.1023/A:1013385125887

Plomin, R., & Kovas, Y. (2005). Generalist genes and learning disabilities. *Psychological Bulletin, 131,* 592–617. DOI: 10.1037/0033-2909.131.4.592

Plomin, R., Lichtenstein, P., Pedersen, N. L., McClearn, G. E., & Nesselroade, J. R. (1990). Genetic influence on life events during the last half of the life span. *Psychology and Aging, 5,* 25–30. DOI: 10.1037/0882-7974.5.1.25

Plomin, R., Loehlin, J. C., & DeFries, J. C. (1985). Genetic and environmental components of "environmental" influences. *Developmental Psychology, 21,* 391–402. DOI: 10.1037/0012-1649.21.3.391

Plomin, R., & McClearn, G. E. (1990). Human behavioral genetics of aging. In J. E. Birren & K. W. Schaie (Eds.), *Handbook of the psychology of aging* (pp. 66–77). New York: Academic Press.

Plomin, R., & McClearn, G. E. (1993). Quantitative trait loci (QTL) analyses and alcohol-related behaviors. *Behavior Genetics, 23,* 197–211. DOI: 10.1007/BF01067425

Plomin, R., Pedersen, N. L., Lichtenstein, P., & McClearn, G. E. (1994). Variability and stability in cognitive abilities are largely genetic later in life. *Behavior Genetics, 24,* 207–215. DOI: 10.1007/BF01067188

Plomin, R., Reiss, D., Hetherington, E. M., & Howe, G. W. (1994). Nature and nurture: Genetic contributions to measures of the family environment. *Developmental Psychology, 30,* 32–43. DOI: 10.1037/0012-1649.30.1.32

Plomin, R., & Schalkwyk, L. C. (2007). Microarrays. *Developmental Science, 10,* 19–23. DOI: 10.1111/j.1467-7687.2007.00558.x

Plomin, R., & Spinath, F. M. (2002). Genetics and general cognitive ability (*g*). *Trends in Cognitive Sciences, 6,* 169–176. DOI: 10.1016/S1364-6613(00)01853-2

Poelmans, G., Buitelaar, J. K., Pauls, D. L., & Franke, B. (2011). A theoretical molecular network for dyslexia: Integrating available genetic findings. *Molecular Psychiatry, 16,* 365–382. DOI: 10.1038/mp.2010.105

Pogue-Geile, M. F., & Rose, R. J. (1985). Developmental genetic studies of adult personality. *Developmental Psychology, 21,* 547–557. DOI: 10.1037/0012-1649.21.3.547

Polanczyk, G., de Lima, M. S., Horta, B. L., Biederman, J., & Rohde, L. A. (2007). The worldwide prevalence of ADHD: A systematic review and metaregression analysis. *American Journal of Psychiatry, 164,* 942–948. DOI: 10.1176/appi.ajp.164.6.942

Pollak, D. D., John, J., Hoeger, H., & Lubec, G. (2006). An integrated map of the murine hippocampal proteome based upon five mouse strains. *Electrophoresis, 27,* 2787–2798. DOI: 10.1002/elps.200600648

Pollak, D. D., John, J., Schneider, A., Hoeger, H., & Lubec, G. (2006). Strain-dependent expression of signaling proteins in the mouse hippocampus. *Neuroscience, 138,* 149–158. DOI: 10.1016/j.neuroscience.2005.11.004

Poorthuis, R. B., Goriounova, N. A., Couey, J. J., & Mansvelder, H. D. (2009). Nicotinic actions on neuronal networks for cognition: General principles and long-term consequences. *Biochemical Pharmacology, 78,* 668–676. DOI: 10.1016/j.bcp.2009.04.031

Posthuma, D., de Geus, E. J. C., Baare, W. F. C., Pol, H. E. H., Kahn, R. S., & Boomsma, D. I. (2002). The association between brain volume and intelligence is of genetic origin. *Nature Neuroscience, 5,* 83–84. DOI: 10.1038/nn0202-83

Posthuma, D., de Geus, E. J. C., & Boomsma, D. I. (2001). Perceptual speed and IQ are associated through common genetic factors. *Behavior Genetics, 31,* 593–602. DOI: 10.1023/A:1013349512683

Posthuma, D., de Geus, E. J. C., Mulder, E. J., Smit, D. J., Boomsma, D. I., & Stam, C. J. (2005). Genetic components of functional connectivity in the brain: The heritability of synchronization likelihood. *Human Brain Mapping, 26,* 191–198. DOI: 10.1002/hbm.20156

Posthuma, D., Luciano, M., Geus, E. J., Wright, M. J., Slagboom, P. E., Montgomery, G. W., ... Martin, N. G. (2005). A genomewide scan for intelligence identifies quantitative trait loci on 2q and 6p. *American Journal of Human Genetics, 77,* 318–326. DOI: 10.1086/432647

Posthuma, D., Neale, M. C., Boomsma, D. I., & de Geus, E. J. (2001). Are smarter brains running faster? Heritability of alpha peak frequency, IQ, and their interrelation. *Behavior Genetics, 31,* 567–579. DOI: 10.1023/A:1013345411774

Prescott, C. A. (2002). Sex differences in the genetic risk for alcoholism. *Alcohol Research and Health, 26,* 264–273.

Prescott, C. A., Sullivan, P. F., Kuo, P. H., Webb, B. T., Vittum, J., Patterson, D. G., ... Kendler, K. S. (2006). Genomewide linkage study in the Irish affected sib pair study of alcohol dependence: Evidence for a susceptibility region for symptoms of alcohol dependence on chromosome 4. *Molecular Psychiatry, 11,* 603–611. DOI: 10.1038/sj.mp.4001811

Price, R. A., Kidd, K. K., Cohen, D. J., Pauls, D. L., & Leckman, J. F. (1985). A twin study of Tourette syndrome. *Archives of General Psychiatry, 42,* 815–820. DOI: 10.1001/archpsyc.1985.01790310077011

Price, T. S., Dale, P. S., & Plomin, R. (2004). A longitudinal genetic analysis of low verbal and nonverbal cognitive abilities in early childhood. *Twin Research, 7,* 139–148. DOI: 10.1375/136905204323016122

Price, T. S., Simonoff, E., Kuntsi, J., Curran, S., Asherson, P., Waldman, I., & Plomin, R. (2005). Continuity and change in preschool ADHD symptoms: Longitudinal genetic analysis with contrast effects. *Behavior Genetics, 35,* 121–132. DOI: 10.1007/s10519-004-1013-x

Profet, M. (1992). Pregnancy sickness as adaptation; a deterrent to maternal ingestion on teratogens. In J. Barkow, L. Cosmides, & J. Tooby (Eds.), *The adapted mind* (pp. 327–366). New York: Oxford University Press.

Propping, P. (1987). Single gene effects in psychiatric disorders. In F. Vogel & K. Sperling (Eds.), *Human genetics: Proceedings of the 7th International Congress, Berlin* (pp. 452–457). New York: Springer.

Purcell, S. (2002). Variance components models for gene-environment interaction in twin analysis. *Twin Research, 5,* 554–571. DOI: 10.1375/136905202762342026

Purcell, S., & Koenen, K. C. (2005). Environmental mediation and the twin design. *Behavior Genetics, 35,* 491–498. DOI: 10.1007/s10519-004-1484-9

Purcell, S. M., Wray, N. R., Stone, J. L., Visscher, P. M., O'Donovan, M. C., Sullivan, P. F., . . . Moran, J. L. (2009). Common polygenic variation contributes to risk of schizophrenia and bipolar disorder. *Nature, 460,* 748–752. DOI: 10.1038/nature08185

Puts, D. A. (2010). Beauty and the beast: Mechanisms of sexual selection in humans. *Evolution and Human Behavior, 31,* 157–175. DOI: 10.1016/j.evolhumbehav.2010.02.005

Puts, D. A., Jones, B. C., & DeBruine, L. M. (2012). Sexual selection on human faces and voices. *Journal of Sex Research, 49,* 227–243. DOI: 10.1080/00224499.2012.658924

Qi, L., Kraft, P., Hunter, D. J., & Hu, F. B. (2008). The common obesity variant near MC4R gene is associated with higher intakes of total energy and dietary fat, weight change and diabetes risk in women. *Human Molecular Genetics, 17,* 3502–3508. DOI: 10.1093/hmg/ddn242

Raiha, I., Kapiro, J., Koskenvuo, M., Rajala, T., & Sourander, L. (1996). Alzheimer's disease in Finnish twins. *Lancet, 347,* 573–578. DOI: 10.1016/S0140-6736(96)91272-6

Raine, A. (1993). *The psychopathology of crime: Criminal behavior as a clinical disorder.* San Diego, CA: Academic Press.

Rakyan, V. K., Down, T. A., Balding, D. J., & Beck, S. (2011). Epigenome-wide association studies for common human diseases. *Nature Reviews Genetics, 12,* 529–541. DOI: 10.1038/nrg3000

Ramachandrappa, S., & Farooqi, I. S. (2011). Genetic approaches to understanding human obesity. *Journal of Clinical Investigation, 121,* 2080–2086. DOI: 10.1172/JCI46044

Rankin, C. H. (2002). From gene to identified neuron to behaviour in Caenorhabditis elegans. *Nature Reviews Genetics, 3,* 622–630. DOI: 10.1038/nrg864

Rankinen, T., Zuberi, A., Chagnon, Y. C., Weisnagel, S. J., Argyropoulos, G., Walts, B., . . . Bouchard, C. (2006). The human obesity gene map: The 2005 update. *Obesity, 14,* 529–644. DOI: 10.1038/oby.2006.71

Rasetti, R., & Weinberger, D. R. (2011). Intermediate phenotypes in psychiatric disorders. *Current Opinion in Genetics and Development, 21,* 340–348. DOI: 10.1016/j.gde.2011.02.003

Rasmussen, E. R., Neuman, R. J., Heath, A. C., Levy, F., Hay, D. A., & Todd, R. D. (2004). Familial clustering of latent class and DSM-IV defined attention-deficit/hyperactivity disorder (ADHD) subtypes. *Journal of Child Psychology and Psychiatry, 45,* 589–598. DOI: 10.1111/j.1469-7610.2004.00248.x

Rasmussen, S. A., & Tsuang, M. T. (1984). The epidemiology of obsessive compulsive disorder. *Journal of Clinical Psychiatry, 45,* 450–457.

Ratcliffe, S. G. (1994). The psychological and psychiatric consequences of sex chromosome abnormalities in children, based on population studies. In F. Poustka (Ed.), *Basic approaches to genetic and molecular-biological developmental psychiatry* (pp. 92–122). Berlin: Quintessenz Library of Psychiatry.

Raymond, F. L. (2010). Monogenic causes of mental retardation. In S. J. L. Knight (Ed.), *Genetics of mental retardation: An overview encompassing learning disability and intellectual disability. Monographs in Human Genetics, 18,* 89–100.

Read, S., Vogler, G. P., Pedersen, N. L., & Johansson, B. (2006). Stability and change in genetic and environmental components of personality in old age. *Personality and Individual Differences, 40,* 1637–1647. DOI: 10.1016/j.paid.2006.01.004

Redon, R., Ishikawa, S., Fitch, K. R., Feuk, L., Perry, G. H., Andrews, T. D., . . . Hurles, M. E. (2006). Global variation in copy number in the human genome. *Nature, 444,* 444–454. DOI: 10.1038/nature05329

Reed, E. W., & Reed, S. C. (1965). *Mental retardation: A family study.* Philadelphia, PA: Saunders.

Reich, T., Edenberg, H. J., Goate, A., Williams, J., Rice, J., Van Eerdewegh, P., . . . Begleiter, H. (1998). Genome-wide search for genes affecting the risk for alcohol dependence. *American Journal of Medical Genetics, 81,* 207–215. DOI: 10.1002/(SICI)1096-8628(19980508)81:3<207::AID-AJMG1>3.0.CO;2-T

Reik, W., & Walter, J. (2001). Genomic imprinting: Parental influence on the genome. *Nature Reviews Genetics, 2,* 21–32. DOI: 10.1038/35047554

Reiss, D., Neiderhiser, J. M., Hetherington, E. M., & Plomin, R. (2000). *The relationship code: Deciphering genetic and social patterns in adolescent development.* Cambridge, MA: Harvard University Press.

Rettew, D. C., Vink, J. M., Willemsen, G., Doyle, A., Hudziak, J. J., & Boomsma, D. I. (2006). The genetic architecture of neuroticism in 3301 Dutch adolescent twins as a function of age and sex: A study from the Dutch Twin Register. *Twin Research and Human Genetics, 9,* 24–29. DOI: 10.1375/twin.9.1.24

Reyes, A., Haynes, M., Hanson, N., Angly, F. E., Heath, A. C., Rohwer, F., & Gordon, J. I. (2010). Viruses in the faecal microbiota of monozygotic twins and their mothers. *Nature, 466,* 334–338. DOI: 10.1038/nature09199

Reynolds, C. A., Prince, J. A., Feuk, L., Brookes, A. J., Gatz, M., & Pedersen, N. L. (2006). Longitudinal memory performance during normal aging: Twin association models of APOE and other Alzheimer candidate genes. *Behavior Genetics, 36,* 185–194. DOI: 10.1007/s10519-005-9027-6

Rhee, S. H., Hewitt, J. K., Young, S. E., Corley, R. P., Crowley, T. J., & Stallings, M. C. (2003). Genetic and environmental influences on substance initiation, use, and problem use in adolescents. *Archives of General Psychiatry, 60,* 1256–1264. DOI: 10.1001/archpsyc.60.12.1256

Rhee, S. H., & Waldman, I. D. (2002). Genetic and environmental influences on antisocial behavior: A meta-analysis of twin and adoption studies. *Psychological Bulletin, 128,* 490–529. DOI: 10.1037/0033-2909.128.3.490

Rhoades, K. A., Leve, L. D., Harold, G. T., Neiderhiser, J. M., Shaw, D. S., & Reiss, D. (2011). Longitudinal pathways from marital hostility to child anger during toddlerhood: Genetic susceptibility and indirect effects via harsh parenting. *Journal of Family Psychology, 25,* 28–291. DOI: 10.1037/a0022886

Rhodes, G. (2006). The evolutionary psychology of facial beauty. *Annual Review of Psychology, 57,* 199–226. DOI: 10.1146/annurev.psych.57.102904.190208

Rice, G., Anderson, C., Risch, N., & Ebers, G. (1999). Male homosexuality: Absence of linkage to microsatellite markers at Xq28. *Science, 284,* 665–667. DOI: 10.1126/science.284.5414.665

Richards, E. J. (2006). Inherited epigenetic variation—revisiting soft inheritance. *Nature Reviews Genetics, 7,* 395–401. DOI: 10.1038/nrg1834

Riemann, R., Angleitner, A., & Strelau, J. (1997). Genetic and environmental influences on personality: A study of twins reared together using the self- and peer report NEO-FFI scales. *Journal of Personality, 65,* 449–476. DOI: 10.1111/j.1467-6494.1997.tb00324.x

Riese, M. L. (1990). Neonatal temperament in monozygotic and dizygotic twin pairs. *Child Development, 61,* 1230–1237. DOI: 10.2307/1130890

Riese, M. L. (1999). Effects of chorion type on neonatal temperament differences in monozygotic pairs. *Behavior Genetics, 29,* 87–94. DOI: 10.1023/A:1021604321243

Rietveld, M. J., Dolan, C. V., van Baal, G. C., & Boomsma, D. I. (2003). A twin study of differentiation of cognitive abilities in childhood. *Behavior Genetics, 33,* 367–381. DOI: 10.1023/A:1025388908177

Rietveld, M. J., Hudziak, J. J., Bartels, M., Van Beijsterveldt, C. E., & Boomsma, D. I. (2004). Heritability of attention problems in children: Longitudinal results from a study of twins, age 3 to 12. *Journal of Child Psychology and Psychiatry, 45,* 577–588. DOI: 10.1111/j.1469-7610.2004.00247.x

Rietveld, M. J., Posthuma, D., Dolan, C. V., & Boomsma, D. I. (2003). ADHD: Sibling interaction or dominance: An evaluation of statistical power. *Behavior Genetics, 33,* 247–255. DOI: 10.1023/A:1023490307170

Rijsdijk, F. V., & Boomsma, D. I. (1997). Genetic mediation of the correlation between peripheral nerve conduction velocity and IQ. *Behavior Genetics, 27,* 87–98. DOI: 10.1023/A:1025600423013

Rijsdijk, F. V., Boomsma, D. I., & Vernon, P. A. (1995). Genetic analysis of peripheral nerve conduction velocity in twins. *Behavior Genetics, 25,* 341–348. DOI: 10.1007/BF02197283

Rijsdijk, F. V., Vernon, P. A., & Boomsma, D. I. (1998). The genetic basis of the relation between speed-of-information-processing and IQ. *Behavioural Brain Research, 95,* 77–84. DOI: 10.1016/S0166-4328(97)00212-X

Rijsdijk, F. V., Vernon, P. A., & Boomsma, D. I. (2002). Application of hierarchical genetic models to Raven and WAIS subtests: A Dutch twin study. *Behavior Genetics, 32,* 199–210. DOI: 10.1023/A:1016021128949

Rijsdijsk, F. V., Viding, E., De Brito, S., Forgiarini, M., Mechelli, A., Jones, A. P., & McCrory, E. (2010). Heritable variations in gray matter concentration as a potential endophenotype for psychopathic traits. *Archives of General Psychiatry, 67,* 406–413. DOI: 10.1001/archgenpsychiatry.2010.20

Riley, B., & Kendler, K. S. (2006). Molecular genetic studies of schizophrenia. *European Journal of Human Genetics, 14,* 669–680. DOI: 10.1038/sj.ejhg.5201571

Rimol, L. M., Panizzon, M. S., Fennema-Notestine, C., Eyler, L. T., Fischl, B., Franz, C.

E., . . . Dale, A. M. (2010). Cortical thickness is influenced by regionally specific genetic factors. *Biological Psychiatry, 67,* 493–499. DOI: 10.1016/j.biopsych.2009.09.032

Ripke, S., Sanders, A. R., Kendler, K. S., Levinson, D. F., Sklar, P., Holmans, P. A., . . . Schizophrenia Psychiatric Genome-wide Association Study (GWAS) Consortium. (2011). Genome-wide association study identifies five new schizophrenia loci. *Nature Genetics, 43,* 969–976. DOI: 10.1038/ng.940

Risch, N., Herrell, R., Lehner, T., Liang, K. Y., Eaves, L., Hoh, J., . . . Merikangas, K. R. (2009). Interaction between the serotonin transporter gene (5-HTTLPR), stressful life events, and risk of depression: A meta-analysis. *Journal of the American Medical Association, 301,* 2462–2471. DOI: 10.1001/jama.2009.878

Risch, N., & Merikangas, K. R. (1996). The future of genetic studies of complex human diseases. *Science, 273,* 1516–1517. DOI: 10.1126/science.273.5281.1516

Risch, N. J. (2000). Searching for genetic determinants in the new millennium. *Nature, 405,* 847–856. DOI: 10.1038/35015718

Ritsner, M. S. (2009). *The handbook of neuropsychiatric biomarkers, endophenotypes and genes: Vol. 1. Neuropsychological endophenotypes and biomarkers.* New York: Springer.

Ritsner, M. S., & Gottesman, I. I. (2011). The schizophrenia construct after 100 years of challenges. In M. S. Ritsner (Ed.), *Handbook of schizophrenia spectrum disorders: Conceptual issues and neurobiological advances.* New York: Springer. DOI: 10.1007/978-94-007-0837-2_1

Robbers, S. C. C., van Oort, F. V. A., Polderman, T. J. C., Bartels, M., Boomsma, D. I., Verhulst, F. C., . . . Huizink, A. C. (2011). Trajectories of CBCL attention problems in childhood. *European Child and Adolescent Psychiatry, 20,* 419–427. DOI: 10.1007/s00787-011-0194-0

Roberts, C. A., & Johansson, C. B. (1974). The inheritance of cognitive interest styles among twins. *Journal of Vocational Behavior, 4,* 237–243. DOI: 10.1016/0001-8791(74)90107-9

Roberts, S. C., & Little, A. C. (2008). Good genes, complementary genes and human mate preferences. *Genetica, 132,* 309–321. DOI: 10.1007/s10709-007-9174-1

Robins, L. N. (1978). Sturdy childhood predictors of adult antisocial behaviour: Replications from longitudinal analyses. *Psychological Medicine, 8,* 611–622. DOI: 10.1017/S0033291700018821

Robins, L. N., & Price, R. K. (1991). Adult disorders predicted by childhood conduct problems: Results from the NIMH Epidemiologic Catchment Area project. *Psychiatry, 54,* 116–132.

Robins, L. N., & Regier, D. A. (1991). *Psychiatric disorders in America.* New York: Free Press.

Robinson, D. G., Woerner, M. G., McMeniman, M., Mendelowitz, A., & Bilder, R. M. (2004). Symptomatic and functional recovery from a first episode of schizophrenia or schizoaffective disorder. *American Journal of Psychiatry, 161,* 473–479. DOI: 10.1176/appi.ajp.161.3.473

Robinson, E. B., Koenen, K. C., McCormick, M. C., Munir, K., Hallett, V., Happé, F., . . . Ronald, A. (2011). Evidence that autistic traits show the same etiology in the general population and at the quantitative extremes (5%, 2.5%, and 1%). *Archives of General Psychiatry, 68,* 1113–1121. DOI: 10.1001/archgenpsychiatry.2011.119

Robinson, J. L., Kagan, J., Reznick, J. S., & Corley, R. (1992). The heritability of inhibited and uninhibited behavior: A twin study. *Developmental Psychology, 28,* 1030–1037. DOI: 10.1037/0012-1649.28.6.1030

Rockman, M. V., & Kruglyak, L. (2006). Genetics of global gene expression. *Nature Reviews Genetics, 7,* 862–872. DOI: 10.1038/nrg1964

Roesti, M., Hendry, A. P., Salzburger, W., & Berner, D. (2012). Genome divergence during evolutionary diversification as revealed in replicate lake–stream stickleback population pairs. *Molecular Ecology, 21,* 2852–2862. DOI: 10.1111/j.1365-294X.2012.05509.x

Roisman, G. I., & Fraley, R. C. (2006). The limits of genetic influence: A behavior-genetic analysis of infant-caregiver relationship quality and temperament. *Child Development, 77,* 1656–1667. DOI: 10.1111/j.1467-8624.2006.00965.x

Roizen, N. J., & Patterson, D. (2003). Down syndrome. *Lancet, 361,* 1281–1289. DOI: 10.1016/S0140-6736(03)12987-X

Romeis, J. C., Grant, J. D., Knopik, V. S., Pedersen, N. L., & Heath, A. C. (2004). The genetics of middle-age spread in middle-class males. *Twin Research, 7,* 596–602. DOI: 10.1375/1369052042663896

Ronald, A., Butcher, L., Docherty, S., Davis, O., Schalkwyk, L., Craig, I., & Plomin, R. (2010). A genome-wide association study of social and non-social autistic-like traits in the general population using pooled DNA, 500k SNP microarrays and an autism replication sample. *Behavior Genetics, 40,* 31–45. DOI: 10.1007/s10519-009-9308-6

Ronald, A., Happé, F., Price, T. S., Baron-Cohen, S., & Plomin, R. (2006). Phenotypic and genetic overlap between autistic traits at the extremes of the general population. *Journal of the American Academy of Child and Adolescent Psychiatry, 45,* 1206–1214. DOI: 10.1097/01.chi.0000230165.54117.41

Ronald, A., & Hoekstra, R. A. (2011). Autism spectrum disorders and autistic traits: A decade of new twin studies. *American Journal of Medical Genetics. B: Neuropsychiatric Genetics, 156B,* 255–274. DOI: 10.1002/ajmg.b.31159

Ronald, A., Larsson, H., Anckarsater, H., & Lichtenstein, P. (2011). A twin study of autism symptoms in Sweden. *Molecular Psychiatry, 16,* 1039–1047. DOI: 10.1038/mp.2010.82

Ronalds, G. A., De Stavola, B. L., & Leon, D. A. (2005). The cognitive cost of being a twin: Evidence from comparisons within families in the Aberdeen children of the 1950s cohort study. *British Medical Journal, 331,* 1306. DOI: 10.1136/bmj.38633.594387.3A

Rooms, L., & Kooy, R. F. (2011). Advances in understanding fragile X syndrome and related disorders. *Current Opinion in Pediatrics, 23,* 601–606. DOI: 10.1097/MOP.0b013e32834c7f1a

Root, T. L., Thornton, L. M., Lindroos, A. K., Stunkard, A. J., Lichtenstein, P., Pedersen, N. L., . . Bulik, C. M. (2010). Shared and unique genetic and environmental influences on binge eating and night eating: A Swedish twin study. *Eating Behaviors, 11,* 92–98. DOI: 10.1016/j.eatbeh.2009.10.004

Rose, R. J., Broms, U., Korhonen, T., Dick, D. M., & Kaprio, J. (2009). Genetics of smoking behavior. In Y.-K. Kim (Ed.), *Handbook of behavior genetics* (pp. 411–432). New York: Springer. DOI: 10.1007/978-0-387-76727-7_28

Rose, R. J., Dick, D. M., Viken, R. J., Pulkkinen, L., & Kaprio, J. (2004). Genetic and environmental effects on conduct disorder and alcohol dependence symptoms and their covariation at age 14. *Alcoholism: Clinical and Experimental Research, 28,* 1541–1548. DOI: 10.1097/01.ALC.0000141822.36776.55

Rosenberg, R. E., Law, J. K., Yenokyan, G., McGready, J., Kaufmann, W. E., & Law, P. A. (2009). Characteristics and concordance of autism spectrum disorders among 277 twin pairs. *Archives of Pediatrics and Adolescent Medicine, 163,* 907–914. DOI: 10.1001/archpediatrics.2009.98

Rosenthal, D., Wender, P. H., Kety, S. S., & Schulsinger, F. (1971). The adopted-away offspring of schizophrenics. *American Journal of Psychiatry, 128,* 307–311.

Rosenthal, D., Wender, P. H., Kety, S. S., Schulsinger, F., Welner, J., & Ostergaard, L. (1968). Schizophrenics' offspring reared in adoptive homes. *Journal of Psychiatric Research, 6,* 377–391. DOI: 10.1016/0022-3956(68)90028-9

Rosenthal, N. E., Sack, D. A., Gillin, J. C., Lewy, A. J., Goodwin, F. K., Davenport, Y., . . . Wehr, T. A. (1984). Seasonal affective disorder: A description of the syndrome and preliminary findings with light therapy. *Archives of General Psychiatry, 41,* 72–80. DOI: 10.1001/archpsyc.1984.01790120076010

Roses, A. D. (2000). Pharmacogenetics and the practice of medicine. *Nature, 405,* 857–865. DOI: 10.1038/35015728

Ross, C. A., & Tabrizi, S. J. (2011). Huntington's disease: From molecular pathogenesis to clinical treatment. *Lancet Neurology, 10,* 83–98. DOI: 10.1016/S1474-4422(10)70245-3

Ross, M. T., Grafham, D. V., Coffey, A. J., Scherer, S., McLay, K., Muzny, D., . . . Yen, J. (2005). The DNA sequence of the human X chromosome. *Nature, 434,* 325–337. DOI: 10.1038/nature03440

Rothe, C., Koszycki, D., Bradwejn, J., King, N., Deluca, V., Tharmalingam, S., . . . Kennedy, J. L. (2006). Association of the Val158Met catechol O-methyltransferase genetic polymorphism with panic disorder. *Neuropsychopharmacology, 31,* 2237–2242. DOI: 10.1038/sj.npp.1301048

Rowe, D. C. (1981). Environmental and genetic influences on dimensions of perceived parenting: A twin study. *Developmental Psychology, 17,* 203–208. DOI: 10.1037/0012-1649.17.2.203

Rowe, D. C. (1983a). A biometrical analysis of perceptions of family environment: A study of twin and singleton sibling relationships. *Child Development, 54,* 416–423. DOI: 10.2307/1129702

Rowe, D. C. (1983b). Biometrical genetic models of self-reported delinquent behavior: A twin study. *Behavior Genetics, 13,* 473–489. DOI: 10.1007/BF01065923

Rowe, D. C. (1987). Resolving the person-situation debate: Invitation to an interdisciplinary dialogue. *American Psychologist, 42,* 218–227. DOI: 10.1037/0003-066X.42.3.218

Rowe, D. C. (1994). *The limits of family influence: Genes, experience, and behaviour.* New York: Guilford Press.

Rowe, D. C., Jacobson, K. C., & van den Oord, E. J. (1999). Genetic and environmental influences on vocabulary IQ: Parental education level as moderator. *Child Development, 70,* 1151–1162. DOI: 10.1111/1467-8624.00084

Rowe, D. C., Vesterdal, W. J., & Rodgers, J. L. (1999). Herrnstein's syllogism: Genetic and shared environmental influences on IQ, education, and income. *Intelligence, 26,* 405–423. DOI: 10.1016/S0160-2896(99)00008-2

Roy, K., Stein, L., & Kaushal, S. (2010). Ocular gene therapy: An evaluation of recombinant adeno-associated virus-mediated gene therapy interventions for the treatment of ocular disease. *Human Gene Therapy, 21,* 915–927. DOI: 10.1089/hum.2010.041

Roy, M. A., Neale, M. C., & Kendler, K. S. (1995). The genetic epidemiology of self-esteem. *British Journal of Psychiatry, 166,* 813–820. DOI: 10.1192/bjp.166.6.813

Roysamb, E., Tambs, K., Reichborn-Kjennerud, T., Neale, M. C., & Harris, J. R. (2003). Happiness and health: Environmental and genetic contributions to the relationship between subjective well-being, perceived health, and somatic illness. *Journal of Personality and Social Psychology, 85,* 1136–1146. DOI: 10.1037/0022-3514.85.6.1136

Rubanyi, G. M. (2001). The future of human gene therapy. *Molecular Aspects of Medicine, 22,* 113–142. DOI: 10.1016/S0098-2997(01)00004-8

Rubinstein, M., Phillips, T. J., Bunzow, J. R., Falzone, T. L., Dziewczapolski, G., Zhang, G., . . . Grandy, D. K. (1997). Mice lacking dopamine D4 receptors are supersensitive to ethanol, cocaine, and methamphetamine. *Cell, 90,* 991–1001. DOI: 10.1016/S0092-8674(00)80365-7

Rush, A. J., & Weissenburger, J. E. (1994). Melancholic symptom features and DSM-IV. *American Journal of Psychiatry, 151,* 489–498.

Rushton, J. P. (2002). New evidence on Sir Cyril Burt: His 1964 speech to the Association of Educational Psychologists. *Intelligence, 30,* 555–567. DOI: 10.1016/S0160-2896(02)00094-6

Rushton, J. P. (2004). Genetic and environmental contributions to pro-social attitudes: A twin study of social responsibility. *Proceedings of the Royal Society of London. B: Biological Sciences, 271,* 2583–2585. DOI: 10.1098/rspb.2004.2941

Rushton, J. P., & Bons, T. A. (2005). Mate choice and friendship in twins: Evidence for genetic similarity. *Psychological Science, 16,* 555–559. DOI: 10.1111/j.0956-7976.2005.01574.x

Rutherford, J., McGuffin, P., Katz, R. J., & Murray, R. M. (1993). Genetic influences on eating attitudes in a normal female twin population. *Psychological Medicine, 23,* 425–436. DOI: 10.1017/S003329170002852X

Rutter, M. (1996). Introduction: Concepts of antisocial behavior, of cause, and of genetic influences. In G. R. Bock & J. A. Goode (Eds.), *Ciba Foundation Symposium 194—Genetics of criminal and antisocial behaviour* (pp. 1–15). Chichester, UK: John Wiley & Sons.DOI: 10.1002/9780470514825.ch1

Rutter, M. (2005a). Aetiology of autism: Findings and questions. *Journal of Intellectual Disability Research, 49,* 231–238. DOI: 10.1111/j.1365-2788.2005.00676.x

Rutter, M. (2005b). Environmentally mediated risks for psychopathology: Research strategies and findings. *Journal of the American Academy of Child and Adolescent Psychiatry, 44,* 3–18. DOI: 10.1097/01.chi.0000145374.45992.c9

Rutter, M. (2006). *Genes and behavior: Nature-nurture interplay explained.* Oxford: Blackwell Publishing.

Rutter, M., Maughan, B., Meyer, J., Pickles, A., Silberg, J., Simonoff, E., & Taylor, E. (1997). Heterogeneity of antisocial behavior: Causes, continuities, and consequences. In R. Dienstbier & D. W. Osgood (Eds.), *Nebraska symposium on motivation: Vol. 44. Motivation and delinquency* (pp. 45–118). Lincoln: University of Nebraska Press.

Rutter, M., Moffitt, T. E., & Caspi, A. (2006). Gene-environment interplay and psychopathology: Multiple varieties but real effects. *Journal of Child Psychology and Psychiatry, 47,* 226–261. DOI: 10.1111/j.1469-7610.2005.01557.x

Rutter, M., & Redshaw, J. (1991). Annotation: Growing up as a twin: Twin-singleton differences in psychological development. *Journal of Child Psychology and Psychiatry, 32,* 885–895. DOI: 10.1111/j.1469-7610.1991.tb01916.x

Rutter, M., Silberg, J., O'Connor, T. G., & Simonoff, E. (1999). Genetics and child psychiatry. II. Empirical research findings. *Journal of Child Psychology and Psychiatry, 40,* 19–55. DOI: 10.1111/1469-7610.00423

Saad, G. (2007). *The evolutionary bases of consumption.* Mahwah, NJ: Lawrence Erlbaum Associates.

Saccone, N. L., Culverhouse, R. C., Schwantes-An, T.-H., Cannon, D. S., Chen, X., Cichon, S., . . . Bierut, L. J. (2010). Multiple independent loci at chromosome 15q25.1 affect smoking quantity: A meta-analysis and comparison with lung cancer and COPD. *PLoS Genetics, 6:* e1001053. DOI: 10.1371/journal.pgen.1001053

Saccone, S. F., Pergadia, M. L., Loukola, A., Broms, U., Montgomery, G. W., Wang, J. C., . . . Madden, P. A. F. (2007). Genetic linkage to chromosome 22q12 for a heavy-smoking quantitative trait in two independent samples. *American Journal of Human Genetics, 80,* 856–866. DOI: 10.1086/513703

Sadler, M. E., Miller, C. J., Christensen, K., & McGue, M. (2011). Subjective wellbeing and longevity: A co-twin control study. *Twin Research and Human Genetics, 14,* 249–256. DOI: 10.1375/twin.14.3.249

Saha, S., Chant, D., Welham, J., & McGrath, J. (2005). A systematic review of the prevalence of schizophrenia. *PLoS Medicine, 2:* e141. DOI: 10.1371/journal.pmed.0020141

Sahoo, T., Theisen, A., Rosenfeld, J. A., Lamb, A. N., Ravnan, J. B., Schultz, R. A., . . . Shaffer, L. G. (2011). Copy number variants of schizophrenia susceptibility loci are associated with a spectrum of speech and developmental delays and behavior problems. *Genetics in Medicine, 13,* 868–880. DOI: 10.1097/GIM.0b013e3182217a06

Salahpour, A., Medvedev, I. O., Beaulieu, J. M., Gainetdinov, R. R., & Caron, M. G. (2007). Local knockdown of genes in the brain using small interfering RNA: A phenotypic comparison with knockout animals. *Biological Psychiatry, 61,* 65–69. DOI: 10.1016/j.biopsych.2006.03.020

Salsberry, P. J., & Reagan, P. B. (2010). Effects of heritability, shared environment, and non shared intrauterine conditions on child and adolescent BMI. *Obesity, 18,* 1775–1780. DOI: 10.1038/oby.2009.485

Samaco, R. C., & Neul, J. L. (2011). Complexities of Rett syndrome and MeCP2. *Journal of Neuroscience, 31,* 7951–7959. DOI: 10.1523/JNEURO SCI.0169-11.2011

Sambandan, D., Yamamoto, A., Fanara, J. J., Mackay, T. F., & Anholt, R. R. (2006). Dynamic genetic interactions determine odor-guided behavior in Drosophila melanogaster. *Genetics, 174,* 1349–1363. DOI: 10.1534/genetics.106.060574

Samuelsson, S., Byrne, B., Olson, R. K., Hulslander, J., Wadsworth, S., Corley, R., . . . DeFries, J. C. (2008). Response to early literacy instruction in the United States, Australia, and Scandinavia: A behavioral-genetic analysis. *Learning and Individual Differences, 18,* 289–295. DOI: 10.1016/j.lindif.2008.03.004

Samuelsson, S., Olson, R., Wadsworth, S., Corley, R., DeFries, J. C., Willcutt, E. . . . Byrne, B. (2007). Genetic and environmental influences on prereading skills and early reading and spelling development in the United States, Australia, and Scandinavia. *Reading and Writing, 20,* 51–75. DOI: 10.1007/s11145-006-9018-x

Saudino, K. J. (2005). Behavioral genetics and child temperament. *Journal of Developmental and Behavioral Pediatrics, 26,* 214–223. DOI: 10.1097/00004703-200506000-00010

Saudino, K. J. (2012). Sources of continuity and change in activity level in early childhood. *Child Development, 83,* 266–281. DOI: 10.1111/j.1467-8624.2011.01680.x

Saudino, K. J., & Eaton, W. O. (1991). Infant temperament and genetics: An objective twin study of motor activity level. *Child Development, 62,* 1167–1174. DOI: 10.2307/1131160

Saudino, K. J., McGuire, S., Reiss, D., Hetherington, E. M., & Plomin, R. (1995). Parent ratings of EAS temperaments in twins, full siblings, half siblings, and step siblings. *Journal of Personality and Social Psychology, 68,* 723–733. DOI: 10.1037/0022-3514.68.4.723

Saudino, K. J., Pedersen, N. L., Lichtenstein, P., McClearn, G. E., & Plomin, R. (1997). Can personality explain genetic influences on life events? *Journal of Personality and Social Psychology, 72,* 196–206. DOI: 10.1037/0022-3514.72.1.196

Saudino, K. J., & Plomin, R. (1997). Cognitive and temperamental mediators of genetic contributions to the home environment during infancy. *Merrill-Palmer Quarterly, 43,* 1–23.

Saudino, K. J., Plomin, R., & DeFries, J. C. (1996). Tester-rated temperament at 14, 20, and 24 months: Environmental change and genetic continuity. *British Journal of Developmental Psychology, 14,* 129–144. DOI: 10.1111/j.2044-835X.1996.tb00697.x

Saudino, K. J., Plomin, R., Pedersen, N. L., & McClearn, G. E. (1994). The etiology of high and low cognitive ability during the second half of the life span. *Intelligence, 19,* 353–371. DOI: 10.1016/0160-2896(94)90007-8

Saudino, K. J., Ronald, A., & Plomin, R. (2005). The etiology of behavior problems in 7-year-old twins: Substantial genetic influence and negligible shared environmental influence for parent ratings and ratings by same and different teachers. *Journal of Abnormal Child Psychology, 33,* 113–130. DOI: 10.1007/s10802-005-0939-7

Saudino, K. J., Wertz, A. E., Gagne, J. R., & Chawla, S. (2004). Night and day: Are siblings as different in temperament as parents say they are? *Journal of Personality and Social Psychology, 87,* 698–706. DOI: 10.1037/0022-3514.87.5.698

Saudou, F., Amara, D. A., Dierich, A., Lemeur, M., Ramboz, S., Segu, L., . . . Hen, R. (1994). Enhanced aggressive-behavior in mice lacking 5-HT1B receptor. *Science, 265,* 1875–1878. DOI: 10.1126/science.8091214

Savelieva, K. V., Caudle, W. M., Findlay, G. S., Caron, M. G., & Miller, G. W. (2002). Decreased ethanol preference and consumption in dopamine transporter female knock-out mice. *Alcoholism: Clinical and Experimental Research, 26,* 758–764. DOI: 10.1111/j.1530-0277.2002.tb02602.x

Scarr, S., & Carter-Saltzman, L. (1979). Twin method: Defense of a critical assumption. *Behavior Genetics, 9,* 527–542. DOI: 10.1007/BF01067349

Scarr, S., & McCartney, K. (1983). How people make their own environments: A theory of genotype → environmental effects. *Child Development, 54,* 424–435.

Scarr, S., & Weinberg, R. A. (1978a). Attitudes, interests, and IQ. *Human Nature, 1,* 29–36.

Scarr, S., & Weinberg, R. A. (1978b). The influence of "family background" on intellectual attainment. *American Sociological Review, 43,* 674–692. DOI: 10.2307/2094543

Scarr, S., & Weinberg, R. A. (1981). The transmission of authoritarianism in familes: Genetic resemblance in social-politial attitudes? In S. Scarr (Ed.), *Race, social class, and individual differences in IQ* (pp. 399–427). Hillsdale, NJ: Erlbaum.

Scerri, T. S., Morris, A. P., Buckingham, L.-L., Newbury, D. F., Miller, L. L., Monaco, A. P.,

... Paracchini, S. (2011). DCDC2, KIAA0319 and CMIP are associated with reading-related traits. *Biological Psychiatry, 70,* 237–245. DOI: 10.1016/j.biopsych.2011.02.005

Schadt, E. E. (2006). Novel integrative genomics strategies to identify genes for complex traits. *Animal Genetics, 37,* 18–23. DOI: 10.1111/j.1365-2052.2006.01473.x

Schafer, W. R. (2005). Deciphering the neural and molecular mechanisms of C. elegans behavior. *Current Biology, 15,* R723–R729.

Scharf, J. M., Moorjani, P., Fagerness, J., Platko, J. V., Illmann, C., Galloway, B., . . . The Tourette Syndrome International Consortium for Genetics (2008). Lack of association between SLITRK1 var321 and Tourette syndrome in a large family-based sample. *Neurology, 70,* 1495–1496. DOI: 10.1212/01.wnl.0000296833.25484.bb

Schermerhorn, A. C., D'Onofrio, B. M., Turkheimer, E., Ganiban, J. M., Spotts, E. L., Lichtenstein, P., . . . Neiderhiser, J. M. (2011). A genetically informed study of associations between family functioning and child psychosocial adjustment. *Developmental Psychology, 47,* 707–725. DOI: 10.1037/a0021362

Scherrer, J. F., True, W. R., Xian, H., Lyons, M. J., Eisen, S. A., Goldberg, J., . . . Tsuang, M. T. (2000). Evidence for genetic influences common and specific to symptoms of generalized anxiety and panic. *Journal of Affective Disorders, 57,* 25–35. DOI: 10.1016/S0165-0327(99)00031-2

Schlaggar, B. L. (2011). Mapping genetic influences on cortical regionalization. *Neuron, 72,* 499–501. DOI: 10.1016/j.neuron.2011.10.024

Schmidt, F. L., & Hunter, J. (2004). General mental ability in the world of work: Occupational attainment and job performance. *Journal of Personality and Social Psychology, 86,* 162–173. DOI: 10.1037/0022-3514.86.1.162

Schmitt, J. E., Prescott, C. A., Gardner, C. O., Neale, M. C., & Kendler, K. S. (2005). The differential heritability of regular tobacco use based on method of administration. *Twin Research and Human Genetics, 8,* 60–62. DOI: 10.1375/1832427053435346

Schmitt, J. E., Wallace, G. L., Lenroot, R. K., Ordaz, S. E., Greenstein, D., Clasen, L., . . . Giedd, J. N. (2010). A twin study of intracerebral volumetric relationships. *Behavior Genetics, 40,* 114–124. DOI: 10.1007/s10519-010-9332-6

Schmitz, S. (1994). Personality and temperament. In J. C. DeFries & R. Plomin (Eds.), *Nature and nurture during middle childhood* (pp. 120–140). Cambridge, MA: Blackwell.

Schoenfeldt, L. F. (1968). The hereditary components of the Project TALENT two-day test battery. *Measurement and Evaluation in Guidance, 1,* 130–140.

Scholz, H., Ramond, J., Singh, C. M., & Heberlein, U. (2000). Functional ethanol tolerance in drosophila. *Neuron, 28,* 261–271. DOI: 10.1016/S0896-6273(00)00101-X

Schousboe, K., Willemsen, G., Kyvik, K. O., Mortensen, J., Boomsma, D. I., Cornes, B. K., . . . Harris, J. R. (2003). Sex differences in heritability of BMI: A comparative study of results from twin studies in eight countries. *Twin Research, 6,* 409–421. DOI: 10.1375/136905203770326411

Schulsinger, F. (1972). Psychopathy: Heredity and environment. *International Journal of Mental Health, 1,* 190–206.

Schulze, T. G., Hedeker, D., Zandi, P., Rietschel, M., & McMahon, F. J. (2006). What is familial about familial bipolar disorder? Resemblance among relatives across a broad spectrum of phenotypic characteristics. *Archives of General Psychiatry, 63,* 1368–1376. DOI: 10.1001/archpsyc.63.12.1368

Schumann, G., Loth, E., Banaschewski, T., Barbot, A., Barker, G., Buchel, C., . . . IMAGEN Consortium. (2010). The IMAGEN study: Reinforcement-related behaviour in normal brain function and psychopathology. *Molecular Psychiatry, 15,* 1128–1139. DOI: 10.1038/mp.2010.4

Schur, E., Afari, N., Goldberg, J., Buchwald, D., & Sullivan, P. F. (2007). Twin analyses of fatigue. *Twin Research and Human Genetics, 10,* 729–733. DOI: 10.1375/twin.10.5.729

Scott, J. P., & Fuller, J. L. (1965). *Genetics and the social behavior of the dog.* Chicago: University of Chicago Press.

Scriver, C. R. (2007). The PAH gene, phenylketonuria, and a paradigm shift. *Human Mutation, 28,* 831–845. DOI: 10.1002/humu.20526

Scriver, C. R., & Waters, P. J. (1999). Monogenetic traits are not simple: Lessons from phenylketonuria. *Trends in Genetics, 15,* 267–272. DOI: 10.1016/S0168-9525(99)01761-8

Seale, T. W. (1991). Genetic differences in response to cocaine and stimulant drugs. In J. C. Crabbe & R. A. Harris (Eds.), *The genetic basis of alcohol and drug actions* (pp. 279–321). New York: Plenum.

Sebat, J., Lakshmi, B., Malhotra, D., Troge, J., Lese-Martin, C., Walsh, T., . . . Wigler, M. (2007). Strong association of de novo copy number mutations with autism. *Science, 316,* 445–449. DOI: 10.1126/science.1138659

Segal, N. L. (1999). *Entwined lives: Twins and what they tell us about human behavior.* New York: Dutton.

Segal, N. L., & MacDonald, K. B. (1998). Behavioral genetics and evolutionary psychology:

Unified perspective on personality research. *Human Biology, 70,* 159–184.

Segurado, R., Detera-Wadleigh, S. D., Levinson, D. F., Lewis, C. M., Gill, M., Nurnberger, J. I., Jr., Akarsu, N. (2003). Genome scan meta-analysis of schizophrenia and bipolar disorder, Part III: Bipolar disorder. *American Journal of Human Genetics 73,* 49–62. DOI: 10.1086/376547

Serretti, A., Calati, R., Mandelli, L., & De Ronchi, D. (2006). Serotonin transporter gene variants and behavior: A comprehensive review. *Current Drug Targets, 7,* 1659–1669. DOI: 10.2174/138945006779025419

Sesardic, N. (2005). *Making sense of heritability.* Cambridge: Cambridge University Press. DOI: 10.1017/CBO9780511487378

Sham, P. C., Cherny, S. S., Purcell, S., & Hewitt, J. (2000). Power of linkage versus association analysis of quantitative traits, by use of variance-components models, for sibship data. *American Journal of Human Genetics, 66,* 1616–1630. DOI: 10.1086/302891

Shapira, N. A., Lessig, M. C., He, A. G., James, G. A., Driscoll, D. J., & Liu, Y. (2005). Satiety dysfunction in Prader-Willi syndrome demonstrated by fMRI. *Journal of Neurology, Neurosurgery and Psychiatry, 76,* 260–262. DOI: 10.1136/jnnp.2004.039024

Sharma, A., Sharma, V. K., Horn-Saban, S., Lancet, D., Ramachandran, S., & Brahmachari, S. K. (2005). Assessing natural variations in gene expression in humans by comparing with monozygotic twins using microarrays. *Physiological Genomics, 21,* 117–123. DOI: 10.1152/physiolgenomics.00228.2003

Sharp, S. I., McQuillin, A., & Gurling, H. M. D. (2009). Genetics of attention-deficit hyperactivity disorder (ADHD). *Neuropharmacology, 57,* 590–600. DOI: 10.1016/j.neuropharm.2009.08.011

Shaw, P., Greenstein, D., Lerch, J., Clasen, L., Lenroot, R., Gogtay, N., . . . Giedd, J. (2006). Intellectual ability and cortical development in children and adolescents. *Nature, 440,* 676–679. DOI: 10.1038/nature04513

Sher, L., Goldman, D., Ozaki, N., & Rosenthal, N. E. (1999). The role of genetic factors in the etiology of seasonal affective disorder and seasonality. *Journal of Affective Disorders, 53,* 203–210. DOI: 10.1016/S0165-0327(98)00194-3

Sherrington, R., Brynjolfsson, J., Petursson, H., Potter, M., Dudleston, K., Barraclough, B., . . . Gurling, H. (1988). Localisation of susceptibility locus for schizophrenia on chromosome 5. *Nature, 336,* 164–167. DOI: 10.1038/336164a0

Shi, J. X., Levinson, D. F., Duan, J. B., Sanders, A. R., Zheng, Y. L., Pe'er, I., . . . Gejman, P. V. (2009). Common variants on chromosome 6p22.1 are associated with schizophrenia. *Nature, 460,* 753–757. DOI: 10.1038/nature08192

Shih, R. A., Belmonte, P. L., & Zandi, P. P. (2004). A review of the evidence from family, twin and adoption studies for a genetic contribution to adult psychiatric disorders. *International Review of Psychiatry, 16,* 260–283. DOI: 10.1080/09540260400014401

Shilyansky, C., Lee, Y. S., & Silva, A. J. (2010). Molecular and cellular mechanisms of learning disabilities: A focus on NF1. *Annual Review of Neuroscience, 33,* 221–243. DOI: 10.1146/annurev-neuro-060909-153215

Siever, L. J., Silverman, K. M., Horvath, T. B., Klar, H., Coccaro, E., Keefe, R. S. E., . . . Davis, K. L. (1990). Increased morbid risk for schizophrenia-related disorders in relatives of schizotypal personality disordered patients. *Archives of General Psychiatry, 47,* 634–640. DOI: 10.1001/archpsyc.1990.01810190034005

Sigvardsson, S., Bohman, M., & Cloninger, C. R. (1996). Replication of Stockholm Adoption Study of alcoholism. *Archives of General Psychiatry, 53,* 681–687. DOI: 10.1001/archpsyc.1996.01830080033007

Silberg, J., Pickles, A., Rutter, M., Hewitt, J., Simonoff, E., Maes, H., . . . Eaves, L. (1999). The influence of genetic factors and life stress on depression among adolescent girls. *Archives of General Psychiatry, 56,* 225–232. DOI: 10.1001/archpsyc.56.3.225

Silberg, J. L., Maes, H., & Eaves, L. J. (2010). Genetic and environmental influences on the transmission of parental depression to children's depression and conduct disturbance: An extended children of twins study. *Journal of Child Psychology and Psychiatry, 51,* 734–744. DOI: 10.1111/j.1469-7610.2010.02205.x

Silberg, J. L., Rutter, M., & Eaves, L. (2001). Genetic and environmental influences on the temporal association between earlier anxiety and later depression in girls. *Biological Psychiatry, 49,* 1040–1049. DOI: 10.1016/S0006-3223(01)01161-1

Silberg, J. L., Rutter, M. L., Meyer, J., Maes, H., Hewitt, J., Simonoff, E., . . . Eaves, L. (1996). Genetic and environmental influences on the covariation between hyperactivity and conduct disturbance in juvenile twins. *Journal of Child Psychology and Psychiatry, 37,* 803–816. DOI: 10.1111/j.1469-7610.1996.tb01476.x

Silva, A. J., Kogan, J. H., Frankland, P. W., & Kida, S. (1998). CREB and memory. *Annual Review of Neuroscience, 21,* 127–148. DOI: 10.1146/annurev.neuro.21.1.127

Silva, A. J., Paylor, R., Wehner, J. M., & Tonegwa, S. (1992). Impaired spatial learning in α-calcium-calmodulin kinase mutant mice. *Science, 257,* 206–211. DOI: 10.1126/science.1321493

Silventoinen, K., Hasselbalch, A. L., Lallukka, T., Bogl, L., Pietilainen, K. H., Heitmann, B. L., . . . Kaprio, J. (2009). Modification effects of physical activity and protein intake on heritability of body size and composition. *American Journal of Clinical Nutrition, 90,* 1096–1103. DOI: 10.3945/ajcn.2009.27689

Silventoinen, K., & Kaprio, J. (2009). Genetics of tracking of body mass index from birth to late middle age: Evidence from twin and family studies. *Obesity Facts, 2,* 196–202. DOI: 10.1159/000219675

Silventoinen, K., Rokholm, B., Kaprio, J., & Sorensen, T. I. A. (2010). The genetic and environmental influences on childhood obesity: A systematic review of twin and adoption studies. *International Journal of Obesity, 34,* 29–40. DOI: 10.1038/ijo.2009.177

Simm, P. J., & Zacharin, M. R. (2006). The psychosocial impact of Klinefelter syndrome—a 10 year review. *Journal of Pediatric Endocrinology and Metabolism, 19,* 499–505.

Singer, J. J., Falchi, M., MacGregor, A. J., Cherkas, L. F., & Spector, T. D. (2006). Genome-wide scan for prospective memory suggests linkage to chromosome 12q22. *Behavior Genetics, 36,* 18–28. DOI: 10.1007/s10519-005-9011-1

Singer, J. J., MacGregor, J. J., Cherkas, L. F., & Spector, T. D. (2006). Genetic influences on cognitive function using The Cambridge Neuropsychological Test Automated Battery. *Intelligence, 34,* 421–428. DOI: 10.1016/j.intell.2005.11.005

Singh, A. L., D'Onofrio, B. M., Slutske, W. S., Turkheimer, E., Emery, R. E., Harden, K. P., . . . Martin, N. G. (2011). Parental depression and offspring psychopathology: A children of twins study. *Psychological Medicine, 41,* 1385–1395. DOI: 10.1017/S0033291710002059

Siontis, K. C. M., Patsopoulos, N. A., & Ioannidis, J. P. A. (2010). Replication of past candidate loci for common diseases and phenotypes in 100 genome-wide association studies. *European Journal of Human Genetics, 18,* 832–837. DOI: 10.1038/ejhg.2010.26

Sison, M., & Gerlai, R. (2010). Associative learning in zebrafish (Danio rerio) in the plus maze. *Behavioural Brain Research, 207,* 99–104. DOI: 10.1016/j.bbr.2009.09.043

Skelly, D. A., Ronald, J., & Akey, J. M. (2009). Inherited variation in gene expression. *Annual Review of Genomics and Human Genetics, 10,* 313–332. DOI: 10.1146/annurev-genom-082908-150121

Skelton, J. A., Irby, M. B., Grzywacz, J. G., & Miller, G. (2011). Etiologies of obesity in children: Nature and nurture. *Pediatric Clinics of North America, 58,* 1333–1354. DOI: 10.1016/j.pcl.2011.09.006

Skodak, M., & Skeels, H. M. (1949). A final follow-up on one hundred adopted children. *Journal of Genetic Psychology, 75,* 84–125.

Skoulakis, E. M. C., & Grammenoudi, S. (2006). Dunces and Da Vincis: The genetics of learning and memory in Drosophila. *Cellular and Molecular Life Sciences, 63,* 975–988. DOI: 10.1007/s00018-006-6023-9

Slater, E., & Cowie, V. (1971). *The genetics of mental disorders.* London: Oxford University Press.

Slof-Op 't Landt, M. C., van Furth, E. F., Meulenbelt, I., Slagboom, P. E., Bartels, M., Boomsma, D. I., & Bulik, C. M. (2005). Eating disorders: From twin studies to candidate genes and beyond. *Twin Research and Human Genetics, 8,* 467–482. DOI: 10.1375/183242705774310114

Smalley, S. L., Asarnow, R. F., & Spence, M. A. (1988). Autism and genetics: A decade of research. *Archives of General Psychiatry, 45,* 953–961. DOI: 10.1001/archpsyc.1988.01800340081013

Smeeth, L., Cook, C., Fombonne, E., Heavey, L., Rodrigues, L. C., Smith, P. G., & Hall, A. J. (2004). MMR vaccination and pervasive developmental disorders: A case-control study. *Lancet, 364,* 963–969. DOI: 10.1016/S0140-6736(04)17020-7

Smith, C. (1974). Concordance in twins: Methods and interpretation. *American Journal of Human Genetics, 26,* 454–466.

Smith, D. L. (2004). *Why we lie: The evolutionary roots of deception and the unconscious mind.* New York: St Martin's Press.

Smith, E. M., North, C. S., McColl, R. E., & Shea, J. M. (1990). Acute postdisaster psychiatric disorders: Identification of persons at risk. *American Journal of Psychiatry, 147,* 202–206.

Smith, J., Cianflone, K., Biron, S., Hould, F. S., Lebel, S., Marceau, S., . . . Marceau, P. (2009). Effects of maternal surgical weight loss in mothers on intergenerational transmission of obesity. *Journal of Clinical Endocrinology and Metabolism, 94,* 4275–4283. DOI: 10.1210/jc.2009-0709

Smith, S. D., Grigorenko, E., Willcutt, E., Pennington, B. F., Olson, R. K., & DeFries, J. C. (2010). Etiologies and molecular mechanisms of communication disorders. *Journal of Developmental and Behavioral Pediatrics, 31,* 555–563. DOI: 10.1097/DBP.0b013e3181ee3d9e

Smith, T. F. (2010). Meta-analysis of the heterogeneity in association of DRD4 7-repeat allele and AD/HD: Stronger association with AD/HD combined type. *American Journal of Medical Genetics. B: Neuropsychiatric Genetics, 153B,* 1189–1199. DOI: 10.1002/ajmg.b.31090

Smits, B. M., & Cuppen, E. (2006). Rat genetics: The next episode. *Trends in Genetics, 22,* 232–240. DOI: 10.1016/j.tig.2006.02.009

Smoller, J. W., Block, S. R., & Young, M. M. (2009). Genetics of anxiety disorders: The complex road from DSM to DNA. *Depression and Anxiety, 26,* 965–975. DOI: 10.1002/da.20623

Smoller, J. W., & Finn, C. T. (2003). Family, twin, and adoption studies of bipolar disorder. *American Journal of Medical Genetics. C: Seminars in Medical Genetics, 123,* 48–58. DOI: 10.1002/ajmg.c.20013

Snyderman, M., & Rothman, S. (1988). *The IQ controversy, the media and publication.* New Brunswick, NJ: Transaction Books.

Sokol, D. K., Moore, C. A., Rose, R. J., Williams, C. J., Reed, T., & Christian, J. C. (1995). Intrapair differences in personality and cognitive ability among young monozygotic twins distinguished by chorion type. *Behavior Genetics, 25,* 457–466. DOI: 10.1007/BF02253374

Sokolowski, M. B. (2001). Drosophila: Genetics meets behaviour. *Nature Reviews Genetics, 2,* 879–890. DOI: 10.1038/35098592

Sora, I., Li, B., Igari, M., Hall, F. S., & Ikeda, K. (2010). Transgenic mice in the study of drug addiction and the effects of psychostimulant drugs. *Annals of the New York Academy of Sciences, 1187,* 218–246. DOI: 10.1111/j.1749-6632.2009.05276.x

Spearman, C. (1904). "General intelligence," objectively determined and measured. *American Journal of Psychology, 15,* 201–292. DOI: 10.2307/1412107

Speliotes, E. K., Willer, C. J., Berndt, S. I., Monda, K. L., Thorleifsson, G., Jackson, A. U., ... Loos, R. J. (2010). Association analyses of 249,796 individuals reveal 18 new loci associated with body mass index. *Nature Genetics, 42,* 937–948. DOI: 10.1038/ng.686

Spence, J. P., Liang, T., Liu, L., Johnson, P. L., Foroud, T., Carr, L. G., & Shekhar, A. (2009). From QTL to candidate gene: A genetic approach to alcoholism research. *Current Drug Abuse Reviews, 2,* 127–134.

Spinath, F. M., Harlaar, N., Ronald, A., & Plomin, R. (2004). Substantial genetic influence on mild mental impairment in early childhood. *American Journal of Mental Retardation, 109,* 34–43. DOI: 10.1352/0895-8017(2004)109<34:SGIOMM>2.0.CO;2

Spitz, H. H. (1988). Wechsler subtest patterns of mentally retarded groups: Relationship to *g* and to estimates of heritability. *Intelligence, 12,* 279–297. DOI: 10.1016/0160-2896(88)90027-X

Spotts, E. L., Neiderhiser, J. M., Towers, H., Hansson, K., Lichtenstein, P., Cederblad, M., ... Reiss, D. (2004). Genetic and environmental influences on marital relationships. *Journal of Family Psychology, 18,* 107–119. DOI: 10.1037/0893-3200.18.1.107

Spotts, E. L., Pederson, N. L., Neiderhiser, J. M., Reiss, D., Lichtenstein, P., Hansson, K., & Cederblad, M. (2005). Genetic effects on women's positive mental health: Do marital relationships and social support matter? *Journal of Family Psychology, 19,* 339–349. DOI: 10.1037/0893-3200.19.3.339

Spotts, E. L., Prescott, C., & Kendler, K. (2006). Examining the origins of gender differences in marital quality: A behavior genetic analysis. *Journal of Family Psychology, 20,* 605–613. DOI: 10.1037/0893-3200.20.4.605

Sprott, R. L., & Staats, J. (1975). Behavioral studies using genetically defined mice—a bibliography. *Behavior Genetics, 5,* 27–82. DOI: 10.1007/BF01067579

Spuhler, J. N. (1968). Assortative mating with respect to physical characterisitcs. *Eugenics Quarterly, 15,* 128–140.

Stallings, M. C., Corley, R. P., Dennehey, B., Hewitt, J. K., Krauter, K. S., Lessem, J. M., ... Crowley, T. J. (2005). A genome-wide search for quantitative trait loci that influence antisocial drug dependence in adolescence. *Archives of General Psychiatry, 62,* 1042–1051. DOI: 10.1001/archpsyc.62.9.1042

Stallings, M. C., Corley, R. P., Hewitt, J. K., Krauter, K. S., Lessem, J. M., Mikulich, S. K., ... Crowley, T. J. (2003). A genome-wide search for quantitative trait loci influencing substance dependence vulnerability in adolescence. *Drug and Alcohol Dependence, 70,* 295–307. DOI: 10.1016/S0376-8716(03)00031-0

Stallings, M. C., Hewitt, J. K., Cloninger, C. R., Heath, A. C., & Eaves, L. J. (1996). Genetic and environmental structure of the tridimensional personality questionnaire: Three or four temperament dimensions? *Journal of Personality and Social Psychology, 70,* 127–140. DOI: 10.1037/0022-3514.70.1.127

Stefansson, H., Ophoff, R. A., Steinberg, S., Andreassen, O. A., Cichon, S., Rujescu, D., ... Collier, D. A. (2009). Common variants conferring risk of schizophrenia. *Nature, 460,* 744–747. DOI: 10.1038/nature08186.

Stefansson, H., Sigurdsson, E., Steinthorsdottir, V., Bjornsdottir, S., Sigmundsson, T., Ghosh, S., ... Stefansson, K. (2002). Neuregulin 1 and susceptibility to schizophrenia. *American Journal of Human Genetics, 71,* 877–892. DOI: 10.1086/342734

Stein, J. L., Medland, S. E., Vasquez, A. A., Hibar, D. P., Senstad, R. E., Winkler, A. M., ... Thompson, P. M. (2012). Identification of common variants associated with human hippocampal and intracranial volumes. *Nature Genetics, 44,* 552–561. DOI: 10.1038/ng.2250

Stein, M. B., Chartier, M. J., Hazen, A. L., Kozak, M. V., Tancer, M. E., Lander, S., ... Walker, J. R. (1998). A direct-interview family study of generalized social phobia. *American Journal of Psychiatry, 155,* 90–97.

Stent, G. S. (1963). *Molecular biology of bacterial viruses.* New York: Freeman.

Steptoe, A., & Wardle, J. (2012). Enjoying life and living longer. *Archives of Internal Medicine, 172,* 273–275. DOI: 10.1001/archinternmed.2011.1028

Stergiakouli, E., Hamshere, M., Holmans, P., Langley, K., Zaharieva, I., Hawi, Z., . . . Nelson, S. (2012). Investigating the contribution of common genetic variants to the risk and pathogenesis of ADHD. *American Journal of Psychiatry, 169,* 186–194. DOI: 10.1176/appi.ajp.2011.11040551

Sternberg, R. J., & Grigorenko, E. L. (1997). *Intelligence: Heredity and environment.* Cambridge: Cambridge University Press.

Stevens, A., & Price, J. (2004). *Evolutionary psychiatry: A new beginning* (2nd ed.) London: Routledge.

Stewart, J. R., & Stringer, C. B. (2012). Human evolution out of Africa: The role of refugia and climate change. *Science, 335,* 1317–1321. DOI: 10.1126/science.1215627

Stewart, S. E., Platko, J., Fagerness, J., Birns, J., Jenike, E., Smoller, J. W., . . . Pauls, D. L. (2007). A genetic family-based association study of OLIG2 in obsessive-compulsive disorder. *Archives of General Psychiatry, 64,* 209–214. DOI: 10.1001/archpsyc.64.2.209

Stoneking, M., & Krause, J. (2011). Learning about human population history from ancient and modern genomes. *Nature Reviews Genetics, 12,* 603–614. DOI: 10.1038/nrg3029

Stoolmiller, M. (1999). Implications of the restricted range of family environments for estimates of heritability and nonshared environment in behavior-genetic adoption studies. *Psychological Bulletin, 125,* 392–409. DOI: 10.1037/0033-2909.125.4.392

Stratakis, C. A., & Rennert, O. M. (2005). Turner syndrome—an update. *Endocrinologist, 15,* 27–36. DOI: 10.1097/01.ten.0000152836.30636.a7

Straub, R. E., Jiang, Y., MacLean, C. J., Ma, Y., Webb, B. T., Myakishev, M. V., . . . Kendler, K. S. (2002). Genetic variation in the 6p22.3 gene DT-NBP1, the human ortholog of the mouse dysbindin gene, is associated with schizophrenia. *American Journal of Human Genetics, 71,* 337–348. DOI: 10.1086/341750

Strine, T. W., Chapman, D. P., Balluz, L. S., Moriarty, D. G., & Mokdad, A. H. (2008). The associations between life satisfaction and health-related quality of life, chronic illness, and health behaviors among US community-dwelling adults. *Journal of Community Health, 33,* 40–50. DOI: 10.1007/s10900-007-9066-4

Stromswold, K. (2001). The heritability of language: A review and metaanalysis of twin, adoption and linkage studies. *Language, 77,* 647–723. DOI: 10.1353/lan.2001.0247

Stromswold, K. (2006). Why aren't identical twins linguistically identical? Genetic, prenatal and postnatal factors. *Cognition, 101,* 333–384. DOI: 10.1016/j.cognition.2006.04.007

Sturm, R. A. (2009). Molecular genetics of human pigmentation diversity. *Human Molecular Genetics, 18,* R9–R17. DOI: 10.1093/hmg/ddp003

Sturtevant, A. H. (1915). A sex-linked character in Drosophila repleta. *American Naturalist, 49,* 189–192. DOI: 10.1086/279474

Sullivan, P. F., Evengard, B., Jacks, A., & Pedersen, N. L. (2005). Twin analyses of chronic fatigue in a Swedish national sample. *Psychological Medicine, 35,* 1327–1336. DOI: 10.1017/S0033291705005222

Sullivan, P. F., Kendler, K. S., & Neale, M. C. (2003). Schizophrenia as a complex trait: Evidence from a meta-analysis of twin studies. *Archives of General Psychiatry, 60,* 1187–1192. DOI: 10.1001/archpsyc.60.12.1187

Sullivan, P. F., Kovalenko, P., York, T. P., Prescott, C. A., & Kendler, K. S. (2003). Fatigue in a community sample of twins. *Psychological Medicine, 33,* 263–281. DOI: 10.1017/S0033291702007031

Sullivan, P. F., Neale, M. C., & Kendler, K. S. (2000). Genetic epidemiology of major depression: Review and meta-analysis. *American Journal of Psychiatry, 157,* 1552–1562. DOI: 10.1176/appi.ajp.157.10.1552

Swan, G. E., Benowitz, N. L., Lessov, C. N., Jacob, P., Tyndale, R. F., & Wilhelmsen, K. (2005). Nicotine metabolism: The impact of CYP2A6 on estimates of additive genetic influence. *Pharmacogenetics and Genomics, 15,* 115–125. DOI: 10.1097/01213011-200502000-00007

Sykes, B. (2007). *Saxons, Vikings and Celts: The genetic roots of Britain and Ireland.* New York: W. W. Norton.

Symons, D. (1979). *The evolution of human sexuality.* New York: Oxford University Press.

Szatmari, P., Paterson, A. D., Zwaigenbaum, L., Roberts, W., Brian, J., Liu, X. Q., . . . Shih, A. (2007). Mapping autism risk loci using genetic linkage and chromosomal rearrangements. *Nature Genetics, 39,* 319–328. DOI: 10.1038/ng1985

Tabakoff, B., Saba, L., Kechris, K., Hu, W., Bhave, S. V., Finn, D. A., . . . Hoffman, P. L. (2008). The genomic determinants of alcohol preference in mice. *Mammalian Genome, 19,* 352–365. DOI: 10.1007/s00335-008-9115-z

Tabakoff, B., Saba, L., Printz, M., Flodman, P., Hodgkinson, C., Goldman, D., . . . WHO/ISBRA

Study on State and Trait Markers of Alcoholism (2009). Genetical genomic determinants of alcohol consumption in rats and humans. *BMC Biology, 7:* 70. DOI: 10.1186/1741-7007-7-70

Tabor, H. K., Risch, N. J., & Myers, R. M. (2002). Candidate-gene approaches for studying complex genetic traits: Practical considerations. *Nature Reviews Genetics, 3*, 391–397. DOI: 10.1038/nrg796

Talbot, C. J., Nicod, A., Cherny, S. S., Fulker, D. W., Collins, A. C., & Flint, J. (1999). High-resolution mapping of quantitative trait loci in outbred mice. *Nature Genetics, 21*, 305–308. DOI: 10.1038/6825

Tambs, K., Czajkowsky, N., Roysamb, E., Neale, M. C., Reichborn-Kjennerud, T., Aggen, S. H., . . . Kendler, K. S. (2009). Structure of genetic and environmental risk factors for dimensional representations of DSM-IV anxiety disorders. *British Journal of Psychiatry, 195*, 301–307. DOI: 10.1192/bjp.bp.108.059485

Tambs, K., Sundet, J. M., & Magnus, P. (1986). Genetic and environmental contribution to the co-variation between the Wechsler Adult Intelligence Scale (WAIS) subtests: A study of twins. *Behavior Genetics, 16*, 475–491. DOI: 10.1007/BF01074266

Tambs, K., Sundet, J. M., Magnus, P., & Berg, K. (1989). Genetic and environmental contributions to the covariance between occupational status, educational attainment, and IQ: A study of twins. *Behavior Genetics, 19*, 209–222. DOI: 10.1007/BF01065905

Tang, Y. P., Shimizu, E., Dube, G. R., Rampon, C., Kerchner, G. A., Zhuo, M., . . . Tsien, J. Z. (1999). Genetic enhancement of learning and memory in mice. *Nature, 401*, 63–69. DOI: 10.1038/43432

Taniai, H., Nishiyama, T., Miyachi, T., Imaeda, M., & Sumi, S. (2008). Genetic influences on the broad spectrum of autism: Study of proband-ascertained twins. *American Journal of Medical Genetics. B: Neuropsychiatric Genetics, 147B*, 844–849. DOI: 10.1002/ajmg.b.30740

Tanzi, R. E., & Bertram, L. (2005). Twenty years of the Alzheimer's disease amyloid hypothesis: A genetic perspective. *Cell, 120*, 545–555. DOI: 10.1016/j.cell.2005.02.008

Tartaglia, N. R., Howell, S., Sutherland, A., Wilson, R., & Wilson, L. (2010). A review of trisomy X (47, XXX). *Orphanet Journal of Rare Diseases, 5*, 8. DOI: 10.1186/1750-1172-5-8

Taubman, P. (1976). The determinants of earnings: Genetics, family and other environments: A study of white male twins. *American Economic Review, 66*, 858–870.

Taylor, E. (1995). Dysfunctions of attention. In D. Cicchetti & D. J. Cohen (Eds.), *Developmental psy-chopathology: Vol. 2. Risk, disorder, and adaptation* (pp. 243–273). New York: Wiley.

Taylor, J., Roehrig, A. D., Hensler, B. S., Connor, C. M., & Schatschneider, C. (2010). Teacher quality moderates the genetic effects on early reading. *Science, 328*, 512–514. DOI: 10.1126/science.1186149

Taylor, J., & Schatschneider, C. (2010). Genetic influence on literacy constructs in kindergarten and first grade: Evidence from a diverse twin sample. *Behavior Genetics, 40*, 591–602. DOI: 10.1007/s10519-010-9368-7

Taylor, J. Y., & Wu, C. Y. (2009). Effects of genetic counseling for hypertension on changes in lifestyle behaviors among African-American women. *Journal of National Black Nurses Association, 20*, 1–10.

Taylor, S. (2011). Etiology of obsessions and compulsions: A meta-analysis and narrative review of twin studies. *Clinical Psychology Review, 31*, 1361–1372. DOI: 10.1016/j.cpr.2011.09.008

Taylor, S., Asmundson, G. J. G., & Jang, K. L. (2011). Etiology of obsessive-compulsive symptoms and obsessive-compulsive personality traits: Common genes, mostly different environments. *Depression and Anxiety, 28*, 863–869. DOI: 10.1002/da.20859

Tellegen, A., Lykken, D. T., Bouchard, T. J., Jr., Wilcox, K., Segal, N., & Rich, A. (1988). Personality similarity in twins reared together and apart. *Journal of Personality and Social Psychology, 54*, 1031–1039. DOI: 10.1037/0022-3514.54.6.1031

Terracciano, A., Sanna, S., Uda, M., Deiana, B., Usala, G., Busonero, F., . . . Costa, P. T., Jr. (2010). Genome-wide association scan for five major dimensions of personality. *Molecular Psychiatry, 15*, 647–656. DOI: 10.1038/mp.2008.113

Tesser, A. (1993). On the importance of heritability in psychological research: The case of attitudes. *Psychological Review, 100*, 129–142. DOI: 10.1037/0033-295X.100.1.129

Tesser, A., Whitaker, D., Martin, L., & Ward, D. (1998). Attitude heritability, attitude change and physiological responsivity. *Personality and Individual Differences, 24*, 89–96. DOI: 10.1016/S0191-8869(97)00137-2

Thakker, D. R., Hoyer, D., & Cryan, J. F. (2006). Interfering with the brain: Use of RNA interference for understanding the pathophysiology of psychiatric and neurological disorders. *Pharmacology and Therapeutics, 109*, 413–438. DOI: 10.1016/j.pharmthera.2005.08.006

Thapar, A., Harold, G., & McGuffin, P. (1998). Life events and depressive symptoms in childhood—shared genes or shared adversity? A research note. *Journal of Child Psychology and Psychiatry, 39,* 1153–1158. DOI: 10.1111/1469-7610.00419

Thapar, A., Langley, K., O'Donovan, M., & Owen, M. (2006). Refining the attention deficit hyperactivity disorder phenotype for molecular genetic studies. *Molecular Psychiatry, 11,* 714–720. DOI: 10.1038/sj.mp.4001831

Thapar, A., & McGuffin, P. (1996). Genetic influences on life events in childhood. *Psychological Medicine, 26,* 813–820. DOI: 10.1017/S0033291700037831

Thapar, A., O'Donovan, M., & Owen, M. J. (2005). The genetics of attention deficit hyperactivity disorder. *Human Molecular Genetics, 14,* R275–R282. DOI: 10.1093/hmg/ddi263

Thapar, A., & Rice, F. (2006). Twin studies in pediatric depression. *Child and Adolescent Psychiatric Clinics of North America, 15,* 869–881. DOI: 10.1016/j.chc.2006.05.007

Thapar, A., & Scourfield, J. (2002). Childhood disorders. In P. McGuffin, M. J. Owen, & I. I. Gottesman (Eds.), *Psychiatric genetics and genomics* (pp. 147–180). Oxford: Oxford University Press.

Theis, S. V. S. (1924). *How foster children turn out.* New York: State Charities Aid Association.

Thielen, A., Klus, H., & Mueller, L. (2008). Tobacco smoke: Unraveling a controversial subject. *Experimental and Toxicologic Pathology, 60,* 141–156. DOI: 10.1016/j.etp.2008.01.014

Tholin, S., Rasmussen, F., Tynelius, P., & Karlsson, J. (2005). Genetic and environmental influences on eating behavior: The Swedish Young Male Twins Study. *American Journal of Clinical Nutrition, 81,* 564–569.

Thomas, D. (2010). Gene-environment-wide association studies: Emerging approaches. *Nature Reviews Genetics, 11,* 259–272. DOI: 10.1038/nrg2764

Thompson, L. A., Detterman, D. K., & Plomin, R. (1991). Associations between cognitive abilities and scholastic achievement: Genetic overlap but environmental differences. *Psychological Science, 2,* 158–165. DOI: 10.1111/j.1467-9280.1991.tb00124.x

Thompson, P. M., Cannon, T. D., Narr, K. L., van Erp, T., Poutanen, V. P., Huttunen, M., ... Toga, A. W. (2001). Genetic influences on brain structure. *Nature Neuroscience, 4,* 1253–1258. DOI: 10.1038/nn758

Thompson, P. M., Martin, N. G., & Wright, M. J. (2010). Imaging genomics. *Current Opinion in Neurology, 23,* 368–373. DOI: 10.1097/WCO.0b013e32833b764c

Thorgeirsson, T. E., Gudbjartsson, D. F., Surakka, I., Vink, J. M., Amin, N., Geller, F., ... Stefansson, K. (2010). Sequence variants at CHRNB3-CHRNA6 and CYP2A6 affect smoking behavior. *Nature Genetics, 42,* 448–453. DOI: 10.1038/ng.573

Tian, Z., Palmer, N., Schmid, P., Yao, H., Galdzicki, M., Berger, B., ... Kohane, I. S. (2009). A practical platform for blood biomarker study by using global gene expression profiling of peripheral whole blood. *PLoS One, 4:* e5157. DOI: 10.1371/journal.pone.0005157

Tienari, P., Wynne, L. C., Sorri, A., Lahti, I., Laksy, K., Moring, J., ... Wahlberg, K. E. (2004). Genotype-environment interaction in schizophrenia-spectrum disorder. Long-term follow-up study of Finnish adoptees. *British Journal of Psychiatry, 184,* 216–222. DOI: 10.1192/bjp.184.3.216

Timberlake, D. S., Rhee, S. H., Haberstick, B. C., Hopfer, C., Ehringer, M., Lessem, J. M., ... Hewitt, J. K. (2006). The moderating effects of religiosity on the genetic and environmental determinants of smoking initiation. *Nicotine and Tobacco Research, 8,* 123–133. DOI: 10.1080/14622200500432054

Tobacco and Genetics Consortium (2010). Genome-wide meta-analyses identify multiple loci associated with smoking behavior. *Nature Genetics, 42,* 441–447. DOI: 10.1038/ng.571

Toga, A. W., & Thompson, P. M. (2005). Genetics of brain structure and intelligence. *Annual Review of Neuroscience, 28,* 1–23. DOI: 10.1146/annurev.neuro.28.061604.135655

Tollefsbol, T. O. (2011). *Handbook of epigenetics: The new molecular and medical genetics.* London: Academic Press.

Toma, D. P., White, K. P., Hirsch, J., & Greenspan, R. J. (2002). Identification of genes involved in Drosophila melanogaster geotaxis, a complex behavioral trait. *Nature Genetics, 31,* 349–353.

Tooley, G. A., Karakis, M., Stokes, M., & Ozanne-Smith, J. (2006). Generalising the Cinderella effect to unintentional childhood fatalities. *Evolution and Human Behavior, 27,* 224–230. DOI: 10.1016/j.evolhumbehav.2005.10.001

Topper, S., Ober, C., & Das, S. (2011). Exome sequencing and the genetics of intellectual disability. *Clinical Genetics, 80,* 117–126. DOI: 10.1111/j.1399-0004.2011.01720.x

Torgersen, S. (1980). The oral, obsessive, and hysterical personality syndromes: A study of hereditary and environmental factors by means of the twin method. *Archives of General Psychiatry, 37,* 1272–1277. DOI: 10.1001/archpsyc.1980.01780240070008

Torgersen, S. (1983). Genetic factors in anxiety disorders. *Archives of General Psychiatry, 40,* 1085–1089. DOI: 10.1001/archpsyc.1983.01790090047007

Torgersen, S. (2009). The nature (and nurture) of personality disorders. *Scandinavian Journal of Psychology, 50,* 624–632. DOI: 10.1111/j.1467-9450.2009.00788.x

Torgersen, S., Edvardsen, J., Oien, P. A., Onstad, S., Skre, I., Lygren, S., & Kringlen, E. (2002). Schizotypal personality disorder inside and outside the schizophrenic spectrum. *Schizophrenia Research, 54,* 33–38. DOI: 10.1016/S0920-9964(01)00349-8

Torgersen, S., Lygren, S., Oien, P. A., Skre, I., Onstad, S., Edvardsen, J., . . . Kringlen, E. (2000). A twin study of personality disorders. *Comprehensive Psychiatry, 41,* 416–425. DOI: 10.1053/comp.2000.16560

Torkamani, A., Dean, B., Schork, N. J., & Thomas, E. A. (2010). Coexpression network analysis of neural tissue reveals perturbations in developmental processes in schizophrenia. *Genome Research, 20,* 403–412. DOI: 10.1101/gr.101956.109

Torrey, E. F. (1990). Offspring of twins with schizophrenia. *Archives of General Psychiatry, 47,* 976–977. DOI: 10.1001/archpsyc.1990.01810220092013

Torrey, E. F., Bowler, A. E., Taylor, E. H., & Gottesman, I. I. (1994). *Schizophrenia and manic-depressive disorder.* New York: Basic Books.

Tourette Syndrome Association International Consortium for Genetics (2007). Genome scan for Tourette disorder in affected-sibling-pair and multigenerational families. *American Journal of Human Genetics, 80,* 265–272. DOI: 10.1086/511052

Trikalinos, T. A., Karvouni, A., Zintzaras, E., Ylisaukko-Oja, T., Peltonen, L., Jarvela, I., & Ioannidis, J. P. (2006). A heterogeneity-based genome search meta-analysis for autism-spectrum disorders. *Molecular Psychiatry, 11,* 29–36. DOI: 10.1038/sj.mp.4001750

Trivers, R. L. (1972). Parental investment and sexual selection. In B. Campbell (Ed.), *Sexual selection and the descent of man: 1871–1971* (pp. 136–179). Chicago: Aldine.

Trivers, R. L. (1985). *Social evolution.* Menlo Park, CA: Benjamin/Cummings.

True, W. R., Rice, J., Eisen, S. A., Heath, A. C., Goldberg, J., Lyons, M. J., & Nowak, J. (1993). A twin study of genetic and environmental contributions to liability for posttraumatic stress symptoms. *Archives of General Psychiatry, 50,* 257–264. DOI: 10.1001/archpsyc.1993.01820160019002

Trut, L., Oskina, I., & Kharlamova, A. (2009). Animal evolution during domestication: The domesticated fox as a model. *Bioessays, 31,* 349–360. DOI: 10.1002/bies.200800070

Trut, L. N. (1999). Early canid domestication: The farm-fox experiment. *American Scientist, 87,* 160–169. DOI: 10.1511/1999.20.813

Trzaskowski, M., Zavos, H., Haworth, C., Plomin, R., & Eley, T. (2011). Stable genetic influence

on anxiety-related behaviours across middle childhood. *Journal of Abnormal Child Psychology, 40,* 85–94. DOI: 10.1007/s10802-011-9545-z

Tsuang, M., & Faraone, S. D. (1990). *The genetics of mood disorders.* Baltimore: John Hopkins University Press.

Tsuang, M. T., Bar, J. L., Harley, R. M., & Lyons, M. J. (2001). The Harvard Twin Study of Substance Abuse: What we have learned. *Harvard Review of Psychiatry, 9,* 267–279.

Tsuang, M. T., Lyons, M. J., Eisen, S. A., True, W. T., Goldberg, J., & Henderson, W. (1992). A twin study of drug exposure and initiation of use. *Behavior Genetics, 22,* 756.

Tucker-Drob, E. M., Rhemtulla, M., Harden, K. P., Turkheimer, E., & Fask, D. (2011). Emergence of a gene × socioeconomic status interaction on infant mental ability between 10 months and 2 years. *Psychological Science, 22,* 125–133. DOI: 10.1177/0956797610392926

Turkheimer, E. (2013). Genetics of personality. *Annual Review of Psychology, 64.* DOI: 10.1146/annurev-psych-113011-143752

Turkheimer, E., Haley, A., Waldron, M., D'Onofrio, B., & Gottesman, I. I. (2003). Socioeconomic status modifies heritability of IQ in young children. *Psychological Science, 14,* 623–628. DOI: 10.1046/j.0956-7976.2003.psci_1475.x

Turkheimer, E., & Waldron, M. (2000). Nonshared environment: A theoretical, methodological, and quantitative review. *Psychological Bulletin, 126,* 78–108. DOI: 10.1037/0033-2909.126.1.78

Turnbaugh, P. J., & Gordon, J. I. (2009). The core gut microbiome, energy balance and obesity. *Journal of Physiology, 587,* 4153–4158. DOI: 10.1113/jphysiol.2009.174136

Turnbaugh, P. J., Hamady, M., Yatsunenko, T., Cantarel, B. L., Duncan, A., Ley, R. E., . . . Gordon, J. I. (2009). A core gut microbiome in obese and lean twins. *Nature, 457,* 480–484. DOI: 10.1038/nature07540

Turner, J. R., Cardon, L. R., & Hewitt, J. K. (1995). *Behavior genetic approaches in behavioral medicine.* New York: Plenum.

Turri, M. G., Henderson, N. D., DeFries, J. C., & Flint, J. (2001). Quantitative trait locus mapping in laboratory mice derived from a replicated selection experiment for open-field activity. *Genetics, 158,* 1217–1226.

Tuttelmann, F., & Gromoll, J. (2010). Novel genetic aspects of Klinefelter's syndrome. *Molecular Human Reproduction, 16,* 386–395. DOI: 10.1093/molehr/gaq019

Tuvblad, C., Grann, M., & Lichtenstein, P. (2006). Heritability for adolescent antisocial behavior differs with socioeconomic status: Gene-environment interaction. *Journal of Child Psychology and Psychiatry, 47,* 734–743. DOI: 10.1111/j.1469-7610.2005.01552.x

Tuvblad, C., Zheng, M., Raine, A., & Baker, L. A. (2009). A common genetic factor explains the covariation among ADHD ODD and CD symptoms in 9–10 year old boys and girls. *Journal of Abnormal Child Psychology, 37,* 153–167. DOI: 10.1007/s10802-008-9278-9

Ulbricht, J. A., & Neiderhiser, J. M. (2009). Genotype-environment correlation and family relationships. In Y.-K. Kim (Ed.), *Handbook of behavior genetics.* New York: Springer. DOI: 10.1007/978-0-387-76727-7_15

Üstün, T. B., Rehm, J., Chatterji, S., Saxena, S., Trotter, R., Room, R., . . . WHO/NIH Joint Project CAR Study Group (1999). Multiple-informant ranking of the disabling effects of different health conditions in 14 countries. *Lancet, 354,* 111–115. DOI: 10.1016/S0140-6736(98)07507-2

Vaisse, C., Clement, K., Guy-Grand, B., & Froguel, P. (1998). A frameshift mutation in human MC4R is associated with a dominant form of obesity. *Nature Genetics, 20,* 113–114. DOI: 10.1038/2407

Valdar, W., Solberg, L. C., Gauguier, D., Burnett, S., Klenerman, P., Cookson, W. O., . . . Flint, J. (2006). Genome-wide genetic association of complex traits in heterogeneous stock mice. *Nature Genetics, 38,* 879–887. DOI: 10.1038/ng1840

Valdar, W., Solberg, L. C., Gauguier, D., Cookson, W. O., Rawlins, J. N. P., Mott, R., & Flint, J. (2006). Genetic and environmental effects on complex traits in mice. *Genetics, 174,* 959–984. DOI: 10.1534/genetics.106.060004

van Baal, G., de Geus, E., & Boomsma, D. I. (1998). Longitudinal study of genetic influences on ERP-P3 during childhood. *Developmental Neuropsychology, 14,* 19–45. DOI: 10.1080/87565649809540699

van Baal, G. C., Boomsma, D. I., & de Geus, E. J. (2001). Longitudinal genetic analysis of EEG coherence in young twins. *Behavior Genetics, 31,* 637–651. DOI: 10.1023/A:1013357714500

Van Beijsterveldt, C. E., Molenaar, P. C., de Geus, E. J., & Boomsma, D. I. (1998). Genetic and environmental influences on EEG coherence. *Behavior Genetics, 28,* 443–453. DOI: 10.1023/A:1021637328512

Van den Oord, J. C. G., & Rowe, D. C. (1997). Continuity and change in children's social maladjustment: A developmental behavior genetic study. *Developmental Psychology, 33,* 319–332. DOI: 10.1037/0012-1649.33.2.319

van der Sluis, S., Dolan, C. V., Neale, M. C., Boomsma, D. I., & Posthuma, D. (2006). Detecting genotype-environment interaction in monozygotic twin data: Comparing the Jinks and Fulker test and a new test based on marginal maximum likelihood estimation. *Twin Research and Human Genetics, 9,* 377–392. DOI: 10.1375/183242706777591218

van der Zwaluw, C. S., Engels, R. C. M. E., Buitelaar, J., Verkes, R. J., Franke, B., & Scholte, R. H. J. (2009). Polymorphisms in the dopamine transporter gene (SLC6A3/DAT1) and alcohol dependence in humans: A systematic review. *Pharmacogenomics, 10,* 853–866. DOI: 10.2217/pgs.09.24

van Ijzendoorn, M. H., Bakermans-Kranenburg, M. J., Belsky, J., Beach, S., Brody, G., Dodge, K. A., . . . Scott, S. (2011). Gene-by-environment experiments: A new approach to finding the missing heritability. *Nature Reviews Genetics, 12,* 881. DOI: 10.1038/nrg2764-c1

van Ijzendoorn, M. H., Moran, G., Belsky, J., Pederson, D., Bakermans-Kranenburg, M. J., & Fisher, K. (2000). The similarity of siblings' attachments to their mother. *Child Development, 71,* 1086–1098. DOI: 10.1111/1467-8624.00211

van Soelen, I. L. C., Brouwer, R. M., van Baal, G. C. M., Schnack, H. G., Peper, J. S., Collins, D. L., . . . Hulshoff Pol, H. E. (2012). Genetic influences on thinning of the cerebral cortex during development. *NeuroImage, 59,* 3871–3880. DOI: 10.1016/j.neuroimage.2011.11.044

van Soelen, I. L. C., Brouwer, R. M., van Leeuwen, M., Kahn, R. S., Pol, H. E. H., & Boomsma, D. I. (2011). Heritability of verbal and performance intelligence in a pediatric longitudinal sample. *Twin Research and Human Genetics, 14,* 119–128. DOI: 10.1375/twin.14.2.119

Vandenberg, S. G. (1971). What do we know today about the inheritance of intelligence and how do we know it? In R. Cancro (Ed.), *Genetic and environmental influences* (pp. 182–218). New York: Grune & Stratton.

Vandenberg, S. G. (1972). Assortative mating, or who marries whom? *Behavior Genetics, 2,* 127–157. DOI: 10.1007/BF01065686

Vanyukov, M. M., Tarter, R. E., Kirillova, G. P., Kirisci, L., Reynolds, M. D., Kreek, M. J., . . . Ridenour, T. A. (2012). Common liability to addiction and "gateway hypothesis": Theoretical, empirical and evolutionary perspective. *Drug and Alcohol Dependence, 123,* S3–S17. DOI: 10.1016/j.drugalcdep.2011.12.018

Venter, J. C., Adams, M. D., Myers, E. W., Li, P. W., Mural, R. J., Sutton, G. G., . . . Nodell, M. (2001). The sequence of the human genome. *Science, 291,* 1304–1351. DOI: 10.1126/science.1058040

Verhulst, B., Eaves, L. J., & Hatemi, P. K. (2012). Correlation not causation: The relationship between personality traits and political ideologies. *American Journal of Political Science, 56,* 34–51. DOI: 10.1111/j.1540-5907.2011.00568.x

Verkerk, A. J., Cath, D. C., van der Linde, H. C., Both, J., Heutink, P., Breedveld, G., . . . Oostra, B. A. (2006). Genetic and clinical analysis of a large Dutch Gilles de la Tourette family. *Molecular Psychiatry, 11,* 954–964. DOI: 10.1038/sj.mp.4001877

Verkerk, A. J. M. H., Pieretti, M., Sutcliffe, J. S., Fu, Y. H., Kuhl, D. P. A., Pizzuti, A., . . . Warren, S. T. (1991). Identification of a gene (FMR-1) containing a CGG repeat coincident with a breakpoint cluster region exhibiting length variation in fragile X syndrome. *Cell, 65,* 905–914. DOI: 10.1016/0092-8674(91)90397-H

Vernon, P. A. (1989). The heritability of measures of speed of information-processing. *Personality and Individual Differences, 10,* 575–576. DOI: 10.1016/0191-8869(89)90040-8

Vernon, P. A., Jang, K. L., Harris, J. A., & McCarthy, J. M. (1997). Environmental predictors of personality differences: A twin and sibling study. *Journal of Personality and Social Psychology, 72,* 177–183. DOI: 10.1037/0022-3514.72.1.177

Verweij, K. J. H., Zietsch, B. P., Lynskey, M. T., Medland, S. E., Neale, M. C., Martin, N. G., . . . Vink, J. M. (2010). Genetic and environmental influences on cannabis use initiation and problematic use: A meta-analysis of twin studies. *Addiction, 105,* 417–430. DOI: 10.1111/j.1360-0443.2009.02831.x

Viding, E. (2004). Annotation: Understanding the development of psychopathy. *Journal of Child Psychology and Psychiatry, 45,* 1329–1337. DOI: 10.1111/j.1469-7610.2004.00323.x

Viding, E., Blair, R. J. R., Moffitt, T. E., & Plomin, R. (2005). Evidence for substantial genetic risk for psychopathy in 7-year-olds. *Journal of Child Psychology and Psychiatry, 46,* 592–597. DOI: 10.1111/j.1469-7610.2004.00393.x

Viding, E., Fontaine, N., Oliver, B., & Plomin, R. (2009). Negative parental discipline, conduct problems and callous-unemotional traits: A monozygotic twin differences study. *British Journal of Psychiatry, 195,* 414–419. DOI: 10.1192/bjp.bp.108.061192

Viding, E., Jones, A. P., Frick, P. J., Moffitt, T. E., & Plomin, R. (2008). Heritability of antisocial behaviour at 9: Do callous-unemotional traits matter? *Developmental Science, 11,* 17–22. DOI: 10.1111/j.1467-7687.2007.00648.x

Villafuerte, S., & Burmeister, M. (2003). Untangling genetic networks of panic, phobia, fear and anxiety. *Genome Biology, 4,* 224. DOI: 10.1186/gb-2003-4-8-224

Vink, J., Willemsen, G., & Boomsma, D. (2005). Heritability of smoking initiation and nicotine dependence. *Behavior Genetics, 35,* 397–406. DOI: 10.1007/s10519-004-1327-8

Vinkhuyzen, A. A. E., Pedersen, N. L., Yang, J., Lee, S. H., Magnusson, P. K. E., Iacono, W. G., . . . Wray, N. R. (2012). Common SNPs explain some of the variation in the personality dimensions of neuroticism and extraversion. *Translational Psychiatry, 2:* e102. DOI: 10.1038/tp.2012.27

Vinkhuyzen, A. A. E., van der Sluis, S., Boomsma, D. I., de Geus, E. J. C., & Posthuma, D. (2010). Individual differences in processing speed and working memory speed as assessed with the Sternberg Memory Scanning task. *Behavior Genetics, 40,* 315–326. DOI: 10.1007/s10519-009-9315-7

Vinkhuyzen, A. A. E., van der Sluis, S., Maes, H. H. M., & Posthuma, D. (2012). Reconsidering the heritability of intelligence in adulthood: Taking assortative mating and cultural transmission into account. *Behavior Genetics, 42,* 187–198. DOI: 10.1007/s10519-011-9507-9

Vinkhuyzen, A. A. E., van der Sluis, S., & Posthuma, D. (2011). Life events moderate variation in cognitive ability (*g*) in adults. *Molecular Psychiatry, 16,* 4–6. DOI: 10.1038/mp.2010.12

Visser, B. A., Ashton, M. C., & Vernon, P. A. (2006). Beyond *g*: Putting multiple intelligences theory to the test. *Intelligence, 34,* 487–502. DOI: 10.1016/j.intell.2006.02.004

Vissers, L., de Ligt, J., Gilissen, C., Janssen, I., Steehouwer, M., de Vries, P., . . . Veltman, J. A. (2010). A de novo paradigm for mental retardation. *Nature Genetics, 42,* 1109–1112. DOI: 10.1038/ng.712

Vitaro, F., Brendgen, M., Boivin, M., Cantin, S., Dionne, G., Tremblay, R. E., . . . Perusse, D. (2011). A monozygotic twin difference study of friends' aggression and children's adjustment problems. *Child Development, 82,* 617–632. DOI: 10.1111/j.1467-8624.2010.01570.x

Vitti, J. J., Cho, M. K., Tishkoff, S. A., & Sabeti, P. C. (2012). Human evolutionary genomics: Ethical and interpretive issues. *Trends in Genetics, 28,* 137–145. DOI: 10.1016/j.tig.2011.12.001

von Gontard, A., Heron, J., & Joinson, C. (2011). Family history of nocturnal enuresis and urinary incontinence: Results from a large epidemiological study. *Journal of Urology, 185,* 2303–2306. DOI: 10.1016/j.juro.2011.02.040

von Gontard, A., Schaumburg, H., Hollmann, E., Eiberg, H., & Rittig, S. (2001). The genetics of enuresis: A review. *Journal of Urology, 166,* 2438–2443. DOI: 10.1016/S0022-5347(05)65611-X

von Knorring, A. L., Cloninger, C. R., Bohman, M., & Sigvardsson, S. (1983). An adoption study of depressive disorders and substance abuse. *Archives of General Psychiatry, 40,* 943–950. DOI: 10.1001/archpsyc.1983.01790080025003 vonHoldt, B. M., Pollinger, J. P., Lohmueller, K. E., Han, E., Parker, H. G., Quignon, P., . . . Wayne, R. K. (2010). Genome-wide SNP and haplotype analyses reveal a rich history underlying dog domestication. *Nature, 464,* 898–902. DOI: 10.1038/nature08837

Voracek, M., & Haubner, T. (2008). Twin-singleton differences in intelligence: A meta-analysis. *Psychological Reports, 102,* 951–962. DOI: 10.2466/pr0.102.3.951-962

Waddell, S., & Quinn, W. G. (2001). What can we teach Drosophila? What can they teach us? *Trends in Genetics, 17,* 719–726. DOI: 10.1016/S0168-9525(01)02526-4

Wade, C. H., & Wilfond, B. S. (2006). Ethical and clinical practice considerations for genetic counselors related to direct-to-consumer marketing of genetic tests. *American Journal of Medical Genetics. C: Seminars in Medical Genetics, 142,* 284–292. DOI: 10.1002/ajmg.c.30110

Wahlsten, D. (1999). Single-gene influences on brain and behavior. *Annual Review of Psychology, 50,* 599–624. DOI: 10.1146/annurev.psych.50.1.599

Wahlsten, D., Bachmanov, A., Finn, D. A., & Crabbe, J. C. (2006). Stability of inbred mouse strain differences in behavior and brain size between laboratories and across decades. *Proceedings of the National Academy of Sciences (USA), 103,* 16364–16369. DOI: 10.1073/pnas.0605342103

Wahlsten, D., Metten, P., Phillips, T. J., Boehm, S. L., Burkhart-Kasch, S., Dorow, J., . . . Crabbe, J. C. (2003). Different data from different labs: Lessons from studies of gene-environment interaction. *Journal of Neurobiology, 54,* 283–311. DOI: 10.1002/neu.10173

Wainwright, M. A., Wright, M. J., Luciano, M., Geffen, G. M., & Martin, N. G. (2005). Multivariate genetic analysis of academic skills of the Queensland Core Skills Test and IQ highlight the importance of genetic *g. Twin Research and Human Genetics, 8,* 602–608. DOI: 10.1375/twin.8.6.602

Wainwright, M. A., Wright, M. J., Luciano, M., Montgomery, G. W., Geffen, G. M., & Martin, N. G. (2006). A linkage study of academic skills defined by the Queensland Core Skills Test. *Behavior Genetics, 36,* 56–64. DOI: 10.1007/s10519-005-9013-z

Waldman, I. D., & Gizer, I. R. (2006). The genetics of attention deficit hyperactivity disorder. *Clinical Psychology Review, 26,* 396–432. DOI: 10.1016/j.cpr.2006.01.007

Walker, S., & Plomin, R. (2005). The nature–nurture question: Teachers' perceptions of how genes and the environment influence educationally relevant behaviour. *Educational Psychology, 25,* 509–516. DOI: 10.1080/0144341050046697

Walker, S. O., & Plomin, R. (2006). Nature, nurture, and perceptions of the classroom environment as they relate to teacher assessed academic achievement: A twin study of 9-year-olds. *Educational Psychology, 26,* 541–561. DOI: 10.1080/01443410500342500

Wallace, B. (2010). *Getting Darwin wrong: Why evolutionary psychology won't work.* Exeter, UK: Imprint Academic.

Wallace, G. L., Schmitt, J. E., Lenroot, R., Viding, E., Ordaz, S., Rosenthal, M. A., . . . Giedd, J. N. (2006). A pediatric twin study of brain morphometry. *Journal of Child Psychology and Psychiatry, 47,* 987–993. DOI: 10.1111/j.1469-7610.2006.01676.x

Waller, K., Kujala, U. M., Kaprio, J., Koskenvuo, M., & Rantanen, T. (2010). Effect of physical activity on health in twins: A 30-yr longitudinal study. *Medicine and Science in Sports and Exercise, 42,* 658–664. DOI: 10.1249/MSS.0b013e3181bdeea3

Waller, N. G., & Shaver, P. R. (1994). The importance of nongenetic influence on romantic love styles: A twin-family study. *Psychological Science, 5,* 268–274. DOI: 10.1111/j.1467-9280.1994.tb00624.x

Walsh, C. M., Zainal, N. Z., Middleton, S. J., & Paykel, E. S. (2001). A family history study of chronic fatigue syndrome. *Psychiatric Genetics, 11,* 123–128. DOI: 10.1097/00041444-200109000-00003

Wang, Z., Gerstein, M., & Snyder, M. (2009). RNA-Seq: A revolutionary tool for transcriptomics. *Nature Reviews Genetics, 10,* 57–63. DOI: 10.1038/nrg2484

Ward, M. J., Vaughn, B. E., & Robb, M. D. (1988). Social-emotional adaptation and infant-mother attachment in siblings: Role of the mother in cross-sibling consistency. *Child Development, 59,* 643–651. DOI: 10.2307/1130564

Wardle, J., Carnell, S., Haworth, C. M. A., & Plomin, R. (2008). Evidence for a strong genetic influence on childhood adiposity despite the force of the obesogenic environment. *American Journal of Clinical Nutrition, 87,* 398–404.

Watson, J. B. (1930). *Behaviorism.* New York: Norton.

Watson, J. D., & Crick, F. H. C. (1953). Genetical implications of the structure of deoxyribonucleic acid. *Nature, 171,* 964–967. DOI: 10.1038/171964b0

Weber, H., Kittel-Schneider, S., Gessner, A., Domschke, K., Neuner, M., Jacob, C. P., . . . Reif, A. (2011). Cross-disorder analysis of bipolar risk genes: Further evidence of *DGKH* as a risk gene for bipolar disorder, but also unipolar depression and adult ADHD. *Neuropsychopharmacology, 36,* 2076–2085. DOI: 10.1038/npp.2011.98

Wedekind, C., Seebeck, T., Bettens, F., & Paepke, A. J. (1995). MHC-dependent mate preferences in humans. *Proceedings of the Royal Society* of London. B: Biological Sciences, 260, 245–249. DOI: 10.1098/rspb.1995.0087

Wedell, N., Kvarnemo, C., Lessels, C. K. M., & Tregenza, T. (2006). Sexual conflict and life histories. Animal Behavior, 71, 999–1011. DOI: 10.1016/j.anbehav.2005.06.023

Weiner, J. (1994). The beak of the finch. New York: Vintage Books.

Weiner, J. (1999). Time, love and memory: A great biologist and his quest for the origins of behavior. New York: Alfred A. Knopf.

Weir, B. S., Anderson, A. D., & Hepler, A. B. (2006). Genetic relatedness analysis: Modern data and new challenges. Nature Reviews Genetics, 7, 771–780. DOI: 10.1038/nrg1960

Weiss, D. S., Marmar, C. R., Schlenger, W. E., Fairbank, J. A., Jordan, B. K., Hough, R. L., & Kulka, R. A. (1992). The prevalence of lifetime and partial posttraumatic stress disorder in Vietnam theater veterans. Journal of Traumatic Stress, 5, 365–376. DOI: 10.1002/jts.2490050304

Weiss, P. (1982). Psychogenetik: Humangenetik in psychologie und psychiatrie. Jena, Germany: Gustav Fisher.

Wender, P. H., Kety, S. S., Rosenthal, D., Schulsinger, F., Ortmann, J., & Lunde, I. (1986). Psychiatric disorders in the biological and adoptive families of adopted individuals with affective disorders. Archives of General Psychiatry, 43, 923–929. DOI: 10.1001/archpsyc.1986.01800100013003

Wender, P. H., Rosenthal, D., Kety, S. S., Schulsinger, F., & Welner, J. (1974). Crossfostering: A research strategy for clarifying the role of genetic and experimental factors in the etiology of schizophrenia. Archives of General Psychiatry, 30, 121–128. DOI: 10.1001/archpsyc.1974.01760070097016

Wheeler, H. E., & Kim, S. K. (2011). Genetics and genomics of human ageing. Philosophical Transactions of the Royal Society. B: Biological Sciences, 366, 43–50. DOI: 10.1098/rstb.2010.0259

Widiger, T. A., & Trull, T. J. (2007). Plate tectonics in the classification of personality disorder: Shifting to a dimensional model. American Psychologist, 62, 71–83. DOI: 10.1037/0003-066X.62.2.71

Widiker, S., Kaerst, S., Wagener, A., & Brockmann, G. A. (2010). High-fat diet leads to a decreased methylation of the MC4R gene in the obese BFMI and the lean B6 mouse lines. Journal of Applied Genetics, 51, 193–197. DOI: 10.1007/BF03195727

Wigman, J. T. W., van Winkel, R., Jacobs, N., Wichers, M., Derom, C., Thiery, E., . . . Van Os, J. (2011). A twin study of genetic and environmental determinants of abnormal persistence of psychotic experiences in young adulthood. American Journal

of Medical Genetics. B: Neuropsychiatric Genetics, 156B, 546–552. DOI: 10.1002/ajmg.b.31193

Wilicox, B. J., Donlon, T. A., He, Q., Chen, R., Grove, J. S., Yano, K., . . . Curb, J. D. (2008). FOXO3A genotype is strongly associated with human longevity. Proceedings of the National Academy of Sciences (USA), 105, 13987–13992. DOI: 10.1073/pnas.0801030105

Wilkins, J. F., & Haig, D. (2003). What good is genomic imprinting: The function of parent-specific gene expression. Nature Reviews Genetics, 4, 359–368. DOI: 10.1038/nrg1062

Willcutt, E. G., Betjemann, R. S., McGrath, L. M., Chhabildas, N. A., Olson, R. K., DeFries, J. C., & Pennington, B. F. (2010). Etiology and neuropsychology of comorbidity between RD and ADHD: The case for multiple-deficit models. Cortex, 46, 1345–1361. DOI: 10.1016/j.cortex.2010.06.009

Willcutt, E. G., Pennington, B. F., Duncan, L., Smith, S. D., Keenan, J. M., Wadsworth, S., . . . Olson, R. K. (2010). Understanding the complex etiologies of developmental disorders: Behavioral and molecular genetic approaches. Journal of Developmental and Behavioral Pediatrics, 31, 533–544. DOI: 10.1097/DBP.0b013e3181ef42a1

Willemsen, R., Levenga, J., & Oostra, B. A. (2011). CGG repeat in the FMR1 gene: Size matters. Clinical Genetics, 80, 214–225. DOI: 10.1111/j.1399-0004.2011.01723.x

Williams, C. A., Driscoll, D. J., & Dagli, A. I. (2010). Clinical and genetic aspects of Angelman syndrome. Genetics in Medicine, 12, 385–395. DOI: 10.1097/GIM.0b013e3181def138

Williams, N. M., Franke, B., Mick, E., Anney, R. J. L., Freitag, C. M., Gill, M., . . . Faraone, S. V. (2012). Genome-wide analysis of copy number variants in attention deficit hyperactivity disorder: The role of rare variants and duplications at 15q13.3. American Journal of Psychiatry, 169, 195–204. DOI: 10.1176/appi.ajp.2011.11060822

Williams, R. W. (2006). Expression genetics and the phenotype revolution. Mammalian Genome, 17, 496–502. DOI: 10.1007/s00335-006-0006-x

Willis-Owen, S. A., & Flint, J. (2007). Identifying the genetic determinants of emotionality in humans; insights from rodents. Neuroscience and Biobehavioral Reviews, 31, 115–124. DOI: 10.1016/j.neubiorev.2006.07.006

Willis-Owen, S. A., Turri, M. G., Munafo, M. R., Surtees, P. G., Wainwright, N. W., Brixey, R. D., & Flint, J. (2005). The serotonin transporter length polymorphism, neuroticism, and depression: A comprehensive assessment of association. Biological Psychiatry, 58, 451–456. DOI: 10.1016/j.biopsych.2005.04.050

Wilmer, J. B., Germine, L., Chabris, C. F., Chatterjee, G., Williams, M., Loken, E., . . . Duchaine, B. (2010). Human face recognition ability is specific and highly heritable. *Proceedings of the National Academy of Sciences (USA), 107,* 5238–5241. DOI: 10.1073/pnas.0913053107

Wilson, E. O. (1975). *Sociobiology: The new synthesis.* Cambridge, MA: Belknap Press.

Wilson, R. K. (1999). How the worm was won: The C. elegans genome sequencing project. *Trends in Genetics, 15,* 51–58. DOI: 10.1016/S0168-9525(98)01666-7

Wilson, R. S. (1983). The Louisville Twin Study: Developmental synchronies in behavior. *Child Development, 54,* 298–316. DOI: 10.2307/1129693

Wilson, R. S., & Matheny, A. P., Jr. (1986). Behavior genetics research in infant temperament: The Louisville Twin Study. In R. Plomin & J. F. Dunn (Eds.), *The study of temperament: Changes, continuities, and challenges* (pp. 81–97). Hillsdale, NJ: Erlbaum.

Winterer, G., & Goldman, D. (2003). Genetics of human prefrontal function. *Brain Research Reviews, 43,* 134–163. DOI: 10.1016/S0165-0173(03)00205-4

Winterer, G., Hariri, A. R., Goldman, D., & Weinberger, D. R. (2005). Neuroimaging and human genetics. *International Review of Neurobiology, 67,* 325–383. DOI: 10.1016/S0074-7742(05)67010-9

Wisdom, N. M., Callahan, J. L., & Hawkins, K. A. (2011). The effects of apolipoprotein E on non-impaired cognitive functioning: A meta-analysis. *Neurobiology of Aging, 32,* 63–74. DOI: 10.1016/j.neurobiolaging.2009.02.003

Wolf, N. (1992). *The beauty myth: How images of beauty are used against women.* New York: Anchor Books.

Wolpert, L. (2007). *Six impossible things before breakfast.* London: Faber & Faber.

Wood, W., & Eagly, A. H. (2002). A cross-cultural analysis of the behavior of women and men: Implications for the origins of sex differences. *Psychological Bulletin, 128,* 699–727. DOI: 10.1037/0033-2909.128.5.699

Wooldridge, A. (1994). *Measuring the mind: Education and psychology in England, c. 1860–c. 1990.* Cambridge: Cambridge University Press. DOI: 10.1017/CBO9780511659997

World Health Organization (2009). *WHO report on the global tobacco epidemic, 2009: Implementing smoke-free environments.* Retrieved from World Health Organization website: http://whqlibdoc.who.int/publications/2009/9789241563918_eng_full.pdf

Wray, N. R., Goddard, M. E., & Visscher, P. M. (2008). Prediction of individual genetic risk of complex disease. *Current Opinion in Genetics and Development, 18,* 257–263. DOI: 10.1016/j.gde.2008.07.006

Wright, D. (2011). QTL mapping of behaviour in the zebrafish. *Neuromethods, 52,* 101–141. DOI: 10.1007/978-1-60761-922-2_5

Wright, S. (1921). Systems of mating. *Genetics, 6,* 111–178.

Wu, C.-Y., Wu, Y.-S., Lee, J.-F., Huang, S.-Y., Yu, L., Ko, H.-C., & Lu, R.-B. (2008). The association between DRD2/ANKK1, 5-HTTLPR gene, and specific personality trait on antisocial alcoholism among Han Chinese in Taiwan. *American Journal of Medical Genetics. B: Neuropsychiatric Genetics, 147B,* 447–453. DOI: 10.1002/ajmg.b.30626

Wu, T., Snieder, H., & de Geus, E. (2010). Genetic influences on cardiovascular stress reactivity. *Neuroscience and Biobehavioral Reviews, 35,* 58–68. DOI: 10.1016/j.neubiorev.2009.12.001

Xu, L.-M., Li, J.-R., Huang, Y., Zhao, M., Tang, X., & Wei, L. (2012). AutismKB: An evidence-based knowledgebase of autism genetics. *Nucleic Acids Research, 40,* D1016–D1022. DOI: 10.1093/nar/gkr1145

Xu, Y. H., Wang, W. J., Song, H. J., Qiu, Z. L., & Luo, Q. Y. (2011). Serum differential proteomics analysis between papillary thyroid cancer patients with 131I-avid and those with non-131I-avid lung metastases. *Hellenic Journal of Nuclear Medicine, 14,* 228–233.

Yalcin, B., Nicod, J., Bhomra, A., Davidson, S., Cleak, J., Farinelli, L., . . . Flint, J. (2010). Commercially available outbred mice for genome-wide association studies. *PLoS Genetics, 6:* e1001085. DOI: 10.1371/journal.pgen.1001085

Yalcin, B., Willis-Owen, S. A., Fullerton, J., Meesaq, A., Deacon, R. M., Rawlins, J. N., . . . Mott, R. (2004). Genetic dissection of a behavioral quantitative trait locus shows that Rgs2 modulates anxiety in mice. *Nature Genetics, 36,* 1197–1202. DOI: 10.1038/ng1450

Yamasaki, C., Koyanagi, K. O., Fujii, Y., Itoh, T., Barrero, R., Tamura, T., . . . Gojobori, T. (2005). Investigation of protein functions through data-mining on integrated human transcriptome database, H-invitational database (H-InvDB). *Gene, 364,* 99–107. DOI: 10.1016/j.gene.2005.05.036

Yang, J., Benyamin, B., McEvoy, B. P., Gordon, S., Henders, A. K., Nyholt, D. R., . . . Visscher, P. M. (2010). Common SNPs explain a large proportion of the heritability for human height. *Nature Genetics, 42,* 565–569. DOI: 10.1038/ng.608

Yang, J., Lee, S. H., Goddard, M. E., & Visscher, P. M. (2011). GCTA: A tool for genome-wide complex trait analysis. *American Journal of Human Genetics, 88,* 76–82. DOI: 10.1016/j.ajhg.2010.11.011

Yang, J., Manolio, T. A., Pasquale, L. R., Boerwinkle, E., Caporaso, N., Cunningham, J. M., . . . Visscher, P. M. (2011). Genome partitioning of genetic variation for complex traits using common SNPs. *Nature Genetics, 43,* 519–525. DOI: 10.1038/ng.823

Yeo, G. S. H., Farooqi, I. S., Aminian, S., Halsall, D. J., Stanhope, R. C., & O'Rahilly, S. (1998). A frameshift mutation in MC4R associated with dominantly inherited human obesity. *Nature Genetics, 20,* 111–112. DOI: 10.1038/2404

Yeo, R. A., Gangestad, S. W., Liu, J., Calhoun, V. D., & Hutchison, K. E. (2011). Rare copy number deletions predict individual variation in intelligence. *PLoS One, 6:* e16339. DOI: 10.1371/journal.pone.0016339

Yi, X., Liang, Y., Huerta-Sanchez, E., Jin, X., Cuo, Z. X. P., Pool, J. E., . . . Wang, J. (2010). Sequencing of 50 human exomes reveals adaptation to high altitude. *Science, 329,* 75–78. DOI: 10.1126/science.1190371

Yin, J. C. P., Delvecchio, M., Zhou, H., & Tully, T. (1995). CREB as a memory modulator—induced expression of a dCREB2 activator isoform enhances long-term-memory in Drosophila. *Cell, 81,* 107–115. DOI: 10.1016/0092-8674(95)90375-5

Young, J. P. R., Fenton, G. W., & Lader, M. H. (1971). The inheritance of neurotic traits: A twin study of the Middlesex Hospital Questionnaire. *British Journal of Psychiatry, 119,* 393–398. DOI: 10.1192/bjp.119.551.393

Young, S. E., Friedman, N. P., Miyake, A., Willcutt, E. G., Corley, R. P., Haberstick, B. C., & Hewitt, J. K. (2009). Behavioral disinhibition: Liability for externalizing spectrum disorders and its genetic and environmental relation to response inhibition across adolescence. *Journal of Abnormal Psychology, 118,* 117–130. DOI: 10.1037/a0014657

Young, S. E., Rhee, S. H., Stallings, M. C., Corley, R. P., & Hewitt, J. K. (2006). Genetic and environmental vulnerabilities underlying adolescent substance use and problem use: General or specific? *Behavior Genetics, 36,* 603–615. DOI: 10.1007/s10519-006-9066-7

Young-Wolff, K. C., Enoch, M.-A., & Prescott, C. A. (2011). The influence of gene-environment interactions on alcohol consumption and alcohol use disorders: A comprehensive review. *Clinical Psychology Review, 31,* 800–816. DOI: 10.1016/j.cpr.2011.03.005

Yuferov, V., Levran, O., Proudnikov, D., Nielsen, D. A., & Kreek, M. J. (2010). Search for genetic markers and functional variants involved in the development of opiate and cocaine addiction and treatment. *Annals of the New York Academy of Sciences, 1187,* 184–207. DOI: 10.1111/j.1749-6632.2009.05275.x

Zahn-Waxler, C., Robinson, J., & Emde, R. N. (1992). The development of empathy in twins. *Developmental Psychology, 28,* 1038–1047. DOI: 10.1037/0012-1649.28.6.1038

Zalsman, G., Huang, Y. Y., Oquendo, M. A., Burke, A. K., Hu, X. Z., Brent, D. A., . . . Mann, J. J. (2006). Association of a triallelic serotonin transporter gene promoter region (5-HTTLPR) polymorphism with stressful life events and severity of depression. *American Journal of Psychiatry, 163,* 1588–1593. DOI: 10.1176/appi.ajp.163.9.1588

Zeggini, E., Weedon, M. N., Lindgren, C. M., Frayling, T. M., Elliott, K. S., Lango, H., . . . Hattersley, A. T. (2007). Replication of genome-wide association signals in UK samples reveals risk loci for type 2 diabetes. *Science, 316,* 1336–1341. DOI: 10.1126/science.1142364

Zhang, L., Chang, S., Li, Z., Zhang, K., Du, Y., Ott, J., & Wang, J. (2012). ADHD gene: A genetic database for attention deficit hyperactivity disorder. *Nucleic Acids Research, 40,* D1003–D1009. DOI: 10.1093/nar/gkr992

Zhang, T. Y., & Meaney, M. J. (2010). Epigenetics and the environmental regulation of the genome and its function. *Annual Review of Psychology, 61,* 439–466. DOI: 10.1146/annurev.psych.60.110707.163625

Zhang, Y., Proenca, R., Maffei, M., Barone, M., Leopold, L., & Friedman, J. M. (1994). Positional cloning of the mouse obese gene and its human homologue. *Nature, 372,* 425–432. DOI: 10.1038/372425a0

Zhong, S., Chew, S. H., Set, E., Zhang, J., Xue, H., Sham, P. C., . . . Israel, S. (2009). The heritability of attitude toward economic risk. *Twin Research and Human Genetics, 12,* 103–107. DOI: 10.1375/twin.12.1.103

Zhou, K., Dempfle, A., Arcos-Burgos, M., Bakker, S. C., Banaschewski, T., Biederman, J., . . . Asherson, P. (2008). Meta-analysis of genome-wide linkage scans of attention deficit hyperactivity disorder. *American Journal of Medical Genetics. B: Neuropsychiatric Genetics, 147B,* 1392–1398. DOI: 10.1002/ajmg.b.30878

Zhou, Y., Litvin, Y., Piras, A. P., Pfaff, D. W., & Kreek, M. J. (2011). Persistent increase in hypothalamic arginine vasopressin gene expression during protracted withdrawal from chronic escalating-dose cocaine in rodents. *Neuropsychopharmacology, 36,* 2062–2075. DOI: 10.1038/npp.2011.97

Zhu, B., Wang, X., & Li, L. (2010). Human gut microbiome: The second genome of human body. *Protein and Cell, 1,* 718–725. DOI: 10.1007/s13238-010-0093-z

Zhu, Q., Song, Y., Hu, S., Li, X., Tian, M., Zhen, Z., . . . Liu, J. (2010). Heritability of the specific cognitive ability of face perception. *Current Biology, 20,* 137–142. DOI: 10.1016/j.cub.2009.11.067

Zombeck, J. A., DeYoung, E. K., Brzezinska, W. J., & Rhodes, J. S. (2011). Selective breeding for increased home cage physical activity in collaborative cross and Hsd:ICR mice. *Behavior Genetics, 41,* 571–582. DOI: 10.1007/s10519-010-9425-2

Zucker, R. A. (2006). The developmental behavior genetics of drug involvement: Overview and comments. *Behavior Genetics, 36,* 616–625. DOI: 10.1007/s10519-006-9070-y

Zucker, R. A., Heitzeg, M. M., & Nigg, J. T. (2011). Parsing the undercontrol-disinhibition pathway to substance use disorders: A multi-level developmental problem. *Child Development Perspectives,* 5, 248–255. DOI: 10.1111/j.1750-8606.2011.00172.x

Zuckerman, M. (1994). *Behavioral expression and biosocial bases of sensation seeking.* New York: Cambridge University Press.

Zwijnenburg, P. J. G., Meijers-Heijboer, H., & Boomsma, D. I. (2010). Identical but not the same: The value of discordant monozygotic twins in genetic research. *American Journal of Medical Genetics. B: Neuropsychiatric Genetics, 153B,* 1134–1149. DOI: 10.1002/ajmg.b.31091

Zyphur, M. J., Narayanan, J., Arvey, R. D., & Alexander, G. J. (2009). The genetics of economic risk preferences. *Journal of Behavioral Decision Making, 22,* 367–377. DOI: 10.1002/bdm.643

Name Index

Subject Index

Note: Page numbers followed by f, t, and b indicate figures, tables, and boxes, respectively.